Variational and Non-variational Methods in Nonlinear Analysis and Boundary Value Problems

Nonconvex Optimization and Its Applications

Volume 67

The titles published in this series are listed at the end of this volume.

Variational and Non-variational Methods in Nonlinear Analysis and Boundary Value Problems

by

D. Motreanu

Department of Mathematics,
University of Perpignan,
Perpignan, France

and

V. Rădulescu

Department of Mathematics,
University of Craiova,
Craiova, Romania

KLUWER ACADEMIC PUBLISHERS
DORDRECHT / BOSTON / LONDON

A C.I.P. Catalogue record for this book is available from the Library of Congress.

ISBN 978-1-4419-5248-6

Published by Kluwer Academic Publishers,
P.O. Box 17, 3300 AA Dordrecht, The Netherlands.

Sold and distributed in North, Central and South America
by Kluwer Academic Publishers,
101 Philip Drive, Norwell, MA 02061, U.S.A.

In all other countries, sold and distributed
by Kluwer Academic Publishers,
P.O. Box 322, 3300 AH Dordrecht, The Netherlands.

Printed on acid-free paper

Contents

Preface

This book reflects a significant part of authors' research activity during the last ten years. The present monograph is constructed on the results obtained by the authors through their direct cooperation or due to the authors separately or in cooperation with other mathematicians. All these results fit in a unitary scheme giving the structure of this work.

The book is mainly addressed to researchers and scholars in Pure and Applied Mathematics, Mechanics, Physics and Engineering.

We are greatly indebted to Viorica Venera Motreanu for the careful reading of the manuscript and helpful comments on important issues.

We are also grateful to our Editors of Kluwer Academic Publishers for their professional assistance.

Our deepest thanks go to our numerous scientific collaborators and friends, whose work was so important for us.

D. Motreanu and V. Rădulescu

Introduction

The present monograph is based on original results obtained by the authors in the last decade. This book provides a comprehensive exposition of some modern topics in nonlinear analysis with applications to the study of several classes of boundary value problems. Our framework includes multivalued elliptic problems with discontinuities, variational inequalities, hemivariational inequalities and evolution problems. The treatment relies on variational methods, monotonicity principles, topological arguments and optimization techniques.

Excepting Sections 1 and 3 in Chapter 1 and Sections 1 and 3 in Chapter 2, the material is new in comparison with any other book, representing research topics where the authors contributed. The outline of our work is the following.

Chapter 1 contains basic facts on nonsmooth analysis in the framework of Clarke's and Degiovanni's theories in Sections 1 and 3, while Section 2 concerns the relationship between the Palais-Smale condition and coerciveness.

Chapter 2 is devoted to modern nonsmooth critical point theories (Chang, Degiovanni, Goeleven-Motreanu-Panagiotopoulos, Szulkin).

Chapter 3 deals with general results ensuring the existence of critical points of nondifferentiable functionals.

Chapter 4 presents an abstract multiplicity theorem of Ljusternik-Schnirelman type and a comparison treatment of a class of stationary Schrödinger equations with lack of compactness.

Chapter 5 studies extremal solutions for initial boundary value problems of parabolic type involving Clarke's gradient.

Chapter 6 focuses on existence results for various classes of boundary value problems expressed by variational, hemivariational or variational-hemivariational inequalities.

Chapter 7 treats simple and double eigenvalue hemivariational inequalities subject to constraints and under the presence of a group of symmetries.

Chapter 8 sets forth multiplicity results for non-symmetric perturbations of the problems treated in Chapter 7.

Chapter 9 discusses the location of critical points for nonsmooth functionals and of solutions to variational-hemivariational inequalities.

Chapter 10 provides existence results for first order evolution quasi-variational inequalities and second order nonlinear evolution equations as well as stability results regarding variational inequalities.

Chapter 11 deals with a variational approach involving a functional arising in the theory of minimal surfaces and which does not satisfy the Palais-Smale condition in BV.

The abstract theory developed in the book is illustrated by various examples and applications.

Chapter 1

ELEMENTS OF NONSMOOTH ANALYSIS

In this Chapter we recall important definitions and results from the theory of generalized gradient for locally Lipschitz functionals due to Clarke [8], different nonsmooth versions of Palais-Smale conditions and basic elements of nonsmooth calculus developed by Degiovanni [9], [10]. A major part in Section 2 is devoted to the relationship between the Palais-Smale condition and the coerciveness in the nonsmooth setting.

1. Generalized Gradients of Locally Lipschitz Functionals

The aim of this Section is to present basic elements of the theory of generalized gradients for locally Lipschitz functionals which will be used throughout the book. For more details and applications we refer to Clarke [8].

Let X be a real Banach space endowed with the norm $\|\cdot\|$. The dual space of X is denoted X^* and is equipped with the dual norm $\|\cdot\|_*$, i.e.

$$\|\zeta\|_* = \sup\{\langle \zeta, v\rangle \, : \, v \in X, \ \|v\| \leq 1\},$$

where the notation $\langle \cdot, \cdot \rangle$ stands for the duality pairing between X^* and X.

We recall that a functional $f : X \to \mathbb{R}$ is called locally Lipschitz if for every $x \in X$ there exist a neighborhood V of x in X and a constant $K > 0$ such that

$$|f(y) - f(z)| \leq K\|y - z\|, \quad \forall y, z \in V.$$

Definition 1.1 (Clarke [8]) The generalized directional derivative of a locally Lipschitz functional $f : X \to \mathbb{R}$ at a point $u \in X$ in the direction

$v \in X$, denoted $f^0(u;v)$, is defined by

$$f^0(u;v) = \limsup_{\substack{x \to u \\ t \downarrow 0}} \frac{f(x+tv) - f(x)}{t}.$$

The locally Lipschitz continuity of f at u ensures that $f^0(u;v) \in \mathbb{R}$, $\forall v \in X$. Moreover, $f^0(u;\cdot) : X \to \mathbb{R}$ is subadditive, positively homogeneous and satisfies the inequality

$$|f^0(u;v)| \leq K\|v\|, \quad \forall v \in X,$$

where $K > 0$ is the Lipschitz constant of f near the point $u \in X$. Moreover, the function $(u,v) \mapsto f^0(u;v)$ is upper semicontinuous.

A basic definition of this Section is the following.

Definition 1.2 (Clarke [8]) The generalized gradient of a locally Lipschitz functional $f : X \to \mathbb{R}$ at a point $u \in X$, denoted $\partial f(u)$, is the subset of X^* defined by

$$\partial f(u) = \{\zeta \in X^* : f^0(u;v) \geq \langle \zeta, v \rangle, \quad \forall v \in X\}.$$

It is readily seen from Definition 1.2 that, by using Hahn-Banach theorem (see, e.g., Brézis [3], p. 1), $\partial f(u) \neq \emptyset$.

The next proposition points out important properties of generalized gradients.

Proposition 1.1 (Clarke [8]) Let $f : X \to \mathbb{R}$ be Lipschitz continuous on a neighborhood of a point $u \in X$.

(*i*) $\partial f(u)$ is a convex, weak $*$ compact subset of X^* and

$$\|\zeta\|_* \leq K, \quad \forall \zeta \in \partial f(u),$$

where $K > 0$ is the Lipschitz constant of f near u.

(*ii*) $f^0(u;v) = \max\{\langle \zeta, v \rangle : \zeta \in \partial f(u)\}$, $\forall v \in X$.

The proof can be found in Clarke [8], Proposition 2.1.2.

The result below states useful properties of the generalized gradient.

Proposition 1.2 (Chang [7]) Let $f : X \to \mathbb{R}$ be a locally Lipschitz functional. Then one has

(*i*) For all $u \in X$, $\varepsilon > 0$ and $v \in X$, there exists $\delta > 0$ such that whenever $w \in \partial f(x)$ with $\|x - u\| < \delta$ one finds $z \in \partial f(u)$ satisfying $|\langle w - z, v \rangle| < \varepsilon$.

(*ii*) The function $\lambda : X \to I\!R$ given by

$$\lambda(x) = \min_{w \in \partial f(x)} \|w\|_*$$

is lower semicontinuous.

(*iii*) If $\varphi \in C^1([a,b]; X)$, with $a < b$ in $I\!R$, and $f : X \to I\!R$ is a locally Lipschitz functional, then the composition $h = f \circ \varphi : [a,b] \to I\!R$ is differentiable a.e. and

$$h'(t) \leq \max\{\langle w, \varphi'(t)\rangle \ : \ w \in \partial f(\varphi(t))\} \ \text{ a.e.}$$

Proof. (*i*) Arguing by contradiction, we admit that there exist $u \in X$, $v \in X$, $\varepsilon_0 > 0$ and sequences $\{x_n\} \subset X$, $\{\xi_n\} \subset X^*$ with $\xi_n \in \partial f(x_n)$ such that

$$\|x_n - u\| < \frac{1}{n},$$

$$|\langle \xi_n - w, v\rangle| \geq \varepsilon_0, \quad \forall w \in \partial f(u). \tag{1.1}$$

Since $x_n \to u$ and $\xi_n \in \partial f(x_n)$ we may suppose $\|\xi_n\|_* \leq K$, where $K > 0$ is the Lipschitz constant around u, and $\xi_n \overset{*}{\rightharpoonup} \xi$ weakly$\,*$ in X^* as $n \to \infty$.

We claim that

$$\xi \in \partial f(u). \tag{1.2}$$

Indeed, the fact that $\xi_n \in \partial f(x_n)$ implies

$$\langle \xi_n, y\rangle \leq f^0(x_n; y), \quad \forall y \in X.$$

Definition 1.1 yields sequences $\lambda_n \downarrow 0$, $h_n \to 0$ such that

$$\frac{f(x_n + h_n + \lambda_n y) - f(x_n + h_n)}{\lambda_n} \geq f^0(x_n; y) - \frac{1}{n} \geq \langle \xi_n, y\rangle - \frac{1}{n}, \ \forall y \in X.$$

Passing to the limit, one finds

$$f^0(u; y) \geq \limsup_{n \to \infty} \frac{f(x_n + h_n + \lambda_n y) - f(x_n + h_n)}{\lambda_n}$$

$$\geq \limsup_{n \to \infty} \left[\langle \xi_n, y\rangle - \frac{1}{n}\right] = \langle \xi, y\rangle, \ \forall y \in X.$$

The Definition 1.2 shows that (1.2) is true.

Letting $n \to \infty$ in (1.1) leads to a contradiction with (1.2). This contradiction establishes property (*i*).

(*ii*) Applying Alaoglu's theorem (see, e.g., Brézis [3], p. 42), in the definition of $\lambda(x)$ the minimum makes sense. In order to show that the function λ is lower semicontinuous, let us suppose, on the contrary, that there exists a sequence $\{x_n\}$ such that $x_n \to u$ and

$$\liminf_{n\to\infty} \lambda(x_n) < \lambda(u).$$

We know that there is $w_n \in \partial f(x_n)$ with $\lambda(x_n) = \|w_n\|_*$. Therefore we can choose a subsequence of $\{w_n\}$, denoted again $\{w_n\}$, and an element $z \in \partial f(u)$ such that $w_n \stackrel{*}{\rightharpoonup} z$ weakly $*$. Then one obtains

$$\liminf_{n\to\infty} \|w_n\|_* \geq \|z\|_* \geq \lambda(u).$$

This contradiction shows that assertion (*ii*) is valid.

(*iii*) Since the function $h : [a,b] \to I\!R$ is locally Lipschitz, it is differentiable a.e. Then at any point $t_0 \in [a,b]$ where h is differentiable we may write

$$h'(t_0) = \lim_{\lambda\to 0} \frac{1}{\lambda} [f(\varphi(t_0+\lambda)) - f(\varphi(t_0))]$$

$$= \lim_{\lambda\to 0} \frac{1}{\lambda} [f(\varphi(t_0) + \varphi'(t_0)\lambda + o(\lambda)) - f(\varphi(t_0))]$$

$$= \lim_{\lambda\to 0} \left\{ \frac{1}{\lambda} [f(\varphi(t_0) + \varphi'(t_0)\lambda) - f(\varphi(t_0))] \right.$$

$$\left. + \frac{1}{\lambda} [f(\varphi(t_0) + \varphi'(t_0)\lambda + o(\lambda)) - f(\varphi(t_0) + \varphi'(t_0)\lambda)] \right\}$$

$$= \lim_{\lambda\to 0} \frac{1}{\lambda} [f(\varphi(t_0) + \varphi'(t_0)\lambda) - f(\varphi(t_0))]$$

$$\leq \limsup_{\substack{h\to 0\\ \lambda\downarrow 0}} \frac{1}{\lambda} [f(\varphi(t_0) + h + \varphi'(t_0)\lambda) - f(\varphi(t_0)+h)] = f^0(\varphi(t_0); \varphi'(t_0))$$

$$= \max\{\langle w, \varphi'(t_0)\rangle \, : \, w \in \partial f(\varphi(t_0))\},$$

with the notation $o(\lambda)$ meaning $\lim_{\lambda\to 0} \frac{o(\lambda)}{\lambda} = 0$. This completes the proof. ■

For making a comparison with other concept of differentiability, we recall the following definition.

Definition 1.3 A mapping $F : X \to Y$ between the Banach spaces X and Y is said to be strictly differentiable at $u \in X$ if there exists an element $D_sF(u) \in L(X,Y)$ such that

$$\lim_{\substack{u'\to u\\ t\downarrow 0}} \frac{F(u'+tv) - F(u')}{t} = DF_s(u)v, \;\; \forall v \in X,$$

where convergence is uniform for v in compact sets.

We state now the characterization of strict differentiability.

Proposition 1.3 (Clarke [8], p. 31) Let $F : X \to Y$, $u \in X$ and $\zeta \in L(X,Y)$. The following assertions are equivalent:

(a) F is strictly differentiable at u and $D_sF(u) = \zeta$.

(b) F is Lipschitz continuous near u and

$$\lim_{\substack{u' \to u \\ t \downarrow 0}} \frac{F(u' + tv) - F(u')}{t} = \zeta(v), \quad \forall v \in X.$$

Proof. $(a) \Rightarrow (b)$ Since the equality in (b) follows directly from (a), we need only to check that F is Lipschitz continuous near u. Supposing the contrary we assume that there exist sequences $\{u_i\}$ and $\{u_i'\}$ such that $\|u_i - u\|_X < \frac{1}{i}$, $\|u_i' - u\|_X < \frac{1}{i}$ and

$$\|F(u_i') - F(u_i)\|_Y > i\|u_i' - u_i\|_X. \tag{1.3}$$

For every i, let us take $t_i > 0$ and $v_i \in X$ such that $u_i' = u_i + t_i v_i$ and $\|v_i\|_X = i^{-1/2}$. This choice is possible in view of (1.3). We also note that $t_i \to 0$ as $i \to \infty$.

Using the uniform convergence in the definition of strict differentiability on the compact set $\{v_i\} \cup \{0\}$, we find that for every $\varepsilon > 0$ there exists n_ε such that for all $i \geq n_\varepsilon$ and $j \in I\!N$ we have

$$\left\| \frac{F(u_i + t_i v_j) - F(u_i)}{t_i} - D_sF(u)v_j \right\|_Y < \varepsilon. \tag{1.4}$$

Inequality (1.3) yields

$$\left\| \frac{F(u_i + t_i v_i) - F(u_i)}{t_i} \right\|_Y > i\|v_i\|_X = i^{1/2} \to +\infty \quad \text{as } i \to \infty. \tag{1.5}$$

On the other hand, there is a constant $C > 0$ such that

$$\|D_sF(u)v_j\|_Y \leq \|D_sF(u)\|_{L(X,Y)} \, \|v_j\|_X \leq C, \quad \forall j \in I\!N.$$

Combining with (1.4), we arrive at

$$\left\| \frac{F(u_i + t_i v_i) - F(u_i)}{t_i} \right\|_Y \leq \left\| \frac{F(u_i + t_i v_i) - F(u_i)}{t_i} - D_sF(u)v_i \right\|_Y$$
$$+ \|DF_s(u)v_i\|_Y < \varepsilon + C,$$

which contradicts (1.5). Assertion (b) follows.

(b) $\Rightarrow$ (a) It is sufficient to prove that the limit in assertion (b) is uniform with respect to v in compact sets. Let V be a compact subset of X and let an arbitrary number $\varepsilon > 0$. Property (b) provides for every $v \in V$ a number $\delta(v) > 0$ such that

$$\left\| \frac{F(u' + tv) - F(u')}{t} - \zeta(v) \right\|_Y < \varepsilon \tag{1.6}$$

whenever $\|u' - u\| \leq \delta$ and $t \in]0, \delta[$. For a possible smaller δ, the Lipschitz continuity of F near u implies

$$\left\| \frac{F(u' + tv') - F(u')}{t} - \frac{F(u' + tv) - F(u')}{t} \right\|_Y$$

$$= \frac{\|F(u' + tv') - F(u' + tv)\|_Y}{t} \leq K\|v' - v\|_X.$$

Then, by (1.6), we deduce that

$$\left\| \frac{F(u' + tv') - F(u')}{t} - \zeta(v') \right\|_Y$$

$$\leq \left\| \frac{F(u' + tv') - F(u')}{t} - \frac{F(u' + tv) - F(u')}{t} \right\|_Y$$

$$+ \left\| \frac{F(u' + tv) - F(u')}{t} - \zeta(v') \right\|_Y \leq K\|v' - v\| + \varepsilon < K\delta + \varepsilon$$

$$< 2\varepsilon, \text{ if } \|u' - u\| < \delta, \ \|v' - v\| < \delta, \ t \in]0, \delta[,$$

taking $\delta < \varepsilon/K$.

Using the compactness of V it is straightforward to get the desired conclusion. The proof is thus complete. ∎

Corollary 1.1 (Clarke [8], p. 33) If $f : X \to \mathbb{R}$ is strictly differentiable at $u \in X$, then f is Lipschitz continuous near u and $\partial f(u) = \{D_s f(u)\}$. Conversely, if $f : X \to \mathbb{R}$ is Lipschitz continuous near $u \in X$ and $\partial f(u) = \{\zeta\}$, then f is strictly differentiable at u and $D_s f(u) = \zeta$.

Proof. Assume that f is strictly differentiable at $u \in X$. Proposition 1.3 implies that f is Lipschitz continuous on a neighborhood of u and, by Definitions 1.1 and 1.3, one has

$$\langle D_s f(u), v\rangle = \lim_{\substack{u' \to u \\ t \downarrow 0}} \frac{f(u' + tv) - f(u')}{t} = f^0(u; v), \quad \forall v \in X.$$

Since in this case $f^0(u;\cdot)$ is linear on X, it results $\partial f(u) = \{D_s f(u)\}$.

Conversely, assume that f is Lipschitz continuous on a neighborhood of u and $\partial f(u) = \{\zeta\}$. Applying part (*ii*) of Proposition 1.1, then we may write

$$\langle \zeta, v\rangle = f^0(u;v) := \limsup_{\substack{u'\to u\\ t\downarrow 0}} \frac{f(u'+tv)-f(u')}{t}, \quad \forall v \in X. \tag{1.7}$$

In order to use part (*b*) of Proposition 1.3, it is sufficient to show that

$$\liminf_{\substack{u'\to u\\ t\downarrow 0}} \frac{f(u'+tv)-f(u')}{t} = f^0(u;v), \quad \forall v \in X. \tag{1.8}$$

On the basis of (1.7) it follows that

$$\liminf_{\substack{u'\to u\\ t\downarrow 0}} \frac{f(u'+tv)-f(u')}{t} = -\limsup_{\substack{u'\to u\\ t\downarrow 0}} \frac{f(u'+tv-tv)-f(u'+tv)}{t}$$

$$= -\limsup_{\substack{y\to u\\ t\downarrow 0}} \frac{f(y+t(-v))-f(y)}{t} = -f^0(u;-v) = -\langle \zeta, -v\rangle = \langle \zeta, v\rangle$$

$$= f^0(u;v) = \limsup_{\substack{u'\to u\\ t\downarrow 0}} \frac{f(u'+tv)-f(u')}{t}, \quad \forall v \in X,$$

which is just (1.8). Hence part (*b*) of Proposition 1.3 is verified. Applying Proposition 1.3 we complete the proof. ■

The notion of regularity in the sense of Clarke [8] is now recalled.

Definition 1.4 (Clarke [8], p. 39) A function $f : X \to \mathbb{R}$ is said to be regular at a point $u \in X$ if f is Lipschitz near u and

(i) there exists the usual directional derivative $f'(u;v)$, for every $v \in X$;

(ii) $f^0(u;v) = f'(u;v)$, $\forall v \in X$.

Significant classes of regular functions are pointed out in the next result.

Proposition 1.4 (Clarke [8], p. 40) Let $f : X \to \mathbb{R}$ be Lipschitz continuous on a neighborhood of a point $u \in X$.

(*a*) If f is strictly differentiable at u, then f is regular at u. In particular, if f is continuously differentiable on a neighborhood of u, then f is regular at u.

(*b*) If f is convex, then f is regular at u.

(*c*) If f is Gâteaux differentiable and regular at the point u, then one has $\partial f(u) = \{Df(u)\}$, where $Df(u)$ stands for the Gâteaux differential of f at u.

Proof. (*a*) This follows directly from Definitions 1.3 and 1.4. It is straightforward to check that a real-valued function f which is continuously differentiable near u is strictly differentiable at u.

(*b*) It is known that a convex function which is Lipschitz continuous near u admits the directional derivative $f'(u;\cdot)$, which coincides with $f^0(u;\cdot)$ (see, e.g., Clarke [8], Proposition 2.2.7). This ensures that f is regular at u.

(*c*) Let $\zeta \in \partial f(u)$. Then, from Definitions 1.2 and 1.4 in conjunction with the assumption that f is Gâteaux differentiable at u, it turns that

$$\langle \zeta, v\rangle \le f^0(u;v) = f'(u;v) = \langle Df(u), v\rangle, \quad \forall v \in X.$$

This yields the equality $\zeta = Df(u)$. ∎

The result below deals with Lebourg's mean value theorem.

Theorem 1.1 (Lebourg [18]) Given the points x and y in X and a real-valued function f which is Lipschitz continuous on an open set containing the segment $[x, y] = \{(1-t)x + ty \ : \ t \in [0,1]\}$, there exist $u = x + t_0(y-x)$, with $0 < t_0 < 1$, and $x^* \in \partial f(u)$ such that

$$f(y) - f(x) = \langle x^*, y - x\rangle.$$

Proof. Consider the function $\theta : [0,1] \to I\!R$ defined by

$$\theta(t) = f(x + t(y-x)) + t[f(x) - f(y)], \quad \forall t \in [0,1].$$

The continuity of θ combined with the equalities $\theta(0) = \theta(1) = f(x)$ yields a point $t_0 \in]0,1[$ where θ assumes a local minimum or maximum. A direct verification ensures that $0 \in \partial\theta(t_0)$ and

$$\partial\theta(t_0) \subset \langle \partial f(x + t_0(y-x)), y - x\rangle + [f(x) - f(y)]$$

(see for details Lebourg [18] or Clarke [8], p. 41). It follows that some $x^* \in \partial f(x + t_0(y-x))$ can be found such that the conclusion of Theorem 1.1 holds. ∎

Another important result in the calculus with generalized gradients is the chain rule.

Theorem 1.2 (Clarke [8], p. 45) Let X, Y be Banach spaces, let $F : X \to Y$ be a mapping which is strictly differentiable at a point

$u \in X$ and let $g : Y \to \mathbb{R}$ be a function which is Lipschitz continuous on a neighborhood of $F(u)$. Then the functional $g \circ F : X \to \mathbb{R}$ is Lipschitz continuous on a neighborhood of u, and the formula below holds

$$\partial(g \circ F)(u) \subset \partial g(F(u)) \circ D_s F(u), \tag{1.9}$$

in the sense that every $z \in \partial(g \circ F)(u)$ can be expressed as

$$z = D_s F(u)^* \zeta, \quad \text{for some } \zeta \in \partial g(F(u)),$$

where $D_s F(u)^*$ denotes the adjoint of the operator $D_s F(u)$. If g (or $-g$) is regular at $F(u)$, then $g \circ F$ (or $-g \circ F$) is regular at u and equality holds in (1.9). If F maps every neighborhood of u onto a set which is dense in a neighborhood of $F(u)$, then equality holds in (1.9).

Proof. Proposition 1.3 ensures that F is Lipschitz continuous near u, so the same is true for $g \circ F$.

Inclusion (1.9) is equivalent to the following inequality

$$\begin{aligned}(g \circ F)^0(u; v) &\le \max\{\langle z, D_s F(u) v\rangle \,:\, z \in \partial g(F(u))\rangle\} \\ &= g^0(F(u); D_s F(u) v), \quad \forall v \in X, \end{aligned} \tag{1.10}$$

where Proposition 1.1 has been used. The proof of inequality (1.10) follows the argument in Theorem 2.3.9 of Clarke [8] and is omitted.

Suppose that g is regular at $F(u)$. The case where $-g$ is regular can then be derived because $\partial(-g)(F(u)) = -\partial g(F(u))$.

Using the regularity of g at $F(u)$, the equalities below are valid

$$\begin{aligned} g^0(F(u); D_s F(u) v) &= g'(F(u); D_s F(u) v) \\ &= \lim_{t \downarrow 0} \frac{g(F(u) + t D_s F(u) v) - g(F(u))}{t} \\ = \lim_{t \downarrow 0} \Big[&\frac{g(F(u) + t D_s F(u) v) - g(F(u + tv))}{t} + \frac{g(F(u + tv)) - g(F(u))}{t} \Big] \\ &= (g \circ F)'(u; v) \le (g \circ F)^0(u; v), \quad \forall v \in X. \end{aligned}$$

We infer that one has equality in (1.10), so equality holds in (1.9).

Assume that F maps an arbitrary neighborhood of u onto a dense subset of a neighborhood of $F(u)$. This allows to write the equalities

$$\begin{aligned} g^0(F(u); D_s F(u) v) &= \limsup_{\substack{y \to F(u) \\ t \downarrow 0}} \frac{g(y + t D_s F(u) v) - g(y)}{t} \\ &= \limsup_{\substack{x \to u \\ t \downarrow 0}} \frac{g(F(x) + t D_s F(u) v) - g(F(x))}{t} \end{aligned}$$

$$= \limsup_{\substack{x \to u \\ t \downarrow 0}} \frac{g(F(x+tv)) - g(F(x))}{t} = (g \circ F)^0(u;v), \quad \forall v \in X.$$

Therefore, (1.10) holds with equality, so one has equality in (1.9), too. This completes the proof. ■

Corollary 1.2 (Chang [7], Theorem 2.2, Clarke [8], p. 47) Assume that the Banach space X is continuously and densely embedded in the Banach space Y. If $g : Y \to \mathbb{R}$ is Lipschitz continuous near $u \in X$, then

$$\partial(g|_X)(u) = \partial g(u),$$

in the sense that every element $z \in \partial(g|_X)(u)$ admits a unique extension to an element of $\partial g(u)$.

Proof. One applies the last assertion in Theorem 1.2 for F equal to the embedding of X into Y. ■

We close this Section with the important Aubin-Clarke [2] result concerning subdifferentiation of integral functionals.

Let $m \geq 1$, $1 < p < +\infty$ and let T be a positive complete measure space with $|T| < \infty$, where $|T|$ stands for the measure of T. Let $j : T \times \mathbb{R}^m \to \mathbb{R}$ be a function such that $j(\cdot, y) : T \to \mathbb{R}$ is measurable whenever $y \in \mathbb{R}^m$, and satisfies either

$$|j(x, y_1) - j(x, y_2)| \leq k(x)|y_1 - y_2|, \quad \forall x \in T, \ y_1, y_2 \in \mathbb{R}^m, \tag{1.11}$$

for a function $k \in L^q(T)$, with $1/p + 1/q = 1$, or, $j(x, \cdot) : \mathbb{R}^m \to \mathbb{R}$ is locally Lipschitz for almost all $x \in T$ and there is a constant $c > 0$ such that

$$|z| \leq c(1 + |y|^{p-1}), \quad \forall x \in T, \ y \in \mathbb{R}^m, \ z \in \partial_y j(x, y). \tag{1.12}$$

The notation $\partial_y j(x, y)$ in (1.12) means the generalized gradient of j with respect to the second variable $y \in \mathbb{R}^m$, i.e., $\partial j(x, \cdot)(y)$. We introduce the functional $J : L^p(T; \mathbb{R}^m) \to \mathbb{R}$ by

$$J(v) = \int_T j(x, v(x))dx, \quad \forall v \in L^p(T; \mathbb{R}^m). \tag{1.13}$$

We state the following main result.

Theorem 1.3 (Aubin and Clarke [2], Clarke [8], p. 83) Assume (1.11) or (1.12). Then the functional $J : L^p(T; \mathbb{R}^m) \to \mathbb{R}$ in (1.13) is Lipschitz continuous on bounded sets and satisfies

$$\partial J(u) \subset \int_T \partial_y j(x, u(x))dx, \quad \forall u \in L^p(T; \mathbb{R}^m) \tag{1.14}$$

in the sense that for every $z \in \partial J(u)$ there is $w \in L^q(T;\mathbb{R}^m)$ such that

$$w(x) \in \partial_y j(x,u(x)) \quad \text{for a.e. } x \in T,$$

and

$$\langle z, v\rangle = \int_T \langle w(x), v(x)\rangle \, dx, \quad \forall v \in L^p(T;\mathbb{R}^m).$$

Moreover, if $j(x,\cdot)$ is regular at $u(x)$ for almost all $x \in T$, then J is regular at u and (1.14) holds with equality.

Proof. Using Hölder's inequality, it is straightforward to check that, under assumption (1.11) or (1.12), J is Lipschitz continuous on bounded subsets of $L^p(T;\mathbb{R}^m)$.

Definition 1.1 shows that the map $x \mapsto j_y^0(x,u(x);v(x))$ is measurable on T.

Let us check the inequality

$$J^0(u;v) \le \int_T j_y^0(x,u(x);v(x))dx, \quad \forall u,v \in L^p(T;\mathbb{R}^m), \tag{1.15}$$

where the subscript y indicates that the generalized directional derivative j^0 is taken with respect to the second variable.

If (1.11) is assumed, then Fatou's lemma leads directly to (1.15). Admit now (1.12). On the basis of Theorem 1.1, it is permitted to write

$$\frac{j(x,u(x)+\lambda v(x)) - j(x,u(x))}{\lambda} = \langle \zeta_x, v(x)\rangle$$

for some $\zeta_x \in \partial j(x,u^*(x))$ and for $u^*(x)$ lying in the open segment with endpoints $u(x)$ and $u(x)+\lambda v(x)$. Then Fatou's lemma implies (1.15).

We observe that, in view of (1.15), any $z \in \partial J(u)$ belongs to the subdifferential at $0 \in L^p(T;\mathbb{R}^m)$ (in the sense of convex analysis) of the convex function on $L^p(T;\mathbb{R}^m)$ mapping

$$v \in L^p(T;\mathbb{R}^m) \mapsto \int_T j_y^0(x,u(x);v(x))dx \in \mathbb{R}. \tag{1.16}$$

The subdifferentiation rule in Ioffe and Levin [14] applied to (1.16) yield (1.14).

Finally, assume that $j(x,\cdot)$ is regular at $u(x)$ for almost all $x \in T$. Then, under either assumption (1.11) or (1.12), we may apply Fatou's lemma to get

$$\liminf_{\lambda \downarrow 0} \frac{1}{\lambda}(J(u+\lambda v) - J(u)) \ge \int_T j_y'(x,u(x);v(x))dx$$

$$= \int_T j_y^0(x, u(x); v(x))dx, \quad \forall v \in L^p(T; \mathbb{R}^m).$$

Combining with (1.15) it follows that there exists the directional derivative $J'(u; v)$ and one has $J'(u; v) = J^0(u; v)$, for every $v \in L^p(T; \mathbb{R}^m)$, thus one obtains the regularity of J at u and the equality

$$J^0(u; v) = \int_T j_y'(x, u(x); v(x))dx, \quad \forall v \in L^p(T; \mathbb{R}^m).$$

This relation and the regularity assumption for $j(x, \cdot)$ show that

$$\langle z, v \rangle = \int_T \langle z(x), v(x) \rangle dx \le J^0(u; v), \quad \forall v \in L^p(T; \mathbb{R}^m),$$

whenever z belongs to the right-hand side of (1.14). It turns out that every such element z satisfies $z \in \partial J(u)$. Consequently, (1.14) becomes an equality. This completes the proof. ∎

Remark 1.1 For applications of the theory of generalized gradients in nonsmooth and nonconvex optimization we refer to [1], [2], [8], [11], [12], [22], [23].

2. Palais-Smale Condition and Coerciveness for a Class of Nonsmooth Functionals

In this Section we deal with the class of nonsmooth functions which can be written as a sum $\Phi + \Psi$ of a locally Lipschitz functional Φ and a proper, convex, lower semicontinuous functional Ψ on a Banach space.

Precisely, we are concerned with an appropriate version of Palais-Smale condition and its relationship with the coerciveness property. The contents of this Section is taken from D. Motreanu and V. V. Motreanu [19].

Throughout this Section we denote by X a real Banach space endowed with the norm $\|\cdot\|$. The notation X^* stands for the dual space of X.

We recall three basic definitions of Palais-Smale conditions for nonsmooth functionals.

Definition 1.5 (Chang [7]) The locally Lipschitz functional $\Phi : X \to \mathbb{R}$ satisfies the Palais-Smale condition if every sequence $\{u_n\} \subset X$ with $\Phi(u_n)$ bounded and for which

$$\lambda(u_n) := \min_{w \in \partial\Phi(u_n)} \|w\|_* \to 0 \text{ as } n \to \infty, \tag{1.17}$$

has a (strongly) convergent subsequence in X.

The notation $\partial\Phi$ in (1.17) means the generalized gradient of the locally Lipschitz functional Φ (see Definition 1.2).

Definition 1.6 (Szulkin [24]) Let $\Phi : X \to I\!R$ be a continuously differentiable function and let $\Psi : X \to I\!R \cup \{+\infty\}$ be a proper (i.e., $\not\equiv +\infty$), convex and lower semicontinuous function. The functional $\Phi + \Psi : X \to I\!R \cup \{+\infty\}$ satisfies the Palais-Smale condition if every sequence $\{u_n\} \subset X$ with $\Phi(u_n) + \Psi(u_n)$ bounded and for which there exists a sequence $\{\varepsilon_n\} \subset I\!R^+$, $\varepsilon_n \downarrow 0$, such that

$$\Phi'(u_n)(v - u_n) + \Psi(v) - \Psi(u_n) \geq -\varepsilon_n \|v - u_n\|, \ \forall v \in X, \tag{1.18}$$

contains a (strongly) convergent subsequence in X.

Definition 1.7 (Motreanu and Panagiotopoulos [21], p. 64) Let $\Phi : X \to I\!R$ be a locally Lipschitz functional and let $\Psi : X \to I\!R \cup \{+\infty\}$ be a proper, convex and lower semicontinuous function. The functional $\Phi + \Psi : X \to I\!R \cup \{+\infty\}$ satisfies the Palais-Smale condition if every sequence $\{u_n\} \subset X$ with $\Phi(u_n) + \Psi(u_n)$ bounded and for which there exists a sequence $\{\varepsilon_n\} \subset I\!R^+$, $\varepsilon_n \downarrow 0$, such that

$$\Phi^0(u_n; v - u_n) + \Psi(v) - \Psi(u_n) \geq -\varepsilon_n \|v - u_n\|, \ \forall v \in X, \tag{1.19}$$

contains a (strongly) convergent subsequence in X.

The comparison between Definitions 1.5, 1.6 and 1.7 relies on the following result.

Lemma 1.1 (Szulkin [24]) Let $\chi : X \to I\!R \cup \{+\infty\}$ be a lower semicontinuous, convex function with $\chi(0) = 0$. If

$$\chi(x) \geq -\|x\|, \ \forall x \in X,$$

then there exists some $z \in X^*$ such that $\|z\|_* \leq 1$ and

$$\chi(x) \geq \langle z, x \rangle, \ \forall x \in X.$$

Proof. Consider the following convex subsets A and B of $X \times I\!R$:

$$A = \{(x,t) \in X \times I\!R \,:\, \|x\| < -t\} \text{ and } B = \{(x,t) \in X \times I\!R \,:\, \chi(x) \leq t\}.$$

Notice that A is an open set and due to the condition $\chi(x) \geq -\|x\|$, one has $A \cap B = \emptyset$. A well-known separation result (Brézis [3], p. 5) yields the existence of numbers $\alpha, \beta \in I\!R$ and $w \in X^*$ such that $(w, \alpha) \neq (0,0)$,

$$\langle w, x \rangle - \alpha t \geq \beta, \ \forall (x,t) \in \bar{A}$$

and

$$\langle w, x \rangle - \alpha t \leq \beta, \ \forall (x,t) \in B.$$

We see that $\beta = 0$ since $(0,0) \in \bar{A} \cap B$. Set $t = -\|x\|$ in the first inequality above. It follows that $\langle w, x\rangle \geq -\alpha\|x\|$, $\forall x \in X$, which implies $\alpha > 0$ and $\|w\|_* \leq \alpha$.

Set $z = \alpha^{-1}w$ and $t = \chi(x)$ in the second equality above. We deduce that $\langle z, x\rangle \leq \chi(x)$, $\forall x \in X$. Since $\|w\|_* \leq \alpha$ we obtain $\|z\|_* \leq 1$. The conclusion is achieved. ■

The result below discusses the relationship between Definitions 1.5, 1.6 and 1.7.

Theorem 1.4 *(i)* If $\Psi = 0$, Definition 1.7 reduces to Definition 1.5. *(ii)* If $\Phi \in C^1(X; I\!R)$, Definition 1.7 coincides with Definition 1.6.

Proof. *(i)* Let $\Psi = 0$ in Definition 1.7. It is sufficient to show the equivalence between relations (1.17) and (1.19). Suppose that property (1.17) holds. It is known from part *(ii)* of Proposition 1.2 that an element $z_n \in \partial\Phi(u_n)$ can be found such that $\lambda(u_n) = \|z_n\|_*$. Then part *(ii)* in Proposition 1.1 ensures that

$$\Phi^0(u_n; v) \geq \langle z_n, v\rangle \geq -\|z_n\|_* \, \|v\|, \ \forall v \in X.$$

Therefore inequality (1.19) (with $\Psi = 0$) is verified for $\varepsilon_n = \|z_n\|$.

Conversely, we admit that (1.19) is satisfied. We can apply Lemma 1.1 for $\chi = \frac{1}{\varepsilon_n}\Phi^0(u_n; \cdot)$ since χ is continuous, convex and (1.19) is satisfied (with $\psi = 0$). Lemma 1.1 yields an element $w_n \in X^*$ with $\|w_n\|_* \leq 1$ and

$$\frac{1}{\varepsilon_n}\Phi^0(u_n; x) \geq \langle w_n, x\rangle, \ \forall x \in X.$$

Choosing $z_n = \varepsilon_n w_n$ we get (1.17).

(ii) This assertion follows from the fact that Φ^0 is equal to the Fréchet differential Φ' if the functional $\Phi : X \to I\!R$ is of class C^1. Therefore, in this case the inequalities (1.18) and (1.19) coincide. The proof is complete. ■

In the following we need the next version of celebrated Ekeland's variational principle [11], [12].

Theorem 1.5 (Ekeland [11], [12]) Let M be a complete metric space endowed with the distance d and let $f : M \to I\!R \cup \{+\infty\}$ be a function which is proper, lower semicontinuous and bounded from below. Then for every number $\varepsilon > 0$ and every point $x_0 \in M$ there exists $v_0 \in M$ such that

$$f(v_0) \leq f(x_0) - \varepsilon d(v_0, x_0), \tag{1.20}$$

$$f(x) > f(v_0) - \varepsilon d(v_0, x), \ \forall x \in M \setminus \{v_0\}. \tag{1.21}$$

The proof can be found in Ekeland [11], [12].

Now we describe the asymptotic behavior of a large class of nonsmooth functions.

Theorem 1.6 Let $\Phi : X \to I\!R$ be a locally Lipschitz functional and let $\Psi : X \to I\!R \cup \{+\infty\}$ be a proper, convex, lower semicontinuous function. For the function

$$f = \Phi + \Psi \tag{1.22}$$

we suppose that

$$\alpha := \liminf_{\|v\| \to \infty} f(v) \in I\!R. \tag{1.23}$$

Then for every sequence $\{\varepsilon_n\} \subset I\!R^+$ with $\varepsilon_n \downarrow 0$, there exists a sequence $\{u_n\} \subset X$ satisfying

$$\|u_n\| \to \infty \text{ as } n \to \infty, \tag{1.24}$$

$$f(u_n) \to \alpha \text{ as } n \to \infty \tag{1.25}$$

and property (1.19).

Proof. As in the proof of Proposition 1 in [4] we denote, for each $r > 0$,

$$m(r) = \inf_{\|u\| \geq r} f(u). \tag{1.26}$$

Assumption (1.23) in conjunction with (1.26) leads to

$$\alpha = \lim_{r \to \infty} m(r) \in I\!R. \tag{1.27}$$

Assertion (1.27) ensures that for each $\varepsilon > 0$ there exists $r_\varepsilon > 0$ satisfying

$$\alpha - \varepsilon^2 \leq m(r), \; \forall r \geq r_\varepsilon. \tag{1.28}$$

For any fixed $\varepsilon > 0$, let us choose a number $\overline{r}_\varepsilon$ with

$$\overline{r}_\varepsilon \geq \max\{r_\varepsilon, 2\varepsilon\}. \tag{1.29}$$

Using assumption (1.23), we can fix some $u_0 = u_0(\varepsilon) \in X$ such that

$$\|u_0\| \geq 2\overline{r}_\varepsilon \;\text{ and }\; f(u_0) < \alpha + \varepsilon^2. \tag{1.30}$$

The set $M = M(\varepsilon) \subset X$ given by

$$M = \{x \in X : \|x\| \geq \overline{r}_\varepsilon\} \tag{1.31}$$

is a closed subset of X, so M is a complete metric space with respect to the metric induced on M by the norm $\|\cdot\|$. The function $f : X \to$

$\mathbb{R} \cup \{+\infty\}$ expressed in (1.22) is lower semicontinuous on X, thus on M. By (1.26), (1.28) and (1.29) we derive that

$$f(u) \geq m(\|u\|) \geq \alpha - \varepsilon^2, \ \forall u \in X, \ \|u\| \geq \overline{r}_\varepsilon. \tag{1.32}$$

Estimate (1.32) ensures that the function f is bounded from below on M. From (1.31) and the first inequality in (1.30) it is seen that $u_0 \in M$. Hence by the second relation in (1.30) we know that the function f is proper on M. Since all the assumptions of Theorem 1.5 are fulfilled for the functional $f|_M : M \to \mathbb{R} \cup \{+\infty\}$, it is allowed to apply Theorem 1.5 where the fixed number $\varepsilon > 0$ and the point $x_0 = u_0$ are the data entering relations (1.28), (1.29) and (1.30). Consequently, we find some $v_\varepsilon \in M$ such that

$$f(v_\varepsilon) \leq f(u_0) - \varepsilon\|v_\varepsilon - u_0\|, \tag{1.33}$$

and

$$f(x) > f(v_\varepsilon) - \varepsilon\|v_\varepsilon - x\|, \ \forall x \neq v_\varepsilon, \ \|x\| \geq \overline{r}_\varepsilon \tag{1.34}$$

(see (1.20), (1.21)).

Since $v_\varepsilon \in M$, using relations (1.28), (1.29), (1.31), (1.26), (1.33) and the second inequality in (1.30), we have

$$\alpha - \varepsilon^2 \leq m(\overline{r}_\varepsilon) \leq f(v_\varepsilon) \leq f(u_0) - \varepsilon\|v_\varepsilon - u_0\| < \alpha + \varepsilon^2 - \varepsilon\|v_\varepsilon - u_0\|.$$

This implies that

$$\|v_\varepsilon - u_0\| < 2\varepsilon. \tag{1.35}$$

Combining (1.35), the first inequality in (1.30) and (1.29) we deduce that

$$\|v_\varepsilon\| \geq \|u_0\| - \|v_\varepsilon - u_0\| > 2\overline{r}_\varepsilon - 2\varepsilon \geq \overline{r}_\varepsilon. \tag{1.36}$$

From (1.36) it is clear that v_ε is an interior point of M defined in (1.31). This guaranties that for an arbitrary $v \in X$, $v \neq v_\varepsilon$, it is true that $x = v_\varepsilon + t(v - v_\varepsilon)$ belongs to the interior of M in (1.31) whenever $t > 0$ is sufficiently small. It is thus permitted to use a point x as above in (1.34). By means of (1.22) and (1.34) we can write

$$\Phi(v_\varepsilon + t(v - v_\varepsilon)) + \Psi(v_\varepsilon + t(v - v_\varepsilon))$$

$$> \Phi(v_\varepsilon) + \Psi(v_\varepsilon) - \varepsilon t\|v - v_\varepsilon\|, \ \forall v \in X \setminus \{v_\varepsilon\}, \tag{1.37}$$

for all $t > 0$ sufficiently small. On the other hand, we observe from inequality (1.33) and the second relation in (1.30) that $\Psi(v_\varepsilon) < +\infty$. On the basis of the convexity of $\Psi : X \to \mathbb{R} \cup \{+\infty\}$, the inequality (1.37) yields

$$\Phi(v_\varepsilon + t(v - v_\varepsilon)) - t\Psi(v_\varepsilon) + t\Psi(v) > \Phi(v_\varepsilon) - \varepsilon t\|v - v_\varepsilon\|, \ \forall v \in X \setminus \{v_\varepsilon\},$$

for $t > 0$ small enough. Passing to the limit one obtains that

$$\limsup_{t\downarrow 0} \frac{1}{t}(\Phi(v_\varepsilon + t(v - v_\varepsilon)) - \Phi(v_\varepsilon)) + \Psi(v) - \Psi(v_\varepsilon)$$

$$> -\varepsilon\|v - v_\varepsilon\|, \ \forall v \in X \setminus \{v_\varepsilon\}.$$

Taking into account Definition 1.1 we deduce that

$$\Phi^0(v_\varepsilon; v - v_\varepsilon) + \Psi(v) - \Psi(v_\varepsilon) > -\varepsilon\|v - v_\varepsilon\|, \ \forall v \in X \setminus \{v_\varepsilon\}. \quad (1.38)$$

Consider now a sequence $\{\varepsilon_n\} \subset I\!R^+$ with $\varepsilon_n \downarrow 0$. Corresponding to it we may choose a sequence of positive numbers

$$r_{\varepsilon_n} \to +\infty \ \text{ as } \ n \to \infty$$

satisfying (1.28) with $\varepsilon = \varepsilon_n$. We denote $u_n = v_{\varepsilon_n}$, where we recall that $v_{\varepsilon_n} \in M = M(\varepsilon_n)$ is the point satisfying (1.38) with $\varepsilon = \varepsilon_n$, i.e., property (1.19) holds true. Since $\|u_n\| \geq \overline{r}_{\varepsilon_n} \geq r_{\varepsilon_n}$ (cf. (1.31) and (1.29)), we obtain that property (1.24) is satisfied. In order to check relation (1.25), we notice that (1.33) and the second inequality in (1.30) imply

$$f(u_n) \leq f(u_0) - \varepsilon_n\|u_n - u_0\| \leq f(u_0) < \alpha + \varepsilon_n^2.$$

This combined with (1.24) and (1.23) expresses that

$$\alpha \leq \liminf_{n\to\infty} f(u_n) \leq \limsup_{n\to\infty} f(u_n) \leq \alpha,$$

which establishes (1.25). The proof of Theorem 1.6 is complete. ■

Adding a suitable Palais-Smale condition, one obtains from Theorem 1.6 important coerciveness results.

Corollary 1.3 Assume that the functional $f : X \to I\!R \cup \{+\infty\}$ satisfies the structure hypothesis (1.22), with Φ and Ψ as in the statement of Theorem 1.6, together with

$$\alpha > -\infty, \quad (1.39)$$

where α is defined in (1.23), and

$$f \text{ verifies the Palais-Smale condition in Definition 1.7.} \quad (1.40)$$

Then f is coercive on X, i.e.,

$$f(u) \to +\infty \ \text{ as } \ \|u\| \to \infty. \quad (1.41)$$

Proof. Arguing by contradiction we admit that the functional f in (1.22) is not coercive. Since (1.41) does not hold, there exists a sequence $\{v_n\} \subset X$ satisfying $\|v_n\| \to \infty$ and

$$\alpha \leq \liminf_{n \to \infty} f(v_n) < +\infty. \tag{1.42}$$

From (1.39) and (1.42) one obtains that

$$\alpha = \liminf_{\|v\| \to \infty} f(v) \in \mathbb{R}.$$

Consequently, we may apply Theorem 1.6 to the functional $f : X \to \mathbb{R} \cup \{+\infty\}$ for a fixed sequence $\{\varepsilon_n\} \subset \mathbb{R}^+$ with $\varepsilon_n \downarrow 0$. In this way a sequence $\{u_n\} \subset X$ is found fulfilling the properties (1.24), (1.25) and (1.19). According to assumption (1.40) it results that $\{u_n\}$ possesses a convergent subsequence denoted again by $\{u_n\}$, say

$$u_n \to u \quad \text{as } n \to \infty,$$

for some $u \in X$. This contradicts assertion (1.24), which accomplishes the proof. ■

Corollary 1.4 Let $\Phi : X \to \mathbb{R}$ be a locally Lipschitz functional which satisfies the Palais-Smale condition in the sense of Definition 1.5 and

$$\liminf_{\|v\| \to \infty} \Phi(v) > -\infty.$$

Then Φ is coercive on X, i.e. $\Phi(u) \to +\infty$ as $\|u\| \to \infty$.

Proof. Let us apply Corollary 1.3 with $\Psi = 0$. Then condition (1.39) with $\Psi = 0$ is satisfied (for α introduced in (1.23)). By part (*i*) in Theorem 1.4 the requirement (1.40) is satisfied for $f = \Phi$. Then Corollary 1.3 leads to the desired result. ■

Corollary 1.5 Let $\Phi : X \to \mathbb{R}$ be a function of class C^1 and let $\Psi : X \to \mathbb{R} \cup \{+\infty\}$ be a proper, convex, lower semicontinuous function. Assume that the functional $f = \Phi + \Psi : X \to \mathbb{R} \cup \{+\infty\}$ satisfies the Palais-Smale condition in the sense of Definition 1.6 and fulfils also (1.39), where α is introduced in (1.23). Then f is coercive on X.

Proof. Let us apply Corollary 1.3 for $f = \Phi + \Psi : X \to \mathbb{R} \cup \{+\infty\}$, with Φ and Ψ as in the statement of Corollary 1.5. Since we supposed that property (1.39) holds, it remains to check (1.40). In turn, this follows from part (*ii*) in Theorem 1.4. The proof is thus complete. ■

Remark 1.2 If $\Phi \in C^1(X; \mathbb{R})$ and $\Psi = 0$ in (1.22), Theorem 1.6 reduces to Proposition 1 of Brézis and Nirenberg [4].

Remark 1.3 The case in (1.22) where Φ is Gâteaux differentiable and lower semicontinuous has been studied in Caklovic, Li and Willem [5] (with $\Psi = 0$) and in Goeleven [13]. Our Corollary 1.3 provides, in particular, nondifferentiable versions of these results. Precisely, Corollary 1.3 covers the nondifferentiable situation where, in (1.22), $\Phi : X \to \mathbb{R}$ is locally Lipschitz and $\Psi : X \to \mathbb{R} \cup \{+\infty\}$ is proper, convex and lower semicontinuous. Therefore Corollary 1.3 deals with different situations with respect to [5] and [13]. Corollary 1.4 treats the purely locally Lipschitz case, i.e. $\Psi = 0$ in (1.22). It extends Corollary 1 in [4] and allows to extend the main result in [5] to the locally Lipschitz functionals. It overlaps with the main result in [5] if $\Phi \in C^1(X; \mathbb{R})$ and Φ is bounded from below. Corollary 1.5 represents the version of Corollary 1.3 in the case where $\Phi \in C^1(X; \mathbb{R})$. Under the assumption that $\Phi \in C^1(X; \mathbb{R})$ is bounded from below, Corollary 1.5 has been obtained in [13].

Remark 1.4 Corollaries 1.3, 1.4 and 1.5 correspond to the three concepts of Palais-Smale conditions in Definitions 1.7, 1.5 and 1.6, respectively.

Remark 1.5 The coerciveness property for the same class of nonsmooth functionals as introduced in (1.22) is studied in D. Motreanu, V. V. Motreanu and D. Paşca [20] by using a more general Palais-Smale condition, inspired by Zhong [25].

3. Nonsmooth Analysis in the Sense of Degiovanni

Let X be a metric space endowed with the metric d and let $f : X \to \overline{\mathbb{R}}$ be a function. We denote by $B_r(u)$ the open ball of centre u and radius r and we set

$$\operatorname{epi}(f) = \{(u, \lambda) \in X \times \mathbb{R} :\ f(u) \leq \lambda\}$$

(the epigraph of f). In the following, $X \times \mathbb{R}$ will be endowed with the metric

$$d\left((u, \lambda), (v, \mu)\right) = \left(d(u, v)^2 + (\lambda - \mu)^2\right)^{\frac{1}{2}}$$

and $\operatorname{epi}(f)$ with the induced metric.

Definition 1.8 For every $u \in X$ with $f(u) \in \mathbb{R}$, we denote by $|df|\,(u)$ the supremum of the σ's in $[0, +\infty[$ such that there exist $\delta > 0$ and a continuous map

$$\mathcal{H} : (B_\delta(u, f(u)) \cap \operatorname{epi}(f)) \times [0, \delta] \to X$$

satisfying

$$d(\mathcal{H}((w,\mu),t),w) \le t\,, \quad f(\mathcal{H}((w,\mu),t)) \le \mu - \sigma t\,,$$

whenever $(w,\mu) \in B_\delta(u, f(u)) \cap \mathrm{epi}(f)$ and $t \in [0,\delta]$.

The extended real number $|df|\,(u)$ is called the weak slope of f at u.

Remark 1.6 It is seen from Definition 1.8 that always the weak slope $|df|(u)$ exists (as a number in $[0,+\infty]$). However, this is far from well established theories. For instance, the function $f : \mathbb{R} \to \mathbb{R} \cup \{+\infty\}$ given by $f(x) = -\sqrt{x}$ for $x \ge 0$ and $f(x) = +\infty$ for $x < 0$ is convex, lower semicontinuous, proper with empty subdifferential at zero, that is $\partial f(0) = \emptyset$, so no notion of derivative is usual for such a function at the point 0.

The above notion has been introduced in [10], following an equivalent approach. When f is continuous, it has been independently introduced in [17], while a variant has been considered in [15], [16]. The above version appeared in [6].

Define $\mathcal{G}_f : \mathrm{epi}(f) \to \mathbb{R}$ by $\mathcal{G}_f(u,\lambda) = \lambda$. Clearly, $\mathcal{G}_f$ is Lipschitz continuous of constant 1.

The next basic property of the weak slope is due to Campa and Degiovanni (Proposition 2.3 in [6]).

Proposition 1.5 For every $u \in X$ with $f(u) \in \mathbb{R}$, we have

$$|df|\,(u) = \begin{cases} \dfrac{|d\mathcal{G}_f|\,(u,f(u))}{\sqrt{1-|d\mathcal{G}_f|\,(u,f(u))^2}} & \text{if } |d\mathcal{G}_f|\,(u,f(u)) < 1\,, \\ +\infty & \text{if } |d\mathcal{G}_f|\,(u,f(u)) = 1\,. \end{cases}$$

Proof. We claim that

$$|df|\,(u) \ge \begin{cases} \dfrac{|d\mathcal{G}_f|\,(u,f(u))}{\sqrt{1-|d\mathcal{G}_f|\,(u,f(u))^2}} & \text{if } |d\mathcal{G}_f|\,(u,f(u)) < 1\,, \\ +\infty & \text{if } |d\mathcal{G}_f|\,(u,f(u)) = 1\,. \end{cases}$$

If $|d\mathcal{G}_f|\,(u,f(u)) = 0$, the assertion is obvious. Otherwise, let $0 < \sigma < |d\mathcal{G}_f|\,(u,f(u))$. Since $\mathcal{G}_f$ is continuous, there exists a continuous mapping

$$\mathcal{H} : (B_\delta(u,f(u)) \cap \mathrm{epi}\,(f)) \times [0,\delta] \to \mathrm{epi}\,(f)$$

such that

$$d(\mathcal{H}((\xi,\mu),t),(\xi,\mu)) \le t, \quad f(\mathcal{H}((\xi,\mu),t)) \le f(\xi,\mu) - \sigma t,$$

whenever $(\xi, \mu) \in B_\delta(u, f(u)) \cap \mathrm{epi}\,(f)$ and $t \in [0, \delta]$. Let $\delta' > 0$ be such that $\delta' < \delta\sqrt{1-\sigma^2}$ and let

$$\mathcal{K} : (B_{\delta'}(u, f(u)) \cap \mathrm{epi}\,(f)) \times [0, \delta'] \to X$$

be defined by

$$\mathcal{K}((\xi, \mu), t) = \mathcal{H}_1 \left((\xi, \mu), \frac{t}{\sqrt{1-\sigma^2}} \right),$$

where $\mathcal{H}_1$ is the first component of $\mathcal{H}$. The map $\mathcal{K}$ is continuous and

$$d(\mathcal{K}((\xi, \mu), t), \xi)^2 = d\left(\mathcal{H}_1 \left((\xi, \mu), \frac{t}{\sqrt{1-\sigma^2}} \right), \xi \right)^2$$

$$\leq \frac{t^2}{1-\sigma^2} - \left| \mathcal{H}_2 \left((\xi, \mu), \frac{t}{\sqrt{1-\sigma^2}} \right) - \mu \right|^2 \leq \frac{t^2}{1-\sigma^2} - \frac{\sigma^2 t^2}{1-\sigma^2} = t^2,$$

where $\mathcal{H}_2$ is the second component of $\mathcal{H}$. Moreover, we have

$$f(\mathcal{K}((\xi, \mu), t)) = f\left(\mathcal{H}_1 \left((\xi, \mu), \frac{t}{\sqrt{1-\sigma^2}} \right) \right) \leq \mathcal{H}_2 \left((\xi, \mu), \frac{t}{\sqrt{1-\sigma^2}} \right)$$

$$= \mathcal{G}_f \left(\mathcal{H} \left((\xi, \mu), \frac{t}{\sqrt{1-\sigma^2}} \right) \right) \leq \mathcal{G}_f(\xi, \mu) - \frac{\sigma}{\sqrt{1-\sigma^2}} t = \mu - \frac{\sigma}{\sqrt{1-\sigma^2}} t.$$

Hence

$$|df|\,(u) \geq \frac{\sigma}{\sqrt{1-\sigma^2}}$$

and the claim follows since σ is arbitrary.

We prove in what follows the opposite inequality. If $|df|\,(u) = 0$ or $|d\mathcal{G}_f|\,(u, f(u)) = 1$, the assertion is evident. Otherwise, let $0 < \sigma < |df|\,(u)$ and let

$$\mathcal{H} : (B_\delta(u, f(u)) \cap \mathrm{epi}\,(f)) \times [0, \delta] \to X$$

be such that

$$d(\mathcal{H}(\xi, t), \xi) \leq t, \quad f(\mathcal{H}(\xi, t)) \leq f(\xi) - \sigma t,$$

whenever $\xi \in B_\delta(u, f(u)) \cap \mathrm{epi}\,(f)$ and $t \in [0, \delta]$. Define

$$\mathcal{K} : (B_\delta(u, f(u)) \cap \mathrm{epi}\,(f)) \times [0, \delta] \to \mathrm{epi}\,(f)$$

by

$$\mathcal{K}((\xi, \mu), t) = \left(\mathcal{H} \left((\xi, \mu), \frac{t}{\sqrt{1+\sigma^2}} \right), \mu - \frac{\sigma}{\sqrt{1+\sigma^2}} t \right).$$

Since

$$f\left(\mathcal{H}\left((\xi,\mu),\frac{t}{\sqrt{1+\sigma^2}}\right)\right)\leq \mu-\frac{\sigma}{\sqrt{1+\sigma^2}}t\,,$$

we actually have $\mathcal{K}((\xi,\mu),t) \in \mathrm{epi}\,(f)$. It is seen that $\mathcal{K}$ is continuous and

$$d(\mathcal{K}((\xi,\mu),t),(\xi,\mu))$$

$$=\left(d\left(\mathcal{H}\left((\xi,\mu),\frac{t}{\sqrt{1+\sigma^2}}\right),\xi\right)^2+\left(\frac{\sigma}{\sqrt{1+\sigma^2}}t\right)^2\right)^{1/2}$$

$$\leq\left(\frac{t^2}{1+\sigma^2}+\frac{\sigma^2t^2}{1+\sigma^2}\right)^{1/2}=t\,.$$

Moreover, we have

$$\mathcal{G}_f(\mathcal{K}((\xi,\mu),t))=\mu-\frac{\sigma}{\sqrt{1+\sigma^2}}t=\mathcal{G}_f(\xi,\mu)-\frac{\sigma}{\sqrt{1+\sigma^2}}t\,.$$

Therefore

$$|d\mathcal{G}_f|\,(u,f(u))\geq\frac{\sigma}{\sqrt{1+\sigma^2}}\,,$$

namely

$$\sigma\leq\frac{|d\mathcal{G}_f|\,(u,f(u))}{\sqrt{1-|d\mathcal{G}_f|\,(u,f(u))^2}}\,.$$

Since σ is arbitrary, our assertion follows. ■

Proposition 1.5 allows us to reduce, at some extent, the study of the weak slope of the general function f to that of the continuous function $\mathcal{G}_f$.

Definition 1.8 can be simplified, when f is continuous.

Proposition 1.6 ([6], Proposition 2.2) Let $f : X \to \mathbb{R}$ be continuous. Then $|df|\,(u)$ is the supremum of the σ's in $[0,+\infty[$ such that there exist $\delta > 0$ and a continuous map

$$\mathcal{H} : B_\delta(u)\times[0,\delta]\to X$$

satisfying

$$d(\mathcal{H}(w,t),w)\leq t\,,\quad f(\mathcal{H}(w,t))\leq f(w)-\sigma t\,, \tag{1.43}$$

whenever $w \in B_\delta(u)$ and $t \in [0,\delta]$.

Proof. If

$$\mathcal{H} : (B_\delta(u,f(u))\cap\mathrm{epi}\,(f))\times[0,\delta]\to X$$

is a map as in Definition 1.8, taking into account the continuity of f we may define

$$\mathcal{K} : B_{\delta'}(u) \times [0, \delta'] \to X$$

by $\mathcal{K}(\xi, t) = \mathcal{H}((\xi, f(\xi)), t)$ for some small $\delta' > 0$. It is easy to observe that $\mathcal{K}$ has the properties required in the statement of the proposition.

Conversely, let

$$\mathcal{K} : B_\delta(u) \times [0, \delta] \to X$$

be a map as in the statement of the proposition. Then

$$\mathcal{H} : (B_\delta(u, f(u)) \cap \operatorname{epi}(f)) \times [0, \delta] \to X$$

defined by $\mathcal{H}((\xi, \mu), t) = \mathcal{K}(\xi, t)$ has the properties required by Definition 1.8, as

$$f(\mathcal{H}((\xi, \mu), t)) = f(\mathcal{K}(\xi, t)) \le f(\xi) - \sigma t \le \mu - \sigma t.$$

The proof is complete. ■

Remark 1.7 The notion of weak slope in Definition 1.8 and Proposition 1.6 does not always reflect the slope in mathematical analysis. For example, if $f : \mathbb{R} \to \mathbb{R}$ is the continuous function expressed by $f(x) = \sqrt{|x|}$, $\forall x \in \mathbb{R}$, we have that $|df|(0) = 0$ because 0 is a minimum point, while the slope of the tangent to the graph of f at the point 0 is $+\infty$.

We need also, in a particular case, the notion of equivariant weak slope.

Definition 1.9 Let X be a normed space and $f : X \to \overline{\mathbb{R}}$ an even function with $f(0) < +\infty$. For every $(0, \lambda) \in \operatorname{epi}(f)$ we denote by $\left|d_{\mathbb{Z}_2}\mathcal{G}_f\right|(0, \lambda)$ the supremum of the σ's in $[0, +\infty[$ such that there exist $\delta > 0$ and a continuous map

$$\mathcal{H} = (\mathcal{H}_1, \mathcal{H}_2) : (B_\delta(0, \lambda) \cap \operatorname{epi}(f)) \times [0, \delta] \to \operatorname{epi}(f)$$

satisfying

$$d\,(\mathcal{H}((w, \mu), t), (w, \mu)) \le t\,, \quad \mathcal{H}_2((w, \mu), t) \le \mu - \sigma t\,,$$

$$\mathcal{H}_1((-w, \mu), t) = -\mathcal{H}_1((w, \mu), t)\,,$$

whenever $(w, \mu) \in B_\delta(0, \lambda) \cap \operatorname{epi}(f)$ and $t \in [0, \delta]$.

Remark 1.8 In Proposition 1.6, if there exist $\varrho > 0$ and a continuous map $\mathcal{H}$ satisfying

$$d(\mathcal{H}(w, t), w) \le \varrho t\,, \quad f(\mathcal{H}(w, t)) \le f(w) - \sigma t\,,$$

instead of (1.43), we can deduce that $|df|(u) \geq \sigma/\varrho$. A similar remark applies to Definition 1.9.

By means of the weak slope, we can now introduce the two main notions of critical point theory in this framework.

Definition 1.10 We say that $u \in X$ is a (lower) critical point of f if $f(u) \in \mathbb{R}$ and $|df|(u) = 0$. We say that $c \in \mathbb{R}$ is a (lower) critical value of f, if there exists a (lower) critical point $u \in X$ of f with $f(u) = c$.

Definition 1.11 Let $c \in \mathbb{R}$. A sequence $\{u_h\}$ in X is said to be a Palais-Smale sequence at level c ($(PS)_c$-sequence, for short) for f if $f(u_h) \to c$ and $|df|(u_h) \to 0$. We say that f satisfies the Palais-Smale condition at level c ($(PS)_c$, for short), if every $(PS)_c$-sequence $\{u_h\}$ for f admits a convergent subsequence $\{u_{h_k}\}$ in X.

The main feature of the weak slope is that it allows to prove results in critical point theory for general continuous functions defined on complete metric spaces. Moreover, in the context of critical points one can try to reduce the study of a lower semicontinuous function f to that of the continuous function $\mathcal{G}_f$. Actually, Proposition 1.5 suggests to exploit the bijective correspondence between the set where f is finite and the graph of f. This approach can be successful, if we can ensure that the remaining part of $\mathrm{epi}(f)$ does not carry much information. The next notion turns out to be useful for this purpose.

Definition 1.12 Let $c \in \mathbb{R}$. We say that f satisfies condition $(epi)_c$ if there exists $\varepsilon > 0$ such that

$$\inf\{|d\mathcal{G}_f|(u,\lambda) : f(u) < \lambda,\ |\lambda - c| < \varepsilon\} > 0\,.$$

The next two results may help in dealing with condition $(epi)_c$.

Proposition 1.7 Let $(u,\lambda) \in \mathrm{epi}(f)$. Assume that there exist constants $\varrho, \sigma, \delta, \varepsilon > 0$ and a continuous map

$$\mathcal{H} : \{w \in B_\delta(u) : f(w) < \lambda + \delta\} \times [0,\delta] \to X$$

satisfying

$$d(\mathcal{H}(w,t), w) \leq \varrho t\,, \quad f(\mathcal{H}(w,t)) \leq \max\{f(w) - \sigma t, \lambda - \varepsilon\}$$

whenever $w \in B_\delta(u)$, $f(w) < \lambda + \delta$ and $t \in [0,\delta]$. Then we have

$$|d\mathcal{G}_f|(u,\lambda) \geq \frac{\sigma}{\sqrt{\varrho^2 + \sigma^2}}\,.$$

If, moreover, X is a normed space, f is even, $u = 0$ and $\mathcal{H}(-w,t) = -\mathcal{H}(w,t)$, then we have

$$\left|d_{\mathbb{Z}_2}\mathcal{G}_f\right|(0,\lambda) \geq \frac{\sigma}{\sqrt{\varrho^2+\sigma^2}}\,.$$

Proof. Let $\delta' \in]0,\delta]$ be such that $\delta' + \sigma\delta' \leq \varepsilon$ and let

$$\mathcal{K} : (B_{\delta'}(u,\lambda) \cap \mathrm{epi}(f)) \times [0,\delta'] \to \mathrm{epi}(f)$$

be defined by $\mathcal{K}((w,\mu),t) = (\mathcal{H}(w,t), \mu - \sigma t)$. If $(w,\mu) \in B_{\delta'}(u,\lambda) \cap \mathrm{epi}(f)$ and $t \in [0,\delta']$, we have

$$\lambda - \varepsilon \leq \lambda - \delta' - \sigma\delta' < \mu - \sigma t\,, \quad f(w) - \sigma t \leq \mu - \sigma t\,,$$

hence

$$f(\mathcal{H}(w,t)) \leq \max\{f(w) - \sigma t, \lambda - \varepsilon\} \leq \mu - \sigma t\,.$$

Therefore $\mathcal{K}$ actually takes its values in $\mathrm{epi}(f)$. Furthermore, one has

$$d(\mathcal{K}((w,\mu),t),(w,\mu)) \leq \sqrt{\varrho^2+\sigma^2}\,t\,,$$

$$\mathcal{G}_f(\mathcal{K}((w,\mu),t)) = \mu - \sigma t = \mathcal{G}_f(w,\mu) - \sigma t\,.$$

Taking into account Proposition 1.6 and Remark 1.8, the first assertion follows.

In the symmetric case, $\mathcal{K}$ automatically satisfies the further condition required in Definition 1.9. ∎

Corollary 1.6 Let $(u,\lambda) \in \mathrm{epi}(f)$ with $f(u) < \lambda$. Assume that for every $\varrho > 0$ there exist $\delta > 0$ and a continuous map

$$\mathcal{H} : \{w \in B_\delta(u) : f(w) < \lambda + \delta\} \times [0,\delta] \to X$$

satisfying

$$d(\mathcal{H}(w,t),w) \leq \varrho t\,, \quad f(\mathcal{H}(w,t)) \leq f(w) + t(f(u) - f(w) + \varrho)$$

whenever $w \in B_\delta(u)$, $f(w) < \lambda + \delta$ and $t \in [0,\delta]$. Then we have $|d\mathcal{G}_f|(u,\lambda) = 1$. If, moreover, X is a normed space, f is even, $u = 0$ and $\mathcal{H}(-w,t) = -\mathcal{H}(w,t)$, then we have $\mathbb{Z}_2\mathcal{G}_f(0,\lambda) = 1$.

Proof. Let $\varepsilon > 0$ with $\lambda - 2\varepsilon > f(u)$, let $0 < \varrho < \lambda - f(u) - 2\varepsilon$ and let δ and $\mathcal{H}$ be as in the hypothesis. By reducing δ, we may also assume that

$$\delta \leq 1\,, \quad \delta(|\lambda - 2\varepsilon| + |f(u) + \varrho|) \leq \varepsilon\,.$$

Now consider $w \in B_\delta(u)$ with $f(w) < \lambda + \delta$ and $t \in [0, \delta]$. If $f(w) \leq \lambda - 2\varepsilon$, we have

$$\begin{aligned} f(w) + t(f(u) - f(w) + \varrho) &= (1-t)f(w) + t(f(u) + \varrho) \\ &\leq (1-t)(\lambda - 2\varepsilon) + t(f(u) + \varrho) \\ &\leq \lambda - 2\varepsilon + t|\lambda - 2\varepsilon| + t|f(u) + \varrho| \leq \lambda - \varepsilon\,, \end{aligned}$$

while, if $f(w) > \lambda - 2\varepsilon$, we have

$$f(w) + t(f(u) - f(w) + \varrho) \leq f(w) - (\lambda - f(u) - 2\varepsilon - \varrho)t\,.$$

In any case it follows

$$f(\mathcal{H}(w,t)) \leq \max\left\{f(w) - (\lambda - f(u) - 2\varepsilon - \varrho)t, \lambda - \varepsilon\right\}\,.$$

From Proposition 1.7 we get

$$|d\mathcal{G}_f|\,(u,\lambda) \geq \frac{\lambda - f(u) - 2\varepsilon - \varrho}{\sqrt{\varrho^2 + (\lambda - f(u) - 2\varepsilon - \varrho)^2}}$$

and the first assertion follows by the arbitrariness of ϱ.

The same proof works also in the symmetric case. ∎

Now assume that X is a normed space over $\mathbb{R}$ and $f : X \to \overline{\mathbb{R}}$ a function.

Definition 1.13 For every $u \in X$ with $f(u) \in \mathbb{R}$, $v \in X$ and $\varepsilon > 0$, let $f^\circ_\varepsilon(u; v)$ be the infimum of r's in $\overline{\mathbb{R}}$ such that there exist $\delta > 0$ and a continuous map

$$\mathcal{V} : (B_\delta(u, f(u)) \cap \mathrm{epi}(f)) \times]0, \delta] \to B_\varepsilon(v)$$

satisfying

$$f(z + t\,\mathcal{V}((z,\mu),t)) \leq \mu + rt$$

whenever $(z,\mu) \in B_\delta(u, f(u)) \cap \mathrm{epi}(f)$ and $t \in]0,\delta]$. Then let

$$f^\circ(u; v) = \sup_{\varepsilon > 0} f^\circ_\varepsilon(u; v)\,.$$

The function $f^\circ(u; \cdot)$ is convex, lower semicontinuous and positively homogeneous of degree 1 (see [6], Corollary 4.6).

Definition 1.14 For every $u \in X$ with $f(u) \in \mathbb{R}$, we set

$$\partial f(u) = \{u^* \in X^* \;:\; \langle u^*, v\rangle \leq f^\circ(u; v), \quad \forall v \in X\}\,.$$

It turns out from Definition 1.13 that $f^\circ(u;v)$ is greater than or equal to the generalized directional derivative in the sense of Rockafellar ([8], [23]). Consequently, $\partial f(u)$ as introduced in Definition 1.14 contains the generalized gradient of f at u in the sense of Clarke. These modified notions of $f^\circ(u;v)$ and $\partial f(u)$ have been introduced in [6], [9], because they are better related with the notion of weak slope and hence more suitable for critical point theory in the framework of this Section, as the next result shows.

Theorem 1.7 If $u \in X$ and $f(u) \in \mathbb{R}$, the following facts hold:

(*a*) $|df|(u) < +\infty \iff \partial f(u) \neq \emptyset$;

(*b*) $|df|(u) < +\infty \implies |df|(u) \geq \min\{\|u^*\| : u^* \in \partial f(u)\}$.

Proof. See [6], Theorem 4.13. ∎

If $f : X \to \mathbb{R}$ is locally Lipschitz, it is proved in [6], Corollary 4.10, that these notions agree with those of Clarke as we considered in Definitions 1.1 and 1.2. Thus, in such a case, $f^\circ(u;\cdot)$ is also Lipschitz continuous and we have that

$$\forall u, v \in X : f^\circ(u;v) = \limsup_{\substack{z\to u,\, w\to v \\ t\to 0+}} \frac{f(z+tw)-f(z)}{t}, \tag{1.44}$$

$$f^\circ(\cdot;\cdot) \text{ is upper semicontinuous on } X \times X. \tag{1.45}$$

Properties (1.44), (1.45) will be essentially used in Chapter 11.

References

[1] S. Aizicovici, D. Motreanu and N. H. Pavel, Nonlinear programming problems associated with closed range operators, *Appl. Math. Optimization* **40** (1999), 211-228.

[2] J. P. Aubin and F. H. Clarke, Shadow Prices and Duality for a Class of Optimal Control Problems, *SIAM J. Control Optimization* **17** (1979), 567-586.

[3] H. Brézis, *Analyse Fonctionnelle - Théorie et Applications*, Masson, Paris, 1983.

[4] H. Brézis and L. Nirenberg, Remarks on finding critical points, Commun. Pure Appl. Math. **44** (1991), 939-963.

[5] L. Caklovic, S. Li and M. Willem, A note on Palais-Smale condition and coercivity, *Differ. Integral Equ.* **3** (1990), 799-800.

[6] I. Campa and M. Degiovanni, Subdifferential calculus and nonsmooth critical point theory, *SIAM J. Optim.* **10** (2000), 1020-1048.

[7] K.-C. Chang, Variational methods for non-differentiable functionals and their applications to partial differential equations, *J. Math. Anal. Appl.* **80** (1981), 102-129.

[8] F. H. Clarke, *Optimization and Nonsmooth Analysis*, New York, John Wiley-Interscience, 1983.

[9] M. Degiovanni, Nonsmooth critical point theory and applications, *Second World Congress of Nonlinear Analysts* (Athens, 1996), *Nonlinear Anal.* **30** (1997), 89-99.

[10] M. Degiovanni and M. Marzocchi, A critical point theory for nonsmooth functionals, *Ann. Mat. Pura Appl. IV. Ser.* **167** (1994), 73-100.

[11] I. Ekeland, On the variational principle, *J. Math. Anal. Appl.* **47** (1974), 324-353.

[12] I. Ekeland, Nonconvex minimization problems, *Bull. Amer. Math. Soc. (New Series)* **1** (1979), 443-474.

[13] D. Goeleven, A note on Palais-Smale condition in the sense of Szulkin, *Differ. Integral Equ.* **6** (1993), 1041-1043.

[14] A. D. Ioffe and V. L. Levin, Subdifferentials of Convex Functions, *Trans. Mosc. Math. Soc.* **26** (1972), 1-72.

[15] A. Ioffe and E. Schwartzman, Metric critical point theory I. Morse regularity and homotopic stability of a minimum, *J. Math. Pures Appl.* (9) **75** (1996), 125-153.

[16] A. Ioffe and E. Schwartzman, Metric critical point theory II. Deformation techniques. *New Results in Operator Theory and its Applications*, 131-144, *Oper. Theory Adv. Appl.*, **98**, Birkhäuser, Basel, 1997.

[17] G. Katriel, Mountain pass theorems and global homeomorphism theorems, *Ann. Inst. Henri Poincaré Anal. Non Linéaire* **11** (1994), 189-209.

[18] G. Lebourg, Valeur moyenne pour gradient généralisé, *C. R. Acad. Sci.* Paris **281** (1975), 795-797.

[19] D. Motreanu and V. V. Motreanu, Coerciveness Property for a Class of Nonsmooth Functionals, *Z. Anal. Anwend.* **19** (2000), 1087-1093.

[20] D. Motreanu, V. V. Motreanu and D. Paşca, A version of Zhong's coercivity result for a general class of nonsmooth functionals, *Abstr. Appl. Anal.*, to appear.

[21] D. Motreanu and P. D. Panagiotopoulos, *Minimax Theorems and Qualitative Properties of the Solutions of Hemivariational Inequalities and Applications*, Kluwer Academic Publishers, Nonconvex Optimization and Its Applications, Vol. 29, Dordrecht/Boston/London, 1999.

[22] D. Motreanu and N. H. Pavel, *Tangency, Flow-Invariance for Differential Equations and Optimization Problems*, Marcel Dekker, Inc., New York, Basel, 1999.

[23] R. T. Rockafellar, Generalized directional derivatives and subgradients of non-convex functions, *Can. J. Math.* **32** (1980), 257-280.

[24] A. Szulkin, Minimax principles for lower semicontinuous functions and applications to nonlinear boundary value problems, *Ann. Inst. Henri Poincaré Anal. Non Linéaire* **3** (1986), 77-109.

[25] C.-K. Zhong, A generalization of Ekeland's variational principle and application to the study of the relation between the weak P. S. condition and coercivity, *Nonlinear Anal.* **29** (1997), 1421-1431.

Chapter 2

CRITICAL POINTS FOR NONSMOOTH FUNCTIONALS

The present Chapter deals with critical point theory for three different nonsmooth situations. First, we set forth the critical point theory for locally Lipschitz functionals, following the approach of Chang [4] and Clarke [5]. Then a critical point theory is described for nonsmooth functionals expressed as a sum of a locally Lipschitz function and a convex, proper and lower semicontinuous function, using the development in Motreanu and Panagiotopoulos [26]. Finally, the critical point theory for continuous functionals defined on a complete metric space as introduced by Degiovanni and Marzocchi [7] is presented.

1. Critical Point Theory for Locally Lipschitz Functionals

The critical point theory for the locally Lipschitz functionals have been developed by Chang [4] using the calculus with generalized gradients as constructed by Clarke [5]. Related results can be found in [1], [19], [21], [22], [27], [28], [32], [34], [36], [37]. Applications can be found in [12], [24], [25], [30].

Throughout this Section we assume that X is a real Banach space and X^* is its topological dual. We denote by $\langle \cdot, \cdot \rangle$ the duality pairing between X^* and X. Let $\mathrm{Lip}_{loc}(X; \mathbb{R})$ be the set of all locally Lipschitz functionals $f : X \to \mathbb{R}$.

Definition 2.1 (Chang [4]) A point $x \in X$ is said to be a critical point of the locally Lipschitz functional $f : X \to \mathbb{R}$ if $0 \in \partial f(x)$, that is $f^0(x; v) \geq 0$, for any $v \in X$. The real number c is a critical value of f provided that there exists a critical point $x \in X$ such that $f(x) = c$.

We observe that a local minimum point is a critical point. Indeed, if x local minimum point of $f \in \mathrm{Lip}_{loc}(X; \mathbb{R})$, then for any $v \in X$ we have

$$0 \leq \limsup_{\lambda \searrow 0} \frac{f(x + \lambda v) - f(x)}{\lambda} \leq f^0(x; v).$$

We now introduce a compactness condition for locally Lipschitz functionals. In the C^1 framework this notion was given by Palais and Smale (in global variant, see [31]) and by Brézis, Coron and Nirenberg (in local variant, see [3]).

Throughout this Section we use the Palais-Smale condition for a locally Lipschitz functional $f : X \to \mathbb{R}$ at a level $c \in \mathbb{R}$ (in short, $(\mathrm{PS})_c$) as introduced in Definition 1.5. We say that the mapping f satisfies the Palais-Smale condition (in short, (PS)) if any sequence $\{x_n\}$ such that $\sup_n |f(x_n)| < +\infty$ and $\lambda(x_n) \to 0$, where

$$\lambda(x) = \inf\{\|x^*\|_* \,:\, x^* \in \partial f(x)\},$$

has a convergent subsequence.

Let $f : X \to \mathbb{R}$ be a locally Lipschitz functional. Assume that K is a compact metric space and K^* is a nonempty, closed subset of K. If $p^* : K^* \to X$ is a fixed continuous map, set

$$\mathcal{P} = \{p \in C(K; X) \,:\, p = p^* \text{ on } K^*\}.$$

By the Dugundji theorem, $\mathcal{P}$ is nonempty.

Define

$$c = \inf_{p \in \mathcal{P}} \max_{t \in K} f(p(t)).$$

Obviously, $c \geq \max_{t \in K^*} f(p^*(t))$.

The following result generalizes the celebrated Mountain Pass Theorem of Ambrosetti and Rabinowitz [2], [33] in the locally Lipschitz framework.

Theorem 2.1 Assume that $c > \max_{t \in K^*} f(p^*(t))$. Then there exists a sequence $\{x_n\}$ in X such that

(i) $\lim_{n \to \infty} f(x_n) = c$.

(ii) $\lim_{n \to \infty} \lambda(x_n) = 0$.

In the proof of this theorem we need the following auxiliary result.

Lemma 2.1 Let M be a compact metric space and let $\varphi : M \to 2^{X^*}$ be a multivalued mapping which is upper semicontinuous and such that for any $t \in M$, the set $\varphi(t)$ is convex and $\sigma(X^*, X)$-compact. For any $t \in M$, denote

$$\gamma(t) = \inf\{\|x^*\| \ : \ x^* \in \varphi(t)\}$$

and

$$\gamma = \inf_{t \in M} \gamma(t).$$

Then, for any $\varepsilon > 0$, there exists a continuous function $v : M \to X$ such that for any $t \in M$ and $x^* \in \varphi(t)$,

$$\|v(t)\| \leq 1 \quad \text{and} \quad \langle x^*, v(t)\rangle \geq \gamma - \varepsilon.$$

Proof. Without loss of generality we can assume that $\gamma > 0$ and $0 < \varepsilon < \gamma$. If B_r denotes the open ball in X^* centered in the origin and radius r, then for any $t \in M$ we have

$$B_{\gamma - \frac{\varepsilon}{2}} \cap \varphi(t) = \emptyset.$$

Since $\varphi(t)$ and $B_{\gamma-\frac{\varepsilon}{2}}$ are convex, disjoint and $\sigma(X^*, X)$-compact, we can apply the separation theorem in locally convex spaces (Theorem 3.4 in [38]). More precisely, we apply this theorem to the space X^* endowed with the $\sigma(X^*, X)$-topology and we use the fact that the dual of this space is X. It follows that for any $t \in M$ there exists $v_t \in X$ such that

$$\|v_t\| = 1 \quad \text{and} \quad \langle \xi, v_t\rangle \leq \langle x^*, v_t\rangle,$$

for any $\xi \in B_{\gamma-\frac{\varepsilon}{2}}$ and $x^* \in \varphi(t)$. Thus for any $x^* \in \varphi(t)$,

$$\langle x^*, v_t\rangle \geq \sup_{\xi \in B_{\gamma-\frac{\varepsilon}{2}}} \langle \xi, v_t\rangle = \gamma - \frac{\varepsilon}{2}\cdot$$

Since φ is upper semicontinuous, there exists an open neighborhood $V(t)$ of t such that for any $t' \in V(t)$ and $x^* \in \varphi(t')$,

$$\langle x^*, v_t\rangle > \gamma - \varepsilon.$$

Since M is compact and $M = \bigcup_{t \in M} V(t)$, we can find a finite subcovering $\{V_1, ..., V_n\}$ of M. Let $v_1, ..., v_n$ on the unit sphere of X such that

$$\langle x^*, v_i\rangle > \gamma - \varepsilon,$$

for any $1 \leq i \leq n$, $t \in V_i$ and $x^* \in \varphi(t)$. If $\rho_i(t) = \text{dist}(t, \partial V_i)$, define

$$\zeta_i(t) = \frac{\rho_i(t)}{\sum_{j=1}^n \rho_j(t)} \quad \text{and} \quad v(t) = \sum_{i=1}^n \zeta_i(t) v_i.$$

It is obvious that v satisfies the conclusion of the lemma. ■

Proof of Theorem 2.1. We apply Ekeland's variational principle [9], [10] to the functional

$$\psi(p) = \max_{t \in K} f(p(t))$$

defined on $\mathcal{P}$, which becomes a complete metric space if it is endowed with the usual metric. The mapping ψ is continuous and bounded from below because for any $p \in \mathcal{P}$,

$$\psi(p) \geq \max_{t \in K} f(p^*(t)).$$

Since

$$c = \inf_{p \in \mathcal{P}} \psi(p),$$

it follows that for any $\varepsilon > 0$ there exists $p \in \mathcal{P}$ such that

$$\psi(q) - \psi(p) + \varepsilon d(p, q) \geq 0, \quad \text{for any} \quad q \in \mathcal{P} \tag{2.1}$$

and

$$c \leq \psi(p) \leq c + \varepsilon.$$

Set

$$B(p) = \{t \in K \; : \; f(p(t)) = \psi(p)\}.$$

We observe that for concluding our proof it is enough to establish the existence of some $t' \in B(p)$ such that

$$\lambda(p(t')) \leq 2\varepsilon.$$

Indeed, our conclusion follows by choosing $\varepsilon = \frac{1}{n}$ and $x_n = p(t')$.

We apply Lemma 2.1 for $M = B(p)$ and $\varphi(t) = \partial f(p(t))$. Thus we obtain a continuous function $v : B(p) \to X$ such that for any $t \in B(p)$ and $x^* \in \partial f(p(t))$ we have

$$\|v(t)\| \leq 1 \quad \text{and} \quad \langle x^*, v(t) \rangle \geq \gamma - \varepsilon,$$

where

$$\gamma = \inf_{t \in B(p)} \lambda(p(t)).$$

It follows that for any $t \in B(p)$,

$$f^0(p(t); -v(t)) = \max\{\langle x^*, -v(t) \rangle \; : \; x^* \in \partial f(p(t))\}$$

$$= -\min\{\langle x^*, v(t) \rangle \; : \; x^* \in \partial f(p(t))\} \leq -\gamma + \varepsilon.$$

Our assumption implies that $B(p) \cap K^* = \emptyset$. Hence there exists a continuous extension $w : K \to X$ of v such that $w = 0$ on K^* and, for any $t \in K$,

$$\|w(t)\| \leq 1.$$

In the role of q in (2.1) we choose small variations of the path p:

$$q_h(t) = p(t) - hw(t),$$

where $h > 0$ is sufficiently small.

By (2.1) it follows that for any $h > 0$,

$$-\varepsilon \leq -\varepsilon \|w\|_\infty \leq \frac{\psi(q_h) - \psi(p)}{h}. \tag{2.2}$$

In what follows, $\varepsilon > 0$ is fixed, while $h \to 0$. Let $t_h \in K$ be such that $f(q_h(t_h)) = \psi(q_h)$. We can choose $h_n \to 0$ such that the sequence $\{t_{h_n}\}$ converges to some t_0, and it is obvious that $t_0 \in B(p)$. We also observe that

$$\frac{\psi(q_h) - \psi(p)}{h} = \frac{\psi(p - hw) - \psi(p)}{h} \leq \frac{f(p(t_h) - hw(t_h)) - f(p(t_h))}{h}.$$

This relation combined with (2.2) yields

$$-\varepsilon \leq \frac{f(p(t_h) - hw(t_h)) - f(p(t_h))}{h} \leq \frac{f(p(t_h) - hw(t_0)) - f(p(t_h))}{h}$$
$$+ \frac{f(p(t_h) - hw(t_h)) - f(p(t_h) - hw(t_0))}{h}.$$

Since f is locally Lipschitz and $t_{h_n} \to t_0$, we obtain that

$$\lim_{n \to \infty} \frac{f(p(t_{h_n}) - h_n w(t_{h_n})) - f(p(t_{h_n}) - h_n w(t_0))}{h_n} = 0.$$

Hence

$$-\varepsilon \leq \limsup_{n \to \infty} \frac{f(p(t_0) + z_n - h_n w(t_0)) - f(p(t_0) + z_n)}{h_n},$$

where $z_n = p(t_{h_n}) - p(t_0)$. Therefore

$$-\varepsilon \leq f^0(p(t_0); -w(t_0)) = f^0(p(t_0); -v(t_0)) \leq -\gamma + \varepsilon.$$

It follows that

$$\gamma = \inf\{\|x^*\| \,:\, x^* \in \partial f(p(t)),\ t \in B(p)\} \leq 2\varepsilon.$$

Taking into account the lower semicontinuity of λ we obtain the existence of some $t' \in B(p)$ such that

$$\lambda(p(t')) = \inf\{\|x^*\| \; : \; x^* \in \partial f(p(t'))\} \leq 2\varepsilon.$$

The proof is complete. ∎

Corollary 2.1 If f satisfies the $(\mathrm{PS})_c$ condition and the assumptions of Theorem 2.1, then c is a critical value of f that corresponds to a critical point which does not belong to $p^*(K^*)$.

Proof. The proof of this result follows by Theorem 2.1 combined with the lower semicontinuity of λ. ∎

Corollary 2.2 Let $f : X \to I\!\!R$ be a locally Lipschitz functional. Assume that there exists a subset S of X such that for any $p \in \mathcal{P}$,

$$p(K) \cap S \neq \emptyset.$$

If

$$\inf_{x \in S} f(x) > \max_{t \in K^*} f(p^*(t)),$$

then the conclusion of Theorem 2.1 holds.

Proof. It is enough to observe that

$$\inf_{p \in \mathcal{P}} \max_{t \in K} f(p(t)) \geq \inf_{x \in S} f(x) > \max_{t \in K^*} f(p^*(t)).$$

The proof is complete. ∎

Corollary 2.3 (Saddle Point Theorem) Let $f : X \to I\!\!R$ be a locally Lipschitz function. Assume that $X = Y \oplus Z$, where Z is a finite dimensional subspace of X and for some $z_0 \in Z$ there exists $R > \|z_0\|$ such that

$$\inf_{y \in Y} f(y + z_0) > \max\{f(z) \; : \; z \in Z, \; \|z\| = R\}.$$

Let

$$K = \{z \in Z \; : \; \|z\| \leq R\}$$

and

$$\mathcal{P} = \{p \in C(K; X) \; : \; p(x) = x \text{ if } \|x\| = R\}.$$

If f satisfies the $(\mathrm{PS})_c$ condition, then c is a critical value of f.

Proof. It suffices to apply Corollary 2.2 for $S = z_0 + Y$. In this respect we have to prove that for every $p \in \mathcal{P}$,

$$p(K) \cap (z_0 + Y) \neq \emptyset.$$

If $P : X \to Z$ is the canonical projection, the above condition is equivalent to the fact that, for each $p \in \mathcal{P}$, there exists $x \in K$ such that

$$P(p(x) - z_0) = P(p(x)) - z_0 = 0.$$

This follows easily by a topological degree argument. Indeed, for some fixed $p \in \mathcal{P}$, we have

$$P \circ p = id \quad \text{on} \quad K^* = \partial K.$$

Hence we have the following equalities involving the Brouwer degree

$$d(P \circ p, \operatorname{Int} K, 0) = d(P \circ p, \operatorname{Int} K, z_0) = d(id, \operatorname{Int} K, z_0) = 1.$$

By the existence property of the Brouwer degree we find some $x \in \operatorname{Int} K$ such that $(P \circ p)(x) - z_0 = 0$, which concludes our proof. ■

Theorem 2.1 enables us to prove the following result which is due to Brézis, Coron and Nirenberg (see Theorem 2 in [3]).

Corollary 2.4 Let $f : X \to I\!R$ be a Gâteaux differentiable functional such that $f' : (X, \|\cdot\|) \to (X^*, \sigma(X^*, X))$ is continuous. Assume that $c > \max_{t \in K^*} f(p^*(t))$. Then there exists a sequence $\{x_n\}$ in X such that

(i) $\lim_{n\to\infty} f(x_n) = c$

(ii) $\lim_{n\to\infty} \|f'(x_n)\| = 0.$

Moreover, if f satisfies the $(\mathrm{PS})_c$ condition, then there exists $x \in X$ such that $f(x) = c$ and $f'(x) = 0$.

Proof. We first observe that f' is locally bounded. Indeed, given a sequence $\{x_n\}$ which converges to x_0, we have

$$\sup_n |\langle f'(x_n), v\rangle| < \infty,$$

for every $v \in X$. So, by the Banach-Steinhaus theorem,

$$\limsup_{n\to\infty} \|f'(x_n)\| < \infty.$$

If $\lambda > 0$ and $h \in X$ are sufficiently small we have

$$|f(x_0 + h + \lambda v) - f(x_0 + h)| = |\lambda\langle f'(x_0 + h + \lambda\theta v), v\rangle| \leq C\|\lambda v\|, \quad (2.3)$$

where $\theta \in]0,1[$. Hence, by the continuity of f', we obtain that $f \in \mathrm{Lip}_{loc}(X; I\!R)$ and $f^0(x_0; v) = \langle f'(x_0), v\rangle$. In (2.3) the existence of C

follows by the local boundedness of f'. Since f^0 is linear in v, it follows that

$$\partial f(x) = \{f'(x)\}$$

and it remains to apply Theorem 2.1 and Corollary 2.1. ■

It is natural to ask what happens if the assumption $c > \max_{t \in K^*} f(p^*(t))$ fails, that is if $c = \max_{t \in K^*} f(p^*(t))$. The following example shows that in this case the conclusion of Theorem 2.1 is no more valid.

Example 2.1. Let us choose $X = \mathbb{R}^2$, $K = [0,1] \times \{0\}$, $K^* = \{(0,0),(1,0)\}$ and denote by p^* the identity mapping of K^*. The functional $f : X \to \mathbb{R}$ defined by $f(x,y) = x + |y|$ is locally Lipschitz. In this case,

$$c = \max_{t \in K^*} f(p^*(t)) = 1.$$

A straightforward computation shows that

$$\partial f(x,y) = \begin{pmatrix} 1 \\ 1 \end{pmatrix}, \text{ if } y > 0$$

$$\partial f(x,y) = \begin{pmatrix} 1 \\ -1 \end{pmatrix}, \text{ if } y < 0$$

$$\partial f(x,0) = \left\{ \begin{pmatrix} 1 \\ a \end{pmatrix} : -1 \leq a \leq 1 \right\}.$$

It follows that f satisfies the Palais-Smale condition. However, f has no critical points.

The next result gives a sufficient condition which ensures the conclusion of Theorem 2.1 if $c = \max_{t \in K^*} f(p^*(t))$. In the sequel we suppose that the previous equality holds.

Theorem 2.2 Assume that for every $p \in \mathcal{P}$ there exists $t \in K \setminus K^*$ such that $f(p(t)) \geq c$. Then there exists a sequence $\{x_n\}$ in X such that

(i) $\lim_{n \to \infty} f(x_n) = c$

(ii) $\lim_{n \to \infty} \lambda(x_n) = 0$.

Moreover, if f satisfies the $(\mathrm{PS})_c$ condition, then c is a critical value of f. Furthermore, if $\{p_n\}$ is an arbitrary sequence in $\mathcal{P}$ satisfying

$$\lim_{n \to \infty} \max_{t \in K} f(p_n(t)) = c,$$

then there exists a sequence $\{t_n\}$ in K such that

$$\lim_{n\to\infty} f(p_n(t_n)) = c \quad \text{and} \quad \lim_{n\to\infty} \lambda(p_n(t_n)) = 0.$$

Proof. For every $\varepsilon > 0$ we apply Ekeland's variational principle to the perturbed functional $\psi_\varepsilon : \mathcal{P} \to I\!R$ defined by

$$\psi_\varepsilon(p) = \max_{t\in K}(f(p(t)) + \varepsilon d(t)),$$

where

$$d(t) = \min\{\text{dist}(t, K^*), 1\}.$$

If

$$c_\varepsilon = \inf_{p\in\mathcal{P}} \psi_\varepsilon(p),$$

then

$$c \le c_\varepsilon \le c + \varepsilon.$$

Applying Ekeland's variational principle, we find a path $p \in \mathcal{P}$ such that for every $q \in \mathcal{P}$,

$$\psi_\varepsilon(q) - \psi_\varepsilon(p) + \varepsilon d(p,q) \ge 0, \tag{2.4}$$

$$c \le c_\varepsilon \le \psi_\varepsilon(p) \le c_\varepsilon + \varepsilon \le c + 2\varepsilon.$$

Set

$$B_\varepsilon(p) = \{t \in K \;:\; f(p(t)) + \varepsilon d(t) = \psi_\varepsilon(p)\}.$$

It remains to show that there exists $t' \in B_\varepsilon(p)$ such that $\lambda(p(t')) \le 2\varepsilon$. Choosing $\varepsilon = \frac{1}{n}$ and $x_n = p(t')$, the first part of the conclusion of the theorem follows.

Applying Lemma 2.1 with $M = B_\varepsilon(p)$ and $\varphi(t) = \partial f(p(t))$, we find a continuous mapping $v : B_\varepsilon(p) \to X$ such that for any $t \in B_\varepsilon(p)$ and $x^* \in \partial f(p(t))$,

$$\|v(t)\| \le 1 \quad \text{and} \quad \langle x^*, v(t)\rangle \ge \gamma_\varepsilon - \varepsilon,$$

where

$$\gamma_\varepsilon = \inf_{t\in B_\varepsilon(p)} \lambda(p(t)).$$

On the other hand, our assumption implies that

$$\psi_\varepsilon(p) > \max_{t\in K^*} f(p(t)).$$

Hence

$$B_\varepsilon(p) \cap K^* = \emptyset.$$

So there exists a continuous extension w of v which is defined on K and such that

$$w = 0 \quad \text{on} \quad K^* \qquad \text{and} \quad \|w(t)\| \leq 1, \quad \text{for any} \quad t \in K.$$

We now replace in (2.4) q by

$$q_h(t) = p(t) - hw(t),$$

for $h > 0$ sufficiently small. In what follows $\varepsilon > 0$ is fixed, while $h \to 0$.

Let $t_h \in B_\varepsilon(p)$ be such that

$$f(q_h(t_h)) + \varepsilon d(t_h) = \psi_\varepsilon(q_h).$$

There exists a sequence $\{h_n\}$ which converges to 0 such that $\{t_{h_n}\}$ converges to some $t_0 \in B_\varepsilon(p)$. It follows that

$$-\varepsilon \leq -\varepsilon\|w\|_\infty \leq \frac{\psi_\varepsilon(q_h) - \psi_\varepsilon(p)}{h} = \frac{f(q_h(t_h)) + \varepsilon d(t_h) - \psi_\varepsilon(p)}{h}$$

$$\leq \frac{f(q_h(t_h)) - f(p(t_h))}{h} = \frac{f(p(t_h) - hw(t_h)) - f(p(t_h))}{h}.$$

With the same reasoning as in the proof of Theorem 2.1 we obtain $t' \in B_\varepsilon(p)$ such that

$$\lambda(p(t')) \leq 2\varepsilon.$$

Moreover, if f satisfies the $(\mathrm{PS})_c$ condition, then c is a critical value of f. This follows by the lower semicontinuity of λ.

For the second part of the proof, applying again Ekeland's variational principle, we obtain the existence of a sequence of paths $\{q_n\}$ in $\mathcal{P}$ such that for every $q \in \mathcal{P}$,

$$\psi_{\varepsilon_n^2}(q) - \psi_{\varepsilon_n^2}(q_n) + \varepsilon_n d(q, q_n) \geq 0,$$

$$\psi_{\varepsilon_n^2}(q_n) \leq \psi_{\varepsilon_n^2}(p_n) - \varepsilon_n d(p_n, q_n),$$

where $\{\varepsilon_n\}$ is a sequence of positive numbers convergent to 0, while $\{p_n\}$ are paths in $\mathcal{P}$ such that

$$\psi_{\varepsilon_n^2}(p_n) \leq c + 2\varepsilon_n^2.$$

Applying the preceding argument to q_n in place of p, we find $t_n \in K$ such that

$$c - \varepsilon_n^2 \leq f(q_n(t_n)) \leq c + 2\varepsilon_n^2,$$

$$\lambda(q_n(t_n)) \leq 2\varepsilon_n.$$

The sequence $\{t_n\}$ is the desired one. Indeed, according to the $(\mathrm{PS})_c$ condition, the sequence $\{q_n(t_n)\}$ contains a subsequence converging to a critical point. The corresponding subsequence of the sequence $\{p_n(t_n)\}$ has the same limit. A standard limit based on the continuity of f and the lower semicontinuity of λ shows that for the whole sequence we have

$$\lim_{n\to\infty} f(p_n(t_n)) = c$$

and

$$\lim_{n\to\infty} \lambda(p_n(t_n)) = 0.$$

The proof is complete. ∎

Corollary 2.5 Let $f : X \to I\!R$ be a locally Lipschitz functional satisfying the Palais-Smale condition. If f has two different local minimum points, then f possesses a third critical point.

Proof. Let x_0 and x_1 be two different local minimum points of f.

Case 1. $f(x_0) = f(x_1) = a$. Choose $0 < R < \frac{1}{2}\|x_1 - x_0\|$ such that $f(x) \geq a$, for any $x \in B(x_0, R) \cup B(x_1, R)$. Set $A = \overline{B}(x_0, \frac{R}{2}) \cup \overline{B}(x_1, \frac{R}{2})$.

Case 2. $f(x_0) > f(x_1)$. Choose $0 < R < \|x_1 - x_0\|$ such that $f(x) \geq f(x_0)$, for any $x \in B(x_0, R)$. Set $A = \overline{B}(x_0, \frac{R}{2}) \cup \{x_1\}$.

In both cases we fix $p^* \in C([0,1]; X)$ such that $p^*(0) = x_0$ and $p^*(1) = x_1$. Setting $K^* = (p^*)^{-1}(A)$ and applying Theorem 2.2, we obtain a critical point of f which is different from x_0 and x_1, as observed after examining the proof of Theorem 2.2. ∎

The following result is a strengthened variant of Theorems 2.1 and 2.2.

Theorem 2.3 Let $f : X \to I\!R$ be a locally Lipschitz functional and let F be a nonempty closed subset of X, which is disjoint of $p^*(K^*)$. Assume that

$$f(x) \geq c, \quad \text{for any } \ x \in F$$

and

$$p(K) \cap F \neq \emptyset\,, \quad \text{for any } \ p \in \mathcal{P}. \tag{2.5}$$

Then there exists a sequence $\{x_n\}$ in X such that

(i) $\lim\limits_{n\to\infty} \mathrm{dist}\,(x_n, F) = 0$,

(ii) $\lim\limits_{n\to\infty} f(x_n) = c$,

(iii) $\lim\limits_{n\to\infty} \lambda(x_n) = 0$.

Proof. Fix $\varepsilon > 0$ such that

$$\varepsilon < \min\{1;\ \text{dist}\,(p^*(K^*), F)\}.$$

Let $p \in \mathcal{P}$ be such that

$$\max_{t \in K} f(p(t)) \leq c + \frac{\varepsilon^2}{4}\,.$$

The set

$$K_0 = \{t \in K\ :\ \text{dist}\,(p(t), F) \geq \varepsilon\}$$

is closed and contains K^*. Define

$$\mathcal{P}_0 = \{q \in C(K; X)\ :\ q = p \ \text{ on } \ K_0\}.$$

Let

$$\eta : X \to I\!R, \quad \eta(x) = \max\{0;\ \varepsilon^2 - \varepsilon\ \text{dist}\,(x, F)\}.$$

Define $\psi : \mathcal{P}_0 \to I\!R$ by

$$\psi(q) = \max_{t \in K}\ (f + \eta)(q(t)).$$

The functional ψ is continuous and bounded from below, so by Ekeland's variational principle, there exists $p_0 \in \mathcal{P}_0$ such that for any $q \in \mathcal{P}_0$,

$$\psi(p_0) \leq \psi(q),$$

$$d(p_0, q) \leq \frac{\varepsilon}{2}\,, \tag{2.6}$$

$$\psi(p_0) \leq \psi(q) + \frac{\varepsilon}{2}\, d(q, p_0). \tag{2.7}$$

The set

$$B(p_0) = \{t \in K\ :\ (f + \eta)(p_0(t)) = \psi(p_0)\}$$

is nonempty and closed.

For concluding the proof, it is enough to show that there exists $t \in B(p_0)$ such that

$$\text{dist}\,(p_0(t), F) \leq \frac{3\varepsilon}{2}\,, \tag{2.8}$$

$$c \leq f(p_0(t)) \leq c + \frac{5\varepsilon^2}{4}\,, \tag{2.9}$$

$$\lambda(p_0(t)) \leq \frac{5\varepsilon}{2}\,. \tag{2.10}$$

Proof of (2.8). The definition of $\mathcal{P}_0$ and relation (2.5) imply that for any $q \in \mathcal{P}_0$, we have

$$q(K \setminus K_0) \cap F \neq \emptyset,$$

hence

$$\psi(q) \geq c + \varepsilon^2.$$

On the other hand, we can write

$$\psi(p) \leq c + \frac{\varepsilon^2}{4} + \varepsilon^2 = c + \frac{5\varepsilon^2}{4}.$$

It follows that

$$c + \varepsilon^2 \leq \psi(p_0) \leq \psi(p) \leq c + \frac{5\varepsilon^2}{4}. \tag{2.11}$$

So, for any $t \in B(p_0)$, we obtain that

$$c + \varepsilon^2 \leq \psi(p_0) = (f + \eta)(p_0(t)).$$

If in addition $t \in K_0$, then

$$(f + \eta)(p_0(t)) = (f + \eta)(p(t)) = f(p(t)) \leq c + \frac{\varepsilon^2}{4}.$$

This implies that

$$B(p_0) \subset K \setminus K_0.$$

The definition of K_0 shows that for any $t \in B(p_0)$ we have

$$\text{dist } (p(t), F) < \varepsilon.$$

Using now (2.6), we obtain

$$\text{dist } (p_0(t), F) < \frac{3\varepsilon}{2}.$$

Proof of (2.9). For every $t \in B(p_0)$ we have

$$\psi(p_0) = (f + \eta)(p_0(t)).$$

Taking into account that $0 \leq \eta \leq \varepsilon^2$ and using (2.11) we obtain

$$c \leq f(p_0(t)) \leq c + \frac{5\varepsilon^2}{4}.$$

Proof of (2.10). Applying Lemma 2.1 for $\varphi(t) = \partial f(p_0(t))$, we find a continuous mapping $v : B(p_0) \to X$ such that for every $t \in B(p_0)$,

$$\|v(t)\| \leq 1.$$

In addition, for any $t \in B(p_0)$ and $x^* \in \partial f(p_0(t))$ we have

$$\langle x^*, v(t) \rangle \geq \gamma - \varepsilon,$$

where

$$\gamma = \inf_{t \in B(p_0)} \lambda(p_0(t)).$$

So, for any $t \in B(p_0)$,

$$f^0(p_0(t); -v(t)) = \max\{\langle x^*, -v(t)\rangle \, : \, x^* \in \partial f(p_0(t))\}$$

$$= -\min\{\langle x^*, v(t)\rangle \, : \, x^* \in \partial f(p_0(t))\} \leq -\gamma + \varepsilon.$$

Since $B(p_0) \cap K_0 = \emptyset$, there exists a continuous extension w of v to the set K such that $w = 0$ on K_0 and $\|w(t)\| \leq 1$, for any $t \in K$. It follows now by (2.7) that for any $\lambda > 0$, we have

$$-\frac{\varepsilon}{2} \leq -\frac{\varepsilon}{2} \|w\|_\infty \leq \frac{\psi(p_0 - \lambda w) - \psi(p_0)}{\lambda}. \tag{2.12}$$

For any n, there exists $t_n \in K$ such that

$$\psi\left(p_0 - \frac{1}{n}w\right) = (f + \eta)(p_0(t_n) - \frac{1}{n}w(t_n)).$$

Passing eventually to a subsequence we can assume that $\{t_n\}$ converges to $t_0 \in B(p_0)$. On the other hand, for any $t \in K$ and $\lambda > 0$ we have

$$f(p_0(t) - \lambda w(t)) \leq f(p_0(t)) + \lambda\varepsilon.$$

So we obtain

$$n[f(p_0 - \frac{1}{n}w) - \psi(p_0)] \leq n[f(p_0(t_n) - \frac{1}{n}w(t_n)) + \frac{\varepsilon}{n} - f(p_0(t_n))].$$

This inequality and (2.12) yield

$$-\frac{3\varepsilon}{2} \leq n[f(p_0(t_n) - \frac{1}{n}w(t_n)) - f(p_0(t_n))]$$

$$= n[f(p_0(t_n) - \frac{1}{n}w(t_0)) - f(p_0(t_n))]$$

$$+n[f(p_0(t_n) - \frac{1}{n}w(t_n)) - f(p_0(t_n) - \frac{1}{n}w(t_0))].$$

Since f is locally Lipschitz and $t_n \to t_0$ we conclude that

$$\limsup_{n\to\infty} n\left[f(p_0(t_n) - \frac{1}{n}w(t_n)) - f(p_0(t_n) - \frac{1}{n}w(t_0))\right] = 0.$$

Therefore,

$$-\frac{3\varepsilon}{2} \leq \limsup_{n\to\infty} n\left[f(p_0(t_0) + z_n - \frac{1}{n}w(t_0)) - f(p_0(t_0) + z_n)\right],$$

where $z_n = p_0(t_n) - p_0(t_0)$. Thus we derive

$$-\frac{3\varepsilon}{2} \leq f^0(p_0(t_0); -w(t_0)) \leq -\gamma + \varepsilon.$$

It follows that

$$\gamma = \inf\{\|x^*\| \,:\, x^* \in \partial f(p_0(t)),\ t \in B(p_0)\} \leq \frac{5\varepsilon}{2}.$$

By the lower semicontinuity of λ, there exists $t \in B(p_0)$ such that

$$\lambda(p_0(t)) = \inf_{x^* \in \partial f(p_0(t))} \|x^*\| \leq \frac{5\varepsilon}{2}.$$

This completes the proof. ■

Corollary 2.6 (Ghoussoub-Preiss Theorem [13]) Let $f : X \to I\!R$ be a Gâteaux differentiable function such that $f' : (X, \|\cdot\|) \to (X^*, \sigma(X^*, X))$ is continuous. Let a and b in X be such that

$$c = \inf_{p \in \mathcal{P}} \max_{t \in [0,1]} f(p(t)),$$

where $\mathcal{P}$ denotes the set of all continuous paths joining a and b. Let F be a nonempty closed subset of X which does not contain a and b such that $f(x) \geq c$, for any $x \in F$. In addition, we assume that for any $p \in \mathcal{P}$,

$$p([0,1]) \cap F \neq \emptyset.$$

Then there exists a sequence $\{x_n\}$ in X such that

(i) $\lim_{n\to\infty} \operatorname{dist}(x_n, F) = 0$,

(ii) $\lim_{n\to\infty} f(x_n) = c$,

(iii) $\lim_{n\to\infty} \|f'(x_n)\| = 0$.

Moreover, if f satisfies the $(PS)_c$ condition, then there exists $x \in F$ such that $f(x) = c$ and $f'(x) = 0$.

Proof. With the same arguments as in the proof of Corollary 2.4, we deduce that f is locally Lipschitz and

$$\partial f(x) = \{f'(x)\}.$$

Then we apply Theorem 2.3 for $K = [0,1]$, $K^* = \{0,1\}$, $p^*(0) = a$ and $p^*(1) = b$. The last part of the conclusion follows directly from Theorem 2.3. ■

2. Critical Point Theory for Convex Perturbations of Locally Lipschitz Functionals

This Section deals with the class of nonsmooth functionals which can be written as a sum of a locally Lipschitz function and a convex, proper and lower semicontinuous functional (possibly, taking the value $+\infty$). Namely, we are concerned with functionals $f : X \to]-\infty, +\infty]$, where $(X, \|\cdot\|)$ is a real reflexive Banach space, satisfying the structure hypothesis

(H_f) $f = \Phi + \Psi$, *where* $\Phi : X \to \mathbb{R}$ *is locally Lipschitz and* $\Psi : X \to]-\infty, +\infty]$ *is convex, proper (i.e.,* $\not\equiv +\infty$*), lower semicontinuous.*

Definition 2.2 An element $u \in X$ is said to be a critical point of functional $f : X \to]-\infty, +\infty]$ satisfying assumption (H_f) if

$$\Phi^0(u; x-u) + \Psi(x) - \Psi(u) \geq 0, \quad \forall x \in X.$$

Remark 2.1 This definition has been formulated in Motreanu and Panagiotopoulos [26]. If $\Psi = 0$, Definition 2.2 becomes the notion of critical point for a locally Lipschitz functional as introduced by Chang [4]. In particular, if $\Psi = 0$ and Φ is continuously differentiable, one obtains the usual concept of critical point (see, e.g., Ambrosetti and Rabinowitz [2]). In the case where Φ is continuously differentiable and Ψ is convex, l.s.c., proper, Definition 2.2 reduces with the notion of critical point in the sense of Szulkin [40].

Remark 2.2 An equivalent formulation of Definition 2.2 is that $u \in X$ is a critical point of $f : X \to]-\infty, +\infty]$ if and only if

$$0 \in \partial\Phi(u) + \partial\Psi(u),$$

where $\partial\Phi(u)$ denotes the generalized gradient of Φ in the sense of Definition 1.2 and $\partial\Psi(u)$ is the subdifferential of Ψ in the sense of convex analysis. This can be seen by using Definition 1.1.

Given a real number c, we denote as usually

$$K_c(f) = \{u \in X : f(u) = c,\ u \text{ is a critical point of } f\}.$$

We say that the number $c \in \mathbb{R}$ is a critical value of the functional $f : X \to]-\infty, +\infty]$ satisfying (H_f) if $K_c(f) \neq \emptyset$.

The following proposition provides critical points in the sense of Definition 2.2.

Proposition 2.1 Let the function f satisfy hypothesis (H_f). Then each local minimum of f is a critical point of f in the sense of Definition 2.2.

Proof. Suppose that u is a local minimum of f and fix $v \in X$. Using the convexity of Ψ yields

$$0 \leq f((1-t)u+tv) - f(u) \leq \Phi(u+t(v-u)) - \Phi(u) + t(\Psi(v) - \Psi(u))$$

for all small $t > 0$. Dividing by t and letting $t \to 0^+$ we infer that u is a critical point of f in the sense of Definition 2.2. ∎

The appropriate Palais-Smale condition for the function $f : X \to]-\infty, +\infty]$ in (H_f) at the level $c \in I\!R$ is stated below.

Definition 2.3 The function $f : X \to]-\infty, +\infty]$ satisfying assumption (H_f) is said to verify the Palais-Smale condition at the level $c \in I\!R$ if the next property is true

$(PS)_{f,c}$ *Each sequence* $\{x_n\} \subset X$ *such that* $f(x_n) \to c$ *and*

$$\Phi^0(x_n; x - x_n) + \Psi(x) - \Psi(x_n) \geq -\varepsilon_n \|x_n - x\|, \quad \forall n \in I\!N,\ x \in X,$$

where $\varepsilon_n \to 0^+$, *possesses a strongly convergent subsequence.*

The following proposition expresses significant aspects related to the Palais-Smale condition in Definition 2.3.

Proposition 2.2 (i) Any limit point u of a (Palais-Smale) sequence $\{x_n\}$ entering Definition 2.3 belongs to $K_c(f)$.

(ii) The inequality in Definition 2.3 is equivalent to

$$\Phi^0(x_n; x - x_n) + \Psi(x) - \Psi(x_n) \geq \langle z_n, x - x_n \rangle, \quad \forall n \in I\!N,\ \forall x \in X,$$

for some sequence $\{z_n\} \subset X^*$ with $z_n \to 0$.

Proof. (i) Passing to a subsequence we may admit that $x_n \to u$ strongly in X. Then, letting $n \to \infty$, the upper semicontinuity of Φ^0 (see the property stated below Definition 1.1) and the lower semicontinuity of Ψ allow us to derive that u is a critical point of f in the sense of Definition 2.2. Taking $x = u$ in the inequality in $(PS)_{f,c}$ we obtain that

$$\Psi(u) \leq \liminf_{n\to\infty} \Psi(x_n) \leq \limsup_{n\to\infty} \Psi(x_n) \leq \Phi^0(u; u-u) + \Psi(u) = \Psi(u).$$

This ensures that $f(u) = \lim_{n\to\infty} f(x_n) = c$, so $u \in K_c(f)$.

(ii) Assume that the inequality in Definition 2.3 is true. We note that the left-hand side of this inequality is, with respect to the variable $w := x - x_n$, convex, lower semicontinuous and vanishes at $w = 0$. Then Lemma 1.1 provides the existence of $z_n \in X^*$ for which $z_n \to 0$ in X^* and the inequality in part (ii) of Proposition 2.1 is valid. The converse assertion holds because one can take $\varepsilon_n = \|z_n\|_*$. ■

In using the Palais-Smale condition given in Definition 2.3 we start by presenting a basic minimization result for nonsmooth functionals of type (H_f).

Theorem 2.4 Assume that the function $f : X \to]-\infty, +\infty]$ satisfies hypothesis (H_f) on the Banach space X, is bounded from below and verifies condition $(\mathrm{PS})_{f,m}$, with $m = \inf_X f \in \mathbb{R}$. Then there exists $u \in X$ such that $f(u) = m$ and u is a critical point of f in the sense of Definition 2.2.

Proof. Using the definition of m, we find a (minimizing) sequence $\{u_n\} \subset X$ such that

$$f(u_n) < m + \varepsilon_n^2,$$

for a sequence $\{\varepsilon_n\}$ of positive numbers, with $\varepsilon_n \downarrow 0$. Since the function f is proper, lower semicontinuous and bounded from below on X, we may apply Ekeland's variational principle [9], [10] (for an equivalent form, see Theorem 1.5). Then there exists a sequence $\{v_n\} \subset X$ such that

$$f(v_n) < m + \varepsilon_n^2,$$

and

$$f(v) \geq f(v_n) - \varepsilon_n \|v_n - v\|, \ \forall v \in X, \ \forall n \in \mathbb{N}.$$

Setting $v = (1-t)v_n + tw$ in the inequality above, for arbitrary $0 < t < 1$ and $w \in X$, we obtain

$$\Phi((1-t)v_n + tw) + \Psi((1-t)v_n + tw)$$

$$\geq \Phi(v_n) + \Psi(v_n) - \varepsilon_n t \|w - v_n\|, \ \forall w \in X, \ \forall t \in (0,1).$$

The convexity of $\Psi : X \to \mathbb{R} \cup \{+\infty\}$ yields

$$\Phi((1-t)v_n + tw) - t\Psi(v_n) + t\Psi(w)$$

$$\geq \Phi(v_n) - \varepsilon_n t \|w - v_n\|, \ \forall w \in X, \ \forall t \in (0,1).$$

Passing to lim sup with respect to t one gets

$$\limsup_{t \downarrow 0} \frac{1}{t}(\Phi(v_n + t(w - v_n)) - \Phi(v_n)) + \Psi(w) - \Psi(v_n)$$

$$\geq -\varepsilon_n \|w - v_n\|, \ \forall w \in X.$$

Taking into account Definition 1.1 we deduce that

$$\Phi^0(v_n; w - v_n) + \Psi(w) - \Psi(v_n) \geq -\varepsilon_n \|w - v_n\|, \ \forall w \in X.$$

On the other hand, from the choice of $\{v_n\}$ we have that

$$\Phi(v_n) + \Psi(v_n) \to m \ \text{ as } n \to \infty.$$

Then the hypothesis of f to satisfy $(\mathrm{PS})_m$ implies that the sequence $\{v_n\}$ has a convergent subsequence $\{v_{n_k}\}$, say $v_{n_k} \to u$, for some $u \in X$. The lower semicontinuity of f yields

$$m \leq f(u) \leq \liminf_{k \to \infty} f(v_{n_k}) \leq m,$$

so $f(u) = m$ as well as

$$\limsup_{k \to \infty} \Phi^0(v_{n_k}; w - v_{n_k}) + \Psi(w) \geq \limsup_{k \to \infty} \Psi(v_{n_k})$$

$$\geq \liminf_{k \to \infty} \Psi(v_{n_k}) \geq \Psi(u), \ \forall w \in X,$$

i.e. u is a critical point of f. The proof is complete. ■

In order to present a minimax principle for the functionals complying with (H_f), we need the following deformation result.

Lemma 2.2 Let the functional $f : X \to]-\infty, +\infty]$ and the number $c \in \mathbb{R}$ be such that conditions (H_f) and $(\mathrm{PS})_{f,c}$ are satisfied. Let U be a neighborhood of $K_c(f)$ and let $\bar{\varepsilon} > 0$ be a fixed number. Then there exists a number $\varepsilon \in]0, \bar{\varepsilon}[$ such that for every compact subset $A \subset X \setminus U$ with

$$c \leq \sup_A f \leq c + \varepsilon \tag{2.13}$$

there exist a closed neighborhood W of A in X and a deformation $h_A : W \times [0, \bar{s}] \to X$, $\bar{s} > 0$, i.e. a continuous map with $h_A(\cdot, 0) = id_W$, satisfying

$$\|u - h_A(u, s)\| \leq s, \ \ \forall u \in W, \ \ \forall s \in [0, \bar{s}]; \tag{2.14}$$

$$f(h_A(u, s)) - f(u) \leq Ms, \ \ \forall u \in W, \ \ \forall s \in [0, \bar{s}], \tag{2.15}$$

with a constant $M > 0$ independent of u and s;

$$f(h_A(u, s)) - f(u) \leq -2\varepsilon s, \tag{2.16}$$

for all $u \in W$ with $f(u) \geq c - \varepsilon$, $\ \forall s \in [0, \bar{s}]$;

$$\sup_{u \in A} f(h_A(u, s)) - \sup_{u \in A} f(u) \leq -2\varepsilon s, \ \ \forall s \in [0, \bar{s}]. \tag{2.17}$$

Furthermore, if W_0 is a closed subset of X which does not contain critical points of f, then W and h_A can be chosen so that

$$f(h_A(u,s)) - f(u) \leq 0, \quad \forall u \in W \cap W_0, \quad \forall s \in [0, \bar{s}].$$

Proof. For the proof we refer to Theorem 3.1 in Motreanu and Panagiotopoulos [26], p. 69. ■

On the basis of Lemma 2.2 we may state the next minimax principle.

Theorem 2.5 Let the functional $f : X \to]-\infty, +\infty]$ on the Banach space X satisfy assumption (H_f). Let Q be a compact topological submanifold of X with nonempty boundary ∂Q (in the sense of manifolds with boundary) such that

$$\sup_Q f \in \mathbb{R}. \tag{2.18}$$

Consider the numbers $a := \sup_{\partial Q} f$ and

$$c := \inf_{\varphi \in \Gamma} \sup_{x \in Q} f(\varphi(x)), \tag{2.19}$$

where

$$\Gamma = \{\varphi \in C(Q; X) : \ \varphi|_{\partial Q} = id_{\partial Q}\}. \tag{2.20}$$

If

$$c > a \tag{2.21}$$

and condition $(\mathrm{PS})_{f,c}$ is satisfied, then c in (2.19) is a critical value of f, i.e. $K_c(f)$ is nonempty.

Proof. Relation (2.18) yields $c < \infty$ by putting $\varphi = id_Q$ in (2.19), (2.20).

We have to show that c is a critical value of f. Arguing by contradiction we assume that $K_c(f) = \emptyset$. Then we can take $U = \emptyset$ as the neighborhood of $K_c(f)$ in Lemma 2.2. Applying Lemma 2.2 with

$$\bar{\varepsilon} = c - a, \tag{2.22}$$

which is a positive number by (2.21), we obtain an $\varepsilon \in]0, \bar{\varepsilon}[$ as stated in Lemma 2.2. Notice that

$$\partial Q \subset f_{c-\frac{\varepsilon}{2}} := \left\{x \in X : f(x) \leq c - \frac{\varepsilon}{2}\right\}. \tag{2.23}$$

Indeed, (2.23) follows from (2.21) and (2.22).

It is worth to notice that the class Γ in (2.20) does not generally fulfill the property $h_A(\varphi(\cdot), s) \in \Gamma$ whenever $\varphi \in \Gamma$, where h_A is the deformation in Lemma 2.2 corresponding to a compact set $A \subset X$ satisfying (2.13). We enlarge Γ by defining

$$\Gamma_1 := \{\varphi \in C(Q;X) : \ \varphi|_{\partial Q} \text{ and } id_{\partial Q} \text{ are homotopic as}$$

$$\text{maps from } \partial Q \text{ into } f_{c-\frac{\varepsilon}{4}} \text{ and } \varphi(\partial Q) \subset f_{c-\frac{\varepsilon}{2}}\}, \tag{2.24}$$

where

$$f_{c-\frac{\varepsilon}{4}} := \left\{x \in X : f(x) \leq c - \frac{\varepsilon}{4}\right\}.$$

Taking into account (2.23), the family (2.24) is well defined and nonempty (containing at least id_Q). Corresponding to the family Γ_1 in (2.24) we introduce the minimax value

$$c_1 = \inf_{\varphi \in \Gamma_1} \sup_{x \in Q} f(\varphi(x)). \tag{2.25}$$

We claim that

$$c = c_1 \,. \tag{2.26}$$

Since $\Gamma \subset \Gamma_1$, from (2.19), (2.25) it is clear that $c_1 \leq c$. Assuming by contradiction that $c_1 < c$, there would exist $\varphi \in \Gamma_1$ such that

$$\sup_{x \in Q} f(\varphi(x)) < c.$$

Recalling that $\varphi|_{\partial Q}$ and $id_{\partial Q}$ are homotopic in $f_{c-\frac{\varepsilon}{4}}$, the (absolute) Homotopy Extension Property (see Proposition I.9.2 of Hu [18] or Spanier [39]) guarantees that $id_{\partial Q}$ can be extended to some $\psi \in C(Q;X)$, so $\psi \in \Gamma$, such that

$$\sup_{x \in Q} f(\psi(x)) < c.$$

This happens because $(Q, \partial Q)$ is a (finitely) triangulable pair in view of the fact that Q is a compact topological manifold with boundary ∂Q (see Munkres [29]). This fact is in contradiction with the definition of c in (2.19). Hence the claim in (2.26) is verified.

The next step in the proof is to show that Γ_1 in (2.24) is a closed subset of the Banach space $C(Q;X)$ with respect to the uniform norm (recall that Q is compact in X), i.e.

$$\|\varphi\| = \sup_{x \in Q} \|\varphi(x)\|, \quad \forall \varphi \in C(Q;X).$$

To check this assertion let us take a sequence $\{\varphi_n\} \subset \Gamma_1$ which converges strongly in $C(Q;X)$, say $\varphi_n \to \varphi$. By (2.24) one has

$$f(\varphi_n(x)) \le c - \frac{\varepsilon}{2}, \quad \forall x \in \partial Q, \quad \forall n \ge 1.$$

Since f is lower semicontinuous, we derive

$$f(\varphi(x)) \le \liminf_{n\to\infty} f(\varphi_n(x)) \le c - \frac{\varepsilon}{2}, \quad \forall x \in \partial Q. \tag{2.27}$$

Using Lebourg's mean value theorem for Φ and the convexity of Ψ it is seen that

$$\begin{aligned} & f(t\varphi_n(x) + (1-t)\varphi(x)) \\ &= \Phi(t\varphi_n(x) + (1-t)\varphi(x)) + \Psi(t\varphi_n(x) + (1-t)\varphi(x)) \\ &\le \Phi(\varphi(x)) + \bar{t}\Phi^0(\varphi(x); \varphi_n(x) - \varphi(x)) \\ &+t\Psi(\varphi_n(x)) + (1-t)\Psi(\varphi(x)), \quad \forall x \in \partial Q, \ \forall t \in [0,1], \end{aligned}$$

for some $\bar{t} \in]0,1[$. Since $\varphi(\partial Q)$ is compact, the sequence $\{\varphi_n\}$ converges uniformly to φ and Φ being locally Lipschitz, it follows that Φ is Lipschitz continuous near $\varphi(\partial Q)$ of Lipschitz constant $K > 0$. It turns out that

$$f(t\varphi_n(x) + (1-t)\varphi(x)) \le \Phi(\varphi(x)) + K\|\varphi_n(x) - \varphi(x)\|$$

$$+t\Psi(\varphi_n(x)) + (1-t)\Psi(\varphi(x)), \quad \forall x \in \partial Q, \quad \forall t \in [0,1].$$

Taking into account the Lipschitz continuity of Φ around $\varphi(\partial Q)$ we obtain that

$$\Phi(\varphi(x)) \le \Phi(\varphi_n(x)) + K\|\varphi_n(x) - \varphi(x)\|, \quad \forall x \in \partial Q.$$

It results that

$$f(t\varphi_n(x) + (1-t)\varphi(x)) \le t\Phi(\varphi_n(x)) + (1-t)\Phi(\varphi(x))$$

$$\begin{aligned} &+2K\|\varphi_n(x) - \varphi(x)\| + t\Psi(\varphi_n(x)) + (1-t)\Psi(\varphi(x)) \\ &= tf(\varphi_n(x)) + (1-t)f(\varphi(x)) + 2K\|\varphi_n(x) - \varphi(x)\| \end{aligned}$$

for all $x \in \partial Q$ and $t \in [0,1]$. Then, using that $\varphi_n \in \Gamma_1$ and (2.27), we obtain

$$f(t\varphi_n(x) + (1-t)\varphi(x)) \le c - \frac{\varepsilon}{2} + 2K\|\varphi_n(x) - \varphi(x)\|, \quad \forall x \in \partial Q.$$

Since $\{\varphi_n\}$ converges uniformly to φ it follows

$$f(t\varphi_n(x) + (1-t)\varphi(x)) \le c - \frac{\varepsilon}{4}, \quad \forall x \in \partial Q\,, \tag{2.28}$$

provided that n is sufficiently large. From (2.28) we deduce that $\varphi_n|_{\partial Q}$ and $\varphi|_{\partial Q}$ are homotopic in $f_{c-\frac{\varepsilon}{4}}$. Hence, by (2.24) and $\varphi_n \in \Gamma_1$, we conclude that $\varphi|_{\partial Q}$ and $id_{\partial Q}$ are homotopic in $f_{c-\frac{\varepsilon}{4}}$. This property combined with (2.27) entails that $\varphi \in \Gamma_1$. Consequently, Γ_1 is a closed subset of $C(Q;X)$, so a complete metric space.

Let us introduce the functional $\Pi : C(Q;X) \to]-\infty, +\infty]$ by

$$\Pi(\varphi) = \sup_{x\in Q} f(\varphi(x)), \quad \forall \varphi \in C(Q;X). \tag{2.29}$$

A direct verification yields that Π is lower semicontinuous. This is the consequence of the lower semicontinuity of f. Therefore we know that $\Pi : \Gamma_1 \to]-\infty, +\infty]$ is lower semicontinuous. It is known that Γ_1 is a complete metric space because it is a closed subset of $C(Q;X)$. Thus it is allowed to apply to $\Pi : \Gamma_1 \to]-\infty, +\infty]$ Ekeland's variational principle (see Ekeland [9], [10] or, in an equivalent form, Theorem 1.5). Notice that Π on Γ_1 is bounded from below by $c \in \mathbb{R}$ because (2.25), (2.26) hold. Ekeland's variational principle gives rise to some $\varphi \in \Gamma_1$ satisfying

$$c \leq \Pi(\varphi) \leq c + \varepsilon \tag{2.30}$$

and

$$\Pi(\psi) - \Pi(\varphi) \geq -\varepsilon\|\psi - \varphi\|, \quad \forall \psi \in \Gamma_1. \tag{2.31}$$

Now we invoke Lemma 2.2 that provides the deformation $h_\varphi : W \times [0, \bar{s}] \to X$ corresponding to the compact set $A = \varphi(Q)$, where W is a closed neighborhood of A in X and $\bar{s}$ is a positive number. We point out that the set A satisfies the required condition (2.13) due to (2.30). Let us show that for $\bar{s} > 0$ small enough we have

$$h_\varphi(\varphi(\cdot), s) \in \Gamma_1, \quad \forall s \in [0, \bar{s}]. \tag{2.32}$$

In view of (2.24), we see that in order to prove (2.32) it suffices to establish that $h_\varphi(\varphi(\cdot), s)|_{\partial Q}$ and $\varphi|_{\partial Q}$ are homotopic in $f_{c-\frac{\varepsilon}{2}}$. Clearly, a homotopy in X between the involved mappings is the following one $(x,t) \in \partial Q \times [0,1] \to h_\varphi(\varphi(x), ts) \in X$. Therefore, it is sufficient to show that

$$f(h_\varphi(\varphi(x), s)) \leq c - \frac{\varepsilon}{2}, \quad \forall x \in \partial Q, \ \forall\, 0 \leq s \leq \bar{s}. \tag{2.33}$$

If one has for $x \in \partial Q$ that

$$f(\varphi(x)) \in [c - \varepsilon, c - \frac{\varepsilon}{2}],$$

then by (2.16) we get

$$f(h_\varphi(\varphi(x), s)) \leq f(\varphi(x)) - 2\varepsilon s \leq c - \frac{\varepsilon}{2} - 2\varepsilon s < c - \frac{\varepsilon}{2}.$$

If for $x \in \partial Q$ one has

$$f(\varphi(x)) < c - \varepsilon,$$

then (2.15) implies

$$f(h_\varphi(\varphi(x), s)) \leq f(\varphi(x)) + Ms < c - \varepsilon + Ms \leq c - \frac{\varepsilon}{2}$$

provided that $\bar{s} > 0$ is sufficiently small. We thus justified that (2.33) is valid, so (2.32) holds true.

From (2.17), (2.29), (2.31), (2.32) and (2.14) we then deduce

$$\begin{aligned} -2\varepsilon s &\geq \Pi(h_\varphi(\varphi(\cdot), s)) - \Pi(\varphi) \\ &\geq -\varepsilon \| h_\varphi(\varphi(\cdot), s) - \varphi \| \geq -\varepsilon s, \quad \forall\, 0 \leq s \leq \bar{s}. \end{aligned}$$

Thus we arrived at a contradiction, which proves that our initial assumption that c in (2.19) is not a critical value of f is false. Thus the proof of Theorem 2.5 is complete. ■

The inequality $c > a$ in relation (2.21), with c and a as given in the statement of Theorem 2.5, is difficult to be checked. Using the notion of linking in Definition 2.4 below we obtain a verifiable sufficient condition to have inequality (2.21). In particular we obtain the main result of Chapter 3 in Motreanu and Panagiotopoulos ([26], Theorem 3.2).

Definition 2.4 In a real reflexive Banach space X, let Q be a compact topological submanifold of X with nonempty boundary ∂Q (in the sense of manifolds) and let S be a nonempty closed subset of X. We say that Q links with S provided $\partial Q \cap S = \emptyset$ and $\gamma(Q) \cap S \neq \emptyset$ for every $\gamma \in \Gamma$, where Γ is the class of mappings described in (2.20).

Corollary 2.7 (Motreanu and Panagiotopoulos [26], Theorem 3.2) Let the functional $f : X \to]-\infty, +\infty]$ on the Banach space X satisfy assumption (H_f). Let S and Q link in the sense of Definition 2.4. Assume that

$$\sup_Q f \in I\!R, \quad b := \inf_S f \in I\!R, \quad a := \sup_{\partial Q} f < b. \tag{2.34}$$

Consider the number $c \in I\!R$ introduced in (2.19), with Γ in (2.20), and suppose that condition $(\mathrm{PS})_{f,c}$ is satisfied. Then the number c in (2.19) is a critical value of f with

$$c \geq b. \tag{2.35}$$

This means in particular that $K_c(f)$ is nonempty.

Proof. According to the linking property of Definition 2.4 we know that $\gamma(Q) \cap S \neq \emptyset$ for all $\gamma \in \Gamma$. By (2.34) and (2.19) this ensures that (2.35) is true. In addition, the first relation in (2.34) yields $c < \infty$ by putting $\gamma = id_Q \in \Gamma$ in (2.19). Combining (2.35) and the last inequality in (2.34) one obtains that relation (2.21) is true. We are now in the setting of Theorem 2.5. The application of Theorem 2.5 yields the desired result. ■

For the applications of the abstract results in this Section we refer to [8], [11], [14]-[17], [20], [23].

3. A Critical Point Theory in Metric Spaces

In this Section we present some basic facts related to a critical point theory for continuous functionals defined on metric spaces. This theory is essentially due to Degiovanni (see [6], [7]). It is based on a generalized notion of norm of the derivative (weak slope) which is denoted by $|df|(u)$ (see Chapter 1, Section 3). The notion of critical point by means of weak slope is stated in Definition 1.10. The corresponding Palais-Smale condition in this framework is formulated in Definition 1.11.

A first result is the following easy consequence of Ekeland's variational principle [9], [10].

Lemma 2.3 Let X be a complete metric space endowed with the metric ρ and let $f : X \to I\!R \cup \{+\infty\}$ be a bounded from below, lower semicontinuous, proper function. Let $r > 0$, $\sigma > 0$ and let E be a nonempty subset of X such that

$$\inf_E f < \inf_X f + r\sigma.$$

Then there exists $v \in X$ such that

$$f(v) < \inf_X f + r\sigma, \quad \rho(v, E) < r, \quad |df|(v) < \sigma.$$

Proof. Let $u \in E$ and $0 < \sigma' < \sigma$ be such that $f(u) < \inf_X f + r\sigma'$. Applying Ekeland's variational principle we find $v \in X$ such that $f(v) \leq f(u)$, $\rho(v, u) < r$ and

$$f(w) \geq f(v) - \sigma' \rho(v, w), \quad \forall w \in X.$$

It follows that $f(v) < \inf_X f + r\sigma$, $\rho(v, E) < r$ and

$$|df|(v) \leq \sigma' < \sigma.$$

The proof is complete. ■

The result below expresses the deformation property for the nonsmooth critical point theory treated here.

Lemma 2.4 Let X be a metric space endowed with the metric ρ and let $f : X \to \mathbb{R}$ be a continuous function. Let K be a compact subset of X and $\sigma > 0$ such that

$$\inf\{|df|(u) \,:\, u \in K\} > \sigma.$$

Then there exists a neighborhood U of K in X, $\delta > 0$ and a continuous map $\mathcal{K} : X \times [0, \delta] \to X$ such that

(i) $\rho(\mathcal{K}(u,t), u) \leq t$, for any $(u,t) \in X \times [0, \delta]$;

(ii) $f(\mathcal{K}(u,t)) \leq f(u)$, for any $(u,t) \in X \times [0, \delta]$;

(iii) $f(\mathcal{K}(u,t)) \leq f(u) - \sigma t$, for any $(u,t) \in U \times [0, \delta]$.

Proof. For any $u \in K$ we choose $\delta_u > 0$ and

$$\mathcal{K}_u : B(u, \delta_u) \times [0, \delta_u] \to X$$

according to Proposition 1.6. The compactness of $\mathcal{K}$ implies the existence of $u_1, \ldots, u_n \in K$ such that

$$K \subset \bigcup_{j=1}^{n} B(u_j, \frac{\delta_{u_j}}{2}).$$

We set $\delta_j = \delta_{u_j}$, $\mathcal{K}_j = \mathcal{K}_{u_j}$ and, in view of the fact that $\delta_j > 0$, $1 \leq j \leq n$, choose arbitrarily

$$0 < \delta < \min\left\{\frac{\delta_1}{2}, \ldots, \frac{\delta_n}{2}\right\}.$$

Let us take a neighborhood U of K in X and continuous functions $\theta_j : X \to [0,1]$ $(1 \leq j \leq n)$ such that

$$\operatorname{supp} \theta_j \subset B(u_j, \frac{\delta_j}{2}),$$

$$\sum_{j=1}^{n} \theta_j(v) \leq 1, \quad \forall v \in X,$$

$$\sum_{j=1}^{n} \theta_j(v) = 1, \quad \forall v \in U.$$

We claim that for every $j = 1, \ldots, n$ there exists a continuous map

$$\mathcal{H}_j : X \times [0, \delta] \to X$$

such that for any $(u,t) \in X \times [0,\delta]$ we have

$$\rho(\mathcal{H}_j(u,t),u) \leq \left(\sum_{k=1}^{j} \theta_k(u)\right) t$$

and

$$f(\mathcal{H}_j(u,t)) \leq f(u) - \sigma \left(\sum_{k=1}^{j} \theta_k(u)\right) t\,.$$

For this purpose we first define

$$\mathcal{H}_1(u,t) = \begin{cases} \mathcal{K}_1(u,\theta_1(u)t) & \text{if } u \in \overline{B(u_1,\frac{\delta_1}{2})} \\ u & \text{if } u \notin B(u_1,\frac{\delta_1}{2}). \end{cases}$$

The mapping $\mathcal{H}_1$ satisfies the requested conditions.

Let now $2 \leq j \leq n$ and suppose we have defined the mapping $\mathcal{H}_{j-1}$. Since

$$\rho(\mathcal{H}_{j-1}(u,t),u) \leq \left(\sum_{k=1}^{j-1} \theta_k(u)\right) t \leq \delta < \frac{\delta_j}{2}\,,$$

it follows that

$$\mathcal{H}_{j-1}(u,t) \in B(u_j,\delta_j), \quad \forall u \in \overline{B(u_j,\frac{\delta_j}{2})}\,.$$

Set

$$\mathcal{H}_j(u,t) = \begin{cases} \mathcal{K}_j(\mathcal{H}_{j-1}(u,t),\theta_j(u)t) & \text{if } u \in \overline{B(u_j,\frac{\delta_j}{2})} \\ \mathcal{H}_{j-1}(u,t) & \text{if } u \notin B(u_j,\frac{\delta_j}{2}). \end{cases}$$

By the inductive hypothesis one can verify that $\mathcal{H}_j$ satisfies the requested conditions.

To conclude the proof, it is now enough to set $\mathcal{K} = \mathcal{H}_n$. ■

In what follows we denote by $\mathcal{R}$ the family of all compact nonempty subsets of X. The set $\mathcal{R}$ is endowed with the Hausdorff metric

$$\varrho(A,B) = \max\{\max_{a\in A} \rho(a,B),\ \max_{b\in B} \rho(b,A)\}, \quad \forall A,B \in \mathcal{R}.$$

A classical result asserts that if (X,ρ) is complete, then $(\mathcal{R},\varrho)$ is complete, too.

Given a continuous function $f : X \to \mathbb{R}$, we define a function $\mathcal{F} : \mathcal{R} \to \mathbb{R}$ by the formula

$$\mathcal{F}(K) = \max_K f, \quad \forall K \in \mathcal{R}.$$

The function $\mathcal{F}$ is continuous with respect to the metric ϱ.

Another classical result asserts that the set

$$\Gamma_n = \{K \in \mathcal{R} : \operatorname{Cat}_X K \geq n\}$$

is closed in $(\mathcal{R}, \varrho)$, for any integer $n \geq 1$. We recall that the Ljusternik-Schnirelman category of K in X, denoted $\operatorname{Cat}_X K$, is the smallest $k \in \mathbb{N} \cup \{+\infty\}$ such that K can be covered by k closed and contractible sets in X. For more details we refer to Chapters 4 and 8.

Lemma 2.5 Let X be a metric space and let $f : X \to \mathbb{R}$ be a continuous function. For any integer $n \geq 1$ let $\mathcal{F}_n = \mathcal{F}|_{\Gamma_n}$. Let $K \in \Gamma_n$, $\rho > 0$, $\sigma > 0$ be such that, for every $u \in K$,

$$f(u) \geq \max_K f - \rho \Rightarrow |df|(u) \geq \sigma.$$

Then $|d\mathcal{F}_n|(K) \geq \sigma$.

Proof. Choose arbitrarily $0 < \sigma' < \sigma$ and let U and $\mathcal{K} : X \times [0, \delta] \to X$ be obtained applying Lemma 2.4 to the compact set

$$\{u \in K : f(u) \geq \max_K f - \rho\}$$

and to σ'. Let $\mathcal{V}$ be a neighborhood of K in Γ_n such that

$$\max_A f \geq \max_K f - \frac{\rho}{2}, \quad \forall A \in \mathcal{V},$$

$$\{u \in A : f(u) \geq \max_K f - \rho\} \subset U, \quad \forall A \in \mathcal{V}.$$

Let $\delta' = \min\{\rho/(2\sigma'), \delta\}$ and let $\mathcal{H} : \mathcal{V} \times [0, \delta'] \to \Gamma_n$ be defined by

$$\mathcal{H}(A, t) = \mathcal{K}(A \times \{t\}).$$

The function $\mathcal{H}$ is continuous and, by part (i) of Lemma 2.4,

$$\varrho(\mathcal{H}(A, t), A) \leq t.$$

Let now $A \in \mathcal{V}$. If $u \in A$ and $f(u) \leq \max_K f - \rho$, then, using part (ii) of Lemma 2.4, for any $t \in [0, \delta']$ we have

$$f(\mathcal{K}(u, t)) \leq f(u) \leq \max_K f - \rho \leq \max_A f - \frac{\rho}{2} \leq \mathcal{F}_n(A) - \sigma' t.$$

Otherwise, if $u \in A$ and $f(u) \geq \max_K f - \rho$, then, taking into account part (iii) of Lemma 2.4, for any $t \in [0, \delta']$ we have

$$f(\mathcal{K}(u, t)) \leq f(u) - \sigma' t \leq \mathcal{F}_n(A) - \sigma' t.$$

In any case we have

$$\mathcal{F}_n(\mathcal{H}(A,t)) \leq \mathcal{F}_n(A) - \sigma' t, \quad \forall A \in \mathcal{V}, \ \forall t \in [0, \delta'],$$

hence

$$|d\mathcal{F}_n|(K) \geq \sigma'.$$

Since the parameter σ' is chosen arbitrarily in $]0, \sigma[$, our assertion follows. ∎

We are now ready to prove a result of Ljusternik-Schnirelman type.

Theorem 2.6 (Degiovanni and Marzocchi [7]) Let X be a complete metric space and $f : X \to \mathbb{R}$ a continuous function. For any integer

$$1 \leq h \leq \sup\{\mathrm{Cat}_X K \ : \ K \text{ is a compact subset of } X\}$$

let

$$c_h = \inf_{\Gamma_h} \mathcal{F} = \inf_{K \in \Gamma_h} \max_K f.$$

Assume that for some $h \geq 1$ and $m \geq 1$ we have

$$-\infty < c_h = \ldots = c_{h+m-1} \, .$$

If f satisfies the Palais-Smale condition $(\mathrm{PS})_{c_h}$, then

$$\mathrm{Cat}_X \{u \in X \ : \ |df|(u) = 0, \ f(u) = c_h\} \geq m.$$

In particular, c_h is a critical value of f.

Proof. Let $c = c_h$ and set

$$K_c = \{u \in X \ : \ |df|(u) = 0, \ f(u) = c\}.$$

Arguing by contradiction, we assume that there exists a neighborhood U of K_c with $\mathrm{Cat}_X \overline{U} \leq m - 1$. For any $\varepsilon > 0$ let

$$\mathcal{N}_\varepsilon(K_c) = \{u \in X \ : \ \rho(u, K_c) < \varepsilon\}.$$

Since the Palais-Smale condition at level c holds, K_c is compact. Therefore we can suppose that $U = \mathcal{N}_{2r}(K_c)$, for some $r > 0$.

There exists $\sigma > 0$ such that

$$u \notin \mathcal{N}_r(K_c) \ \text{ and } \ c - \sigma \leq f(u) \leq c + \sigma \Rightarrow |df|(u) \geq \sigma.$$

For every $c' > c$ there exists $A_1 \in \Gamma_{h+m-1}$ such that $\mathcal{F}(A_1) < c'$. Let

$$A_2 = \overline{A_1 \setminus \overline{\mathcal{N}_{2r}(K_c)}}.$$

Then

$$\mathrm{Cat}_X A_2 \geq \mathrm{Cat}_X A_1 - \mathrm{Cat}_X \overline{\mathcal{N}_{2r}(K_c)} \geq h + m - 1 - (m-1) = h,$$

$\mathcal{F}(A_2) < c'$ (since $\mathcal{F}(A_1) < c'$) and $A_2 \cap \mathcal{N}_{2r}(K_c) = \emptyset$.

Setting

$$E = \{A \in \Gamma_h \; : \; A \cap \mathcal{N}_{2r}(K_c) = \emptyset\},$$

then

$$\inf_E \mathcal{F}_h = \inf_{\Gamma_h} \mathcal{F}_h.$$

Since $c_h > -\infty$, it follows that $\mathcal{F}_h$ is bounded from below. Then by Lemma 2.3 we obtain $A \in \Gamma_h$ such that

$$\mathcal{F}_h(A) < c + r\sigma, \quad \varrho(A, E) < r, \quad |d\mathcal{F}_h|(A) < \sigma.$$

Then $A \cap \mathcal{N}_{2r}(K_c) = \emptyset$ and $\max_A f \geq c$, so, for every $u \in A$,

$$f(u) \geq \max_A f - \sigma \Rightarrow |df|(u) \geq \sigma.$$

Thus, by Lemma 2.5 we obtain $|d\mathcal{F}_h|(A) \geq \sigma$, which is a contradiction.■

Now we prove a result that will be useful for establishing a sufficient condition of existence of critical points.

Theorem 2.7 (Degiovanni and Marzocchi [7]) Let X be a metric space and $f : X \to \mathbb{R}$ a continuous function. Assume D is a compact metric space and S is a nonempty closed subset of D. Fix $\psi : S \to X$ a continuous map and consider the set

$$\mathcal{P} = \{p \in C(D; X) \; : \; p|_S = \psi\},$$

which is endowed with the uniform metric ϱ. Let $\mathcal{F} : \mathcal{P} \to \mathbb{R}$ be the continuous function defined by

$$\mathcal{F}(p) = \max_D (f \circ p), \quad \forall p \in \mathcal{P}.$$

Let $p \in \mathcal{P}$, $\rho > 0$, $\sigma > 0$ be such that

$$\max_S (f \circ \psi) < \max_D (f \circ p)$$

and, for every $\xi \in D$,

$$f(p(\xi)) \geq \max_D (f \circ p) - \rho \Rightarrow |df|(p(\xi)) \geq \sigma.$$

Then $|d\mathcal{F}|(p) \geq \sigma$.

Proof. Without loss of generality we can assume

$$\max_S (f \circ \psi) \le \max_D (f \circ p) - 3\rho.$$

Fix $0 < \sigma' < \sigma$ and let U and $\mathcal{K} : X \times [0, \delta] \to X$ be obtained applying Lemma 2.4 to the compact set

$$\{p(\xi) \ : \ f(p(\xi)) \ge \max_D (f \circ p) - \rho\}$$

and to σ'. Furthermore, we can assume that $f(u) > \max_D(f \circ p) - 2\rho$ for any $u \in U$.

We can also suppose that $\mathcal{K}(u, t) = u$ whenever $f(u) \le \max_D(f \circ p) - 3\rho$. Otherwise, we substitute $\mathcal{K}(u, t)$ with $\mathcal{K}(u, t\lambda(u))$, where $\lambda : X \to [0, 1]$ is a continuous function such that $\lambda(u) = 0$ for $f(u) \le \max_D(f \circ p) - 3\rho$ and $\lambda(u) = 1$ for $f(u) \ge \max_D(f \circ p) - 2\rho$.

Let $\mathcal{V}$ be a neighborhood of p in $\mathcal{P}$ such that for any $\eta \in \mathcal{V}$ we have

$$\max_D (f \circ \eta) \ge \max_D (f \circ p) - \frac{\rho}{2},$$

$$\{\eta(\xi) \ : \ f(\eta(\xi)) \ge \max_D (f \circ p) - \rho\} \subset U.$$

Let $\delta' = \min\{\rho/(2\sigma'), \delta\}$ and let $\mathcal{H} : \mathcal{V} \times [0, \delta'] \to \mathcal{P}$ be defined by

$$\mathcal{H}(\eta, t)(\xi) = \mathcal{K}(\eta(\xi), t), \quad \forall \xi \in D.$$

The function $\mathcal{H}$ is continuous and $\varrho(\mathcal{H}(\eta, t), \eta) \le t$.

Let now $\eta \in \mathcal{V}$ and $\xi \in D$. If $f(\eta(\xi)) \le \max_D(f \circ p) - \rho$, then, using part (ii) in Lemma 2.4 and the fact that $\eta \in \mathcal{V}$, for any $t \in [0, \delta']$ we have

$$\begin{aligned} f(\mathcal{H}(\eta, t)(\xi)) &= f(\mathcal{K}(\eta(\xi), t)) \le f(\eta(\xi)) \\ &\le \max_D (f \circ p) - \rho \le \max_D (f \circ \eta) - \frac{\rho}{2} < \mathcal{F}(\eta) - \sigma' t. \end{aligned}$$

Instead, if $f(\eta(\xi)) \ge \max_D(f \circ p) - \rho$, then for any $t \in [0, \delta']$ we have

$$f(\mathcal{H}(\eta, t)(\xi)) = f(\mathcal{K}(\eta(\xi), t)) \le f(\eta(\xi)) - \sigma' t \le \mathcal{F}(\eta) - \sigma' t.$$

Then

$$\mathcal{F}(\mathcal{H}(\eta, t)) \le \mathcal{F}(\eta) - \sigma' t, \quad \forall t \in [0, \delta'].$$

Therefore, $|d\mathcal{F}|(p) \ge \sigma'$.

Since σ' is arbitrary chosen in $]0, \sigma[$, our assertion follows. ■

We state below the main result of this Section.

Theorem 2.8 (Degiovanni and Marzocchi [7]) Let X be a complete metric space and $f : X \to \mathbb{R}$ a continuous function. Assume D is a compact metric space and S is a nonempty closed subset of D. Fix $\psi : S \to X$ a continuous map and consider the set

$$\mathcal{P} = \{p \in C(D; X) \, : \, p|_S = \psi\},$$

which is endowed with the Hausdorff metric ϱ. Assume that $\mathcal{P} \neq \emptyset$ and

$$\max_S (f \circ \psi) < \max_D (f \circ p), \quad \forall p \in \mathcal{P}.$$

If f verifies the Palais-Smale condition at level

$$c = \inf_{p \in \mathcal{P}} \max_D (f \circ p),$$

then c is a critical value of f.

Proof. Arguing by contradiction, let us suppose that c is not a critical value of f. By hypothesis we have that $c \in \mathbb{R}$. Since f has the Palais-Smale condition at level c, there exists $\sigma > 0$ such that

$$c - \sigma \leq f(u) \leq c + \sigma \Rightarrow |df|(u) \geq \sigma.$$

Let us define the mapping $\mathcal{F} : \mathcal{P} \to \mathbb{R}$ as in Theorem 2.7. Since $\mathcal{F}$ is bounded from below, Lemma 2.3 yields the existence of $p \in \mathcal{P}$ such that $\mathcal{F}(p) < c + \sigma$ and $|d\mathcal{F}|(p) < \sigma$. By the definition of c and the choice of σ, it follows that, for every $\xi \in D$,

$$f(p(\xi)) \geq \max_D (f \circ p) - \sigma \Rightarrow |df|(p(\xi)) \geq \sigma.$$

Applying Theorem 2.7 we obtain $|d\mathcal{F}|(p) \geq \sigma$, which is a contradiction. ■

Remark 2.3 Generally the critical points in the sense of Definition 1.10 do not coincide with the critical points in the sense of Definition 2.1 (so, a fortiori, in the sense of Definition 2.2). For instance, the locally Lipschitz function $f : \mathbb{R}^2 \to \mathbb{R}$ given by $f(x, y) = |y - |x|| - x$, $\forall (x, y) \in \mathbb{R}^2$, admits $0 \in \mathbb{R}^2$ as a critical point in the sense of Definition 2.1 (i.e., $0 \in \partial f(0)$), but $|df|(0) > 0$, which means that 0 is not a critical point as introduced in Definition 1.10. For examples of this type we refer to [6], [35].

References

[1] S. Adly, G. Buttazzo and M. Théra, Critical points for nonsmooth energy functions and applications, *Nonlinear Anal.* **32** (1998), 711-718.

[2] A. Ambrosetti and P. H. Rabinowitz, Dual variational methods in critical point theory and applications, *J. Funct. Anal.* **14** (1973), 349-381.

[3] H. Brézis, J.-M. Coron and L. Nirenberg, Free vibrations for a nonlinear wave equation and a theorem of P. Rabinowitz, *Commun. Pure Appl. Math.* **33** (1980), 667-684.

[4] K.-C. Chang, Variational methods for non-differentiable functionals and their applications to partial differential equations, *J. Math. Anal. Appl.* **80** (1981), 102-129.

[5] F. H. Clarke, *Optimization and Nonsmooth Analysis*, New York, John Wiley-Interscience, 1983.

[6] M. Degiovanni, Nonsmooth critical point theory and applications, *Second World Congress of Nonlinear Analysts* (Athens, 1996), *Nonlinear Anal.* **30** (1997), 89-99.

[7] M. Degiovanni and M. Marzocchi, A critical point theory for nonsmooth functionals, *Ann. Mat. Pura Appl., IV. Ser.* **167** (1994), 73-100.

[8] G. Dincă, P. Jebelean and D. Motreanu, Existence and approximation for a general class of differential inclusions, *Houston J. Math.* **28** (2002), 193-215.

[9] I. Ekeland, On the variational principle, *J. Math. Anal. Appl.* **47** (1974), 324-353.

[10] I. Ekeland, Nonconvex minimization problems, *Bull. (New Series) Amer. Math.* **1** (1979), 443-474.

[11] M. Fundo, Hemivariational inequalities in subspaces of $L^p(\Omega)(p \geq 3)$, *Nonlinear Anal.* **33** (1998), 331-340.

[12] L. Gasiński and N. S. Papageorgiou, Solutions and multiple solutions for quasilinear hemivariational inequalities at resonance, *Proc. R. Soc. Edinb.* **131** (2001), 1091-1111.

[13] N. Ghoussoub and D. Preiss, A general mountain pass principle for locating and classifying critical points, *Ann. Inst. Henri Poincaré, Anal. Non Linéaire* **6** (1989), 321-330.

[14] D. Goeleven and D. Motreanu, *Minimax methods of Szulkin's type unilateral problems*, Functional analysis. Selected topics. Dedicated to the memory of the late Professor P. K. Kamthan (Jain, Pawan K. (ed.)), Narosa Publishing House, New Delhi, 158-172, 1998.

[15] D. Goeleven, D. Motreanu, Y. Dumont and M. Rochdi, *Variational and Hemivariational Inequalities, Theory, Methods and Applications*, Volume I: Unilateral Analysis and Unilateral Mechanics, Kluwer Academic Publishers, Dordrecht / Boston / London, to appear.

[16] D. Goeleven and D. Motreanu, *Variational and Hemivariational Inequalities, Theory, Methods and Applications*, Volume II: Unilateral Problems, Kluwer Academic Publishers, Dordrecht / Boston / London, to appear.

[17] J. Haslinger and D. Motreanu, Hemivariational inequalities with a general growth condition: existence and approximation, submitted.

[18] L. T. Hu, *Homotopy Theory*, Academic Press, New York, 1959.

[19] N. C. Kourogenis and N. S. Papageorgiou, Nonsmooth critical point theory and nonlinear elliptic equations at resonance, *Kodai Math. J.* **23** (2000), 108-135.

[20] S. Marano and D. Motreanu, Infinitely many critical points of non-differentiable functions and applications to a Neumann-type problem involving the p-Laplacian, *J. Differ. Equations* **182** (2002), 108-120.

[21] J. Mawhin and M. Willem, *Critical point theory and Hamiltonian systems*, Applied Mathematical Sciences **74**, Springer-Verlag, New York, 1989.

[22] D. Motreanu, Existence and critical points in a general setting, *Set-Valued Anal.* **3** (1995), 295-305.

[23] D. Motreanu, Eigenvalue problems for variational-hemivariational inequalities in the sense of P. D. Panagiotopoulos, *Nonlinear Anal.* **47** (2001), 5101-5112.

[24] D. Motreanu and Z. Naniewicz, Discontinuous semilinear problems in vector-valid function spaces, *Differ. Integral Equ.* **9** (1996), 581-598.

[25] D. Motreanu and Z. Naniewicz, A topological approach to hemivariational inequalities with unilateral growth condition, *J. Appl. Anal.* **7** (2001), 23-41.

[26] D. Motreanu and P. D. Panagiotopoulos, *Minimax Theorems and Qualitative Properties of the Solutions of Hemivariational Inequalities*, Nonconvex Optimization and its Applications **29**, Kluwer Academic Publishers, Dordrecht/Boston/London, 1998.

[27] D. Motreanu and Cs. Varga, Some critical points results for locally Lipschitz functionals, *Commun. Appl. Nonlinear Anal.* **4** (1997), 17-33.

[28] D. Motreanu and Cs. Varga, A nonsmooth equivariant minimax principle, *Commun. Appl. Anal.* **3** (1999), 115-130.

[29] J. R. Munkres, *Elementary Differential Topology*, Princeton University Press, Princeton, 1966.

[30] Z. Naniewicz and P. D. Panagiotopoulos, *Mathematical theory of hemivariational inequalities and applications*, Pure and Applied Mathematics, Marcel Dekker **188**, New York, 1994.

[31] R. S. Palais and S. Smale, A generalized Morse theory, *Bull. Am. Math. Soc.* **70**, (1964) 165-172.

[32] J.-P. Penot, Well-behaviour, well-posedness and nonsmooth analysis, *PLISKA, Stud. Math. Bulg.* **12** (1998), 141-190.

[33] P. H. Rabinowitz, *Minimax methods in critical point theory with applications to differential equations*, CBMS Reg. Conf. Ser. in Math. **65**, Amer. Math. Soc., Providence, 1986.

[34] V. Rădulescu, Mountain pass theorems for non-differentiable functions and applications, *Proc. Japan Acad., Ser. A* **69** (1993), 193-198.

[35] N. K. Ribarska, Ts. Y. Tsachev, M. I. Krastanov, Speculating about mountains, *Serdica Math. J.* **22** (1996), 341-358.

[36] B. Ricceri, Existence of three solutions for a class of elliptic eigenvalue problems. Nonlinear operator theory, *Math. Comput. Modelling* **32** (2000), 1485-1494.

[37] B. Ricceri, *New results on local minima and their applications*, in: Equilibrium Problems and Variational Models (editors: F. Giannessi, A. Maugeri and P. M. Pardalos), Taormina, 2-5 December 1998, Kluwer Academic Publishers, Dordrecht/Boston/London, 2001, pp. 255-268.

[38] W. Rudin, *Functional Analysis*, McGraw-Hill Series in Higher Mathematics, New York, 1973.

[39] E. H. Spanier, *Algebraic Topology,* McGraw-Hill, New York, 1966.

[40] A. Szulkin, Minimax principles for lower semicontinuous functions and applications to nonlinear boundary value problems, *Ann. Inst. Henri Poincaré. Anal. Non Linéaire* **3** (1986), 77-109.

Chapter 3

VARIATIONAL METHODS

This Chapter deals with general techniques for studying the existence and multiplicity of critical points of nondifferentiable functionals in the so-called limit case (see Remark 3.1). There are proved nonsmooth versions of several celebrated results like: Deformation Lemma, Mountain Pass Theorem, Saddle Point Theorem, Generalized Mountain Pass Theorem. First, we present a general deformation result for nonsmooth functionals which can be expressed as a sum of a locally Lipschitz function and a concave, proper, upper semicontinuous function. Then we give a general minimax principle for nonsmooth functionals which can be expressed as a sum of a locally Lipschitz function and a convex, proper, lower semicontinuous functional. Here we are concerned with the limit case (i.e. the equality $c = a$, see Remark 3.1), obtaining results which are complementary to the minimax principles in Section 2 of Chapter 2. These general results are applied in the second Section of this Chapter for proving existence, multiplicity and location of solutions to various boundary value and unilateral problems with discontinuous nonlinearities.

1. Critical Point Theory for Convex Perturbations of Locally Lipschitz Functionals in the Limit Case

First, we study in this Section the deformation properties of a nonsmooth functional $g : X \to [-\infty, +\infty[$ on a real reflexive Banach space $(X, \|\cdot\|)$ fulfilling the structure hypothesis:

(H_g) *$g = \Psi + \beta$, where $\Psi : X \to \mathbb{R}$ is locally Lipschitz and $\beta : X \to [-\infty, +\infty[$ is concave, proper (i.e., $\not\equiv -\infty$) and upper semicontinuous.*

Definition 3.1 An element $u \in X$ is called a critical point of functional $g : X \to [-\infty, +\infty[$ satisfying (H_g) if the below inequality holds

$$\Psi^0(u; u - x) + \beta(u) - \beta(x) \geq 0, \quad \forall x \in X.$$

Here Ψ^0 stands for the generalized directional derivative in the sense of Clarke [5] for the locally Lipschitz functional Ψ. The notation $\partial\Psi$ will stand for the generalized gradient of Ψ in the sense of Clarke [5] (see also [4]).

Given a real number c we denote

$$K_c(g) = \{u \in X : g(u) = c, \quad u \text{ is a critical point of } g \text{ in the sense of Definition 3.1}\}.$$

We denote

$$D_\beta = \{x \in X : \beta(x) > -\infty\}.$$

Let us introduce

$$\partial\beta(x) = -\partial(-\beta)(x), \quad x \in X. \tag{3.1}$$

Relation (3.1) makes sense because $-\beta : X \to]-\infty, +\infty]$ is a convex function, so in the right-hand side of (3.1), $\partial(-\beta)$ means the subdifferential of $-\beta$ in the sense of convex analysis.

Definition 3.2 A nonsmooth functional $g : X \to [-\infty, +\infty[$ satisfying condition (H_g) is said to verify the Palais-Smale condition around a subset $B \subset X$ at the level $c \in \mathbb{R}$ if the property below holds:

$(PS)_{g,B,c}$ *Each sequence $\{x_n\} \subset X$ such that $d(x_n, B) \to 0$, $g(x_n) \to c$, and*

$$\Psi^0(x_n; x_n - x) + \beta(x_n) - \beta(x) \geq -\varepsilon_n \|x_n - x\|, \quad \forall n \in \mathbb{N},\ x \in X,$$

where $\varepsilon_n \to 0^+$, possesses a strongly convergent subsequence.

In the sequel we make use of the following assumption regarding two given subsets A and B of X:

(g) *A and B are two nonempty closed subsets of X such that*

$$A \cap B = \emptyset, \quad A \subset g^c, \quad B \subset g_c, \quad K_c(g) \cap B = \emptyset.$$

Moreover, there exists $\varepsilon_0 > 0$ satisfying $N_{\varepsilon_0}(B) \subset \text{int}(D_\beta)$.

In the statement of (g) we used the following notations:

$$g_c = \{x \in X : g(x) \leq c\},$$

$$g^c = \{x \in X : c \leq g(x)\},$$

$$N_{\varepsilon_0}(B) = \{z \in X : d(z, B) \leq \varepsilon_0\}.$$

In a series of lemmas we establish some preliminary facts which are necessary for our deformation result.

Lemma 3.1 Assume that conditions (H_g), (g) and $(\mathrm{PS})_{g,B,c}$ hold. Then there exist $\varepsilon_1 \in]0, \varepsilon_0[$ and $\sigma > 0$ such that for every $x \in N_{\varepsilon_1}(B) \cap g^{c-\varepsilon_1} \cap g_{c+\varepsilon_1}$, $x^* \in \partial\Psi(x)$, $z^* \in \partial\beta(x)$ one has $\|x^* + z^*\|_{X^*} \geq \sigma$.

Proof. Arguing by contradiction, we suppose that the conclusion were false. Then one could construct three sequences $\{x_n\} \subset X$, $\{x_n^*\}$, $\{z_n^*\} \subset X^*$ having the following properties:

$$d(x_n, B) \to 0; \tag{3.2}$$

$$g(x_n) \to c; \tag{3.3}$$

$$x_n^* \in \partial\Psi(x_n) \text{ and } z_n^* \in \partial\beta(x_n), \quad \forall n \in I\!N; \tag{3.4}$$

$$\|x_n^* + z_n^*\|_{X^*} \to 0. \tag{3.5}$$

By (3.1) and (3.4) it is seen that

$$\Psi^0(x_n; x_n - x) + \beta(x_n) - \beta(x) \geq -\|x_n^* + z_n^*\|_{X^*}\|x_n - x\|, \quad x \in X. \tag{3.6}$$

Setting $\varepsilon_n = \|x_n^* + z_n^*\|_{X^*}$ and using $(\mathrm{PS})_{g,B,c}$ in conjunction with (3.2), (3.3), (3.5), we deduce from (3.6) that there exists a subsequence of $\{x_n\}$, denoted again by $\{x_n\}$, such that $x_n \to u$ strongly in X for some $u \in X$. Since Ψ^0 and β are upper semicontinuous, we derive that u is a critical point of g in the sense of Definition 3.1. By (3.2) and (g) we then infer $u \in \mathrm{int}(D_\beta)$, thus $\beta(x_n) \to \beta(u)$ as $n \to \infty$. According to (3.3) this leads to $u \in K_c(g)$. Since (3.2) ensures that $u \in B$, we arrive at a contradiction with assumption (g). This completes the proof. ■

Lemma 3.2 Assume that the function g satisfies (H_g), (g) and $(\mathrm{PS})_{g,B,c}$. Let the positive numbers ε_1 and σ as in Lemma 3.1. Then corresponding to every point $x \in N_{\varepsilon_1}(B) \cap g^{c-\varepsilon_1} \cap g_{c+\varepsilon_1}$ there exists some $\xi_x \in X$ such that

$$\|\xi_x\| = 1, \quad \langle x^* + z^*, \xi_x\rangle \geq \sigma, \quad \forall x^* \in \partial\Psi(x),\ z^* \in \partial\beta(x). \tag{3.7}$$

Proof. Fix $x \in N_{\varepsilon_1}(B) \cap g^{c-\varepsilon_1} \cap g_{c+\varepsilon_1} \subset D_{\partial\beta}$. Since $\partial\Psi(x)$ and $\partial\beta(x)$ are nonempty and convex (see (3.1)), one has that $\partial\Psi(x) + \partial\beta(x)$ is a convex subset of X. We show that it is also closed. To this end let $\{x_n^*\} \subset \partial\Psi(x)$ and $\{z_n^*\} \subset \partial\beta(x)$ be sequences such that $x_n^* + z_n^* \to u^*$ in X^*. The reflexivity of X and Proposition 2.1.2 in [5] yield $x^* \in \partial\Psi(x)$ with along a subsequence $x_n^* \rightharpoonup x^*$ weakly. Hence, $z_n^* \rightharpoonup u^* - x^*$ weakly. The choice of $\{z_n^*\}$ implies $u^* - x^* \in \partial\beta(x)$, from which the claim follows.

Lemma 3.1 ensures that $0 \notin \partial\Psi(x) + \partial\beta(x)$. Applying Corollary III.20 in [3] we obtain $u^* \in \partial\Psi(x)$, $v^* \in \partial\beta(x)$ fulfilling

$$B_{\delta^*} \cap (\partial\Psi(x) + \partial\beta(x)) = \emptyset\,,$$

where $\delta^* = \|u^* + v^*\|_{X^*}$ and $B_{\delta^*} = \{v \in X \,:\, \|v\| \leq \delta^*\}$. The Hahn-Banach Theorem (see [3],Theorem I.6) provides a point $\xi_x \in X$ with the properties $\|\xi_x\| = 1$ and

$$\langle x^* + z^*, \xi_x \rangle \geq \langle w^*, \xi_x \rangle, \quad \forall w^* \in B_{\delta^*}$$

whenever $x^* \in \partial\Psi(x)$, $z^* \in \partial\beta(x)$. Taking into account that

$$\|u^* + v^*\|_{X^*} = \|u^* + v^*\|_{X^*}\|\xi_x\| = \max\left\{ \langle w^*, \xi_x \rangle : w^* \in \overline{B}_{\delta^*} \right\},$$

the above inequality and Lemma 3.1 lead to

$$\langle x^* + z^*, \xi_x \rangle \geq \|u^* + v^*\|_{X^*} \geq \sigma, \quad \forall x^* \in \partial\Psi(x),\ z^* \in \partial\beta(x)\,,$$

as claimed. ∎

Lemma 3.3 Assume that conditions (H_g), (g), $(\mathrm{PS})_{g,B,c}$ are satisfied. Let ε_1 and σ be positive numbers like in Lemma 3.1. Then for every $x \in N_{\varepsilon_1}(B) \cap g^{c-\varepsilon_1} \cap g_{c+\varepsilon_1}$ there exists $\delta_x > 0$ such that

$$\langle x^* + z^*, \xi_x \rangle > \frac{\sigma}{2}, \quad \forall x^* \in \partial\Psi(x'),\ z^* \in \partial\beta(x''),\ x', x'' \in B(x, \delta_x)\,,$$

where ξ_x is given by Lemma 3.2.

Proof. If the conclusion were not true, we could find $x \in N_{\varepsilon_1}(B) \cap g^{c-\varepsilon_1} \cap g_{c+\varepsilon_1}$, $\{x'_n\}, \{x''_n\} \subset X$ and $\{x_n^*\}, \{z_n^*\} \subset X^*$ fulfilling the following conditions:

$$x'_n \to x\,, \quad x_n^* \in \partial\Psi(x'_n), \quad \forall n \in I\!N; \tag{3.8}$$

$$x''_n \to x\,, \quad z_n^* \in \partial\beta(x''_n), \quad \forall n \in I\!N; \tag{3.9}$$

$$\langle x_n^* + z_n^*, \xi_x \rangle \leq \frac{\sigma}{2}, \quad \forall n \in I\!N. \tag{3.10}$$

Due to the reflexivity of X and (3.8), Proposition 2.1.2 in [5] yields $x^* \in X^*$ such that $x_n^* \rightharpoonup x^*$ weakly in X^*, where a subsequence is considered when necessary. Proposition 2.1.5 in [5] implies $x^* \in \partial\Psi(x)$. From (3.10) we thus get

$$\limsup_{n\to+\infty}\langle z_n^*, \xi_x\rangle \le \frac{\sigma}{2} - \langle x^*, \xi_x\rangle . \tag{3.11}$$

Since the multifunction $\partial(-\beta) : \mathrm{int}(D_{\partial\Psi}) \to 2^{X^*}$ is locally bounded, it follows that the sequence $\{z_n^*\}$ is bounded. Then along a relabelled subsequence $\{z_{r_n}^*\}$ of $\{z_n^*\}$ we have that $z_{r_n}^* \rightharpoonup z^*$ weakly in X^* because X^* is reflexive. In addition, we infer that $z^* \in \partial\beta(x)$ and

$$\langle z^*, \xi_x\rangle = \lim_{n\to+\infty}\langle z_{r_n}^*, \xi_x\rangle .$$

By (3.11) this yields $\langle x^* + z^*, \xi_x\rangle \le \sigma/2$ which contradicts (3.7). The proof is thus complete. ■

We are now in a position to formulate our main deformation result of this Section.

Theorem 3.1 (Marano and Motreanu [12]) Assume the function $g : X \to [-\infty, +\infty[$ satisfies (H_g), (g), $(\mathrm{PS})_{g,B,c}$ and the set $N_{\varepsilon_1}(B) \cap g^{c-\varepsilon_1} \cap g_{c+\varepsilon_1}$, with ε_1 like in Lemma 3.1, is closed. Then there exist a number $\varepsilon > 0$ and a homeomorphism $\eta : X \to X$ having the following properties:

(i) $\eta(x) = x$ for every $x \in A$;

(ii) $\eta(B) \subset g_{c-\varepsilon}$.

Proof. The family of balls $\mathcal{B} = \{B(x, \delta_x) : x \in N_{\varepsilon_1}(B) \cap g^{c-\varepsilon_1} \cap g_{c+\varepsilon_1}\}$ constructed in Lemma 3.3 is an open covering of

$$M := N_{\varepsilon_1}(B) \cap g^{c-\varepsilon_1} \cap g_{c+\varepsilon_1}.$$

Then the family $\mathcal{B} \cup \{X \setminus M\}$ is an open covering of X. Since X is paracompact, $\mathcal{B} \cup \{X \setminus M\}$ possesses an open refinement $\mathcal{O} = \{O_k : k \in K\}$ which is a locally finite covering of X (cf. [7], Theorem VIII.2.4). Let $\mathcal{U} = \{U_k : k \in K\}$ be a locally finite covering of X satisfying $\overline{U_k} \subset O_k$ (see [7], Theorems VIII.2.2 and VII.6.1).

Since $\mathcal{O}$ is a refinement of $\mathcal{B} \cup \{X \setminus M\}$, for each $k \in K$ either there is $x_k \in M$ such that $O_k \subset B(x_k, \delta_{x_k})$ or $O_k \subset X \setminus M$. Denote

$$I := \{k \in K : O_k \subset B(x_k, \delta_{x_k}) \text{ for some } x_k \in M\}.$$

Consider the families $\mathcal{V} = \{V_i : i \in I\}$ and $\mathcal{W} = \{W_i : i \in I\}$ given by $V_i = O_i$ and $W_i = U_i$ for $i \in I$.

We remark that $\mathcal{V}$ is a covering of M. Indeed, let $x \in M$. Since $\mathcal{O}$ covers X, we have that $x \in O_k$ for some $k \in K$. Then either $O_k \subset B(x_k, \delta_{x_k})$ for some $x_k \in M$ or $O_k \subset X \setminus M$. As $x \in M$, we have only the first situation, thus $k \in I$ and $x \in O_k = V_k$. Consequently, $M \subset \bigcup_{i \in I} V_i$, i.e. $\mathcal{V}$ is a covering of M.

For each $i \in I$, we define

$$d_i(x) = d\,(x, X \setminus W_i)\;, \quad x \in X\,.$$

The function d_i is Lipschitz continuous. Define the function $\rho_i : W \to [0,1]$ by

$$\rho_i(x) = \frac{d_i(x)}{\sum_{j \in I} d_j(x)}\,, \quad x \in W,$$

where

$$W := \bigcup_{j \in I} W_j.$$

Let us show that the functions ρ_i are well-defined. First, we see that for every $x \in W$ we find $j \in I$ such that $x \in W_j$, i.e. $d_j(x) > 0$. We have to prove that for each $x \in W$, the sum $\sum_{j \in I} d_j(x)$ is finite. Let $x \in W$. Since $\mathcal{U}$ is a locally finite covering for X, there exist a neighborhood N_x of x in X and a finite subset J_x of K such that

$$N_x \cap U_k \neq \emptyset, \quad \forall k \in J_x$$

and

$$N_x \cap U_k = \emptyset, \quad \forall k \in K \setminus J_x.$$

We note that $I = (J_x \cap I) \cup ((K \setminus J_x) \cap I)$, with $(J_x \cap I) \cap ((K \setminus J_x) \cap I) = \emptyset$. Thus if $i \in I$, then $U_i = W_i$ and either $i \in J_x \cap I$ or $i \in (K \setminus J_x) \cap I$. Therefore we obtain a neighborhood N_x of x in X and a finite subset $J_x \cap I$ of I such that

$$N_x \cap W_i \neq \emptyset, \quad \forall i \in J_x \cap I$$

and

$$N_x \cap W_i = \emptyset, \quad \forall i \in (K \setminus J_x) \cap I.$$

Consequently, if $i \in (K \setminus J_x) \cap I$ then $x \notin W_i$, i.e. $d_i(x) = 0$. This means that $\sum_{j \in I} d_j(x) = \sum_{j \in J_x \cap I} d_j(x)$ is finite. Here $J_x \cap I \neq \emptyset$, since if this were not true, we would have that $N_x \cap W_i = \emptyset$, $\forall i \in I$, that would contradict $x \in W$. We conclude that the functions ρ_i are well-defined for each $x \in W$ and $i \in I$.

Moreover, the functions $\rho_i : W \to [0,1]$, $i \in I$, have the following properties:

$$\operatorname{supp} \rho_i = \overline{W_i} \subset V_i \quad \forall i \in I\,; \quad \sum_{i \in I} \rho_i(x) = 1 \quad \forall x \in W.$$

The family $\{\operatorname{supp} \rho_i : i \in I\}$ is locally finite. Indeed, let $x \in X$. Since $\mathcal{O}$ is a locally finite recovering of X, there is a neighborhood N_x of x in X and a finite subset of J_x of K such that

$$N_x \cap O_k \neq \emptyset, \quad \forall k \in J_x$$

and

$$N_x \cap O_k = \emptyset, \quad \forall k \in K \setminus J_x.$$

Using again $I = (J_x \cap I) \cup ((K \setminus J_x) \cap I)$ we deduce that

$$N_x \cap V_i \neq \emptyset, \quad \forall i \in J_x \cap I$$

and

$$N_x \cap V_i = \emptyset, \quad \forall i \in (K \setminus J_x) \cap I.$$

Since $\overline{W_i} \subset V_i$ one has

$$N_x \cap \overline{W_i} = \emptyset, \quad \forall i \in (K \setminus J_x) \cap I,$$

thus $N_x \cap \overline{W_i} \neq \emptyset$ only for a finite number of indices i in $J_x \cap I$, hence the family $\{\operatorname{supp} \rho_i : i \in I\}$ is locally finite.

For each $i \in I$ we find $x_i \in M$ such that $V_i \subset B(x_i, \delta_{x_i})$. In view of Lemma 3.2, it corresponds $\xi_i \in X$ such that $\|\xi_i\| = 1$ and

$$\langle x^* + z^*, \xi_i \rangle > \frac{\sigma}{2}, \quad \forall x^* \in \partial\Psi(x'),\ z^* \in \partial\beta(x''),\ x', x'' \in V_i\,. \tag{3.12}$$

Let $\theta : W \to X$ be given by

$$\theta(x) = \sum_{i \in I} \rho_i(x) \xi_i\,, \quad x \in W\,.$$

The function θ is well defined due to the fact that the family $\{\operatorname{supp} \rho_i : i \in I\}$ is locally finite in X. Clearly, θ is locally Lipschitz and $\|\theta(x)\| \leq 1$ in W.

Set, for every $x \in X$,

$$\Theta(x) = \begin{cases} -\varepsilon_1 \theta(x) & \text{if } x \in W, \\ 0 & \text{otherwise.} \end{cases} \tag{3.13}$$

The function $\Theta : X \to X$ turns out locally Lipschitz. To see this, we simply note that the set $\bigcup_{i \in I} \operatorname{supp} \rho_i$ is closed, which comes from the local finiteness of the family $\{\operatorname{supp} \rho_i : i \in I\}$, while $\Theta(x) = 0$ in $X \setminus \bigcup_{i \in I} \operatorname{supp} \rho_i$. Furthermore, one has $\|\Theta(x)\| \leq \varepsilon_1$ for all $x \in X$.

The existence and uniqueness theorem for ordinary differential equations in Banach spaces provides a function $\gamma \in C(I\!R \times X, X)$ satisfying

$$\frac{d\gamma(t,x)}{dt} = \Theta(\gamma(t,x)), \quad \gamma(0,x) = x, \quad \forall (t,x) \in I\!R \times X. \tag{3.14}$$

Next, define $B_1 = \gamma([0,1] \times B)$. If $x \in B$ then

$$\|\gamma(t,x) - x\| = \|\int_0^t \frac{d\gamma(\tau,x)}{d\tau} d\tau\| = \|\int_0^t \Theta(\gamma(\tau,x)) d\tau\| \leq \varepsilon_1, \quad \forall t \in [0,1].$$

Consequently, it is true that

$$B_1 \subset N_{\varepsilon_1}(B). \tag{3.15}$$

Let us verify that the set B_1 is closed. To this end, pick a sequence $\{y_n\} \subset B_1$ converging to some $y \in X$. Since $y_n = \gamma(t_n, x_n)$ with $(t_n, x_n) \in [0,1] \times B$, by eventually taking a subsequence, we can suppose $t_n \to t$ in $[0,1]$. Write $z_n = \gamma(t, x_n)$, $n \in I\!N$, and observe that

$$\|y_n - z_n\| = \|\int_t^{t_n} \frac{d\gamma(\tau, x_n)}{d\tau} d\tau\| \leq \varepsilon_1 |t_n - t|, \quad \forall n \in I\!N.$$

Therefore, $z_n \to y$ in X. Through the properties of γ we thus achieve

$$x_n = \gamma(-t, z_n) \to \gamma(-t, y).$$

Setting $x = \gamma(-t, y)$ one has $x_n \to x$, the point x lies in B because B is closed, while $y = \gamma(t,x) \in \gamma([0,1] \times B) = B_1$. The claim is thus verified.

Our next goal is to show that

$$\forall x \in B \text{ the function } t \mapsto g(\gamma(t,x)) \text{ is decreasing on } [0,1]. \tag{3.16}$$

The claim is proved once we see that to each $x_0 \in B$, $t_0 \in [0,1]$ it corresponds $\delta_0 > 0$ fulfilling

$$\frac{g(\gamma(t,x_0)) - g(\gamma(t_0,x_0))}{t - t_0} \leq 0, \quad \forall t \in [0,1] \cap B(t_0, \delta_0) \setminus \{t_0\}. \tag{3.17}$$

So, fix $(t_0, x_0) \in [0,1] \times B$. If $\gamma(t_0, x_0) \notin \bigcup_{i \in I} \operatorname{supp} \rho_i$ then one can find $\delta_0 > 0$ such that

$$\gamma(t, x_0) \notin \bigcup_{i \in I} \operatorname{supp} \rho_i, \quad \forall t \in [0,1] \cap B(t_0, \delta_0).$$

This implies $\Theta(\gamma(t,x_0)) = 0$ and hence $\gamma(t,x_0) = \gamma(t_0,x_0)$ in $[0,1] \cap B(t_0,\delta_0)$, from which (3.17) follows. Suppose $\gamma(t_0,x_0) \in \bigcup_{i \in I} \operatorname{supp} \rho_i$.

Since the family $\{\operatorname{supp} \rho_i : i \in I\}$ is locally finite, there exists $\delta' > 0$ satisfying

$$\operatorname{supp} \rho_i \cap B(\gamma(t_0, x_0), \delta') \neq \emptyset$$

for a finite number of $i \in I$, say $i_1, \ldots, i_p$. Consequently,

$$\operatorname{supp} \rho_i \cap B(\gamma(t_0, x_0), \delta') \begin{cases} \neq \emptyset & \text{if } i \in \{i_1, \ldots, i_p\}, \\ = \emptyset & \text{otherwise.} \end{cases} \tag{3.18}$$

Let $i'_1, \ldots, i'_q$ be the elements in $\{i_1, \ldots, i_p\}$ such that $\gamma(t_0, x_0) \in \operatorname{supp} \rho_{i'_j}$ whenever $j = 1, \ldots, q$. One has

$$\begin{aligned} &\gamma(t_0, x_0) \in \operatorname{supp} \rho_{i'_j}, \quad \forall j = 1, \ldots, q, \\ &\delta_i = d(\gamma(t_0, x_0), \operatorname{supp} \rho_i) > 0, \quad \forall i \in \{i_1, \ldots, i_p\} \setminus \{i'_1, \ldots, i'_q\}. \end{aligned} \tag{3.19}$$

Choose $\delta'' \in]0, \delta'[$ with the following properties:

$$\delta'' < \delta_i, \quad \forall i \in \{i_1, \ldots, i_p\} \setminus \{i'_1, \ldots, i'_q\};$$

$$B(\gamma(t_0, x_0), \delta'') \subset V_{i'_j}, \quad \forall j = 1, \ldots, q.$$

Thanks to (3.18) and (3.19) we get

$$\begin{aligned} &\operatorname{supp} \rho_i \cap B(\gamma(t_0, x_0), \delta'') = \emptyset \quad \forall i \in I \setminus \{i'_1, \ldots, i'_p\}, \\ &B(\gamma(t_0, x_0), \delta'') \subset V_{i'_j}, \quad \forall j = 1, \ldots, q. \end{aligned} \tag{3.20}$$

Finally, choose $\delta_0 > 0$ such that

$$\gamma(t, x_0) \in B(\gamma(t_0, x_0), \delta''), \quad \forall t \in [0, 1] \cap B(t_0, \delta_0). \tag{3.21}$$

Let $t \in [0, 1] \cap B(t_0, \delta_0) \setminus \{t_0\}$. Suppose $t > t_0$. Since $x_0 \in B$, inclusion (3.15) and assumption (g) imply

$$\gamma(\tau, x_0) \in N_{\varepsilon_1}(B) \subset D_{\partial\beta}, \quad \forall \tau \in [t_0, t].$$

Exploiting Theorem 2.3.7 in [5] as well as the definition of $\partial\beta$ (see (3.1)) we have, for suitable $x \in [\gamma(t_0, x_0), \gamma(t, x_0)]$, $x^* \in \partial\Psi(x)$ and $z^* \in \partial\beta(\gamma(t_0, x_0))$,

$$g(\gamma(t, x_0)) - g(\gamma(t_0, x_0)) \leq \langle x^* + z^*, \gamma(t, x_0) - \gamma(t_0, x_0) \rangle$$

$$= \langle x^* + z^*, \int_{t_0}^{t} \frac{d\gamma(\tau, x_0)}{d\tau} d\tau \rangle = \int_{t_0}^{t} \langle x^* + z^*, \Theta(\gamma(\tau, x_0)) \rangle d\tau. \tag{3.22}$$

On account of (3.21), (3.20) it results

$$\Theta(\gamma(\tau, x_0)) = -\varepsilon_1 \sum_{j=1}^{q} \rho_{i'_j}(\gamma(\tau, x_0)) \xi_{i'_j}, \quad \forall \tau \in [t_0, t].$$

Using that $x, \gamma(t_0, x_0) \in V_{i'_j}$, $j = 1, \ldots, q$, inequality (3.12) can be applied to obtain

$$\langle x^* + z^*, \Theta(\gamma(\tau, x_0)) \rangle \leq -\frac{\varepsilon_1 \sigma}{2} \sum_{j=1}^{q} \rho_{i'_j}(\gamma(\tau, x_0)) = -\frac{\varepsilon_1 \sigma}{2} \tag{3.23}$$

for every $\tau \in [t_0, t]$. Hence, by (3.22), we can write

$$\frac{g(\gamma(t, x_0)) - g(\gamma(t_0, x_0))}{t - t_0} \leq -\frac{\varepsilon_1 \sigma}{2} < 0 .$$

Now, suppose $t < t_0$. Combining (3.22), with exchanged t_0 and t, and (3.23) yields

$$g(\gamma(t_0, x_0)) - g(\gamma(t, x_0)) \leq \int_t^{t_0} \langle x^* + z^*, \Theta(\gamma(\tau, x_0)) \rangle d\tau \leq -\frac{\varepsilon_1 \sigma}{2}(t_0 - t) ,$$

which leads to the same conclusion as for $t > t_0$. Thus (3.17) is completely achieved.

We next claim that

$$A \cap B_1 = \emptyset . \tag{3.24}$$

Indeed, if (3.24) were false one could find $(t_0, x_0) \in]0, 1] \times B$ fulfilling $\gamma(t_0, x_0) \in A$. Because of assumption (g) and (3.16) this implies

$$g(\gamma(t, x_0)) = c, \quad \forall t \in [0, t_0] . \tag{3.25}$$

Hence, due to (3.15), $\gamma(t, x_0) \in N_{\varepsilon_1}(B) \cap g^{c-\varepsilon_1} \cap g_{c+\varepsilon_1}$ for all $t \in [0, t_0]$ and, in particular, $\gamma(t_0, x_0) \in \bigcup_{i \in I} \operatorname{supp} \rho_i$. Arguing as above gives $\delta_0 > 0$ such that

$$\frac{g(\gamma(t, x_0)) - g(\gamma(t_0, x_0))}{t - t_0} \leq -\frac{\varepsilon_1 \sigma}{2}, \quad \forall t \in [0, 1] \cap B(t_0, \delta_0) \setminus \{t_0\} .$$

Through (3.25) we then obtain, whenever $t \in]t_0 - \delta_0, t_0[\cap [0, 1]$,

$$\begin{aligned} g(\gamma(t_0, x_0)) &= \frac{g(\gamma(t_0, x_0)) - g(\gamma(t, x_0))}{t_0 - t}(t_0 - t) + g(\gamma(t, x_0)) \\ &\leq -\frac{\varepsilon_1 \sigma}{2}(t_0 - t) + c < c , \end{aligned}$$

which contradicts (3.25) written for $t = t_0$.

Note that from (3.24) it follows $d(x, A) + d(x, B_1) > 0$ at each point $x \in X$. Let $A_1 = \{x \in X : \zeta_1(x) \leq 1/2\}$, where

$$\zeta_1(x) = \frac{d(x, A)}{d(x, A) + d(x, B_1)} , \quad x \in X .$$

Since the function ζ_1 is continuous, the set A_1 is closed. Moreover, one has $A \subset \text{int}(A_1)$ as well as $A_1 \cap B_1 = \emptyset$. Putting

$$\zeta(x) = \frac{d(x, A_1)}{d(x, A_1) + d(x, B_1)}, \quad \forall x \in X$$

provides a locally Lipschitz function $\zeta : X \to [0, 1]$ such that

$$\zeta|_{A_1} \equiv 0, \quad \zeta|_{B_1} \equiv 1. \tag{3.26}$$

Thanks to the properties of Θ the function $\Lambda : X \to X$ given by

$$\Lambda(x) = \zeta(x)\Theta(x), \quad x \in X, \tag{3.27}$$

is bounded and locally Lipschitz. Denote by $\chi : \mathbb{R} \times X \to X$ the solution of the Cauchy problem

$$\frac{d\chi(t, x)}{dt} = \Lambda(\chi(t, x)), \quad \chi(0, x) = x,$$

and define

$$\varepsilon = \varepsilon_1 \min\left\{\frac{\sigma}{2}, 1\right\}, \quad \eta(x) = \chi(1, x), \quad \forall x \in X. \tag{3.28}$$

Classical results concerning ordinary differential equations in Banach spaces ensure that $\eta : X \to X$ is a homeomorphism. If $x \in A$ then $x \in \text{int}(A_1)$ and, because of (3.26), $\Lambda \equiv 0$ on some neighbourhood of x. This implies $\eta(x) = x$, thus showing assertion (i).

Finally, the proof is accomplished once we verify (ii). Suppose on the contrary that there exists $x_0 \in B$ satisfying

$$g(\eta(x_0)) > c - \varepsilon. \tag{3.29}$$

Through (3.27) and (3.26) we obtain

$$\Lambda(\gamma(t, x_0)) = \Theta(\gamma(t, x_0)), \quad \forall t \in [0, 1],$$

from which it follows, due to (3.14),

$$\frac{d\gamma(t, x_0)}{dt} = \Lambda(\gamma(t, x_0)) \quad \text{in} \quad [0, 1], \quad \gamma(0, x_0) = x_0.$$

By uniqueness of solutions to the same Cauchy problem we thus have

$$\gamma(t, x_0) = \chi(t, x_0), \quad \forall t \in [0, 1]. \tag{3.30}$$

Fix $t_0 \in [0, 1]$. Since $x_0 \in B$, relations (3.29), (3.28), (3.30), (3.16) and (g) lead to

$$c - \varepsilon < g(\gamma(t, x_0)) = g(\chi(t, x_0)) < c + \varepsilon, \quad t \in [0, 1],$$

while gathering (3.27) and (3.13) together yields

$$\|\chi(t, x_0) - x_0\| = \| \int_0^t \frac{d\chi(\tau, x_0)}{d\tau} d\tau\| = \| \int_0^t \Lambda(\chi(\tau, x_0)) d\tau\| \leq \varepsilon_1 t$$

$$\leq \varepsilon_1, \quad \forall t \in [0,1].$$

Therefore,

$$\gamma(t, x_0) = \chi(t, x_0) \in N_{\varepsilon_1}(B) \cap g^{c-\varepsilon_1} \cap g_{c+\varepsilon_1}, \quad \forall t \in [0,1].$$

Using the compactness of $[0,1]$ and the fact that $\mathcal{W}$ is a locally finite covering of $N_{\varepsilon_1}(B) \cap g^{c-\varepsilon_1} \cap g_{c+\varepsilon_1}$ we can find a decomposition $0 = t_0 < t_1 < \ldots < t_{p-1} < t_p = 1$ of $[0,1]$ such that to every $j \in \{1, \ldots, p\}$ it corresponds a finite family $I_j \subset I$ for which one has

$$[\gamma(t_{j-1}, x_0), \gamma(t_j, x_0)] \subset W_i \subset \operatorname{supp} \rho_i \subset V_i,$$

$$\gamma(\tau, x_0) \in W_i, \ \forall \tau \in [t_{j-1}, t_j]$$

whenever $i \in I_j$. By [5, Theorem 2.3.7] and the definition of $\partial\beta$ in (3.1) there exist $x_j \in [\gamma(t_{j-1}, x_0), \gamma(t_j, x_0)]$, $x_j^* \in \partial\Psi(x_j)$ and $z_j^* \in \partial\beta(\gamma(t_{j-1}, x_0))$ fulfilling

$$g(\gamma(t_j, x_0)) - g(\gamma(t_{j-1}, x_0)) \leq \langle x_j^* + z_j^*, \gamma(t_j, x_0) - \gamma(t_{j-1}, x_0)\rangle$$
$$= \langle x_j^* + z_j^*, \int_{t_{j-1}}^{t_j} \frac{d\gamma(\tau, x_0)}{d\tau} d\tau\rangle = \int_{t_{j-1}}^{t_j} \langle x_j^* + z_j^*, \Theta(\gamma(\tau, x_0))\rangle d\tau.$$

Due to (3.13) this inequality becomes

$$g(\gamma(t_j, x_0)) - g(\gamma(t_{j-1}, x_0)) \leq -\varepsilon_1 \sum_{i \in I_j} \langle x_j^* + z_j^*, \xi_i\rangle \int_{t_{j-1}}^{t_j} \rho_i(\gamma(\tau, x_0))\, d\tau.$$

Now, since $x_j, \gamma(t_{j-1}, x_0) \in V_i$ for all $i \in I_j$, using (3.12) we get

$$g(\gamma(t_j, x_0)) - g(\gamma(t_{j-1}, x_0)) \leq -\frac{\varepsilon_1 \sigma}{2} \int_{t_{j-1}}^{t_j} \sum_{i \in I_j} \rho_i(\gamma(\tau, x_0))\, d\tau$$

$$= -\frac{\varepsilon_1 \sigma}{2}(t_j - t_{j-1}).$$

Hence, as j was arbitrary,

$$g(\gamma(1, x_0)) - g(\gamma(0, x_0)) = \sum_{j=1}^{p} [g(\gamma(t_j, x_0)) - g(\gamma(t_{j-1}, x_0))] \leq -\frac{\varepsilon_1 \sigma}{2}.$$

By virtue of (3.28), (3.30)and (g) one finally has

$$g(\eta(x_0)) \leq -\frac{\varepsilon_1 \sigma}{2} + g(x_0) \leq c - \varepsilon\,,$$

which contradicts (3.29). The proof is thus complete. ■

In the rest of this Section we deal with a minimax principle in the limit case $c = a$ (see Remark 3.1 below) for the class of nonsmooth functionals $f : X \to]-\infty, +\infty]$, on a real reflexive Banach space $(X, \|\cdot\|)$, which satisfies the structure hypothesis

(H_f) $f = \Phi + \alpha$, *where* $\Phi : X \to \mathbb{R}$ *is locally Lipschitz and* $\alpha : X \to]-\infty, +\infty]$ *is convex, proper (i.e.,* $\not\equiv +\infty$*), lower semicontinuous,*

which has been introduced at the beginning of Section 2 in Chapter 2.

We recall from Definition 2.2 that a critical point of functional $f : X \to]-\infty, +\infty]$ satisfying assumption (H_f) is a point $u \in X$ such that

$$\Phi^0(u; x - u) + \alpha(x) - \alpha(u) \geq 0, \quad \forall x \in X.$$

If $\Phi \in C^1(X)$ one obtains the notion of critical point of Szulkin [20]. Given a real number c, we denote as usually

$$K_c(f) = \{u \in X : f(u) = c,\ u \text{ is a critical point of } f\}\,.$$

We also recall that the number $c \in \mathbb{R}$ is a critical value of the functional $f : X \to]-\infty, +\infty]$ satisfying (H_f) if $K_c(f) \neq \emptyset$.

We state now a suitable Palais-Smale condition for the function $f : X \to]-\infty, +\infty]$ in (H_f) around a set $S \subset X$ at the level $c \in \mathbb{R}$.

Definition 3.3 The function $f : X \to]-\infty, +\infty]$ satisfying (H_f) is said to verify the Palais-Smale condition around the set $S \subset X$ at the level $c \in \mathbb{R}$ if the following property holds

$(\mathrm{PS})_{f,S,c}$ *Each sequence* $\{x_n\} \subset X$ *such that* $d(x_n, S) \to 0$, $f(x_n) \to c$ *and*

$$\Phi^0(x_n; x - x_n) + \alpha(x) - \alpha(x_n) \geq -\varepsilon_n \|x_n - x\|, \quad \forall n \in \mathbb{N},\ x \in X\,,$$

where $\varepsilon_n \to 0^+$, *possesses a strongly convergent subsequence.*

Comparing with Definitions 2.4 and 3.3 we see that the Palais-Smale condition $(\mathrm{PS})_{f,c}$ means the Palais-Smale condition $(\mathrm{PS})_{f,S,c}$ with $S = X$.

On the basis of deformation result formulated in Theorem 3.1 we may state now a new minimax principle (see Theorem 3.2 below), which is different from the minimax principle formulated in Theorem 2.5. In this respect, the following remark is important.

Remark 3.1 By the definition of the minimax value in (2.19) and using relation (2.18) we deduce that $c \geq a$ in $\mathbb{R}$. Therefore two situations arise: $c > a$ and $c = a$. The case $c > a$ has been treated in Theorem 2.5. The case $c = a$ called the limit case is the object of Theorem 3.2 below.

In the following we deal with the case of equality $c = a$, for the notations in (2.21), which is not covered by Theorem 2.5. Weakening the strict inequality $c > a$ assumption to allow also equality in the setting of our hypothesis (H_f) can be considered as a continuation of the study initiated in Du [6] for the smooth case and then extended in Motreanu and Varga [15] to the locally Lipschitz setting. Classical results on the same subject in the framework of continuously differentiable functionals are those by Pucci and Serrin [17, Theorem 1], Rabinowitz [19, Theorem 2.13], Ghoussoub and Preiss [9, Theorem 1.bis]. The extension to the functionals of type (H_f) is given below.

Theorem 3.2 (Marano and Motreanu [12]) Suppose Q and S link in the sense of Definition 2.4 while the function $f : X \to]-\infty, +\infty]$ satisfies the following assumptions in addition to (H_f):

(f_1) $\sup_{x\in Q} f(x) < +\infty$.

(f_2) $\partial Q \subset f_a$ and $S \subset f^a$ for some $a \in \mathbb{R}$.

(f_3)

$$a = \inf_{\gamma\in\Gamma} \sup_{z\in\gamma(Q)} f(z),$$

for Γ in (2.20), and condition $(\mathrm{PS})_{f,S,a}$. Further, there exists $\varepsilon_0 > 0$ such that

($\mathrm{f}_{3.1}$) $N_{\varepsilon_0}(S) \subset D_{\partial\alpha}$,

($\mathrm{f}_{3.2}$) the set $N_\delta(S) \cap f^{a-\delta} \cap f_{a+\delta}$, $\delta \in]0, \varepsilon_0[$, is closed.

Then one has $K_a(f) \cap S \neq \emptyset$.

Proof. First we note that $a < +\infty$. This is true because the function $\gamma = id|_Q$ lies in Γ while (f_1) gives $\sup_{z\in\gamma(Q)} f(z) < +\infty$. In order to achieve the conclusion we argue by contradiction, supposing that $K_a(f) \cap S = \emptyset$. Define $A = \partial Q$, $B = S$, $g = -f$. We observe that the function g fulfils condition $(\mathrm{PS})_{g,B,-a}$ while

$$A \cap B = \emptyset, \quad A \subset g^{-a}, \quad B \subset g_{-a}, \quad K_{-a}(g) \cap B = \emptyset .$$

We note that for any $\delta > 0$ we have

$$N_\delta(B) \cap f^{a-\delta} \cap f_{a+\delta} = N_\delta(B) \cap g^{-a-\delta} \cap g_{-a+\delta}\,.$$

Consequently, by (f_3), the data g and $-a$ satisfy the hypotheses of Theorem 3.1. Thus, there exist $\varepsilon > 0$ and a homeomorphism $\eta : X \to X$ such that

$$\eta(x) = x \;\; \forall x \in \partial Q\,, \quad a + \varepsilon \le f(\eta(x)) \;\; \forall x \in S\,. \tag{3.31}$$

The formula of a in (f_3) produces, for some $\gamma_\varepsilon \in \Gamma$,

$$f(\gamma_\varepsilon(x)) < a + \varepsilon\,, \quad x \in Q\,. \tag{3.32}$$

Since Q links with S while $\eta^{-1} \circ \gamma_\varepsilon \in \Gamma$, we can find a point x_ε in Q fulfilling $\eta^{-1}(\gamma_\varepsilon(x_\varepsilon)) \in S$. So, due to (3.31) and (3.32), one obtains

$$a + \varepsilon \le f(\eta(\eta^{-1}(\gamma_\varepsilon(x_\varepsilon)))) = f(\gamma_\varepsilon(x_\varepsilon)) < a + \varepsilon\,.$$

This contradiction completes the proof. ■

Remark 3.2 When $\alpha \equiv 0$ the preceding result gives Theorem 2.1 by Motreanu and Varga [15], but with the notion of linking as introduced in Definition 2.4.

Important particular cases of the general minimax principles formulated in Theorems 2.4 and 3.3 are stated in the sequel. For the corresponding smooth versions we refer to Mountain Pass Theorem [2, Theorem 2.1], the Saddle Point Theorem [18, Theorem 4.6] and Generalized Mountain Pass Theorem [18, Theorem 5.3].

Corollary 3.1 (Mountain Pass Theorem) Let f be like in (H_f). Suppose that

(f_4) there exist $x_1 \in X$, $r > 0$ and $a \in \mathbb{R}$ satisfying $\|x_1\| > r$ in addition to

$$\max\{f(0), f(x_1)\} \le a \le f(x), \quad \forall x \in \partial B_r\,;$$

(f_5) setting $Q = [0, x_1]$ (the segment joining the points 0 and x_1), $S = \partial B_r$ and

$$c = \inf_{\gamma \in \Gamma} \sup_{z \in \gamma(Q)} f(z),$$

either $(PS)_{f,c}$ or $(PS)_{f,S,c}$ holds according to whether $c > a$ or $c = a$. Further, if $c = a$, there exists $\varepsilon_0 > 0$ such that ($f_{3.1}$), ($f_{3.2}$) are verified for S as above.

Then $c \ge a$ and $K_c(f) \setminus \{0, x_1\} \ne \emptyset$.

Proof. Corollary 2.7 and Theorem 3.2 can be applied for $Q = [0, x_1]$ and $S = \partial B_r$ observing that we are in a condition of linking as described in Definition 2.4. ■

Corollary 3.2 (Saddle Point Theorem) Let $X = V \oplus E$, where $V \neq \{0\}$ is finite dimensional. Assume that

(f_6) there are two real numbers $r > 0$ and a such that

$$\sup_{x \in V \cap \overline{B}_r} f(x) < +\infty, \quad \partial(V \cap \overline{B}_r) \subset f_a, \quad E \subset f^a,$$

(f_7) setting $Q = V \cap \overline{B}_r$, $S = E$ and

$$c = \inf_{\gamma \in \Gamma} \sup_{z \in \gamma(Q)} f(z),$$

either $(PS)_{f,c}$ or $(PS)_{f,S,c}$ holds according to whether $c > a$ or $c = a$. Further, if $c = a$, there exists $\varepsilon_0 > 0$ such that ($f_{3.1}$), ($f_{3.2}$) are verified for S as above.

Then $c \geq a$ and $K_c(f) \neq \emptyset$.

Proof. We apply Corollary 2.7 and Theorem 3.2 for $Q = V \cap \overline{B}_r$ and $S = E$ noticing that we have the linking property as described in Definition 2.4. ■

Corollary 3.3 (Generalized Mountain Pass Theorem) Let $X = V \oplus E$, where $V \neq \{0\}$ is finite dimensional. Let $Q = (V \cap \overline{B}_r) \oplus \{te : t \in [0, r]\}$ and $S = E \cap \partial B_\rho$, for some constants $\rho > 0$, $r > \rho$ and a point $e \in E$, $\|e\| = 1$. Assume that

(f_8) there exists a constant $a > 0$ such that

$$\sup_{x \in Q} f(x) < +\infty, \quad \partial Q \subset f_a, \quad S \subset f^a;$$

(f_9) setting

$$c = \inf_{\gamma \in \Gamma} \sup_{z \in \gamma(Q)} f(z),$$

either $(PS)_{f,c}$ or $(PS)_{f,S,c}$ holds according to whether $c > a$ or $c = a$. Further, if $c = a$, there exists $\varepsilon_0 > 0$ such that ($f_{3.1}$), ($f_{3.2}$) are verified for S as above.

Then $c \geq a$ and $K_c(f) \neq \emptyset$.

Proof. Corollary 2.7 and Theorem 3.2 can be applied for $Q = (V \cap \overline{B}_r) \oplus \{te : t \in [0, r]\}$ and $S = E \cap \partial B_\rho$. This is possible since Q and S satisfy the linking property in Definition 2.4. ■

2. Examples

This Section is devoted to the presentation of some concrete examples of boundary value problems which illustrate the application of the nonsmooth variational results given in Section 2. We follow the lines in [12], [13].

We start with an application to an elliptic variational-hemivariational inequality in the sense of Panagiotopoulos [16], [14], where the strict inequality $c > a$ (see Theorem 2.5 and Remark 3.1) occurs.

Example 3.1. Let Ω be a bounded domain of the real Euclidean N-space $(\mathbb{R}^N, |\cdot|)$, $N \geq 3$, having a smooth boundary $\partial\Omega$. Consider the Sobolev space $H_0^1(\Omega)$ which is the closure of $C_0^\infty(\Omega)$ with respect to the norm

$$\|u\| := \left(\int_\Omega |\nabla u(x)|^2 dx\right)^{1/2}.$$

As usually, denote by 2^* the critical exponent for the Sobolev embedding $H_0^1(\Omega) \subset L^p(\Omega)$, i.e.

$$2^* = \frac{2N}{N-2}.$$

If $p \in [1, 2^*]$ then there exists a constant $c_p > 0$ fulfilling

$$\|u\|_{L^p(\Omega)} \leq c_p \|u\|, \quad \forall u \in H_0^1(\Omega).$$

The embedding is compact if $p \in [1, 2^*[$. Let $\{\lambda_n\}$ be the sequence of eigenvalues of the operator $-\Delta$ on $H_0^1(\Omega)$, with

$$0 < \lambda_1 < \lambda_2 \leq \ldots \leq \lambda_n \leq \ldots$$

and let $\{\varphi_n\}$ be a corresponding sequence of eigenfunctions normalized as follows:

$$\|\varphi_n\|^2 = 1 = \lambda_n \|\varphi_n\|^2_{L^2(\Omega)}, \quad n \in \mathbb{N}; \tag{3.33}$$

$$\int_\Omega \nabla\varphi_m(x) \cdot \nabla\varphi_n(x)\, dx = \int_\Omega \varphi_m(x)\varphi_n(x)\, dx = 0 \quad \text{provided } m \neq n.$$

Consider a function $j : \Omega \times \mathbb{R} \to \mathbb{R}$ satisfying the conditions

(j_1) *j is measurable with respect to each variable separately,*

(j_2) *there exist $a_1 > 0$, $p \in]2, 2^*[$ such that*

$$|j(x,t)| \leq a_1 \left(1 + |t|^{p-1}\right), \quad \forall (x,t) \in \Omega \times \mathbb{R}.$$

The function $J : \Omega \times \mathbb{R} \to \mathbb{R}$ given by

$$J(x,\xi) = \int_0^\xi -j(x,t)\, dt\,, \quad (x,\xi) \in \Omega \times \mathbb{R},$$

is well defined, $J(\cdot,\xi)$ is measurable and $J(x,\cdot)$ is locally Lipschitz. So it makes sense to consider its generalized directional derivative with respect to the variable ξ that will be denoted by J^0.

Let us further assume that

(j_3) $\lambda_k < \lambda_{k+1}$ *for some* $k \in \mathbb{N}$, $\lambda \in [\lambda_k, \lambda_{k+1}[$, *besides*

$$J(x,\xi) \le \min\{0, a_2(1 - |\xi|^q)\} \quad \text{in } \Omega \times \mathbb{R}\,,$$

where $a_2 > 0$, $q \in]2,p]$;

(j_4) *there exists* $r > 0$ *fulfilling*

$$\inf_{\|u\|=r} \frac{1}{r^2} \int_\Omega J(x,u(x))\, dx \ge -\frac{1}{2}\left(1 - \frac{\lambda}{\lambda_{k+1}}\right).$$

Given $R \in]0,+\infty[$, consider the following elliptic variational-hemivariational inequality problem:

(P_R) *Find* $u \in \overline{B}_R \subset H_0^1(\Omega)$ *such that*

$$-\int_\Omega \nabla u(x)\cdot\nabla(v-u)(x)\, dx + \lambda\int_\Omega u(x)(v(x)-u(x))\, dx$$
$$\le \int_\Omega J^0(x,u(x); v(x)-u(x))\, dx$$

for all $v \in \overline{B}_R$.

We state the following result for solving problem (P_R).

Theorem 3.3 Suppose that conditions (j_1)-(j_4) hold. Then there exists $R_0 \in]0,+\infty[$ such that problem (P_R) possesses a solution for every real number $R > R_0$.

Proof. Set $X = H_0^1(\Omega)$, $V = \text{span}\{\varphi_1,\ldots,\varphi_k\}$ and $E = V^\perp$. One has $X = V \oplus E$ as well as $\dim(V) = k < +\infty$. Introduce

$$\Phi(u) = \frac{1}{2}\left(\|u\|^2 - \lambda\|u\|^2_{L^2(\Omega)}\right) + \int_\Omega J(x,u(x))\, dx\,, \quad u \in X\,.$$

By (j_2) the function Φ is locally Lipschitz. Set $W = \text{span}\{\varphi_1,\ldots,\varphi_{k+1}\}$. It results, for a suitable $\rho > r$,

$$\Phi(u) \le 0, \quad \forall u \in W \setminus B_\rho. \tag{3.34}$$

To see this we first note that W is a (finite dimensional) subspace of $L^q(\Omega)$ because $q \le p < 2^*$. Hence, there exists $a_3 > 0$ fulfilling

$$\|u\|_{L^q(\Omega)} \ge a_3 \|u\|\,, \quad u \in W\,.$$

Denote by $|\Omega|$ the Lebesgue measure of Ω. Since each $u \in W$ can be written as $u = \sum_{i=1}^{k+1} t_i \varphi_i$, where $t_1, \ldots, t_{k+1} \in I\!R$, through (3.33) and ($j_3$) we obtain

$$\begin{aligned}
\Phi(u) &= \frac{1}{2}\sum_{i=1}^{k+1} t_i^2 \left(1 - \frac{\lambda}{\lambda_i}\right) + \int_\Omega J(x, u(x))\, dx \\
&\le \frac{1}{2} t_{k+1}^2 \left(1 - \frac{\lambda}{\lambda_{k+1}}\right) + a_2 \left(|\Omega| - \|u\|^q_{L^q(\Omega)}\right) \\
&\le \frac{1}{2}\|u\|^2 \left(1 - \frac{\lambda}{\lambda_{k+1}}\right) + a_2\, |\Omega| - a_2 a_3^q \|u\|^q,
\end{aligned}$$

which, by assumption $q > 2$, leads to (3.34).

Now, choose

$$Q = \left(V \cap \overline{B}_\rho\right) \oplus [0, \rho\varphi_{k+1}]\,, \quad S = \partial B_r \cap E\,, \quad R_0 = 2\rho.$$

It is known that the compact topological manifold Q links with the closed set S (see Rabinowitz [18]). Further, one has

$$Q \subset B_R\,, \quad S \subset B_{R/2}\,, \quad \forall R > R_0\,. \tag{3.35}$$

Fix a real number $R > R_0$. If for every $u \in X$ we define

$$\alpha(u) = \begin{cases} 0 & \text{when } u \in \overline{B}_R \\ +\infty & \text{otherwise} \end{cases}$$

and

$$f(u) = \Phi(u) + \alpha(u)\,,$$

then hypothesis (H_f) in Section 2 is satisfied. The first inclusion in (3.35) yields $f|_Q = \Phi|_Q$, which implies the first condition in (f_8) of Corollary 3.3.

Let us next verify the second condition in (f_8) with $a = 0$. From (3.35) and (3.34) it follows $f(u) \le 0$, $u \in W \cap B_R \setminus B_\rho$. Since each $u \in V \cap B_R$ can be written as $u = \sum_{i=1}^{k} t_i \varphi_i$, where $t_1, \ldots, t_k \in I\!R$, exploiting ($j_3$) we get

$$f(u) = \Phi(u) \le \frac{1}{2}\sum_{i=1}^{k} t_i^2 \left(1 - \frac{\lambda}{\lambda_i}\right) \le 0\,.$$

This implies $\partial Q \subset f_0$. Taking now $u \in S$ it results $u = \sum_{i=k+1}^{+\infty} t_i \varphi_i$ for suitable $t_i \in I\!R$ in addition to $\|u\| = r$. Through (3.33) and (j_4) we thus achieve

$$
\begin{aligned}
f(u) = \Phi(u) &= \frac{1}{2} \sum_{i=k+1}^{+\infty} t_i^2 \left(1 - \frac{\lambda}{\lambda_i}\right) + \int_\Omega J(x, u(x))\, dx \\
&\geq \frac{1}{2}\left(1 - \frac{\lambda}{\lambda_{k+1}}\right) \|u\|^2 + \int_\Omega J(x, u(x))\, dx \\
&\geq r^2 \left[\frac{1}{2}\left(1 - \frac{\lambda}{\lambda_{k+1}}\right) + \inf_{\|u\|=r} \frac{1}{r^2} \int_\Omega J(x, u(x))\, dx\right] \geq 0.
\end{aligned}
$$

Consequently $S \subset f^0$, which completes the proof of condition (f_8) of Corollary 3.3.

Let us finally show (f_3). To this end, pick a sequence $\{u_n\} \subset X$ such that $f(u_n) \to c$ and

$$
\Phi^0(u_n; v - u_n) + \alpha(v) - \alpha(u_n) \geq -\varepsilon_n \|v - u_n\|, \quad \forall n \in I\!N,\ v \in X, \tag{3.36}
$$

where $\varepsilon_n \to 0^+$. One has

$$
\{u_n\} \subset D_\alpha = \overline{B}_R.
$$

So, passing to a subsequence if necessary, we may suppose $u_n \rightharpoonup u$ weakly in X as well as $u_n \to u$ strongly in $L^2(\Omega)$. Exploiting (3.36) with $v = u$ and taking account formula (2) at p. 77 in Clarke [5] yields

$$
\begin{aligned}
&\int_\Omega \nabla u_n(x) \cdot \nabla u(x)\, dx - \lambda \int_\Omega u_n(x)\,(u(x) - u_n(x))\, dx \\
&\quad + \int_\Omega J^0(x, u_n(x); u(x) - u_n(x))\, dx \\
&\quad + \varepsilon_n \|u_n - u\| \geq \int_\Omega |\nabla u_n(x)|^2\, dx, \quad n \in I\!N.
\end{aligned}
$$

By the upper semicontinuity of J^0 we then have

$$
\limsup_{n \to +\infty} \|u_n\|^2 \leq \limsup_{n \to +\infty} \int_\Omega J^0(x, u_n(x); u(x) - u_n(x))\, dx + \|u\|^2 = \|u\|^2,
$$

thus $u_n \to u$ in X. Hence, either $(PS)_{f,c}$ or $(PS)_{f,S,c}$ holds according to whether $c > 0$ or $c = 0$. Owing to the second inclusion in (3.35) there exists $\varepsilon_0 > 0$ fulfilling

$$
N_{\varepsilon_0}(S) \subset B_R = \operatorname{int}(D_{\partial\alpha}).
$$

Finally, since for every $\delta \in]0, \varepsilon_0[$ it results

$$N_\delta(S) \cap f^{c-\delta} \cap f_{c+\delta} = N_\delta(S) \cap \left\{x \in \overline{B}_R : c - \delta \leq \Phi(x) \leq c + \delta\right\},$$

we see that condition (f_9) in Corollary 3.3 is true too. Now, Corollary 3.3 can be applied. We get a point $u \in X$ such that

$$\Phi^0(u; v - u) + \alpha(v) - \alpha(u) \geq 0, \quad \forall v \in X.$$

The choice of α forces $u \in \overline{B}_R$ as well as $\Phi^0(u; v - u) \geq 0$ whenever $v \in \overline{B}_R$. Using formula (2) at p. 77 in [5], we thus have

$$\int_\Omega \nabla u(x) \cdot \nabla(v - u)(x)\, dx - \lambda \int_\Omega u(x)(v(x) - u(x))\, dx$$
$$+ \int_\Omega J^0(x, u(x); v(x) - u(v))\, dx \geq 0, \quad \forall v \in \overline{B}_R,$$

i.e. the function u is a solution to problem (P_R). ■

Remark 3.3 The proof of Theorem 3.3 yields the following value for the constant R_0 entering the statement of Theorem 3.3:

$$R_0 = 2 \inf\left\{\rho > 0 : \frac{1}{2}\left(1 - \frac{\lambda}{\lambda_{k+1}}\right)\rho^2 + a_4|\Omega| - a_4 a_5^q \rho^q \leq 0\right\}.$$

The next example treats an nonsmooth elliptic boundary value problem at resonance in the case $c = a$ of Theorem 3.2.

Example 3.2. We are concerned with the following resonant problem (at the k th eigenvalue λ_k of $-\Delta$ on $H_0^1(\Omega)$) stated in form of a variational-hemivariational inequality

(P_1) *Find $u \in D_\alpha \subset H_0^1(\Omega)$ such that*

$$-\int_\Omega \nabla u(x) \cdot \nabla(v - u)(x)\, dx + \lambda_k \int_\Omega u(x)(v(x) - u(x))\, dx$$
$$\leq \int_\Omega J^0(x, u(x); v(x) - u(x))\, dx + \alpha(v) - \alpha(u), \quad \forall v \in D_\alpha.$$

In the statement of problem (P_1) the meaning of the used data is the following. The set $\Omega \subset I\!R^N$ is as in Example 3.1. The positive integer k is fixed such that $\lambda_k < \lambda_{k+1}$ (see the sequence above (3.33)). The function $J : \Omega \times I\!R \to I\!R$ is measurable with respect to the first variable and locally Lipschitz with respect to the second variable whose generalized gradient $\partial J(x,t)$ (with respect to the second variable $t \in \mathbb{R}$) satisfies the growth condition

$$|z| \le c_1(1+|t|^{p-1}), \quad \forall z \in \partial J(x,t) \ \text{ a.e. } x \in \Omega, \ \forall t \in \mathbb{R}, \qquad (3.37)$$

for constants $c_1 \ge 0$ and $2 < p < 2^*$. The notation J^0 stands for the generalized directional derivative of J with respect to the second variable. The function $\alpha : H_0^1(\Omega) \to \mathbb{R} \cup \{+\infty\}$ entering (P_1) is convex, l.s.c., proper. Let us denote

$$V = \operatorname{span}\{\varphi_1, \dots, \varphi_k\}, \ V^\perp = \{w \in H_0^1(\Omega) : \ (w,v) = 0, \ \forall v \in V\}.$$

Suppose that

(k$_1$) *there exists* $\delta > 0$ *such that*

$$A_\delta := \{x_1 + x_2 \in H_0^1(\Omega) : \ x_1 \in V, \ x_2 \in V^\perp, \ \|x_1\| < \delta\} \subset D_{\partial\alpha},$$

(k$_2$) D_α *is closed,*

(k$_3$) *there exists* $0 < \rho \le \delta$, *for* $\delta > 0$ *given in* (k$_1$), *such that*

$$\int_\Omega J(x, v(x))\, dx + \alpha(v) \le 0, \quad \forall v \in V, \ \|v\| \le \rho,$$

(k$_4$) $\displaystyle \frac{1}{2}\left(1 - \frac{\lambda_k}{\lambda_{k+1}}\right)\|v\|^2 + \int_\Omega J(x, v(x))\, dx + \alpha(v) \ge 0, \quad \forall v \in V^\perp,$

(k$_5$)

$$\liminf_{\substack{\|v_2\| \to +\infty \\ v_2 \in V^\perp}} \frac{1}{\|v_2\|^2}\left[\int_\Omega J(x, v_1(x) + v_2(x))\, dx + \alpha(v_1 + v_2)\right]$$

$$> -\frac{1}{2}\left(1 - \frac{\lambda_k}{\lambda_{k+1}}\right)$$

uniformly with respect to $v_1 \in V$ *on bounded sets in* V.

Our result concerning problem (P_1) is the following.

Theorem 3.4 Assume that conditions (k$_1$)-(k$_5$) are fulfilled for J and α in problem (P_1). Then problem (P_1) has at least a solution $u \in H_0^1(\Omega)$ satisfying $u \in V^\perp$.

Proof. The method of proof is to apply Theorem 3.2 for a suitable functional $f : H_0^1(\Omega) \to \mathbb{R} \cup \{+\infty\}$ associated to problem (P_1). To this end we introduce the nonsmooth functional $\Phi : H_0^1(\Omega) \to \mathbb{R}$ by

$$\Phi(u) = \frac{1}{2}\left(\|u\|^2 - \lambda_k \|u\|^2_{L^2(\Omega)}\right) + \int_\Omega J(x, u(x))\, dx, \quad \forall u \in H_0^1(\Omega). \quad (3.38)$$

Due to the growth condition (3.37) for ∂J we have that Φ in (3.38) is locally Lipschitz. Then the functional $f = \Phi + \alpha : H_0^1(\Omega) \to I\!R \cup \{+\infty\}$ has the form required in (H_f).

In order to apply Theorem 3.2, we verify that the assumptions therein are satisfied.

Towards this, we define

$$Q = \overline{B}_\rho \cap V, \quad S = V^\perp,$$

with $\rho > 0$ in (k_3). Since V is finite dimensional, Q is a compact topological manifold which links with the closed set S (see Rabinowitz [18], p. 24).

Each $u \in Q$ can be expressed as $u = \sum_{i=1}^{k} t_i \varphi_i$, with $t_1, \ldots, t_k \in I\!R$. By (3.38) and ($\mathrm{k}_3$), we have

$$f(u) = \Phi(u) + \alpha(u) = \frac{1}{2}\sum_{i=1}^{k}\left(1 - \frac{\lambda_k}{\lambda_i}\right) t_i^2 + \int_\Omega J(x, u(x))\, dx + \alpha(u)$$

$$\leq \int_\Omega J(x, u(x))\, dx + \alpha(u) \leq 0, \quad \forall u \in Q.$$

Thus (f_1) in Theorem 3.2 holds true.

Next we check that assumption (f_2) is satisfied with $a = 0$. The previous inequality shows that $Q \subset f_0$, so $\partial Q \subset f_0$.

Every $u \in S$ can be written as $u = \sum_{i=k+1}^{+\infty} t_i \varphi_i$, with $t_i \in I\!R$, $\forall i \geq k+1$. Using (3.38) and (k_4), it results that

$$f(u) = \Phi(u) + \alpha(u) = \frac{1}{2}\sum_{i=k+1}^{+\infty}\left(1 - \frac{\lambda_k}{\lambda_i}\right) t_i^2 + \int_\Omega J(x, u(x))\, dx + \alpha(u)$$

$$\geq \frac{1}{2}\left(1 - \frac{\lambda_k}{\lambda_{k+1}}\right)\|u\|^2 + \int_\Omega J(x, u(x))\, dx + \alpha(u) \geq 0, \quad \forall u \in S.$$

It follows that $S \subset f^0$, hence (f_2) is verified.

Moreover, by virtue of linking property, it is seen that

$$0 \leq \inf_{\gamma \in \Gamma} \sup_{z \in \gamma(Q)} f(z) \leq \sup_{z \in Q} f(z) \leq 0,$$

consequently the first equality in (f_2) is valid with $a = 0$.

Let us now check condition $(\mathrm{PS})_{f,S,a}$ at level $a = 0$. Let $\{u_n\} \subset H_0^1(\Omega)$ be a sequence such that $d(u_n, S) \to 0$, $f(u_n) \to 0$ and

$$\Phi^0(u_n; v - u_n) + \alpha(v) - \alpha(u_n) \geq -\varepsilon_n \|v - u_n\|, \quad \forall n \geq 1,\ v \in D_\alpha, \tag{3.39}$$

where $\varepsilon_n \to 0^+$. Consider the decomposition $u_n = u_n^1 + u_n^2$ with $u_n^1 \in V$ and $u_n^2 \in V^\perp$. The property $d(u_n, S) \to 0$ implies that the sequence $\{u_n^1\}$ is bounded in $H_0^1(\Omega)$. Then, by (3.38) and the variational characterization of λ_{k+1}, we infer that

$$f(u_n) = \frac{1}{2}\left[(\|u_n^1\|^2 - \lambda_k\|u_n^1\|^2_{L^2(\Omega)}) + (\|u_n^2\|^2 - \lambda_k\|u_n^2\|^2_{L^2(\Omega)})\right]$$

$$+ \int_\Omega J(x, u_n(x))\, dx + \alpha(u_n) \geq -C + \frac{1}{2}\left(1 - \frac{\lambda_k}{\lambda_{k+1}}\right)\|u_n^2\|^2$$

$$+ \int_\Omega J(x, u_n(x))\, dx + \alpha(u_n), \quad \forall n \geq 1, \tag{3.40}$$

for some constant $C > 0$. Inequality (3.40) in conjunction with (k_5) implies the boundedness of $\{u_n^2\}$ in $H_0^1(\Omega)$. If not, there would exist a subsequence of $\{u_n^2\}$, denoted again $\{u_n^2\}$, such that $\|u_n^2\| \to +\infty$ as $n \to +\infty$. By (3.40) we deduce that

$$\frac{1}{\|u_n^2\|^2}(f(u_n) + C) - \frac{1}{2}\left(1 - \frac{\lambda_k}{\lambda_{k+1}}\right)$$

$$\geq \frac{1}{\|u_n^2\|^2}\left[\int_\Omega J(x, u_n(x))\, dx + \alpha(u_n)\right], \quad \forall n \geq 1.$$

Taking the limit inferior as $n \to +\infty$ in the inequality above and using the fact that $f(u_n) \to 0$ as $n \to +\infty$ we obtain that

$$\liminf_{n\to+\infty} \frac{1}{\|u_n^2\|^2}\left[\int_\Omega J(x, u_n(x))\, dx + \alpha(u_n)\right] \leq -\frac{1}{2}\left(1 - \frac{\lambda_k}{\lambda_{k+1}}\right),$$

which contradicts (k_5). This proves that $\{u_n^2\}$ is bounded in $H_0^1(\Omega)$, thus the sequence $\{u_n\}$ is bounded in $H_0^1(\Omega)$. Consequently, passing eventually to a subsequence of $\{u_n\}$, denoted again $\{u_n\}$, we may admit that $u_n \rightharpoonup u$ weakly in $H_0^1(\Omega)$, $u_n \to u$ strongly in $L^2(\Omega)$ and $u_n(x) \to u(x)$ a.e. $x \in \Omega$.

Since $f(u_n) \to 0$ as $n \to +\infty$ it follows that $u_n \in D_\alpha$. Since D_α is convex and closed (cf. (k_2)), D_α is weakly closed, so $u \in D_\alpha$. Setting $v = u$ in (3.39) and taking into account relation (2) in [5], p. 77, we deduce

$$\int_\Omega \nabla u_n(x) \cdot \nabla u(x)\, dx - \lambda_k \int_\Omega u_n(x)\,(u(x) - u_n(x))\, dx$$

$$+ \int_\Omega J^0(x, u_n(x); u(x) - u_n(x))\, dx + \alpha(u) - \alpha(u_n)$$

$$\geq -\varepsilon_n\|u_n - u\| + \int_\Omega |\nabla u_n(x)|^2\, dx, \quad \forall n \geq 1.$$

It turns out that

$$\|u\|^2 + \limsup_{n\to+\infty} \int_\Omega J^0(x, u_n(x); u(x) - u_n(x))\, dx$$

$$+\alpha(u) - \liminf_{n\to+\infty} \alpha(u_n) \geq \limsup_{n\to+\infty} \|u_n\|^2.$$

By the upper semicontinuity of J^0 and the lower semicontinuity of α we get $\limsup_{n\to+\infty} \|u_n\| \leq \|u\|$. This combined with $u_n \rightharpoonup u$ weakly in $H_0^1(\Omega)$ implies $u_n \to u$ strongly in $H_0^1(\Omega)$. Thus condition $(\mathrm{PS})_{f,S,a}$ at level $a = 0$ is verified. Thereby, (f_2) in Theorem 3.2 is valid.

We note that

$$S = V^\perp \subset A_{\frac{\delta}{2}} = \{x_1 + x_2 \in H_0^1(\Omega) : \ x_1 \in V, \ x_2 \in V^\perp, \ \|x_1\| < \frac{\delta}{2}\}.$$

Taking $0 < \varepsilon_0 < \frac{\delta}{2}$ one obtains from (k_1) that $N_{\varepsilon_0}(S) \subset A_\delta \subset D_{\partial\alpha}$, so $(\mathrm{f}_{3.1})$ is satisfied.

Since A_δ is open we find that $N_{\varepsilon_0}(S) \subset \mathrm{int}(D_{\partial\alpha})$.

Finally, for each $l \in]0, \varepsilon_0[$ we have

$$N_l(S) \cap f^{-l} \cap f_l = N_l(S) \cap \{x \in D_\alpha : -l \leq \Phi(x) + \alpha(x) \leq l\}.$$

Using the fact that $N_l(S) \subset N_{\varepsilon_0}(S) \subset \mathrm{int}(D_\alpha)$, since Φ is locally Lipschitz on $H_0^1(\Omega)$ and α is continuous on $\mathrm{int}(D_\alpha)$, it results that the set $N_l(S) \cap f^{-l} \cap f_l$ is closed. Condition $(\mathrm{f}_{3.2})$ is satisfied.

All the hypotheses of Theorem 3.2 are verified. Applying Theorem 3.2 we find a critical point u of f fulfilling $u \in K_a(f) \cap S$ at level $a = 0$. It suffices now to notice that any critical point of the functional $f = \Phi + \alpha$ solves problem (P_1). This assertion follows from a basic property of generalized gradient of integral functionals (see Clarke [5], p. 83-85). The proof is thus complete. ■

We provide a specific example where the conditions of Theorem 3.4 are satisfied.

Example 3.2$'$. Let a function $J : \Omega \times I\!R \to I\!R$ be measurable with respect to the first variable, locally Lipschitz with respect to the second variable, satisfies the growth condition (3.37) and

$$-d_1 t^2 \leq J(x, t) \leq 0 \ \text{ a.e. } x \in \Omega, \ \forall t \in I\!R,$$

for some constant $d_1 > 0$. Let $\alpha : H_0^1(\Omega) \to I\!R \cup \{+\infty\}$ be given by

$$\alpha(u) = \begin{cases} d_2\|u_2\|^2 & \text{if } u = u_1 + u_2 \text{ with } u_1 \in \overline{B}_\delta \cap V \text{ and } u_2 \in V^\perp \\ +\infty & \text{otherwise,} \end{cases}$$

with some $\delta > 0$ and for a constant $d_2 > 0$ satisfying

$$\frac{1}{2}\left(1 - \frac{\lambda_k}{\lambda_{k+1}}\right) + d_2 > \frac{d_1}{\lambda_1}.$$

It is clear that α is convex, l.s.c. and proper. We claim that the assumptions of Theorem 3.4 are verified. Indeed, since

$$(B_\delta \cap V) \oplus V^\perp \subset D_{\partial\alpha},$$

with $B_\delta = \{u \in H_0^1(\Omega) : \|u\| < \delta\}$, one sees that ($k_1$) is true. By the definition of α we have that (k_2) is verified, too. Furthermore, (k_3) holds with $\rho = \delta$ because α vanishes on V. The estimate

$$\begin{aligned}
&\frac{1}{2}\left(1 - \frac{\lambda_k}{\lambda_{k+1}}\right)\|v\|^2 + \int_\Omega J(x, v(x))\,dx + \alpha(v) \\
&\geq \frac{1}{2}\left(1 - \frac{\lambda_k}{\lambda_{k+1}}\right)\|v\|^2 - d_1\|v\|^2_{L^2(\Omega)} + d_2\|v\|^2 \\
&\geq \left[\frac{1}{2}\left(1 - \frac{\lambda_k}{\lambda_{k+1}}\right) + d_2 - \frac{d_1}{\lambda_1}\right]\|v\|^2 \geq 0, \quad \forall v \in V^\perp,
\end{aligned}$$

ensures that (k_4) is verified according to the choice of d_2. Moreover, we can write

$$\begin{aligned}
\int_\Omega J(x, v_1(x) + v_2(x))\,dx + \alpha(v_1 + v_2) &\geq -d_1\|v_1 + v_2\|^2_{L^2(\Omega)} + d_2\|v_2\|^2 \\
&\geq -\frac{d_1}{\lambda_1}\|v_1 + v_2\|^2 + d_2\|v_2\|^2 \\
&= -\frac{d_1}{\lambda_1}\|v_1\|^2 + \left(d_2 - \frac{d_1}{\lambda_1}\right)\|v_2\|^2, \quad \forall v_1 \in V,\ v_2 \in V^\perp.
\end{aligned}$$

Thus

$$\begin{aligned}
\liminf_{\substack{\|v_2\| \to +\infty \\ v_2 \in V^\perp}} \frac{1}{\|v_2\|^2}\left[\int_\Omega J(x, v_1(x) + v_2(x))\,dx + \alpha(v_1 + v_2)\right] &\geq d_2 - \frac{d_1}{\lambda_1} \\
&> -\frac{1}{2}\left(1 - \frac{\lambda_k}{\lambda_{k+1}}\right)
\end{aligned}$$

uniformly with respect to v_1 running in bounded subsets of V, which yields (k_5). All the assumptions of Theorem 3.4 are verified.

We deal now with the situation of nonsmooth elliptic boundary value problem with nonresonance and $c = a$ in the framework of Theorem 3.2.

Example 3.3. For a convex, l.s.c., proper functional $\alpha : H_0^1(\Omega) \to \mathbb{R} \cup \{+\infty\}$, a locally Lipschitz functional $g : H_0^1(\Omega) \to \mathbb{R}$ and any number $\lambda \in]\lambda_k, \lambda_{k+1}[$ (with the notation in the sequence above relation (3.33)), consider the following (nonresonant) variational-hemivariational inequality problem:

(P_2) *Find* $u \in D_\alpha \subset H_0^1(\Omega)$ *such that*

$$-\int_\Omega \nabla u(x) \cdot \nabla(v-u)(x)\,dx + \lambda \int_\Omega u(x)(v(x)-u(x))\,dx$$
$$\leq g^0(u; v-u) + \alpha(v) - \alpha(u), \quad \forall v \in D_\alpha.$$

We introduce the spaces (linear subspaces of $H_0^1(\Omega)$)

$$V = \text{span}\{\varphi_1, \ldots, \varphi_k\}, \; W = \text{span}\{\varphi_1, \ldots, \varphi_k, \varphi_{k+1}\},$$
$$V^\perp = \{w \in H_0^1(\Omega) : \; (w, v) = 0, \; \forall v \in V\}.$$

One has

$$H_0^1(\Omega) = V \oplus V^\perp, \; \dim V = k < +\infty, \; \dim W = k+1.$$

We impose for α and g the following conditions:

(h_1) *there exist* $r > 0$ *and* $0 < \varepsilon < r$ *such that*

$$\{u \in H_0^1(\Omega) : \; r - \varepsilon < \|u\| < r + \varepsilon\} \subset D_{\partial\alpha},$$

(h_2) $g(u) + \alpha(u) \geq -\frac{1}{2}\left(1 - \frac{\lambda}{\lambda_{k+1}}\right) r^2, \quad \forall u \in V^\perp, \; \|u\| = r,$

with $r > 0$ *prescribed in* (h_1),

(h_3) *there exists* $\rho > r$, *for* $r > 0$ *in* (h_1), *such that if* $u = u_1 + t\varphi_{k+1}$, $u_1 \in V$, $\|u_1\| \leq \rho$, $t \in [0, \rho]$ *one has*

$$g(u) + \alpha(u) \leq \frac{1}{2}\left(\frac{\lambda}{\lambda_k} - 1\right) \|u_1\|^2 - \frac{1}{2}\left(1 - \frac{\lambda}{\lambda_{k+1}}\right) t^2.$$

(h_4) $\limsup_{n\to\infty} g^0(u_n; u - u_n) \leq 0$ *whenever* $u_n \rightharpoonup u$ *weakly in* $H_0^1(\Omega)$.

Our result in the study of problem (P_2) is the following.

Theorem 3.5 Assume (h_1)-(h_4) together with (k_2) (see Example 3.2) for g and α in problem (P_2). Then problem (P_2) has at least a solution $u \in H_0^1(\Omega)$ satisfying $u \in V^\perp$ and $\|u\| = r$.

Proof. It is sufficient to show the existence of a critical point of the functional $f = \Phi + \alpha : H_0^1(\Omega) \to \mathbb{R} \cup \{+\infty\}$, with $\Phi : H_0^1(\Omega) \to \mathbb{R}$ given by

$$\Phi(u) = \frac{1}{2}\left(\|u\|^2 - \lambda\|u\|^2_{L^2(\Omega)}\right) + g(u), \quad \forall u \in H_0^1(\Omega), \tag{3.41}$$

and α in problem (P_2). It is clear that Φ is locally Lipschitz. Therefore the structure of $f = \Phi + \alpha$ complies with hypothesis (H_f) in this Chapter. In the following we check that the assumptions of Theorem 3.2 are satisfied.

With ρ and r fixed by hypotheses (h_1)-(h_3), we define

$$Q = \left(V \cap \overline{B}_\rho\right) \oplus [0, \rho\,\varphi_{k+1}] \quad \text{and} \quad S = \partial B_r \cap V^\perp. \tag{3.42}$$

Since $r < \rho$, the compact topological manifold Q links with the closed set S (see Ambrosetti [1, Lemma 4.1] or Rabinowitz [18, Proposition 5.9]).

We check the requirements of Theorem 3.2 for the function $f = \Phi + \alpha : H_0^1(\Omega) \to \mathbb{R} \cup \{+\infty\}$, where Φ is introduced in (3.41).

Every $u \in Q$ can be expressed as $u = u_1 + u_2$, with $u_1 = \sum_{i=1}^k t_i\varphi_i \in V$ and $u_2 = t\varphi_{k+1}$, where $t_1, \ldots, t_k \in \mathbb{R}$, $\|u_1\| \le \rho$, $t \in [0, \rho]$. Using (3.41) and (h_3) we have

$$f(u) = \Phi(u) + \alpha(u) = \frac{1}{2}\sum_{i=1}^{k}\left(1 - \frac{\lambda}{\lambda_i}\right)t_i^2 + \frac{1}{2}\left(1 - \frac{\lambda}{\lambda_{k+1}}\right)t^2 + g(u) + \alpha(u)$$

$$\le \frac{1}{2}\left(1 - \frac{\lambda}{\lambda_k}\right)\|u_1\|^2 + \frac{1}{2}\left(1 - \frac{\lambda}{\lambda_{k+1}}\right)t^2 + g(u) + \alpha(u) \le 0.$$

Thus it was shown that $Q \subset f_0$, hence $\partial Q \subset f_0$. This ensures that (f_1) and the first part in (f_2) are verified.

Taking into account (3.42), if $u \in S$ we have that $\|u\| = r$ and $u = \sum_{i=k+1}^{+\infty} t_i\varphi_i$, with $t_i \in \mathbb{R}$, for any $i \ge k+1$. Using (3.41) and (h_2), it results that

$$f(u) = \Phi(u) + \alpha(u) = \frac{1}{2}\sum_{i=k+1}^{+\infty}\left(1 - \frac{\lambda}{\lambda_i}\right)t_i^2 + g(u) + \alpha(u)$$

$$\ge \frac{1}{2}\left(1 - \frac{\lambda}{\lambda_{k+1}}\right)\|u\|^2 + g(u) + \alpha(u) = \frac{1}{2}\left(1 - \frac{\lambda}{\lambda_{k+1}}\right)r^2 + g(u) + \alpha(u) \ge 0.$$

We obtained that $S \subset f^0$, so (f_2) is fulfilled.

In view of the linking property (see Definition 2.4), we find that

$$0 \le \inf_{\gamma \in \Gamma}\ \sup_{z \in \gamma(Q)} f(z) \le \sup_{z \in Q} f(z) \le 0.$$

Consequently, we have that the equality in (f_3) of Theorem 3.2 is satisfied with $a = 0$.

We have to show condition $(\mathrm{PS})_{f,S,a}$ at level $a = 0$. Let a sequence $\{u_n\} \subset H_0^1(\Omega)$ satisfy $d(u_n, S) \to 0$, $f(u_n) \to 0$ and

$$\Phi^0(u_n; v - u_n) + \alpha(v) - \alpha(u_n) \geq -\varepsilon_n \|v - u_n\|, \quad \forall n \geq 1, \ v \in D_\alpha,$$

where $\varepsilon_n \to 0^+$. Since $d(u_n, S) \to 0$ and S is a bounded set, the sequence $\{u_n\}$ is bounded in $H_0^1(\Omega)$. Following the same reasoning as in the proof of Theorem 3.4, on the basis of (k_2) and (h_4), we establish that $\{u_n\}$ possesses a (strongly) convergent subsequence. Thus condition $(\mathrm{PS})_{f,S,a}$ at level $a = 0$ is verified.

Taking $0 < \varepsilon_0 < \varepsilon$ we obtain, from (h_1), that

$$N_{\varepsilon_0}(S) \subset \{u \in H_0^1(\Omega) : \ r - \varepsilon < \|u\| < r + \varepsilon\} \subset \operatorname{int}(D_{\partial\alpha}).$$

This ensures that ($f_{3.1}$) is satisfied.

The verification of assumption ($f_{3.2}$) can be carried out as in the proof of Theorem 3.4.

Summarizing, we have checked all the hypotheses of Theorem 3.2. We complete the proof by pointing out that every critical point of the functional $f = \Phi + \alpha$, with Φ given in (3.41), is a solution to problem (P_2). Such a solution satisfies the additional property $u \in S = \partial B_r \cap V^\perp$. ■

Remark 3.4 The above proof shows that if $\theta \in]0, r[$ then there exists a solution of (P_2) lying in $\partial B_\theta \cap V^\perp$. Therefore, this problem possesses infinitely many nontrivial solutions inside $B_r \cap V^\perp$.

We provide now an example of application of Theorem 3.5.

Example 3.3′. Let $J_1, J_3 : \Omega \times \mathbb{R} \to \mathbb{R}$ be measurable functions, locally Lipschitz with respect to the second variable satisfying $J_1(\cdot, 0) \in L^1(\Omega)$, $J_3(\cdot, 0) \in L^1(\Omega)$,

$$\int_\Omega J_1(x, 0)\, dx = -\int_\Omega J_3(x, 0)\, dx \geq 0, \tag{3.43}$$

$$|z| \leq C(1 + |t|^{p-1}), \quad \forall z \in \partial J_1(x, t) \cup \partial J_3(x, t) \ \text{ a.e. } x \in \Omega, \ \forall t \in \mathbb{R}, \tag{3.44}$$

for some constants $C \geq 0$ and $2 < p < 2^*$,

$$J_1(x, t) \leq \frac{1}{2}\left(\frac{\lambda}{\lambda_k} - 1\right) \lambda_1 t^2 \ \text{ a.e. } x \in \Omega, \ \forall t \in \mathbb{R}, \tag{3.45}$$

$$J_3(x, t) \geq -\frac{1}{2}\left(1 - \frac{\lambda}{\lambda_{k+1}}\right) \lambda_{k+2} t^2 \ \text{ a.e. } x \in \Omega, \ \forall t \in \mathbb{R}. \tag{3.46}$$

Consider the function $g : H_0^1(\Omega) \to I\!R$ defined by

$$g(u) = \int_\Omega J_1(x, u_1(x))\, dx - \frac{1}{2}\left(1 - \frac{\lambda}{\lambda_{k+1}}\right) \|u_2\|^2 + \int_\Omega J_3(x, u_3(x))\, dx,$$

for all $u = u_1 + u_2 + u_3 \in H_0^1(\Omega)$ with $u_1 \in V$, $u_2 \in I\!R\varphi_{k+1}$ and $u_3 \in W^\perp$. Taking into account (3.44), the function $g : H_0^1(\Omega) \to I\!R$ is locally Lipschitz.

Let K be a closed, convex subset of $H_0^1(\Omega)$ such that

$$W \oplus \{u \in W^\perp : \|u\| \le r_0\} \subset K,$$

for some $r_0 > 0$, and let $\alpha = I_K : H_0^1(\Omega) \to I\!R \cup \{+\infty\}$ denote the indicator function of K, i.e.

$$I_K(u) = \begin{cases} 0 & \text{if } u \in K \\ +\infty & \text{otherwise.} \end{cases}$$

We claim that conditions (h_1)-(h_4) and (k_2) in Theorem 3.5 are verified. Clearly, assumption (k_2) holds true.

Fix an arbitrary number $0 < r < r_0$, where $r_0 > 0$ enters the description of the set K. Then, for $0 < \varepsilon < \min\{r_0 - r, r\}$, condition ($h_1$) is satisfied since $\overline{B}_{r+\varepsilon} \subset B_{r_0} \subset K = D_{\partial\alpha}$.

By (3.43), (3.46) and using essentially the variational characterization of λ_{k+2}, it follows that

$$\begin{aligned} g(u) + \alpha(u) &= \int_\Omega J_1(x, 0)\, dx - \frac{1}{2}\left(1 - \frac{\lambda}{\lambda_{k+1}}\right) \|u_2\|^2 \\ &\quad + \int_\Omega J_3(x, u_3(x))\, dx + I_K(u) \\ &\ge -\frac{1}{2}\left(1 - \frac{\lambda}{\lambda_{k+1}}\right) \|u_2\|^2 - \frac{1}{2}\left(1 - \frac{\lambda}{\lambda_{k+1}}\right) \lambda_{k+2} \|u_3\|^2_{L^2(\Omega)} \\ &\ge -\frac{1}{2}\left(1 - \frac{\lambda}{\lambda_{k+1}}\right) \left(\|u_2\|^2 + \|u_3\|^2\right) = -\frac{1}{2}\left(1 - \frac{\lambda}{\lambda_{k+1}}\right) r^2, \end{aligned}$$

for every $u = u_2 + u_3 \in V^\perp$ with $u_2 \in I\!R\varphi_{k+1}$, $u_3 \in W^\perp$, and $\|u\| = r$. This shows that (h_2) is true.

Relations (3.43) and the variational characterization of the first eigenvalue λ_1 imply that for every $u = u_1 + u_2 \in W$ with $u_1 \in V$, $u_2 \in I\!R\varphi_{k+1}$, we have

$$g(u) + \alpha(u) = \int_\Omega J_1(x, u_1(x))\, dx - \frac{1}{2}\left(1 - \frac{\lambda}{\lambda_{k+1}}\right) \|u_2\|^2$$

$$+\int_\Omega J_3(x,0)\,dx + I_K(u)$$

$$\le \frac{1}{2}\left(\frac{\lambda}{\lambda_k}-1\right)\lambda_1 \|u_1\|^2_{L^2(\Omega)} - \frac{1}{2}\left(1-\frac{\lambda}{\lambda_{k+1}}\right)\|u_2\|^2$$

$$\le \frac{1}{2}\left(\frac{\lambda}{\lambda_k}-1\right)\|u_1\|^2 - \frac{1}{2}\left(1-\frac{\lambda}{\lambda_{k+1}}\right)\|u_2\|^2,$$

which shows that (h_3) is verified with an arbitrary $\rho > r$. Using the compactness of the embedding of $H_0^1(\Omega) \subset L^p(\Omega)$, with $p < 2^*$, we derive assumption (h_4). Thus all the assumptions of Theorem 3.5 are satisfied.

Remark 3.5 The nonlinear elliptic boundary value problems considered in this Section are formulated in terms of Dirichlet boundary condition. Boundary value problems of the same type but with Neumann boundary condition are studied in Marano and Motreanu [11]. Elliptic boundary value problems with discontinuous nonlinearities at resonance are treated by using other methods in Gasinski and Papageorgiou [8] and Goeleven, Motreanu and Panagiotopoulos [10].

References

[1] A. Ambrosetti, Critical points and nonlinear variational problems, *Mem. Soc. Math. France* (N.S.) **49**, 1992.

[2] A. Ambrosetti and P. H. Rabinowitz, Dual variational methods in critical point theory and applications, *J. Funct. Anal.* **14** (1973), 349-381.

[3] H. Brézis, *Analyse Fonctionnelle - Théorie et Applications*, Masson, Paris, 1983.

[4] K.-C. Chang, Variational methods for non-differentiable functionals and their applications to partial differential equations, *J. Math. Anal. Appl.* **80** (1981), 102-129.

[5] F. H. Clarke, *Optimization and Nonsmooth Analysis,* Classics Appl. Math. **5**, SIAM, Philadelphia, 1990.

[6] Y. Du, A deformation lemma and some critical point theorems, *Bull. Austral. Math. Soc.* **43** (1991), 161-168.

[7] J. Dugundji, *Topology,* Allyn and Bacon, Boston, 1966.

[8] L. Gasiński and N. S. Papageorgiou, Solutions and multiple solutions for quasilinear hemivariational inequalities at resonance, *Proc. Royal Soc. Edinburgh (Math)* 131A (2001), 1091-1111.

[9] N. Ghoussoub and D. Preiss, A general mountain pass principle for locating and classifying critical points, *Ann. Inst. Henri Poincaré Anal. Non Linéaire* **6** (1989), 321-330.

[10] D. Goeleven, D. Motreanu and P. D. Panagiotopoulos, Eigenvalue problems for variational-hemivariational inequalities at resonance, *Nonlinear Anal.* **33** (1998), 161-180.

[11] S. Marano and D. Motreanu, Infinitely many critical points of non-differentiable functions and applications to a Neumann type problem involving the p-Laplacian, *J. Differ. Equations* 182 (2002), 108-120.

[12] S. Marano and D. Motreanu, A deformation theorem and some critical point results for non-differentiable functions, submitted.

[13] D. Motreanu and V. V. Motreanu, Duality in nonsmooth critical point theory, limit case and applications, submitted.

[14] D. Motreanu and P. D. Panagiotopoulos, *Minimax Theorems and Qualitative Properties of the Solutions of Hemivariational Inequalities*, Nonconvex Optimization and its Applications **29**, Kluwer Academic Publishers, Dordrecht/Boston/London, 1998.

[15] D. Motreanu and C. Varga, Some critical point results for locally Lipschitz functionals, *Comm. Appl. Nonlinear Anal.* **4** (1997), 17-33.

[16] P. D. Panagiotopoulos, *Hemivariational Inequalities. Applications in Mechanics and Engineering*, Springer-Verlag, Berlin, 1993.

[17] P. Pucci and J. Serrin, A mountain pass theorem, *J. Differ. Equations* **60** (1985), 142-149.

[18] P. H. Rabinowitz, *Minimax methods in critical point theory with applications to differential equations*, CBMS Reg. Conf. Ser. in Math. **65**, Amer. Math. Soc., Providence, 1986.

[19] P. H. Rabinowitz, Some aspects of critical point theory, in: Proceedings of the 1982 Changchun Symposium on Differential Geometry and Differential Equations (S.S. Chern, R. Wang and M. Chi (eds)), Science Press, Bijing, 1986, pp. 185-232.

[20] A. Szulkin, Minimax principles for lower semicontinuous functions and applications to nonlinear boundary value problems, *Ann. Inst. Henri Poincaré. Anal. Non Linéaire* **3** (1986), 77-109.

Chapter 4

MULTIVALUED ELLIPTIC PROBLEMS IN VARIATIONAL FORM

In Partial Differential Equations, two important tools for proving existence of solutions are the Mountain Pass Theorem of Ambrosetti and Rabinowitz [1] (and its various generalizations) and the Ljusternik-Schnirelmann Theorem [16]. These results apply to the case when the solutions of the given problem are critical points of an appropriate functional of energy f, which is supposed to be real and C^1, or only differentiable, on a real Banach space X. One may ask what happens if f, which often is associated to the original equation in a canonical way, fails to be differentiable. In this case the gradient of f must be replaced by a generalized one, which is often that introduced by Clarke in the framework of locally Lipschitz functionals. In this setting, Chang [4] was the first who proved a version of the Mountain Pass Theorem, in the case when X is reflexive. For this aim, he used a "Lipschitz version" of the Deformation Lemma. The same result was used for the proof of the Ljusternik-Schnirelmann Theorem in the locally Lipschitz case. As observed by Brézis, the reflexivity assumption on X is not necessary.

1. Multiplicity for Locally Lipschitz Periodic Functionals

The main result in this Section is a multiplicity theorem of the Ljusternik-Schnirelmann type for locally Lipschitz periodic functionals, their set of periods being a discrete subgroup of the space where they are defined. The key arguments in our proofs are Ekeland's Variational Principle and a nonsmooth Pseudo-Gradient Lemma.

Following Chang [4], authors usually impose measurability conditions to some *a priori* unknown functions in order to be able to find ∂f. We first show that these conditions are automatically fulfilled and we then prove the existence of critical points, which are shown to be solutions of a multivalued PDE.

Throughout this chapter, X will be a real Banach space. Let X^* be its topological dual and $\langle x^*, x\rangle$, for $x \in X$, $x \in X^*$, denote the duality pairing between X^* and X. If : $X \to I\!R$ is a locally Lipschitz functional, denote

$$\lambda(x) = \min_{x^* \in \partial f(x)} \|x^*\|_* .$$

We recall (see Chapter 1) that λ is well defined and is lower semicontinuous. We recall that if c is a real number then f is said to satisfy the Palais-Smale condition at the level c (in short $(PS)_c$) if any sequence $\{x_n\}$ in X with the properties $\lim_{n\to\infty} f(x_n) = c$ and $\lim_{n\to\infty} \lambda(x_n) = 0$ has a convergent subsequence. The function f satisfies the Palais-Smale condition (in short (PS)) if each sequence $\{x_n\}$ in X such that $\{f(x_n)\}$ is bounded and $\lim_{n\to\infty} \lambda(x_n) = 0$ has a convergent subsequence (see Definition 1.5).

Let Z be a discrete subgroup of X, that is

$$\inf_{z\in Z\setminus\{0\}} \|z\| > 0 .$$

A function $f : X \to I\!R$ is said to be *Z-periodic* if $f(x+z) = f(x)$, for every $x \in X$ and $z \in Z$.

If f is Z-periodic, then $x \mapsto f^0(x; v)$ is Z-periodic, for all $v \in X$ and ∂f is Z-invariant, that is $\partial f(x+z) = \partial f(x)$, for every $x \in X$ and $z \in Z$. These imply that λ inherits the Z-periodicity property.

If $\pi : X \to X/Z$ is the canonical projection and x is a critical point of f, then $\pi^{-1}(\pi(x))$ contains only critical points. Such a set is called a *critical orbit* of f. We observe that X/Z is a complete metric space endowed with the metric

$$d(\pi(x), \pi(y)) = \inf_{z\in Z} \|x - y - z\| .$$

Definition 4.1 A locally Lipschitz Z-periodic function $f : X \to I\!R$ is said to satisfy the $(PS)_Z$-condition provided that, for each sequence $\{x_n\}$ in X such that $\{f(x_n)\}$ is bounded and $\lambda(x_n) \to 0$, then $\{\pi(x_n)\}$ is relatively compact in X/Z. If c is a real number, then f is said to satisfy the $(PS)_{Z,c}$-condition if, for any sequence $\{x_n\}$ in X such that $f(x_n) \to c$ and $\lambda(x_n) \to 0$, there is a convergent subsequence of $\{\pi(x_n)\}$.

Denote by $K_c(f)$ the set of critical points of the locally Lipschitz function $f : X \to I\!R$ at the level $c \in I\!R$, that is

$$K_c(f) = \{x \in X \;:\; f(x) = c \;\text{ and }\; \lambda(x) = 0\} .$$

We prove the following abstract result.

Theorem 4.1 (Mironescu and Rădulescu [15], [21]) Let $f : X \to \mathbb{R}$ be a bounded below locally Lipschitz Z-periodic function with the $(PS)_Z$-property. Then f has at least $n+1$ distinct critical orbits, where n is the dimension of the vector space generated by the discrete subgroup Z.

We start by recalling the notion of *category* and some of its main properties.

A topological space X is said to be *contractible* if the identity of X is homotopical to a constant map, that is there exist $u_0 \in X$ and a continuous map $F : [0,1] \times X \to X$ such that

$$F(0,\cdot) = id_X \quad \text{and} \quad F(1,\cdot) = u_0 .$$

A subset M of X is said to be *contractible in* X if there exist $u_0 \in X$ and a continuous map $F : [0,1] \times M \to X$ such that

$$F(0,\cdot) = id_M \quad \text{and} \quad F(1,\cdot) = u_0 .$$

If A is a subset of X, we define the category of A in X as follows:

$\text{Cat}_X(A) = 0$, if $A = \emptyset$.

$\text{Cat}_X(A) = n$, if n is the smallest positive integer such that A can be covered by n closed sets which are contractible in X.

$\text{Cat}_X(A) = \infty$, otherwise.

The main properties of this notion are summarized in

Lemma 4.1 Let A and B be subsets of X. Then the following hold:

i) If $A \subset B$, then $\text{Cat}_X(A) \le \text{Cat}_X(B)$.

ii) $\text{Cat}_X(A \cup B) \le \text{Cat}_X(A) + \text{Cat}_X(B)$

iii) Let $h : [0,1] \times A \to X$ be a continuous mapping such that $h(0,x) = x$ for every $x \in A$. If A is closed and $B = h(1,A)$, then $\text{Cat}_X\,(A) \le \text{Cat}_X\,(B)$

iv) If n is the dimension of the vector space generated by the discrete group Z, then, for each $1 \le i \le n+1$, the set

$$\mathcal{A}_i = \{A \subset X \;:\; A \;\text{ is compact and }\; \text{Cat}_{\pi(X)}\pi(A) \ge i\}$$

is nonempty. Obviously, $\mathcal{A}_1 \supset \mathcal{A}_2 \supset \ldots \supset \mathcal{A}_{n+1}$.

Proof. The only nontrivial part is *iv)* , which can be found in [13].

The following two lemmas are proved in [25].

Lemma 4.2 For each $1 \le j \le n+1$, the space $\mathcal{A}_i$ endowed with the Hausdorff metric

$$\rho(A,B) = \max\{\sup_{a\in A}\ \text{dist}(a,B)\ ,\ \sup_{b\in B}\ \text{dist}(b,A)\}$$

is a complete metric space.

Lemma 4.3 If $1 \leq i \leq n+1$ and $f \in C(X; \mathbb{R})$, then the function $\eta : \mathcal{A}_i \to \mathbb{R}$ defined by

$$\eta(A) = \max_{x \in A} f(x)$$

is lower semicontinuous.

If n is the dimension of the vector space generated by the discrete group Z, one sets for each $1 \leq i \leq n+1$

$$c_i = \inf_{A \in \mathcal{A}_i} \eta(A) .$$

For each $c \in \mathbb{R}$ we denote $[f \leq c] = \{x \in X : f(x) \leq c\}$.

Proof of Theorem 4.1. It follows from Lemma 4.1 iv) and the lower boundedness of f that

$$-\infty < c_1 \leq c_2 \leq ... \leq c_{n+1} < +\infty .$$

It is sufficient to show that, if $1 \leq i \leq j \leq n+1$ and $c_i = c_j = c$, then the set $K_c(f)$ contains at least $j-i+1$ distinct critical orbits. We argue by contradiction and suppose that, for some $i \leq j$, $K_c(f)$ has $k \leq j-i$ distinct critical orbits, generated by $x_1, ..., x_k \in X$. We construct first an open neighbourhood of $K_c(f)$ of the form

$$V_r = \bigcup_{l=1}^{k} \bigcup_{z \in Z} B(x_l + z, r) .$$

Moreover, we may suppose that $r > 0$ is chosen such that π is one-to-one on $\overline{B}(x_l, 2r)$. This condition ensures that $\mathrm{Cat}_{\pi(X)}(\pi(\overline{B}(x_l, 2r))) = 1$, for each $l = 1, ..., k$. Here $V_r = \emptyset$ if $k = 0$.

Step 1. We prove that there exists $0 < \varepsilon < \min\{1/4, r\}$ such that, for any $x \in [c - \varepsilon \leq f \leq c + \varepsilon] \setminus V_r$,

$$\lambda(x) > \sqrt{\varepsilon} . \tag{4.1}$$

Indeed, if not, there is a sequence $\{x_m\}$ in $X \setminus V_r$ such that, for each $m \geq 1$,

$$c - \frac{1}{m} \leq f(x_m) \leq c + \frac{1}{m} \quad \text{and} \quad \lambda(x_m) \leq \frac{1}{\sqrt{m}} .$$

Since f satisfies $(PS)_Z$, it follows that, up to a subsequence, $\pi(x_m) \to \pi(x)$ as $m \to \infty$, for some $x \in X \setminus V_r$. By the Z-periodicity of f and λ, we can assume that $x_m \to x$ as $m \to \infty$. The continuity of f and the lower semicontinuity of λ imply $f(x) = c$ and $\lambda(x) = 0$, which is a contradiction, since $x \in X \setminus V_r$.

Step 2. For ε found above and according to the definition of c_j, there exists $A \in \mathcal{A}_j$ such that

$$\max_{x \in A} f(x) < c + \varepsilon^2 .$$

Setting $B = A \setminus V_{2r}$, Lemma 4.1 yields

$$\begin{aligned} j &\le \mathrm{Cat}_{\pi(X)}(\pi(A)) \le \mathrm{Cat}_{\pi(X)}(\pi(B) \cup \pi(\overline{V}_{2r})) \\ &\le \mathrm{Cat}_{\pi(X)}(\pi(B)) + \mathrm{Cat}_{\pi(X)}(\pi(\overline{V}_{2r})) \\ &\le \mathrm{Cat}_{\pi(X)}(\pi(B)) + k \le \mathrm{Cat}_{\pi(X)}(\pi(B)) + j - i . \end{aligned}$$

Hence, $\mathrm{Cat}_{\pi(X)}(\pi(B)) \ge i$, that is $B \in \mathcal{A}_i$.

Step 3. For ε and B as above we apply Ekeland's variational principle (see Theorem 1.5) to the functional η defined in Lemma 4.3. It follows that there exists $C \in \mathcal{A}_i$ such that, for each $D \in \mathcal{A}_i$, $D \ne C$,

$$\begin{array}{ll} \eta(C) & \le \eta(B) \le \eta(A) \le c + \varepsilon^2 , \\ \rho(B,C) & \le \varepsilon , \\ \eta(D) & > \eta(C) - \varepsilon\rho(C,D) . \end{array} \tag{4.2}$$

Since $B \cap V_{2r} = \emptyset$ and $\rho(B,C) \le \varepsilon < r$, it follows that $C \cap V_r = \emptyset$. In particular, the set $F = [c - \varepsilon \le f] \cap C$ is contained in $[c - \varepsilon \le f \le c + \varepsilon]$ and $F \cap V_r = \emptyset$.

Lemma 4.4 Let M be a compact metric space and let $\varphi : M \to 2^{X^*}$ be a set-valued mapping which is upper semicontinuous (in the weak $*$ sense) and with weak $*$ relatively compact convex values. For $t \in M$ denote

$$\gamma(t) = \inf\{\|x^*\| \,:\, x^* \in \varphi(t)\}$$

and

$$\gamma = \inf_{t \in M} \gamma(t) .$$

Then, given $\varepsilon > 0$, there exists a continuous function $v : M \to X$ such that for all $t \in M$ and $x^* \in \varphi(t)$,

$$\|v(t)\| \le 1 \text{ and } \langle x^*, v(t) \rangle \ge \gamma - \varepsilon .$$

Proof. We may suppose $\gamma > 0$ and $0 < \varepsilon < \gamma$. If B_r denotes the open ball in X^* centered at 0 with radius r, then, for each $t \in M$,

$$B_{\gamma - \frac{\varepsilon}{2}} \cap \varphi(t) = \emptyset .$$

Since $\varphi(t)$ and $B_{\gamma-\frac{\varepsilon}{2}}$ are convex, weak $*$ relatively compact and disjoint, we can apply Theorem 3.4 in [24] to the space $(X^*,\ \sigma(X^*,X))$ and we use the fact that the dual space of the above one is X. So, for every $t \in M$, there exists $v_t \in X$, $\|v_t\| = 1$ such that

$$\langle \xi, v_t \rangle \leq \langle x^*, v_t \rangle ,$$

for each $\xi \in B_{\gamma-\frac{\varepsilon}{2}}$ and $x^* \in \varphi(t)$. Therefore, for each $x^* \in \varphi(t)$,

$$\langle x^*, v_t \rangle \geq \sup_{\xi \in B_{\gamma-\frac{\varepsilon}{2}}} \langle \xi, v_t \rangle = \gamma - \frac{\varepsilon}{2} .$$

By the upper semicontinuity of φ, there exists an open neighbourhood $V(t)$ of t such that, for each $t' \in V(t)$ and each $x^* \in \varphi(t')$,

$$\langle x^*, v_t \rangle > \gamma - \varepsilon .$$

Since M is compact and $M = \bigcup_{t \in M} V(t)$, we can find a finite subcovering $\{V_1, ..., V_n\}$ of M. Let $v_1, ..., v_n$ be on the unit sphere of X such that $\langle x^*, v_i \rangle > \gamma - \varepsilon$, for all $1 \leq i \leq n$, $t \in V_i$ and $x^* \in \varphi(t)$.

If $\rho_i(t) = \mathrm{dist}(t, \partial V_i)$, define

$$\zeta_i(t) = \frac{\rho_i(t)}{\sum_{j=1}^n \rho_j(t)} \quad \text{and} \quad v(t) = \sum_{i=1}^n \zeta_i(t) v_i .$$

It follows that the function v is the desired mapping. ■

Proof of Theorem 4.1 continued. Applying Lemma 4.4 to $\varphi = \partial f$ on F, we find a continuous map $v : F \to X$ such that, for all $x \in F$ and $x^* \in \partial f(x)$,

$$\|v(x)\| \leq 1$$

and, by (4.1),

$$\langle x^*, v(x) \rangle \geq \inf_{x \in F} \lambda(x) - \varepsilon \geq \inf_{x \in C} \lambda(x) - \varepsilon \geq \sqrt{\varepsilon} - \varepsilon .$$

It follows that, for each $x \in F$ and $x^* \in \partial f(x)$,

$$f^0(x; -v(x)) = \max_{x^* \in \partial f(x)} \langle x^*, -v(x) \rangle$$

$$= - \min_{x^* \in \partial f(x)} \langle x^*, v(x) \rangle \leq \varepsilon - \sqrt{\varepsilon} < -\varepsilon,$$

from our choice of ε.

By the upper semicontinuity of f^0 and the compactness of F, there exists $\delta > 0$ such that if $x \in F$, $y \in X$, $\|y - x\| \leq \delta$, then

$$f^0(y; -v(x)) < -\varepsilon \,. \tag{4.3}$$

Since $C \cap K_c(f) = \emptyset$ and C is compact, while $K_c(f)$ is closed, there exists a continuous extension $w : X \to X$ of v such that $w|_{K_c(f)} = 0$ and $\|w(x)\| \leq 1$, for all $x \in X$.

Let $\alpha : X \to [0,1]$ be a continuous Z-periodic function such that $\alpha = 1$ on $[f \geq c]$ and $\alpha = 0$ on $[f \leq c - \varepsilon]$. Let $h : [0,1] \times X \to X$ be the continuous mapping defined by

$$h(t,x) = x - t\delta\alpha(x)w(x)\,.$$

If $D = h(1, C)$, it follows from Lemma 4.1 that

$$\mathrm{Cat}_{\pi(X)}(\pi(D)) \geq \mathrm{Cat}_{\pi(X)}(\pi(C)) \geq i$$

which shows that $D \in \mathcal{A}_i$, since D is compact.

Step 4. By Lebourg's mean value theorem we get that, for each $x \in X$, there exists $\theta \in]0,1[$ such that

$$f(h(1,x)) - f(h(0,x)) \in \langle \partial f(h(\theta, x)), -\delta\alpha(x)w(x)\rangle\,.$$

Hence, there is some $x^* \in \partial f(h(\theta, x))$ such that

$$f(h(1,x)) - f(h(0,x)) = \alpha(x)\langle x^*, -\delta w(x)\rangle\,.$$

It follows by (4.3) that, if $x \in F$, then

$$\begin{aligned} f(h(1,x)) - f(h(0,x)) &= \delta\alpha(x)\langle x^*, -w(x)\rangle \\ &\leq \delta\alpha(x) f^0(x - \theta\delta\alpha(x)w(x); -v(x)\rangle \leq -\varepsilon\delta\alpha(x)\,. \end{aligned} \tag{4.4}$$

It follows that, for each $x \in C$,

$$f(h(1,x)) \leq f(x)\,.$$

Let $x_0 \in C$ be such that $f(h(1,x_0)) = \eta(D)$. Therefore

$$c \leq f(h(1,x_0)) \leq f(x_0)\,.$$

By the definition of α and F, it follows that $\alpha(x_0) = 1$ and $x_0 \in F$. So, by (4.4),

$$f(h(1,x_0)) - f(x_0) \leq -\varepsilon\delta\,.$$

Thus

$$\eta(D) + \varepsilon\delta \leq f(x_0) \leq \eta(C)\,. \tag{4.5}$$

Taking into account the definition of D, it follows that

$$\rho(C, D) \leq \delta .$$

Therefore

$$\eta(D) + \varepsilon\rho(C, D) \leq \eta(C) ,$$

so that (4.2) implies $C = D$, which contradicts (4.5). ∎

2. The Multivalued Forced-pendulum Problem

As an application of the abstract results in the preceding Section we solve in what follows the set-valued version of the forced-pendulum problem. More precisely, we consider the periodic problem

$$\begin{cases} x''(t) + f(t) \in [\underline{g}(x(t)), \overline{g}(x(t))] \text{ a.e. } t \in]0,1[\\ x(0) = x(1) , \end{cases} \tag{4.6}$$

where:

$$f \in L^p(0,1) \quad \text{for some } \ p > 1 , \tag{4.7}$$

$$g \in L^\infty(I\!R),\ g(u+T) = g(u) \quad \text{for some } \ T > 0, \text{ a.e. } \ u \in I\!R , \tag{4.8}$$

$$\underline{g}(u) = \lim_{\varepsilon \searrow 0} \operatorname{essinf}\{g(u) \ : \ |u - v| < \varepsilon\}$$

$$\overline{g}(u) = \lim_{\varepsilon \searrow 0} \operatorname{esssup}\{g(u) \ : \ |u - v| < \varepsilon\} ,$$

$$\int_0^T g(u)du = \int_0^1 f(t)dt = 0. \tag{4.9}$$

The following result is a generalization of a theorem due to Mawhin and Willem in the single-valued case (see [12]).

Theorem 4.2 (Mironescu and Rădulescu [15]) Assume that f and g satisfy assumptions (4.7)-(4.9). Then Problem (4.6) has at least two solutions in

$$X := H_p^1(0,1) = \{x \in H^1(0,1) \ : \ \ x(0) = x(1)\} ,$$

which are distinct in the sense that their difference is not an integer multiple of T.

Define the functional ψ in $L^\infty(0,1)$ by

$$\psi(x) = \int_0^1 \left(\int_0^{x(s)} g(u)du \right) ds .$$

It is obvious that ψ is a Lipschitz map on $L^\infty(0,1)$.

Set $G(u) = \int_0^u g(v)dv$, $u \in I\!R$.

The following results show that the description of $\partial\psi$ given in [4] holds without further assumptions on g.

Lemma 4.5 Let g be a locally bounded measurable function defined on $I\!R$ and $\underline{g}$, $\overline{g}$ as above. Then the Clarke subdifferential of G is given by

$$\partial G(u) = [\underline{g}(u), \overline{g}(u)] \qquad u \in I\!R\,.$$

Proof. The required equality is equivalent to $G^0(u;1) = \overline{g}(u)$ and $G^0(u;-1) = \underline{g}(u)$.

As a matter of facts, examining the definitions of G^0, $\overline{g}$ and $\underline{g}$, it follows that $\underline{g}(u) = -(\overline{-g})(u)$ and $G^0(u;-1) = -(-G)^0(u;1)$, so that the second required equality is equivalent to the first one.

Now the inequality $G^0(u;1) \leq \overline{g}(u)$ can be found in [4], so it remains to prove that $G^0(u;1) \geq \overline{g}(u)$. By contradiction, suppose that $G^0(u;1) = \overline{g}(u) - \varepsilon$ for some $\varepsilon > 0$. Let $\delta > 0$ be such that

$$\frac{G(\tau+\lambda) - G(\tau)}{\lambda} < \overline{g}(u) - \frac{\varepsilon}{2},$$

if $|\tau - u| < \delta$ and $0 < \lambda < \delta$. Then

$$\frac{1}{\lambda}\int_\tau^{\tau+\lambda} g(s)ds < \overline{g}(u) - \frac{\varepsilon}{2} \qquad \text{if } |\tau - u| < \delta,\ \lambda > 0 \tag{4.10}$$

We claim that there exist $\lambda_n \searrow 0$ such that

$$\frac{1}{\lambda_n}\int_\tau^{\tau+\lambda_n} g(s)ds \to g(\tau) \qquad a.e.\ \tau \in (u-\delta, u+\delta)\,. \tag{4.11}$$

Suppose for the moment that (4.11) has been proved. Then (4.10) and (4.11) show that

$$g(\tau) \leq \overline{g}(u) - \frac{\varepsilon}{2} \qquad \text{if } \tau \in (u-\delta, u+\delta)\,,$$

so we obtain the contradictory inequalities

$$\overline{g}(u) \leq \operatorname{esssup}\{g(s) \,:\, s \in [u-\delta, u+\delta]\} \leq \overline{g}(u) - \frac{\varepsilon}{2}\,.$$

All it remains to be proved is (4.11). Note that we may cut g in order to suppose that $g \in L^\infty \cap L^1$. Then (4.11) is nothing that the classical fact that for each $\varphi \in L^1(I\!R)$,

$$T_\lambda(\varphi) \to \varphi \quad \text{as } \lambda \searrow 0\,, \tag{4.12}$$

where

$$T_\lambda \varphi(u) = \frac{1}{\lambda}\int_u^{u+\lambda} \varphi(s)ds \quad \text{for } \lambda > 0,\ u \in I\!R,\ \varphi \in L^1(I\!R).$$

Indeed, it can be easily seen that T_λ is linear and continuous in $L^1(I\!R)$ and $\lim_{\lambda \searrow 0} T_\lambda \varphi = \varphi$ in $\mathcal{D}(I\!R)$ for $\varphi \in \mathcal{D}(I\!R)$. Now (4.11) follows by a density argument. ■

Returning to our problem, it follows by Theorem 2.1 in [4] that

$$\partial\psi|_{H_0^1(\Omega)}(x) \subset \partial\psi(x) \tag{4.13}$$

In order to obtain information on $\partial\psi$, we shall need an improvement of the Theorem 2.1. in [4].

Theorem 4.3 If $x \in L^\infty(0,1)$, then

$$\partial\psi(x)(t) \subset [\underline{g}(x(t)), \overline{g}(x(t))] \quad a.e.\ t \in]0,1[,$$

in the sense that if $w \in \partial\psi(x)$ then

$$\underline{g}(x(t)) \leq w(t) \leq \overline{g}(x(t)) \qquad a.e.\ t \in]0,1[. \tag{4.14}$$

Proof. Let h be a Borel function such that $h = g$ a.e. on $I\!R$. It follows that the set

$$A = \{t \in]0,1[\ :\ \underline{g}(x(t)) \neq \underline{h}(x(t))\}$$

is a null set. (A similar reasoning can be done for $\overline{g}$ and $\overline{h}$).

Therefore we may suppose that g is a Borel function. We would like to deal with $\int_0^1 \overline{g}(x(t))dt$, so we have to prove that $\overline{g}$ is a Borel function.

Lemma 4.6 Let $g : I\!R \to I\!R$ be a locally bounded Borel function. Then $\overline{g}$ is a Borel function.

Proof. Since the requirement is local, we may suppose without loss of generality that g is bounded by 1 and it is nonnegative. Since

$$g = \lim_{n\to\infty} \lim_{m\to\infty} g_{m,n},$$

where

$$g_{m,n}(x,t) = \left(\int_{t-\frac{1}{n}}^{t+\frac{1}{n}} |g^m(x,s)|ds\right)^{1/m},$$

it suffices to prove that $g_{m,n}$ is a Borel function.

Set

$$\mathcal{M} = \{g : \Omega \times I\!R \to I\!R : \quad |g| \leq 1 \text{ and } g \text{ is a Borel function}\}$$

$$\mathcal{N} = \{g \in \mathcal{M} : \quad g_{m,n} \text{ is a Borel function}\}.$$

It is known (see [2], p. 178) that $\mathcal{M}$ is the smallest set of functions having the following properties:

i) $\{g \in C(\Omega \times I\!R; I\!R) : \quad |g| \leq 1\} \subset \mathcal{M}$

ii) $g^{(k)} \xrightarrow{k} g$ imply $g \in \mathcal{M}$. We point out that here we have an "each point" convergence.

Since $\mathcal{N}$ contains obviously the continuous functions and ii) is also true for $\mathcal{N}$, by the Lebesgue's Dominated Convergence Theorem, it follows that $\mathcal{M} = \mathcal{N}$. ■

Proof of Theorem 4.3 continued. Let $v \in L^\infty(\Omega)$, $v \geq 0$. Then, for suitable $\lambda_i \searrow 0$ and $h_i \to 0$ in $L^{p+1}(\Omega)$ we have

$$\psi^0(u; v) = \lim_{i\to\infty} \frac{1}{\lambda_i} \int_\Omega \int_{u(x)+h_i(x)}^{u(x)+h_i(x)+\lambda_i v(x)} g(x,s)ds\,dx.$$

We may suppose that $h_i \to 0$ a.e., so that

$$\psi^0(u; v) = \lim_{i\to\infty} \frac{1}{\lambda_i} \int_{[v>0]} \int_{u(x)+h_i(x)}^{u(x)+h_i(x)+\lambda_i v(x)} g(x,s)ds\,dx$$

$$\leq \int_{[v>0]} \left(\limsup_{i\to\infty} \frac{1}{\lambda_i} \int_{u(x)+h_i(x)}^{u(x)+h_i(x)+\lambda_i v(x)} g(x,s)ds \right) dx$$

$$\leq \int_{[v>0]} \overline{g}(x, u(x))v(x)dx.$$

Hence, for any v,

$$\psi^0(u; v) \leq \int_{[v>0]} \overline{g}(x, u(x))v(x)dx. \tag{4.15}$$

Suppose now, by contradiction, that (4.14) is false. So, there exist a set E with $|E| > 0$ and $w \in \partial\psi(u)$ such that

$$w(x) > \overline{g}(x, u(x)) \qquad \text{on } E. \tag{4.16}$$

Choosing $v = id_E$ in (4.15) we obtain

$$\langle w, v\rangle = \int_E w \leq \psi^0(u; v) \leq \int_E \overline{g}(x, u(x))dx,$$

which contradicts (4.16). ■

Proof of Theorem 4.2. Define on the space $X = H^1_p(0,1)$ the locally Lipschitz function

$$\varphi(x) = \frac{1}{2}\int_0^1 x'^2(t)dt - \int_0^1 f(t)x(t)dt + \int_0^1 G(x(t))dt.$$

The critical points of φ are solutions of (4.6). Indeed, it is obvious that

$$\partial\varphi(x) = -x'' - f + \partial\psi|_{H^1_p(0,1)}(x) \quad \text{in} \quad H^{-1}(0,1)\,.$$

If x_0 is a critical point of φ it follows that there exists $w \in \partial\psi|_{H^1_p(0,1)}(x_0)$ such that

$$x'' + f = w \qquad \text{in} \quad H^{-1}(0,1)\,.$$

Since $\varphi(x+T) = \varphi(x)$, we can apply Theorem 4.1. All we have to do is to check the $(PS)_{Z,c}$ condition, for each c, and to prove that (4.6) has a solution x_0 that minimizes φ on $H^1_p(0,1)$. We first observe that every $x \in H^1_p(0,1)$ can be written

$$x(t) = \int_0^1 x(s)ds + \overline{x}(t) \quad \text{with } \overline{x} \in H^1_0(0,1).$$

Hence, by the Poincaré inequality,

$$\begin{aligned}\varphi(x) &= \frac{1}{2}\int_0^1 \overline{x}'^2(t)dt - \int_0^1 f(t)\overline{x}(t)dt + \int_0^1 G(x(t))dt \\ &\geq \frac{1}{2}\,\|\overline{x}'^2\|^2_{L^2} - \|f\|_{L^p}\cdot\|\overline{x}\|_{L^{p'}} - \|G\|_{L^\infty} \\ &\geq \frac{1}{2}\,\|\overline{x}'^2\|^2_{L^2} - C\,\|f\|_{L^p}\cdot\|\overline{x}'\|_{L^2} - \|G\|_{L^\infty} \to +\infty\end{aligned}$$

as $\|\overline{x}\|_{H^1} \to \infty$, where p' denotes the conjugated exponent of p.

We verify in what follows the $(PS)_{Z,c}$ condition, for each real c. Let $(x_n) \subset X$ be such that

$$\varphi(x_n) \to c \tag{4.17}$$

$$\lambda(x_n) \to 0. \tag{4.18}$$

Since $\underline{g} \circ x_n \leq w_n \leq \overline{g} \circ x_n$ and $\underline{g}, \overline{g} \in L^\infty(\mathbb{R})$, let us choose arbitrarily $w_n \in \partial\varphi(x_n) \subset L^\infty(0,1)$ so that

$$\lambda(x_n) = \|x_n'' + f - w_n\|_*.$$

By (4.18) we obtain

$$\int_0^1 (x_n')^2 - \int_0^1 f x_n + \int_0^1 w_n x_n = o(1)\|x_n\|_{H^1_p}$$

and, by (4.17),

$$\frac{1}{2}\int_0^1 (x_n')^2 - \int_0^1 f x_n + \int_0^1 G(x_n) \to c\,.$$

So, there exist positive constants C_1 and C_2 such that

$$\int_0^1 (x_n')^2 \le C_1 + C_2 \|x_n\|_{H_p^1}\,.$$

Note that G is also T-periodic; hence it is bounded.

Replacing x_n by $x_n + kT$ for a suitable integer k, we may suppose that

$$x_n(0) \in [0, T]\,,$$

so that $\{x_n\}$ is bounded in H_p^1.

Let $x \in H_p^1$ be such that, up to a subsequence, $x_n \rightharpoonup x$ and $x_n(0) \to x(0)$. Then

$$\int_0^1 (x_n')^2 = \langle -x_n'' - f + w_n, x_n - x\rangle - \int_0^1 w_n(x_n - x)$$
$$+ \int_0^1 f(x_n - x) + \int_0^1 x_n' x' \to \int_0^1 x'^2\,,$$

because $x_n \to x$ in $L^{p'}$. It follows that $x_n \to x$ in H_p^1. ■

3. Hemivariational Inequalities Associated to Multivalued Problems with Strong Resonance

The literature is very rich in resonant problems, the first who studied such problems in the smooth case being Landesman and Lazer in their celebrated paper [10]. They found sufficient conditions for the existence of solutions for some single-valued equations with Dirichlet conditions. These problems, that arise frequently in Mechanics, were thereafter intensively studied and many applications to concrete situations were given.

We start with an overview on problems at resonance and we do this in the classical C^1-framework, as formulated by Landesman and Lazer in [10].

Let Ω be an open bounded set in $\mathbb{R}^N$ and let $f : \mathbb{R} \to \mathbb{R}$ be a continuous function. Consider the problem

$$\begin{cases} -\Delta u = f(u) & \text{in } \Omega\,,\\ u = 0 & \text{on } \partial\Omega\,.\end{cases}$$

For obtaining information on the existence of solutions, as well as possible estimates on the number of solutions, it is essential to know the asymptotic behavior of the nonlinearity. Assume that f is asymptotic linear at infinity and set

$$a := \lim_{|t|\to\infty} \frac{f(t)}{t} \in I\!R . \tag{4.19}$$

We write $f(t) = at - g(t)$, where

$$\lim_{|t|\to\infty} \frac{g(t)}{t} = 0 .$$

There are several "degrees" of resonance, according to the growth of g at infinity, in the sense that if g has a "smaller" rate of increasing at infinity then its resonance is "stronger". Landesman and Lazer distinguished the following situations:

i) $\lim_{t\to\pm\infty} g(t) = \ell_\pm \in I\!R$ and $(\ell_+, \ell_-) \neq (0,0)$;

ii) $\lim_{t\to\pm\infty} g(t) = 0$ and $\lim_{|t|\to\infty} \int_0^t g(s)ds = \pm\infty$;

iii) $\lim_{t\to\pm\infty} g(t) = 0$ and $\lim_{|t|\to\infty} \int_0^t g(s)ds \in I\!R$.

The last situation corresponds to a problem with *strong resonance at infinity.*

In this Section we solve a nonsmooth problem with strong resonance at infinity. For this purpose we apply basic tools on hemivariational inequalities, as well as results on nonsmooth critical point theory. We remark that it is not natural to consider nonlinearities which are strongly resonant at $+\infty$, but which may not be strongly resonant at $-\infty$.

Let Ω be an open bounded set in $I\!R^N$, $N \geq 2$ and let $V : \Omega \to I\!R$ be a potential satisfying

(V) $V \in L^1_{\rm loc}(\Omega)$, $V^+ = V_1 + V_2 \neq 0$, $V_1 \in L^{N/2}(\Omega)$, and $\limsup_{\substack{x\to y\\ x\in\Omega}} |x - y|^2 V_2(x) = 0$, for every $y \in \overline{\Omega}$.

As usual, we have denoted $V^+(x) := \max\{V(x), 0\}$.

Consider the problem

$$\begin{cases} -\Delta u = \lambda V(x) u & \text{in } \Omega ,\\ u = 0 & \text{on } \partial\Omega . \end{cases} \tag{4.20}$$

Szulkin and Willem proved in [26] that, under assumption (V), problem (4.20) has a sequence of eigenvalues $0 < \lambda_1^V \leq \lambda_2^V \leq \cdots \leq \lambda_n^V \leq \cdots$, with $\lim_{n\to\infty} \lambda_n^V = \infty$. Furthermore, λ_1^V is simple, provided that V fulfills the additional assumption

(V_1) there exist $p > N/2$ and a closed subset S of measure 0 in $\mathbb{R}^N$ such that $\Omega \setminus S$ is connected and $V \in L^p_{\text{loc}}(\Omega \setminus S)$.

Spectral results of this type have been originally proved by Pleijel and Minakshisundaran in [14], [17] under the stronger assumption that $V \in L^\infty(\Omega)$ and there exists $\Omega' \subset \Omega$ with positive measure such that $V(x) \geq 0$ for a.e. $x \in \Omega$, and $V(x) > 0$ if $x \in \Omega'$.

Consider now a function $f \in L^\infty(\mathbb{R})$ and define

$$F(t) := \int_0^t f(s)ds \quad \forall t \in \mathbb{R}.$$

Our aim is to study the following hemivariational problem: find $u \in H^1_0(\Omega) \setminus \{0\}$ such that

$$\int_\Omega (DuDv - \lambda_1^V V(x)uv)dx + \int_\Omega (-F^0)(x, u; v)dx \geq 0\,, \tag{4.21}$$

for all $v \in H^1_0(\Omega)$. Our basic hypothesis on f is

(f_1) $f(+\infty) = F(+\infty) = 0$,

where

$$f(\pm\infty) := \operatorname{ess}\lim_{t\to\pm\infty} f(t)\,, \quad F(\pm\infty) := \operatorname{ess}\lim_{t\to\pm\infty} F(t)\,.$$

We observe that, due to (f_1), the hemivariational inequality (4.21) becomes a Landesman-Lazer type problem with strong resonance at $+\infty$.

We observe that problem (4.21) may be reformulated in the following manner. Set

$$\underline{f}(t) = \lim_{\varepsilon\searrow 0} \operatorname{essinf}\{f(s) \,:\, |t-s| < \varepsilon\}$$

$$\overline{f}(t) = \lim_{\varepsilon\searrow 0} \operatorname{esssup}\{f(s) \,:\, |t-s| < \varepsilon\}\,.$$

Then the hemivariational inequality (4.21) is equivalent to the following multivalued elliptic problem:

$$\begin{cases} -\Delta u - \lambda_1^V V(x)u \in [\,\underline{f}(u(x)), \overline{f}(u(x))\,] & \text{in } \Omega\,, \\ u = 0 & \text{on } \partial\Omega\,, \\ u \not\equiv 0 & \text{in } \Omega\,. \end{cases}$$

We also refer to [19], [20], [22] for the treatment of problems of this type.

Our first result is

Theorem 4.4 (Rădulescu [23]) Assume (V), (V_1), (f_1) and either

(F_1) $F(-\infty) = -\infty$
or $-\infty < F(-\infty) \leq 0$ and there exists $\eta > 0$ such that
(F_2) F is non-negative on $(0, \eta)$ or on $(-\eta, 0)$.
Then problem (4.21) has at least a solution.

For positive values of $F(-\infty)$ it is necessary to impose additional restrictions to f. Our variant in this case is

Theorem 4.5 (Rădulescu [23]) Assume (V), (V_1), (f_1) and $0 < F(-\infty) < +\infty$. Then problem (4.21) has at least a solution, provided that the following conditions are satisfied:

$$f(-\infty) = 0$$

and

$$F(t) \leq \frac{\lambda_2^V - \lambda_1^V}{2} t^2 \qquad \text{for any } t \in \mathbb{R}.$$

We start with some auxiliary results. We first associate to problem (4.21) the energy $E = E_1 - E_2$, where

$$E_1(u) = \frac{1}{2} \int_\Omega (|Du|^2 - \lambda_1^V V(x) u^2) dx \quad \text{and} \quad E_2(u) = \int_\Omega F(u) dx\,,$$

for all $u \in H_0^1(\Omega)$. We observe that E_1 is of class C^1 on $H_0^1(\Omega)$, while E_2 is a Lipschitz functional. Indeed, for any $u, v \in H_0^1(\Omega)$,

$$\begin{aligned} |E_2(u) - E_2(v)| &\leq | \int_\Omega \left(\int_{u(x)}^{v(x)} f(t) dt \right) dx | \leq \\ &\leq \|f\|_{L^\infty} \|u - v\|_{L^1} \leq C \|u - v\|_{H_0^1}. \end{aligned}$$

We also observe that critical points of the locally Lipschitz function E correspond to the solutions of problem (4.21).

Lemma 4.7 Assume that $f \in L^\infty(\mathbb{R})$ and there exist $F(\pm\infty) \in \overline{\mathbb{R}}$. Moreover, suppose that
i) $f(+\infty) = 0$ if $F(+\infty)$ is finite;
and
ii) $f(-\infty) = 0$ if $F(-\infty)$ is finite.
Then

$$\mathbb{R} \setminus \{a|\Omega| \,:\, a = -F(\pm\infty)\} \subset \{c \in \mathbb{R} \,:\, E \text{ satisfies } (PS)_c\}.$$

Proof. We shall assume, without loss of generality, that $F(-\infty) \notin \mathbb{R}$ and $F(+\infty) \in \mathbb{R}$. In this case, if c is a critical value such that E does not satisfy $(PS)_c$, then it is enough to prove that $c = -F(+\infty)|\Omega|$. Let $e_1^V \geq 0$ be an eigenfunction of problem (4.20), corresponding to λ_1^V and let W denote the orthogonal complement of the space spanned by e_1^V with respect to $H_0^1(\Omega)$, that is

$$H_0^1(\Omega) = \mathrm{span}\,\{e_1^V\} \oplus W\,.$$

Since E does not satisfy the condition $(PS)_c$, there exist $t_n \in I\!R$ and $v_n \in W$ such that the sequence $\{u_n\} \subset H_0^1(\Omega)$, where $u_n = t_n e_1^V + v_n$, has no convergent subsequence, while

$$\lim_{n\to\infty} E(u_n) = c, \tag{4.22}$$

$$\lim_{n\to\infty} \lambda_E(u_n) = 0\,. \tag{4.23}$$

Step 1. The sequence $\{v_n\}$ is bounded in $H_0^1(\Omega)$.
By (4.23) and

$$\partial E(u) = -\Delta u - \lambda_1^V V u - \partial E_2(u)\,,$$

it follows that there exists $w_n \in \partial E_2(u_n)$ such that

$$-\Delta u_n - \lambda_1^V V u_n - w_n \to 0 \quad \text{in} \quad H^{-1}(\Omega)\,.$$

So

$$\begin{aligned}&\langle -\Delta u_n - \lambda_1^V V u_n - w_n, v_n\rangle\\ &= \int_\Omega |Dv_n|^2 - \lambda_1^V \int_\Omega V v_n^2 - \int_\Omega g_n(t_n e_1^V + v_n) = o(\|v_n\|_{H_0^1}),\end{aligned}$$

as $n \to \infty$, where $\underline{f} \leq g_n \leq \overline{f}$. Since f is bounded, it follows that

$$\|v_n\|_{H_0^1}^2 - \lambda_1^V \int_\Omega V v_n^2 = O(\|v_n\|_{H_0^1})\,.$$

So, there exists $C > 0$ such that, for every $n \geq 1$, $\|v_n\|_{H_0^1} \leq C$. Now, since $\{u_n\}$ has no convergent subsequence, it follows that the sequence $\{v_n\}$ has no convergent subsequence, too.

Step 2. $t_n \to +\infty$.
Since $\|v_n\|_{H_0^1} \leq C$ and the sequence $\{t_n e_1^V + v_n\}$ has no convergent subsequence, it follows that $|t_n| \to +\infty$.

On the other hand, by Lebourg's Mean Value Theorem, there exist $\theta \in]0,1[$ and $x^* \in \partial F(te_1^V(x) + \theta v(x))$ such that

$$E_2(te_1^V + v) - E_2(te_1^V) = \int_\Omega \langle x^*, v(x)\rangle dx$$

$$\leq \int_\Omega F^0(te_1^V(x)+v(x);v(x))dx$$

$$= \int_\Omega \limsup_{\substack{y\to te_1^V(x)+v(x)\\ \lambda\searrow 0}} \frac{F(y+\lambda v(x))-F(y)}{\lambda}dx$$

$$\leq \|f\|_{L^\infty}\cdot\int_\Omega |v(x)|dx = \|f\|_{L^\infty}\cdot\|v\|_{L^1}\leq C_1\|v\|_{H_0^1}\,.$$

A similar computation for $E_2(te_1^V)-E_2(te_1^V+v)$ together with the above inequality shows that, for every $t\in I\!R$ and for any $v\in V$,

$$|E_2(te_1^V+v)-E_2(te_1^V)|\leq C_2\|v\|_{H_0^1}\,.$$

So, taking into account the boundedness of $\{v_n\}$ in $H_0^1(\Omega)$, we find

$$|E_2(t_ne_1^V+v_n)-E_2(t_ne_1^V)|\leq C\,.$$

Therefore, since $F(-\infty)\notin I\!R$ and

$$E(u_n)=E_1(v_n)-E_2(t_ne_1^V+v_n)\to c,$$

it follows that $t_n\to+\infty$. In this argument we have also used the fact that $E_1(v_n)$ is bounded.

Step 3. $\|v_n\|_{H_0^1}\to 0$ as $n\to\infty$.
By (f_1) and Step 2 it follows that

$$\lim_{n\to\infty}\int_\Omega f(t_ne_1^V+v_n)v_n=0\,.$$

Using now (4.23) and Step 1 we find

$$\lim_{n\to\infty}\|v_n\|_{H_0^1}=0\,.$$

Step 4. We have

$$\lim_{t\to+\infty}E_2(te_1^V+v)=F(+\infty)|\Omega|\,, \tag{4.24}$$

uniformly on the bounded subsets of W.
Assume the contrary. So, there exist $r>0$, $t_n\to+\infty$, $v_n\in W$ with $\|v_n\|\leq r$, such that (4.24) is not fulfilled. Thus there exist $v\in H_0^1(\Omega)$ and $h\in L^2(\Omega)$ such that, up to a subsequence,

$$\begin{aligned} &v_n\rightharpoonup v \quad \text{weakly in } H_0^1(\Omega)\,,\\ &v_n\to v \quad \text{strongly in } L^2(\Omega)\,,\\ &v_n(x)\to v(x) \quad \text{for a.e. } x\in\Omega\,, \end{aligned} \tag{4.25}$$

$$|v_n(x)| \leq h(x) \quad \text{for a.e. } x \in \Omega\,. \tag{4.26}$$

For any $n \geq 1$ we define

$$A_n = \{x \in \Omega \,:\, t_n e_1^V(x) + v_n(x) < 0\}\,,$$

$$h_n(x) = F(t_n e_1^V + v_n)\chi_{A_n}\,,$$

where χ_A represents the characteristic function of the set A. By (4.26) and the choice of t_n it follows that $|A_n| \to 0$ if $n \to \infty$.

Using (4.25) we remark easily that

$$h_n(x) \to 0 \quad \text{for a.e. } x \in \Omega\,.$$

Therefore

$$|h_n(x)| = \chi_{A_n}(x) \cdot \left| \int_0^{t_n e_1^V(x)+v_n(x)} f(s)ds \right|$$

$$\leq \chi_{A_n}(x) \cdot \|f\|_{L^\infty} \cdot |t_n e_1^V(x) + v_n(x)| \leq C|v_n(x)| \leq Ch(x)\,,$$

for a. e. $x \in \Omega$. So, by Lebesgue's Dominated Convergence Theorem,

$$\lim_{n\to\infty} \int_{A_n} F(t_n e_1^V + v_n) = 0\,.$$

On the other hand,

$$\lim_{n\to\infty} \int_{\Omega\setminus A_n} F(t_n e_1^V + v_n) = F(+\infty)|\Omega|\,.$$

So

$$\lim_{n\to\infty} E_2(t_n e_1^V + v_n) = \lim_{n\to\infty} \int_\Omega F(t_n e_1^V + v_n) = F(+\infty)|\Omega|,$$

which contradicts our initial assumption.

Step 5. Taking into account the previous step and the fact that $E(te_1^V + v) = E_1(v) - E_2(te_1^V + v)$, we obtain

$$\lim_{n\to\infty} E(t_n e_1^V + v_n) = \lim_{n\to\infty} E_1(v_n) - \lim_{n\to\infty} E_2(t_n e_1^V + v_n)$$

$$= -F(+\infty)|\Omega|,$$

that is $c = -F(+\infty)|\Omega|$, which concludes our proof. ■

Lemma 4.8 Assume that f fulfills condition (f_1). Then E satisfies $(PS)_c$, whenever $c \neq 0$ and $c < -F(-\infty)|\Omega|$.

Proof. It is enough to show that for every $c \neq 0$ and $\{u_n\} \subset H_0^1(\Omega)$ such that

$$\begin{array}{ll} E(u_n) & \to c\,, \\ \lambda_E(u_n) & \to 0\,, \\ \|u_n\| & \to \infty\,, \end{array} \tag{4.27}$$

we have $c \geq -F(-\infty)|\Omega|$.

Let $t_n \in I\!R$ and $v_n \in W$ be such that, for every $n \geq 1$,

$$u_n = t_n e_1^V + v_n\,.$$

As we have already remarked,

$$E(u_n) = E_1(v_n) - E_2(u_n)\,.$$

Moreover, E_1 is positive and coercive on W. Indeed,

$$E_1(v) = \frac{1}{2}\int_\Omega (|Dv|^2 - \lambda_1^V V v^2) \geq \frac{\lambda_2^V - \lambda_1^V}{2} \cdot \|v\|_{H_0^1}^2 \to +\infty$$

as $\|v\|_{H_0^1} \to \infty$. Also, by (f_1), E_2 is bounded from below. Hence, again by (f_1), we conclude that the sequence $\{v_n\}$ is bounded in $H_0^1(\Omega)$. So, up to a subsequence,

$$\begin{aligned} v_n \rightharpoonup v \quad & \text{weakly in } \; H_0^1(\Omega)\,, \\ v_n \to v \quad & \text{strongly in } \; L^2(\Omega)\,, \\ v_n(x) \to v(x) \quad & \text{for a.e. } \; x \in \Omega\,, \\ |v_n(x)| \leq h(x) \quad & \text{for a.e. } \; x \in \Omega\,, \end{aligned}$$

where $h \in L^2(\Omega)$.

Since $\|u_n\|_{H_0^1} \to \infty$ and $\{v_n\}$ is bounded in $H_0^1(\Omega)$, it follows that $|t_n| \to +\infty$.

Assume for the moment that we have already proved that $\|v_n\|_{H_0^1} \to 0$, if $t_n \to +\infty$. So,

$$E(u_n) = E_1(v_n) - E_2(u_n) \to 0 \quad \text{as} \;\; n \to \infty\,.$$

Here, to prove that $E_2(u_n) \to 0$, we have used (f_1). The last relation yields a contradiction, since $E(u_n) \to c \neq 0$. So, $t_n \to -\infty$.

Moreover, since $E(u) \geq -E_2(u)$ and F is bounded from below, it follows that

$$c = \liminf_{n\to\infty} E(u_n) \geq \liminf_{n\to\infty}(-E_2(u_n)) = -\limsup_{n\to\infty}\int_\Omega F(u_n)$$

$$\geq - \int_\Omega \limsup_{n\to\infty} F(u_n) = -F(-\infty)|\Omega|,$$

which gives the desired contradiction.

So, for concluding the proof, it remains to show that

$$\|v_n\|_{H_0^1} \to 0 \quad \text{as} \quad t_n \to +\infty\,.$$

Since

$$\partial E(u) = -\Delta u - \lambda_1^V V u - \partial E_2(u),$$

it follows from (4.27) that there exists $w_n \in \partial E_2(u_n)$ such that

$$-\Delta u_n - \lambda_1^V V u_n - w_n \to 0 \quad \text{in} \quad H^{-1}(\Omega)\,.$$

Thus

$$\langle -\Delta u_n - \lambda_1^V u_n - w_n, v_n\rangle$$

$$= \int_\Omega |Dv_n|^2 - \lambda_1^V \int_\Omega V v_n^2 - \int_\Omega g_n(t_n e_1^V + v_n)v_n = o(\|v_n\|)$$

as $n \to \infty$, where $\underline{f} \leq g_n \leq \overline{f}$.

Now, for concluding the proof, it is sufficient to show that the last term tends to 0, as $n \to \infty$.

Fix $\varepsilon > 0$. Since $f(+\infty) = 0$, it follows that there exists $T > 0$ such that

$$|f(t)| \leq \varepsilon \qquad \text{for a.e. } \ t \geq T\,.$$

Set

$$A_n = \{x \in \Omega :\ \ t_n e_1^V(x) + v_n(x) \geq T\} \quad \text{and} \quad B_n = \Omega \setminus A_n\,.$$

We remark that for every $x \in B_n$,

$$|t_n e_1^V(x) + v_n(x)| \leq |v_n(x)| + T\,.$$

So, for every $x \in B_n$,

$$|g_n(t_n e_1^V(x) + v_n(x))v_n(x)| \cdot \chi_{B_n}(x) \leq \|f\|_{L^\infty} \cdot h(x)\,.$$

By

$$\chi_{B_n}(x) \to 0 \quad \text{for a.e.} \ \ x \in \Omega$$

and the Lebesgue Dominated Convergence Theorem it follows that

$$\int_{B_n} g_n(t_n e_1^V + v_n)v_n \to 0 \quad \text{as} \ \ n \to \infty\,. \tag{4.28}$$

On the other hand, it is obvious that

$$|\int_{A_n} g_n(t_n e_1^V + v_n)v_n| \le \varepsilon \int_{A_n} |v_n| \le \varepsilon \, \|h\|_{L^1} . \tag{4.29}$$

By (4.28) and (4.29) it follows that

$$\lim_{n\to\infty} \int_\Omega g_n(u_n)v_n = 0,$$

which concludes our proof. ■

Proof of Theorem 4.4. We distinguish two distinct situations:

CASE 1. $F(-\infty)$ is finite, that is $-\infty < F(-\infty) \le 0$. In this case, E is bounded from below since

$$E(u) = \frac{1}{2}\int_\Omega (|Du|^2 - \lambda_1^V V(x)u^2)dx - \int_\Omega F(u)dx$$

and, by our hypothesis on $F(-\infty)$,

$$\sup_{u\in H_0^1(\Omega)} \int_\Omega F(u)dx < +\infty .$$

Therefore

$$-\infty < a := \inf_{u\in H_0^1(\Omega)} E(u) \le 0 = E(0) .$$

Choose c small enough in order to have $F(ce_1^V) < 0$ (note that c may be taken positive if $F > 0$ in $(0, \eta)$ and negative if $F < 0$ in $(-\eta, 0)$). Hence $E(ce_1^V) < 0$, so $a < 0$. It follows now from Lemma 4.7 that E satisfies $(\mathrm{PS})_a$. The proof ends in this case by applying Mountain Pass Theorem in the locally Lipschitz case.

CASE 2. $F(-\infty) = -\infty$. Then, by Lemma 4.8, E satisfies $(PS)_c$ for each $c \neq 0$.

As in the previous Section, let W be the orthogonal complement of the space spanned by e_1^V with respect to $H_0^1(\Omega)$.

For fixed $t_0 > 0$, denote

$$W_0 = \{t_0 e_1^V + w \, : \, w \in W\} \qquad \text{and} \qquad a_0 = \inf_{w\in W_0} E(w) .$$

Note that E is coercive on V. Indeed, if $w \in W$, then

$$E(w) \ge \frac{1}{2}\left(1 - \frac{\lambda_1^V}{\lambda_2^V}\right) \|w\|^2_{H_0^1} - \int_\Omega F(w) \to +\infty$$

as $\|w\|_{H_0^1} \to +\infty$, because the first right-hand term has a quadratic growth at infinity (t_0 being fixed), while $\int_\Omega F(w)$ is uniformly bounded (in w), in view of the behavior of F near $\pm\infty$. Thus, a_0 is attained, because of the coercivity of E on W. From the boundedness of E on $H_0^1(\Omega)$ it follows that $-\infty < a \leq 0 = E(0)$ and $a \leq a_0$.

Again, there are two possibilities:

i) $a < 0$. In this case, by Lemma 4.8, E satisfies $(\mathrm{PS})_a$. Hence $a < 0$ is a critical value of E.

ii) $a = 0 \leq a_0$. Then, either $a_0 = 0$ or $a_0 > 0$. In the first case, as we have already remarked, a_0 is attained. Thus, there is some $w \in W$ such that

$$0 = a_0 = E(t_0 e_1^V + w)\,.$$

Hence, $u = t_0 e_1^V + w \in H_0^1(\Omega) \setminus \{0\}$ is a critical point of E, that is a solution of (4.21).

If $a_0 > 0$, notice that E satisfies $(\mathrm{PS})_b$ for each $b \neq 0$. Since

$$\lim_{t \to +\infty} E(te_1^V) = 0\,,$$

we may apply the Saddle Point Theorem in the locally Lipschitz case to conclude that E has a critical value $c \geq a_0 > 0$. ■

Proof of Theorem 4.5. Set

$$W_+ = \{te_1^V + w \;:\; t > 0,\ w \in W\}\,.$$

It is sufficient to show that the functional E has a non-zero critical point. To do this, we shall make use of two different arguments.

If $u = te_1^V + w \in W_+$ then

$$E(u) = \frac{1}{2}\int_\Omega (\,|Dw|^2 - \lambda_1^V V(x)w^2\,)dx - \int_\Omega F(te_1^V + w)dx\,.$$

In view of the boundedness of F it follows that

$$-\infty < a_+ := \inf_{u \in W_+} E(u) \leq 0\,.$$

We analyze two distinct situations:

CASE 1. $a_+ = 0$.

To prove that E has a critical point, we use the same arguments as in the proof of Theorem 4.5 (the second case). More precisely, for some fixed $t_0 > 0$ we define at the same way W_0 and a_0. Obviously, $a_0 \geq 0 = a_+$, since $W_0 \subset W_+$. The proof follows from now on the same ideas as in Case 2 of Theorem 4.5, by considering the two distinct situations $a_0 > 0$ and $a_0 = 0$.

CASE 2. $a_+ < 0$.

Let $u_n = t_n e_1^V + w_n$ be a minimizing sequence of E in W_+. We observe that the sequences $\{u_n\}$ and $\{w_n\}$ are bounded. Indeed, this is essentially a compactness condition and may be deduced in a similar way in the proof of Lemma 4.7. It follows that there exists $w_0 \in \overline{W}_+$, such that, going eventually to a subsequence,

$$u_n \rightharpoonup w_0 \quad \text{weakly in } H_0^1(\Omega)\,;$$

$$u_n \to w_0 \quad \text{strongly in } L^2(\Omega)\,;$$

$$u_n \to w_0 \quad \text{a.e. in } \Omega\,.$$

Applying the Lebesgue Dominated Convergence Theorem we obtain

$$\lim_{n\to\infty} E_2(u_n) = E_2(w_0)\,.$$

On the other hand,

$$E(w_0) \leq \liminf_{n\to\infty} E_1(u_n) - \lim_{n\to\infty} E_2(u_n) = \liminf_{n\to\infty} E(u_n) = a_+\,.$$

It follows that, necessarily, $E(w_0) = a_+ < 0$. Since the boundary of W_+ is W and

$$\inf_{u\in W} E(u) = 0\,,$$

we conclude that w is a local minimum of E on W_+ and $w \in W_+$. ■

4. A Parallel Nonsmooth Critical Point Theory. Approach to Stationary Schrödinger Type Equations in $\mathbb{R}^n$

In this Section we determine nontrivial solutions of some semilinear and quasilinear elliptic problems on $\mathbb{R}^n$. We make use of two different nonsmooth critical point theories which allow to treat two kinds of nonlinear problems. A comparison between the possible applications of the two theories is also made.

Consider a functional J defined on some Banach space B and having a mountain pass geometry: the celebrated theorem by Ambrosetti and Rabinowitz [1] states that if $J \in C^1(B)$ and J satisfies the Palais-Smale condition (PS condition in the sequel) then J admits a nontrivial critical point. In what follows we drop these two assumptions: in order to determine nontrivial solutions of some nonlinear elliptic equations in $\mathbb{R}^n$ $(n \geq 3)$, we use the mountain pass principle for a class of nonsmooth functionals which do not satisfy the PS condition. More precisely, we consider a model elliptic problem first studied by Rabinowitz [18] with the C^1-theory and we extend his results by means of the nonsmooth critical point theories of Clarke and Degiovanni (see Chapter 2). One of our purposes in this Section is to emphasize some differences between these two theories. This study was inspired by previous work on the existence of standing wave solutions of nonlinear Schrödinger equations. After making a standing wave ansatz, Rabinowitz reduces the problem to that of studying the semilinear elliptic equation

$$-\Delta u + b(x)u = f(x,u) \quad \text{in } \mathbb{R}^n \tag{4.30}$$

under suitable conditions on b and assuming that f is smooth, superlinear and subcritical. Problems of this type appear in the study of several physical phenomena: self channeling of a high-power ultra short laser in matter, in the theory of Heisenberg ferromagnets and magnons, in dissipative quantum mechanics, in condensed matter theory, in plasma physics (e.g., the Kurihara superfluid film equation) etc.

Our goal is to show how variational methods can be used to find existence results for stationary nonlinear Schrödinger equations. The approach we develop in this Section is based on the fact that many nonlinear problems such as those that naturally arise in the study of geodesics, minimal surfaces, harmonic maps, conformal metrics with prescribed curvature, subharmonics of Hamiltonian systems, solutions of boundary value problems and Yang-Mills fields can all be characterized as critical points of some energy functional on an appropriate manifold.

We are concerned with two problems on the existence of critical points and how they relate to the (weak) solutions they represent for the corresponding Euler-Lagrange equations. To explain our results we introduce some functional spaces. We denote by L^p the space of measurable functions u of p-th power absolutely summable on $\mathbb{R}^n$, that is, satisfying

$$\|u\|_p^p := \int_{\mathbb{R}^n} |u|^p < +\infty\,.$$

Let H^1 denote the Sobolev space normed by

$$\|u\|_{H^1}^2 := \int_{\mathbb{R}^n} (|Du|^2 + |u|^2)\,.$$

We assume that the function b in (4.30) is greater than some positive constant. Then we define the Hilbert space E of all functions $u : \mathbb{R}^n \to \mathbb{R}$ with

$$\|u\|_E^2 := \int_{\mathbb{R}^n} (|Du|^2 + b(x)u^2) < +\infty\,.$$

We denote by E^* the dual space of E: as E is continuously embedded in H^1 we also have $H^{-1} \subset E^*$.

We first consider the case where $(-\Delta)$ in (4.30) is replaced by a quasilinear elliptic operator: we seek positive weak solutions $u \in E$ of the problem

$$-\sum_{i,j=1}^{n} D_j(a_{ij}(x,u)D_iu)$$
$$+\frac{1}{2}\sum_{i,j=1}^{n} \frac{\partial a_{ij}}{\partial s}(x,u)D_iuD_ju + b(x)u = f(x,u) \tag{4.31}$$

in $\mathbb{R}^n$. Note that if $a_{ij}(x,s) \equiv \delta_{ij}$, then (4.31) reduces to (4.30). Here and in the sequel, by positive solution we mean a nonnegative nontrivial solution. To determine weak solutions of (4.31) we look for critical points of the functional $J : E \to \mathbb{R}$ defined by

$$J(u) = \frac{1}{2}\int_{\mathbb{R}^n}\sum_{i,j=1}^{n} a_{ij}(x,u)D_iuD_ju + \frac{1}{2}\int_{\mathbb{R}^n} b(x)u^2 - \int_{\mathbb{R}^n} F(x,u),$$

where $F(x,s) = \int_0^s f(x,t)dt$. Under reasonable assumptions on a_{ij}, b, f, the functional J is continuous but not even locally Lipschitz, see [3]. Therefore, we cannot work in the classical framework of critical point theory. Nevertheless, the Gâteaux-derivative of J exists in the smooth directions, i.e. for all $u \in E$ and $\varphi \in C_c^\infty$ we can define

$$J'(u)[\varphi] = \int_{\mathbb{R}^n}\left(\sum_{i,j=1}^{n}\left[a_{ij}(x,u)D_iuD_j\varphi + \frac{1}{2}\frac{\partial a_{ij}}{\partial s}(x,u)D_iuD_ju\varphi\right]\right)$$
$$+\int_{\mathbb{R}^n}(b(x)u\varphi - f(x,u)\varphi)\,.$$

According to the Degiovanni nonsmooth critical point theory, critical points u of J satisfy $J'(u)[\varphi] = 0$ for all $\varphi \in C_c^\infty$ and hence solve (4.31) in distributional sense; moreover, since

$$-\sum_{i,j=1}^{n} D_j(a_{ij}(x,u)D_iu) + b(x)u - f(x,u) \in E^*$$

we also have

$$\frac{1}{2}\sum_{i,j=1}^{n}\frac{\partial a_{ij}}{\partial s}(x,u)D_iuD_j(\cdot) \in E^*$$

and (4.31) is solved in the weak sense. We refer to [3] for the adaptation of this theory to quasilinear equations of the kind of (4.31). Under suitable assumptions on a_{ij}, b, f and by using the above mentioned tools we will prove that (4.31) admits a positive weak solution.

Next, we take into account the case where f is not continuous: let $f(x,\cdot) \in L^\infty_{\rm loc}(\mathbb{R})$ and denote

$$\underline{f}(x,s) = \lim_{\varepsilon \searrow 0} \text{ essinf } \{f(x,t) \, : \, |t-s| < \varepsilon\}$$

$$\overline{f}(x,s) = \lim_{\varepsilon \searrow 0} \text{ esssup } \{f(x,t) \, : \, |t-s| < \varepsilon\}\,.$$

Our aim is to determine $u \in E$ such that

$$-\Delta u + b(x)u \in [\underline{f}(x,u), \overline{f}(x,u)] \quad \text{in } \mathbb{R}^n\,. \tag{4.32}$$

Positive solutions u of (4.32) satisfy $0 \in \partial I(u)$, where

$$I(u) \;=\; \frac{1}{2}\int_{\mathbb{R}^n} (|Du|^2 + b(x)u^2) \;-\; \int_{\mathbb{R}^n} F(x,u^+), \quad \forall u \in E$$

and $\partial I(u)$ stands for the Clarke gradient of the locally Lipschitz energy functional I. This problem may be reformulated, equivalently, in terms of hemivariational inequalities as follows: find $u \in E$ such that

$$\int_{\mathbb{R}^n} (DuDv + b(x)uv) + \int_{\mathbb{R}^n} (-F)^0(x,u;v) \geq 0, \quad \forall v \in E \tag{4.33}$$

where $(-F)^0(x,u;v)$ denotes the Clarke directional derivative of $(-F)$ at $u(x)$ with respect to $v(x)$. So, when $f(x,\cdot)$ is not continuous, Clarke's theory will enable us to prove that (4.33) admits a positive solution.

The two existence results we will state in what follows have several points in common: in both cases we first prove that the corresponding functional has a mountain pass geometry and that a PS sequence can be built at a suitable inf-max level. Then we prove that the PS sequence is bounded and that its weak limit is a solution of the problem considered. The final step is to prove that this solution is not the trivial one. To this end we use the concentration-compactness principle [11] and the behavior of the function b at infinity. However, the construction of a PS sequence and the proof that its weak limit is a solution are definitely different: they highlight the different tools existing in the two theories.

Let us first state our result concerning the problem (4.31). We require the coefficients a_{ij} $(i,j = 1,...,n)$ to satisfy

$$\begin{cases} a_{ij} \equiv a_{ji} \\ a_{ij}(x,\cdot) \in C^1(\mathbb{R}) \quad \text{for a.e. } x \in \mathbb{R}^n \\ a_{ij}(x,s), \ \frac{\partial a_{ij}}{\partial s}(x,s) \in L^\infty(\mathbb{R}^n \times \mathbb{R})\,. \end{cases} \tag{4.34}$$

The matrices $[a_{ij}(x,s)]$ and $[s\frac{\partial a_{ij}}{\partial s}(x,s)]$ are assumed to fulfill the following assumptions: there exists $\nu > 0$ such that

$$\sum_{i,j=1}^{n} a_{ij}(x,s)\xi_i\xi_j \geq \nu|\xi|^2 \quad \text{for a.e. } x \in \mathbb{R}^n,\ \forall s \in \mathbb{R},\ \forall \xi \in \mathbb{R}^n \tag{4.35}$$

and there exist $\mu \in (2, 2^*)$ and $\gamma \in (0, \mu - 2)$ such that

$$0 \leq s \sum_{i,j=1}^{n} \frac{\partial a_{ij}}{\partial s}(x,s)\xi_i\xi_j \leq \gamma \sum_{i,j=1}^{n} a_{ij}(x,s)\xi_i\xi_j\,, \tag{4.36}$$

for a.e. $x \in \mathbb{R}^n$, for all $s \in \mathbb{R}$ and for any $\xi \in \mathbb{R}^n$. We require that $b \in L^\infty_{\text{loc}}(\mathbb{R}^n)$ and that

$$\begin{cases} \exists \underline{b} > 0 \quad \text{such that} \quad b(x) \geq \underline{b} \quad \text{for a.e. } x \in \mathbb{R}^n \\ \text{ess} \lim_{|x|\to\infty} b(x) = +\infty\,. \end{cases} \tag{4.37}$$

Let μ be as in (4.36), assume that $f(x,s) \not\equiv 0$ and

$$\begin{cases} f : \mathbb{R}^n \times \mathbb{R} \to \mathbb{R} \text{ is a Carathéodory function,} \\ f(x,0) = 0 \quad \text{for a.e. } x \in \mathbb{R}^n, \\ 0 \leq \mu F(x,s) \leq sf(x,s), \quad \forall s \geq 0 \text{ and for a.e. } x \in \mathbb{R}^n\,. \end{cases} \tag{4.38}$$

Moreover, we require f to be subcritical in the sense that for any $\varepsilon > 0$ there exists $f_\varepsilon \in L^{\frac{2n}{n+2}}(\mathbb{R}^n)$ such that

$$|f(x,s)| \leq f_\varepsilon(x) + \varepsilon|s|^{\frac{n+2}{n-2}}\,, \quad \forall s \in \mathbb{R} \text{ and for a.e. } x \in \mathbb{R}^n\,. \tag{4.39}$$

Finally, for all $\delta \in (2, 2^*)$ define $q(\delta) = \frac{2n}{2n+(2-n)\delta}$. Then we assume that there exist $C \geq 0$, $\delta \in (2, 2^*)$, and $G \in L^{q(\delta)}(\mathbb{R}^n)$ such that

$$F(x,s) \leq G(x)|s|^\delta + C|s|^{2^*} \quad \forall s \in \mathbb{R} \text{ and for a.e. } x \in \mathbb{R}^n\,. \tag{4.40}$$

In the above assumption we could also consider the case $\delta = 2$. In this situation we also need that $\|G\|_{n/2}$ is sufficiently small.

Our first result is

Theorem 4.6 (Gazzola and Rădulescu [9]) Assume conditions (4.34)-(4.40) are fulfilled. Then (4.31) admits a positive weak solution $\bar{u} \in E$.

Let us turn to the problem (4.32). We assume that $f : \mathbb{R}^n \times \mathbb{R} \to \mathbb{R}$ is a (nontrivial) measurable function such that

$$|f(x,s)| \leq C(|s| + |s|^p) \quad \text{for a.e. } (x,s) \in \mathbb{R}^n \times \mathbb{R} \tag{4.41}$$

where C is a positive constant and $1 < p \leq \frac{n+2}{n-2}$. Here we do not assume that $f(x, \cdot)$ is continuous. Nevertheless, if we define $F(x, s) = \int_0^s f(x, t)dt$ we observe that F is a Carathéodory function which is locally Lipschitz with respect to the second variable. We also observe that the functional

$$\Psi(u) = \int_{\mathbb{R}^n} F(x, u)$$

is locally Lipschitz on E. Indeed, by (4.41), Hölder's inequality and the embedding $E \subset L^{p+1}$,

$$|\Psi(u) - \Psi(v)| \leq C(\|u\|_E, \|v\|_E)\|u - v\|_E$$

where $C(\|u\|_E, \|v\|_E) > 0$ depends only on $\max\{\|u\|_E, \|v\|_E\}$.

We impose to f the following additional assumptions

$$\lim_{\varepsilon \searrow 0} \operatorname{esssup} \left\{ \left| \frac{f(x, s)}{s} \right| \; : \; (x, s) \in \mathbb{R}^n \times] -\varepsilon, \varepsilon[\right\} = 0 \tag{4.42}$$

and there exists $\mu > 2$ such that

$$0 \leq \mu F(x, s) \leq s\underline{f}(x, s) \quad \text{for a.e. } (x, s) \in \mathbb{R}^n \times [0, +\infty[. \tag{4.43}$$

We shall prove

Theorem 4.7 (Gazzola and Rădulescu [9]) Under hypotheses (4.37), (4.41)-(4.43), problem (4.32) has at least a positive solution in E.

Remark 4.1 The couple of assumptions (4.39)-(4.40) is equivalent to the couple (4.41)-(4.42) in the sense that Theorems 4.6 and 4.7 hold under any one of these groups of assumptions.

It seems that it is not possible to use the above mentioned nonsmooth critical point theories to obtain an existence result for the quasilinear operator of (4.31) in the presence of a function f which is discontinuous with respect to the second variable. Indeed, to prove that critical points of J (in the sense of [8]) solve (4.31) in distributional sense, one needs, for all given $\varphi \in C_c^\infty$, the continuity of the map $u \mapsto J'(u)[\varphi]$ (see [3]). Even if $J \notin C^1(E)$, we have at least $J \in C^1(W^{1,p} \cap E)$ for $p \geq \frac{3n}{n+1}$. This smoothness property in a finer topology is in fact the basic (hidden) tool used in Theorem 1.5 in [3]. However, one cannot prove the boundedness of the PS sequences in the $W^{1,p}$ norm. On the other hand, the theory developed in [4]-[7] only applies to locally Lipschitz continuous functionals and therefore it does not allow to manage quasilinear operators as that in (4.31).

Proof of Theorem 4.6. By (4.34) and (4.36) we have

$$u \in E \Longrightarrow \sum_{i,j=1}^{n} \frac{\partial a_{ij}}{\partial s}(x,u) D_i u D_j u u \in L^1(\mathbb{R}^n) . \tag{4.44}$$

Hence $J'(u)[u]$ can be written in integral form. We first remark that positive solutions of (4.31) correspond to critical points of the functional $J_+ : E \to \mathbb{R}$ defined by

$$J_+(u) := \frac{1}{2} \int_{\mathbb{R}^n} \sum_{i,j=1}^{n} a_{ij}(x,u) D_i u D_j u + \frac{1}{2} \int_{\mathbb{R}^n} b(x) u^2 - \int_{\mathbb{R}^n} F(x, u^+) ,$$

where u^+ denotes the positive part of u, i.e. $u^+(x) = \max(u(x), 0)$.

Lemma 4.9 Let $u \in E$ satisfy $J'_+(u)[\varphi] = 0$ for all $\varphi \in C_c^\infty$. Then u is a weak positive solution of (4.31).

For the proof of this result we refer to [3]. Without loss of generality we can therefore suppose that

$$f(x,s) = 0 \quad \forall s \leq 0 \quad \text{for a.e. } x \in \mathbb{R}^n$$

and, from now on, we make this assumption. For simplicity we denote J instead of J_+.

Let us establish the following boundedness criterion which applies, in particular, to Palais-Smale sequences.

Lemma 4.10 Every sequence $\{u_m\} \subset E$ satisfying

$$|J(u_m)| \leq C_1 \quad \text{and} \quad |J'(u_m)[u_m]| \leq C_2 \|u_m\|_E$$

is bounded in E.

Proof. Consider $\{u_m\} \subset E$ such that $|J(u_m)| \leq C_1$, then by (4.38) we get

$$I_m := \frac{1}{2} \int_{\mathbb{R}^n} \sum_{i,j=1}^{n} a_{ij}(x, u_m) D_i u_m D_j u_m$$

$$-\frac{1}{\mu} \int_{\mathbb{R}^n} f(x, u_m) u_m + \frac{1}{2} \int_{\mathbb{R}^n} b(x) u_m^2 \leq C_1 .$$

By (4.44) we can evaluate $J'(u_m)[u_m]$ and by the assumptions we have

$$\left| \int_{\mathbb{R}^n} \sum_{i,j=1}^{n} a_{ij}(x,u_m) D_i u_m D_j u_m \right.$$
$$+ \frac{1}{2} \int_{\mathbb{R}^n} \sum_{i,j=1}^{n} \frac{\partial a_{ij}}{\partial s}(x,u_m) D_i u_m D_j u_m u_m$$
$$\left. - \int_{\mathbb{R}^n} f(x,u_m) u_m + \int_{\mathbb{R}^n} b(x) u_m^2 \right| \leq C_2 \|u_m\|_E .$$

Therefore, by (4.36) and computing $I_m - \frac{1}{\mu} J'(u_m)[u_m]$ we obtain

$$\frac{\mu - 2 - \gamma}{2\mu} \int_{\mathbb{R}^n} \sum_{i,j=1}^{n} a_{ij}(x,u_m) D_i u_m D_j u_m + \frac{\mu - 2}{2\mu} \int_{\mathbb{R}^n} b(x) u_m^2$$
$$\leq C_3 \|u_m\|_E + C_1 .$$

By (4.35) this yields $C_4 > 0$ such that $C_4 \|u_m\|_E^2 \leq C_3 \|u_m\|_E + C_1$ and the result follows. ∎

Let us denote by E_{loc} the space of functions u satisfying $\int_\omega (|Du|^2 + b(x)u^2) < \infty$ for all bounded open set $\omega \subset \mathbb{R}^n$ and by E_{loc}^* its dual space. We establish that the weak limit of a PS sequence solves (4.31).

Lemma 4.11 Let $\{u_m\}$ be a bounded sequence in E satisfying

$$\int_{\mathbb{R}^n} \sum_{i,j=1}^{n} a_{ij}(x,u_m) D_i u_m D_j \varphi$$
$$+ \frac{1}{2} \int_{\mathbb{R}^n} \sum_{i,j=1}^{n} \frac{\partial a_{ij}}{\partial s}(x,u_m) D_i u_m D_j u_m \varphi = \langle \beta_m, \varphi \rangle ,$$

for any $\varphi \in C_c^\infty$, with $\{\beta_m\}$ converging in E_{loc}^* to some $\beta \in E_{\text{loc}}^*$. Then, up to a subsequence, $\{u_m\} \subset E$ converges in E_{loc} to some $u \in E$ satisfying

$$\int_{\mathbb{R}^n} \sum_{i,j=1}^{n} a_{ij}(x,u) D_i u D_j \varphi + \frac{1}{2} \int_{\mathbb{R}^n} \sum_{i,j=1}^{n} \frac{\partial a_{ij}}{\partial s}(x,u) D_i u D_j u \varphi = \langle \beta, \varphi \rangle .$$

for any $\varphi \in C_c^\infty$.

Proof. As b is uniformly positive and locally bounded, for all bounded open set $\omega \subset \mathbb{R}^n$ we have

$$\int_\omega (|Du|^2 + b(x)u^2) < \infty \quad \Longleftrightarrow \quad \int_\omega (|Du|^2 + u^2) < \infty .$$

From now on it is sufficient to follow the same lines as those developed in [3]. ■

The previous results allow us to prove

Proposition 4.1 Assume that $\{u_m\} \subset E$ is a PS sequence for J. Then there exists $\bar{u} \in E$ such that (up to a subsequence)
(i) $u_m \rightharpoonup \bar{u}$ in E
(ii) $u_m \to \bar{u}$ in E_{loc}
(iii) $\bar{u} \geq 0$ and $\bar{u}$ solves (4.31) in weak sense.

Proof. By Lemma 4.10, the sequence $\{u_m\}$ is bounded and (i) follows. To obtain (ii) it suffices to apply Lemma 4.11 with $\beta_m = \alpha_m + f(x, u_m) - b(x)u_m \in E^*$ where $\alpha_m \to 0$ in E^*. Indeed, if $u_m \rightharpoonup u$ in E, then $\beta_m \to \beta$ in E^*_{loc} with $\beta = f(x,u) - b(x)u$. Finally, (iii) follows from Lemmas 4.9 and 4.11. ■

In order to build a Palais-Smale sequence for the functional J we apply the Mountain Pass Theorem for locally Lipschitz functionals. Let us check that J has such a geometrical structure. We first notice that $J(0) = 0$. Since the function F is superquadratic at $+\infty$, we may choose a nonnegative function e such that

$$e \in C_c^\infty \quad e \geq 0 \quad \text{and} \quad J(te) < 0 \quad \forall t > 1\,.$$

Moreover, it is easy to check that there exist $\rho, \beta > 0$ such that $\rho < \|e\|_E$ and $J(u) \geq \beta$ if $\|u\|_E = \rho$. Indeed, by (4.4), we infer that

$$\int_{\mathbb{R}^n} F(x,u) \leq \|G\|_{q(\delta)} \|u\|_{2^*}^{\delta} + C\|u\|_{2^*}^{2^*}\,.$$

Hence, by (4.35),

$$J(u) \geq C_1 \|u\|_E^2 - C_2 \|u\|_E^{\delta} - C_3 \|u\|_E^{2^*}$$

and the existence of ρ, β follows. So, J has a mountain pass geometry. Set

$$\Gamma := \{\gamma \in C([0,1]; E) \,:\, \gamma(0) = 0,\ \gamma(1) = e\}$$

and

$$\alpha := \inf_{\gamma \in \Gamma} \max_{t \in [0,1]} J(\gamma(t))\,.$$

The existence of a Palais-Smale sequence for J at level α follows by the results of Degiovanni and Marzocchi [8]. We have so proved

Proposition 4.2 The functional J admits a PS sequence $\{u_m\}$ at level α.

As we are on an unbounded domain, the problem lacks compactness and we cannot infer that the above Palais-Smale sequence converges strongly. However, by Proposition 4.1, the weak limit $\bar{u}$ of the Palais-Smale sequence is a nonnegative solution of (4.31). The main problem is that it could be $\bar{u} \equiv 0$. To prove that this is not the case we make use of the following technical result.

Lemma 4.12 There exist $p \in (2, 2^*)$ and $C > 0$ such that $\|u_m^+\|_p \geq C$.

Proof. Using the relations $J'(u_m)[u_m] = o(1)$ and $J(u_m) = \alpha + o(1)$, by assumptions (4.36) and (4.38) we have

$$\begin{aligned} 2\alpha &= 2J(u_m) - J'(u_m)[u_m] + o(1) \\ &= \int_{\mathbb{R}^n} [f(x, u_m^+)u_m - 2F(x, u_m^+)] \\ &\quad -\frac{1}{2}\int_{\mathbb{R}^n} \sum_{i,j=1}^{n} \frac{\partial a_{ij}}{\partial s}(x, u_m) D_i u_m D_j u_m u_m + o(1) \\ &\leq \int_{\mathbb{R}^n} f(x, u_m^+)u_m + o(1)\,. \end{aligned}$$

Then, by (4.39), for all $\varepsilon > 0$ there exists $f_\varepsilon \in L^{\frac{2n}{n+2}}(\mathbb{R}^n)$ such that

$$2\alpha \leq \int_{\mathbb{R}^n} |f_\varepsilon(x)u_m^+(x)| + \varepsilon\|u_m^+\|_{2^*}^{2^*}\,.$$

Since $\|u_m\|_{2^*}$ is bounded, one can choose $\varepsilon > 0$ so that

$$\alpha \leq \int_{\mathbb{R}^n} |f_\varepsilon(x)u_m^+(x)|\,. \tag{4.45}$$

Now take $r \in]\frac{2n}{n+2}, 2[$. Then for all $\delta > 0$ there exist $f_\delta \in L^r$ and $f^\delta \in L^{\frac{2n}{n+2}}$ such that

$$f_\varepsilon = f_\delta + f^\delta \quad \text{and} \quad \|f^\delta\|_{\frac{2n}{n+2}} \leq \delta\,.$$

Then, by (4.45) and Hölder's inequality we infer that

$$\alpha \leq \|f_\delta\|_r \|u_m^+\|_p + \delta\|u_m^+\|_{2^*}\,,$$

where $p = \frac{r}{r-1}$. Since $\|u_m\|_{2^*}$ is bounded, one can choose $\delta > 0$ so that

$$\frac{\alpha}{2} \leq \|f_\delta\|_r \|u_m^+\|_p$$

and the result follows. ■

By the previous lemma we deduce that the sequence $\{u_m^+\}$ does not converge strongly to 0 in L^p. Taking into account that $\|u_m^+\|_2$ and

$\|\nabla u_m^+\|_2$ are bounded, by Lemma I.1 p. 231 in [11], we infer that the sequence $\{u_m^+\}$ "does not vanish" in L^2, i.e. there exists a sequence $\{y_m\} \subset I\!R^n$ and $C > 0$ such that

$$\int_{y_m+B_R} |u_m^+|^2 \geq C\,, \tag{4.46}$$

for some R. We claim that the sequence $\{y_m\}$ is bounded. If not, up to a subsequence, it follows by (4.38) that

$$\int_{I\!R^n} b(x)u_m^2 \to +\infty$$

which contradicts $J(u_m) = \alpha + o(1)$. Therefore, by (4.37), there exists an open bounded set $\omega \subset I\!R^n$ such that

$$\int_\omega |u_m|^2 \geq C > 0\,. \tag{4.47}$$

So, consider the Palais-Smale sequence found in Proposition 4.2. by Proposition 4.1, this sequence converges in the L^2_{loc} topology to some nonnegative function $\bar{u}$ which solves (4.31) in weak sense. Finally, (4.47) entails $\bar{u} \not\equiv 0$. ■

Proof of Theorem 4.7. We assume that hypotheses (4.37), (4.41)-(4.43) are fulfilled. Moreover, we set $f(x,s) \equiv 0$ for $s \leq 0$.

To prove Theorem 4.7, it is sufficient to show that the functional I has a critical point $u_0 \in \mathcal{C}$, $\mathcal{C}$ being the cone of positive functions of E. Indeed,

$$\partial I(u) = -\Delta u + b(x)u - \partial\Psi(u) \quad \text{in } E^*$$

and, by Theorem 4.3, we have

$$\partial\Psi(u) \subset [\underline{f}(x,u(x)), \overline{f}(x,u(x))] \quad \text{for a.e. } x \in I\!R^n$$

in the sense that if $w \in \partial\Psi(u)$ then

$$\underline{f}(x,u(x)) \leq w(x) \leq \overline{f}(x,u(x)) \quad \text{for a.e. } x \in I\!R^n\,. \tag{4.48}$$

Thus, if u_0 is a critical point of I, then there exists $w \in \partial\Psi(u_0)$ such that

$$-\Delta u_0 + b(x)u_0 = w \quad \text{in } E^*\,.$$

The existence of u_0 will be justified by the nonsmooth variant of the Mountain Pass Theorem, even if the Palais-Smale condition is not fulfilled. More precisely, we verify the following geometric hypotheses:

$$I(0) = 0 \text{ and } \exists v \in E \text{ such that } I(v) \leq 0, \tag{4.49}$$

$$\exists \beta, \rho > 0 \quad \text{such that} \quad I \geq \beta \quad \text{on} \quad \{u \in E \,:\, \|u\|_E = \rho\}\,. \tag{4.50}$$

Verification of (4.49). It is obvious that $I(0) = 0$. For the second assertion we need

Lemma 4.13 There exist two positive constants C_1 and C_2 such that

$$f(x,s) \geq C_1 s^{\mu-1} - C_2 \quad \text{for a.e. } (x,s) \in \mathbb{R}^n \times [0, +\infty[\,. \tag{4.51}$$

Proof. From the definition we clearly have

$$\underline{f}(x,s) \leq f(x,s) \quad \text{a.e. in } \mathbb{R}^n \times [0,+\infty[\,. \tag{4.52}$$

Then, by (4.43),

$$0 \leq \mu \underline{F}(x,s) \leq s\underline{f}(x,s) \quad \text{for a.e. } (x,s) \in \mathbb{R}^n \times [0,+\infty[\tag{4.53}$$

where

$$\underline{F}(x,s) = \int_0^s \underline{f}(x,t)dt\,.$$

By (4.53), there exist $R > 0$ and $K_1 > 0$ such that

$$\underline{F}(x,s) \geq K_1 s^{\mu} \quad \text{for a.e. } (x,s) \in \mathbb{R}^n \times [R, +\infty)\,. \tag{4.54}$$

The inequality (4.51) follows now by (4.52), (4.53) and (4.54). ■

Verification of (4.49) continued. Choose $v \in C_c^{\infty}(\mathbb{R}^n) \setminus \{0\}$ so that $v \geq 0$ in $\mathbb{R}^n$. We obviously have

$$\int_{\mathbb{R}^n} (|Dv|^2 + b(x)v^2) < +\infty\,.$$

Then, by Lemma 4.13,

$$I(tv) = \frac{t^2}{2} \int_{\mathbb{R}^n} (|Dv|^2 + b(x)v^2) - \Psi(tv)$$

$$\leq \frac{t^2}{2} \int_{\mathbb{R}^n} (|Dv|^2 + b(x)v^2) + C_2 t \int_{\mathbb{R}^n} v - C_1' t^{\mu} \int_{\mathbb{R}^n} v^{\mu} < 0\,,$$

for $t > 0$ large enough. ■

Verification of (4.50). We first observe that (4.41) and (4.42) imply that, for any $\varepsilon > 0$, there exists a constant A_ε such that

$$|f(x,s)| \leq \varepsilon |s| + A_\varepsilon |s|^p \quad \text{for a.e. } (x,s) \in \mathbb{R}^n \times \mathbb{R}\,. \tag{4.55}$$

By (4.55) and Sobolev embeddings it follows that for any $u \in E$

$$\Psi(u) \leq \frac{\varepsilon}{2} \int_{\mathbb{R}^n} u^2 + \frac{A_\varepsilon}{p+1} \int_{\mathbb{R}^n} |u|^{p+1} \leq \varepsilon C_3 \|u\|_E^2 + C_4 \|u\|_E^{p+1},$$

where ε is arbitrary and $C_4 = C_4(\varepsilon)$. Thus, by (4.37),

$$I(u) = \frac{1}{2} \int_{\mathbb{R}^n} (|Du|^2 + b(x)u^2) - \Psi(u)$$

$$\geq C_5 \|u\|_E^2 - \varepsilon C_3 \|u\|_E^2 - C_4 \|u\|_E^{p+1} \geq \beta > 0$$

for $\|u\|_E = \rho$, with ρ, ε and β sufficiently small positive constants. ∎

Denote

$$\Gamma = \{\gamma \in C([0,1], E) \,:\, \gamma(0) = 0,\ \gamma(1) \neq 0 \text{ and } I(\gamma(1)) \leq 0\}$$

and

$$c = \inf_{\gamma \in \Gamma} \max_{t \in [0,1]} I(\gamma(t)) .$$

Set

$$\lambda_I(u) = \min_{\zeta \in \partial I(u)} \|\zeta\|_{E^*} .$$

Then, by the Mountain Pass Theorem for locally Lipschitz functionals, there exists a sequence $\{u_m\} \subset E$ such that

$$I(u_m) \to c \quad \text{and} \quad \lambda_I(u_m) \to 0 . \tag{4.56}$$

Since $I(|u|) \leq I(u)$ for all $u \in E$ we may assume that $\{u_m\} \subset \mathcal{C}$. So, there exists a sequence $\{w_m\} \subset \partial\Psi(u_m) \subset E^*$ such that

$$-\Delta u_m + b(x)u_m - w_m \to 0 \quad \text{in } E^* . \tag{4.57}$$

Note that for all $u \in \mathcal{C}$, by (4.43) we have

$$\Psi(u) \leq \frac{1}{\mu} \int_{\mathbb{R}^n} u(x)\underline{f}(x, u(x)) .$$

Therefore, by (4.48), for every $u \in \mathcal{C}$ and any $w \in \partial\Psi(u)$,

$$\Psi(u) \leq \frac{1}{\mu} \int_{\mathbb{R}^n} u(x)w(x) .$$

Hence, if $\langle \cdot, \cdot \rangle$ denotes the duality pairing between E^* and E, we have

$$I(u_m) = \frac{\mu - 2}{2\mu} \int_{\mathbb{R}^n} (|Du_m|^2 + b(x)u_m^2)$$

$$+\frac{1}{\mu}\langle -\Delta u_m + bu_m - w_m, u_m\rangle + \frac{1}{\mu}\langle w_m, u_m\rangle - \Psi(u_m)$$

$$\geq \frac{\mu-2}{2\mu}\int_{I\!R^n}(|Du_m|^2 + b(x)u_m^2) + \frac{1}{\mu}\langle -\Delta u_m + bu_m - w_m, u_m\rangle$$

$$\geq \frac{\mu-2}{2\mu}\|u_m\|_E^2 - o(1)\|u_m\|_E .$$

This together with (4.56) implies that the Palais-Smale sequence $\{u_m\}$ is bounded in E. Thus it converges weakly (up to a subsequence) in E and strongly in $L^2_{\rm loc}$ to some $u_0 \in \mathcal{C}$. Taking into account that $w_m \in \partial\Psi(u_m)$ for all m, that $u_m \rightharpoonup u_0$ in E and that there exists $w_0 \in E^*$ such that $w_m \rightharpoonup w_0$ in E^* (up to a subsequence), we infer that $w_0 \in \partial\Psi(u_0)$. This follows from the fact that the map $u \mapsto F(x,u)$ is compact from E into L^1. Moreover, if we take $\varphi \in C_c^\infty(I\!R^n)$ and let $\Omega := \text{supp}\varphi$, then by (4.57),

$$\int_\Omega (Du_0 D\varphi + b(x)u_0\varphi - w_0\varphi) = 0 .$$

Since $w_0 \in \partial\Psi(u_0)$, it follows by (4) p. 104 in [4] and by definition of $(-F)^0$ that

$$\int_\Omega (Du_0 D\varphi + b(x)u_0\varphi) + \int_\Omega (-F)^0(x, u_0; \varphi) \geq 0 .$$

By density, this hemivariational inequality holds for all $\varphi \in E$ and (4.33) follows. This means that u_0 solves problem (4.32).

It remains to prove that $u_0 \not\equiv 0$. If w_m is as in (4.57), then by (4.48) (recall that $u_m \in \mathcal{C}$) and (4.56) (for large m) we get

$$\frac{c}{2} \leq I(u_m) - \frac{1}{2}\langle -\Delta u_m + bu_m - w_m, u_m\rangle$$

$$= \frac{1}{2}\langle w_m, u_m\rangle - \int_{I\!R^n} F(x, u_m) \leq \frac{1}{2}\int_{I\!R^n} u_m\overline{f}(x, u_m) . \qquad (4.58)$$

Now, taking into account its definition, one deduces that $\overline{f}$ verifies (4.52), too. So, by (4.55), we obtain

$$\frac{c}{2} \leq \frac{1}{2}\int_{I\!R^n}(\varepsilon|u_m|^2 + A_\varepsilon|u_m|^{p+1}) = \frac{\varepsilon}{2}\|u_m\|_2^2 + \frac{A_\varepsilon}{2}\|u_m\|_{p+1}^{p+1} .$$

Hence $\{u_m\}$ does not converge strongly to 0 in L^{p+1}. Frow now on, with the same arguments as in the proof of Theorem 4.6 (see after Lemma 4.12), we deduce that $u_0 \not\equiv 0$, which concludes our proof. ■

References

[1] A. Ambrosetti and P. Rabinowitz, Dual variational methods in critical point theory and applications, *J. Funct. Anal.* **14** (1973), 349-381.

[2] S. Berberian, *Measure and Integration*, MacMillan, 1967.

[3] A. Canino, Multiplicity of solutions for quasilinear elliptic equations, *Topol. Meth. Nonlin. Anal. 6* (1995), 357-370.

[4] K. C. Chang, Variational methods for non-differentiable functionals and their applications to partial differential equations, *J. Math. Anal. Appl.* **80** (1981), 102-129.

[5] F. H. Clarke, Generalized gradients and applications, *Trans. Amer. Math. Soc.* **205** (1975), 247-262.

[6] F. H. Clarke, Generalized gradients of Lipschitz functionals, *Advances in Mathematics* **40** (1981), 52-67.

[7] F. H. Clarke, *Optimization and Nonsmooth Analysis*, Willey, New York, 1983.

[8] M. Degiovanni and M. Marzocchi, A critical point theory for nonsmooth functionals, *Ann. Mat. Pura Appl.* **167** (1994), 73-100.

[9] F. Gazzola and V. Rădulescu, A nonsmooth critical point theory approach to some nonlinear elliptic problems in $\mathbb{R}^N$, *Differential and Integral Equations* **13** (2000), 47-60.

[10] E. A. Landesman and A. C. Lazer, Nonlinear perturbations of linear elliptic boundary value problems at resonance, *J. Math. Mech.* **19** (1976), 609-623.

[11] P. L. Lions, The concentration-compactness principle in the calculus of variations. The locally compact case, Ann. Inst. H. Poincaré, Analyse Non Linéaire **1** (1984), (I) 109-145, (II) 223-283.

[12] J. Mawhin and M. Willem, Multiple solutions of the periodic boundary value problem for some forced pendulum-type equations, *J. Diff. Equations* **52** (1984), 264-287.

[13] J. Mawhin and M. Willem, *Critical Point Theory and Hamiltonian Systems*, Springer-Verlag, Berlin, 1989.

[14] S. Minakshisundaran and A. Pleijel, Some properties of the eigenfunctions of the Laplace operator on Riemannian manifolds, *Canadian J. Math.* **1** (1949), 242-256.

[15] P. Mironescu and V. Rădulescu, A multiplicity theorem for locally Lipschitz periodic functionals, *J. Math. Anal. Appl.* **195** (1995), 621-637.

[16] R. Palais, Ljusternik-Schnirelmann theory on Banach manifolds, *Topology* **5** (1966), 115-132.

[17] A. Pleijel, On the eigenvalues and eigenfunctions of elastic plates, *Comm. Pure Appl. Math.* **3** (1950), 1-10.

[18] P. H. Rabinowitz, On a class of nonlinear Schrödinger equations, *Zeit. Angew. Math. Phys. (ZAMP)* **43** (1992), 270-291.

[19] V. Rădulescu, Mountain Pass theorems for non-differentiable functions and applications, *Proc. Japan Acad.* **69A** (1993), 193-198.

[20] V. Rădulescu, Locally Lipschitz functionals with the strong Palais-Smale property, *Revue Roum. Math. Pures Appl.* **40** (1995), 355-372.

[21] V. Rădulescu, A Ljusternik-Schnirelmann type theorem for locally Lipschitz functionals with applications to multivalued periodic problems, *Proc. Japan Acad.* **71A** (1995), 164-167.

[22] V. Rădulescu, Nontrivial solutions for a multivalued problem with strong resonance, *Glasgow Math. Journal* **38** (1996), 53-61.

[23] V. Rădulescu, Hemivariational inequalities associated to multivalued problems with strong resonance, in *Nonsmooth/Nonconvex Mechanics: Modeling, Analysis and Numerical Methods*, dedicated to the memory of Professor P.D. Panagiotopoulos, Eds.: D.Y. Gao, R.W. Ogden, G.E. Stavroulakis, Kluwer Academic Publishers, 2000, pp. 333-348.

[24] W. Rudin, *Functional Analysis*, Mc Graw-Hill, 1973.

[25] A. Szulkin, *Critical Point Theory of Ljusternik-Schnirelmann Type and Applications to Partial Differential Equations*, Sémin. Math. Sup., Presses Univ. Montréal, 1989.

[26] A. Szulkin and M. Willem, Eigenvalue problems with indefinite weight, *Studia Mathematica* **135** (1999), 191-201.

Chapter 5

BOUNDARY VALUE PROBLEMS IN NON-VARIATIONAL FORM

This Chapter is devoted to an initial boundary value problem for a parabolic inclusion with a multivalued nonlinearity given by a generalized gradient in the sense of Clarke [13] of some locally Lipschitz function. The elliptic operator is a general quasilinear operator of Leray-Lions type. Of special interest is the case where the multivalued term is described by the usual subdifferential of a convex function. Our main result is the existence of extremal solutions limited by prescribed lower and upper solutions. The main tools used in the proofs are abstract results on nonlinear evolution equations, regularization, comparison, truncation, and special test function techniques as well as the calculus with generalized gradients.

1. The General Setting and Assumptions

Given a bounded domain $\Omega \subset I\!R^N$ with Lipschitz boundary $\partial\Omega$, let us denote $Q = \Omega \times (0,\tau)$ and $\Gamma = \partial\Omega \times (0,\tau)$, for a fixed $\tau > 0$. We study the initial boundary value problem

$$\begin{cases} \dfrac{\partial u}{\partial t} + Au + \partial g(\cdot,\cdot,u) \ni Fu + h & \text{in } Q \\ u(\cdot,0) = 0 & \text{in } \Omega \\ u = 0 & \text{on } \Gamma. \end{cases} \tag{5.1}$$

Here the unknown is $u = u(x,t)$, $(x,t) \in Q$. In the statement of problem (5.1), A stands for a second order quasilinear differential operator in divergence form of Leray-Lions type, namely

$$Au(x,t) = -\sum_{i=1}^{N} \frac{\partial}{\partial x_i} a_i(x,t,u(x,t),\nabla u(x,t)),$$

where $\nabla u = (\frac{\partial u}{\partial x_1}, \ldots, \frac{\partial u}{\partial x_N})$ denotes the gradient. The notation F in (5.1) represents the Nemytski operator associated with a Carathéodory function $f : Q \times \mathbb{R} \to \mathbb{R}$. The function $g : Q \times \mathbb{R} \to \mathbb{R}$ entering (5.1) is supposed to satisfy: for every $s \in \mathbb{R}$, $g(\cdot, \cdot, s) : Q \to \mathbb{R}$ is measurable, and for a.e. $(x,t) \in Q$, $g(x,t,\cdot) : \mathbb{R} \to \mathbb{R}$ is locally Lipschitz. Therefore it is well-defined the generalized gradient in the sense of Clarke (cf. [13]) $\partial(g(x,t,\cdot))$ which is denoted simply by ∂g in (5.1). So the notation ∂g stands for the generalized gradient of g with respect to the third variable. The regularity of the function $h : Q \to \mathbb{R}$ will be specified later.

We considered in (5.1) homogeneous initial and boundary conditions only for the sake of simplicity. Making appropriate translations, a problem of form (5.1) with nonhomogeneous initial and boundary conditions can be reduced to the corresponding problem with homogeneous conditions, without changing the functional setting.

It is clear that problem (5.1) may be considered as a parabolic hemivariational inequality. For other different results concerning evolution hemivariational inequalities and mechanical problems governed by nonconvex, possibly nonsmooth energy functionals, called superpotentials, with nonmonotone, multivalued constitutive laws, we refer to the works [3]-[7], [14], [16]-[20], [24]. General theories of mechanical phenomena can be found in [12], [15], [21]-[23].

A basic qualitative question for problem (5.1) is the existence of extremal solutions. In the papers [5], [6], [9] this problem has been solved under the assumption that the multifunction of the inclusion is either given by the subdifferential of some convex function or Clarke's gradient of so called d.c.-functions. A related result dealing with a quasilinear elliptic problem with multivalued flux boundary conditions is given in [7].

The extension of extremality results to parabolic inclusions with a general Clarke's gradient has been done in Carl and Motreanu [10]. In this Chapter we present the main result in [10]. Precisely, it is shown the existence of extremal solutions to (5.1) within a sector of appropriately defined lower and upper solutions. Moreover, the compactness of the solution set within this sector is shown. A relevant particular case is obtained when the generalized gradient in (5.1) is the subdifferential of a convex function (see Carl and Motreanu [9]).

We precise now the functional setting for initial boundary value problem (5.1).

Let $2 \le p < \infty$ and q satisfy $\frac{1}{p} + \frac{1}{q} = 1$. We set

$$V = L^p(0, \tau; W^{1,p}(\Omega)).$$

The dual space of V is $V^* = L^q(0,\tau;(W^{1,p}(\Omega))^*)$. Denote

$$W = \{w \in V : \frac{\partial w}{\partial t} \in V^*\},$$

where the derivative $\partial/\partial t$ is understood in the sense of vector-valued distributions (see, e.g., [25]). It is known that W is a reflexive, separable Banach space endowed with the norm

$$\|w\|_W = \|w\|_V + \|\frac{\partial w}{\partial t}\|_{V^*}$$

and the embedding $W \subset L^p(Q)$ is compact (see [25]). Further, we introduce the spaces

$$V_0 = L^p(0,\tau;W_0^{1,p}(\Omega)),$$

with the dual $V_0^* = L^q(0,\tau;W^{-1,q}(\Omega))$, and

$$W_0 = \{w \in V_0 : \frac{\partial w}{\partial t} \in V_0^*\}.$$

Suppose $h \in V_0^*$.

We assume the following conditions on the coefficient functions a_i, $i = 1,\ldots,N$, entering the definition of the operator A.

(A1) $a_i : Q \times I\!R \times I\!R^N \to I\!R$ are Carathéodory functions, i.e. $a_i(\cdot,\cdot,s,\xi) : Q \to I\!R$ is measurable for all $(s,\xi) \in I\!R \times I\!R^N$ and $a_i(x,t,\cdot,\cdot) : I\!R \times I\!R^N \to I\!R$ is continuous for a.e. $(x,t) \in Q$. In addition, one has

$$|a_i(x,t,s,\xi)| \le k_0(x,t) + c_0\left(|s|^{p-1} + |\xi|^{p-1}\right)$$

for a.e. $(x,t) \in Q$ and for all $(s,\xi) \in I\!R \times I\!R^N$, for some constant $c_0 > 0$ and some function $k_0 \in L^q(Q)$.

(A2) $\sum_{i=1}^{N}(a_i(x,t,s,\xi) - a_i(x,t,s,\xi'))(\xi_i - \xi_i') > 0$
for a.e. $(x,t) \in Q$, for all $s \in I\!R$ and all $\xi, \xi' \in I\!R^N$ with $\xi \neq \xi'$.

(A3) $\sum_{i=1}^{N} a_i(x,t,s,\xi)\xi_i \ge \nu|\xi|^p - k_1(x,t)$
for a.e. $(x,t) \in Q$ and for all $(s,\xi) \in I\!R \times I\!R^N$, for some constant $\nu > 0$ and some function $k_1 \in L^1(Q)$.

(A4) $|a_i(x,t,s,\xi) - a_i(x,t,s',\xi)|$
$\le [k_2(x,t) + |s|^{p-1} + |s'|^{p-1} + |\xi|^{p-1}]\omega(|s - s'|)$

for a.e. $(x,t) \in Q$, for all $s, s' \in \mathbb{R}$ and all $\xi \in \mathbb{R}^N$, for some function $k_2 \in L^q(Q)$, where $\omega : [0,+\infty[\to [0,+\infty[$ is the modulus of continuity, satisfying

$$\int_{0+} \frac{1}{\omega(r)}\, dr = +\infty.$$

For example, in (A4) we can take $\omega(r) = cr$, with $c > 0$.

Let $\langle \cdot, \cdot \rangle$ denote the duality pairing between the Banach spaces V_0^* and V_0. Condition (A1) allows to associate to the operator A the semilinear form a by

$$\langle Au, \varphi \rangle := a(u,\varphi) = \sum_{i=1}^{N} \int_Q a_i(x,t,u,\nabla u) \frac{\partial \varphi}{\partial x_i}\, dx\, dt, \quad \forall u, \varphi \in V_0. \quad (5.2)$$

The semilinear form a is well-defined, and the operator $A : V_0 \to V_0^*$ is continuous and bounded.

In the following we use the notation $L_+^p(Q)$ for the positive cone of nonnegative elements of $L^p(Q)$, i.e.

$$L_+^p(Q) = \{u \in L^p(Q) \,:\, u \geq 0 \text{ a.e. on } Q\}.$$

The partial ordering in $L^p(Q)$ denoted by $u \leq v$ is defined by $v - u \in L_+^p(Q)$. Given $\underline{u}, \overline{u} \in W_0$ with $\underline{u} \leq \overline{u}$, the order interval formed by $\underline{u}$ and $\overline{u}$ is denoted by $[\underline{u}, \overline{u}]$, thus

$$[\underline{u}, \overline{u}] = \{u \in W_0 \,:\, \underline{u} \leq u \leq \overline{u}\}.$$

We now state the basic definition of (weak) solution to problem (5.1).

Definition 5.1 A function $u \in W_0$ is called a solution of problem (5.1) if $Fu \in L^q(Q)$ and if there is a function $v \in L^q(Q)$ such that

(i) $u(\cdot, 0) = 0$ in Ω,

(ii) $v(x,t) \in \partial g(x,t,u(x,t))$ for a.e. $(x,t) \in Q$,

(iii) $\langle \frac{\partial u}{\partial t}, \varphi \rangle + \langle Au, \varphi \rangle + \int_Q v(x,t)\varphi(x,t)\, dx\, dt$
$= \int_Q (Fu)(x,t)\varphi(x,t)\, dx\, dt + \langle h, \varphi \rangle$, for all $\varphi \in V_0$.

The corresponding extensions for the definitions of the upper and lower solutions in the case of single-valued equations to the multivalued problem (5.1) is given below.

Definition 5.2 A function $\overline{u} \in W$ is called an upper solution of problem (5.1) if $F\overline{u} \in L^q(Q)$ and if there is a function $\overline{v} \in L^q(Q)$ such that

(i) $\overline{u}(x,0) \geq 0$ in Ω and $\overline{u} \geq 0$ on Γ,

(ii) $\overline{v}(x,t) \in \partial g(x,t,\overline{u}(x,t))$ for a.e. $(x,t) \in Q$,

(iii) $$\langle \frac{\partial \overline{u}}{\partial t}, \varphi \rangle + \langle A\overline{u}, \varphi \rangle + \int_Q \overline{v}(x,t)\varphi(x,t)\, dx\, dt$$
$$\geq \int_Q (F\overline{u})(x,t)\, \varphi(x,t)\, dx\, dt + \langle h, \varphi \rangle, \text{ for all } \varphi \in V_0 \cap L^p_+(Q).$$

Similarly, a function $\underline{u} \in W$ is called a lower solution of problem (5.1) if the reversed inequalities hold in Definition 5.2 with $\overline{u}, \overline{v}$ replaced by $\underline{u}, \underline{v}$.

We additionally impose the following hypotheses on the data entering problem (5.1).

(H1) There exist an upper solution $\overline{u}$ and a lower solution $\underline{u}$ of problem (5.1) such that $\underline{u} \leq \overline{u}$.

(H2) The function $g : Q \times I\!R \to I\!R$ satisfies

(i) g is Borel measurable in $Q \times I\!R$,

(ii) $g(x,t,\cdot) : I\!R \to I\!R$ is locally Lipschitz and there exist constants $\alpha > 0$ and $c_1 \geq 0$ such that

$$\xi_1 \leq \xi_2 + c_1(s_2 - s_1)^{p-1}$$

for a.e. $(x,t) \in Q$, for all $\xi_i \in \partial g(x,t,s_i)$, $i = 1,2$, and for all s_1, s_2 with $\underline{u}(x,t) - \alpha \leq s_1 < s_2 \leq \overline{u}(x,t) + \alpha$.

(iii) There is a function $k_3 \in L^q_+(Q)$ such that

$$|z| \leq k_3(x,t)$$

for a.e. $(x,t) \in Q$, for all $s \in [\underline{u}(x,t) - 2\alpha, \overline{u}(x,t) + 2\alpha]$ and all $z \in \partial g(x,t,s)$, where α is the same positive constant as in (ii).

(H3) The function $f : Q \times I\!R \to I\!R$ is Carathéodory and there exists $k_4 \in L^q_+(Q)$ such that

$$|f(x,t,s)| \leq k_4(x,t)$$

for a.e. $(x,t) \in Q$, for all $s \in [\underline{u}(x,t), \overline{u}(x,t)]$.

We end this Section with the definition of extremal solutions for initial boundary value problem (5.1).

Definition 5.3 A solution u^* is said to be the greatest solution within $[\underline{u}, \overline{u}]$ if for any solution $u \in [\underline{u}, \overline{u}]$ we have $u \leq u^*$. Similarly, u_* is said to be the least solution within $[\underline{u}, \overline{u}]$ if for any solution $u \in [\underline{u}, \overline{u}]$ we have $u_* \leq u$. The least and greatest solutions are called the extremal solutions.

2. Extremal Solutions of Quasilinear Parabolic Inclusion (5.1)

Let $\mathcal{S}$ denote the set of the solutions of problem (5.1) enclosed by the lower and upper solutions $\underline{u}$, $\overline{u}$ respectively, which are fixed by assumption (H1). Therefore we have

$$\mathcal{S} = \{u \in W_0 \, : \, u \in [\underline{u}, \overline{u}] \text{ and } u \text{ is a solution of (5.1)}\}.$$

First of all, we have to show that $\mathcal{S}$ is nonempty. This is assured by the following result.

Proposition 5.1 Assume (A1)-(A4) and (H1)-(H3) be satisfied. Then problem (5.1) admits at least one solution u within the order interval $[\underline{u}, \overline{u}]$ formed by the given lower and upper solutions $\underline{u}$ and $\overline{u}$, respectively.

The proof of Proposition 5.1 is given in Section 3.

In view of Proposition 5.1 it follows that $\mathcal{S} \neq \emptyset$. We state the main result of this Chapter ensuring the existence of extremal solutions to problem (5.1).

Theorem 5.1 (Carl and Motreanu [10]) Assume that conditions (A1)-(A4) and (H1)-(H3) are satisfied. Then problem (5.1) admits extremal solutions within the order interval $[\underline{u}, \overline{u}]$ formed by the given lower and upper solutions $\underline{u}$ and $\overline{u}$, respectively.

In the rest of this Chapter we proceed to prove Theorem 5.1. Throughout this Section we assume that the hypotheses of Theorem 5.1 are fulfilled. We only prove the existence of the greatest solution of problem (5.1) because a similar reasoning leads to the existence of the least solution of problem (5.1).

The next lemma expresses that the set $\mathcal{S}$ is upward directed, i.e. whenever $u_1, u_2 \in \mathcal{S}$ there is a $u \in \mathcal{S}$ such that $u_1 \leq u$ and $u_2 \leq u$.

Lemma 5.1 For all $u_1, u_2 \in \mathcal{S}$, there exists a function $u \in \mathcal{S}$ satisfying

$$\max\{u_1, u_2\} \leq u.$$

Proof. The proof of Lemma 5.1 will be done in several steps.

Step 1: *Preliminaries.* Let $u_0 := \max\{u_1, u_2\}$. For $k = 0, 1, 2$ we define the truncation mapping T_k as follows

$$(T_k u)(x,t) = \begin{cases} \overline{u}(x,t) & \text{if} \quad u(x,t) > \overline{u}(x,t) \\ u(x,t) & \text{if} \quad u_k(x,t) \le u(x,t) \le \overline{u}(x,t) \\ u_k(x,t) & \text{if} \quad u(x,t) < u_k(x,t). \end{cases}$$

In addition, with the positive constant α given in (H2) (ii), we introduce the truncation operator T_α by

$$(T_\alpha u)(x,t) = \begin{cases} \overline{u}(x,t) + \alpha & \text{if} \quad u(x,t) > \overline{u}(x,t) + \alpha \\ u(x,t) & \text{if} \quad \underline{u}(x,t) - \alpha \le u(x,t) \le \overline{u}(x,t) + \alpha \\ \underline{u}(x,t) - \alpha & \text{if} \quad u(x,t) < \underline{u}(x,t) - \alpha. \end{cases}$$

It is known that the truncation operators T_k, $k = 0, 1, 2$, and T_α are continuous and bounded from V into V (see, e.g., [8]).

In order to make a regularization process, we fix a mollifier function $\rho : I\!R \to I\!R$, that is a smooth function $\rho \in C_0^\infty((-1,1))$ such that $\rho \ge 0$ and

$$\int_{-\infty}^{+\infty} \rho(s)\, ds = 1.$$

Then, for any $\varepsilon > 0$, we define the regularization (with respect to the third variable) g^ε of the function g given in the statement of problem (5.1), by using the convolution as follows

$$g^\varepsilon(x,t,s) = \frac{1}{\varepsilon} \int_{-\infty}^{+\infty} g(x,t,s-\zeta)\rho(\frac{\zeta}{\varepsilon})\, d\zeta.$$

Notice that the function g^ε admits the derivative with respect to the third variable $s \in I\!R$, which will be denoted by $(g^\varepsilon)'$. Let us then define the nonlinear operator $G_\alpha^\varepsilon : L^p(Q) \to L^q(Q)$ by

$$G_\alpha^\varepsilon u := (g^\varepsilon)'(\cdot,\cdot,(T_\alpha u)(\cdot,\cdot)). \tag{5.3}$$

The definition in (5.3) makes sense since, by (H2) (iii), $k_3 \in L^q(Q)$ and we have that

$$|(G_\alpha^\varepsilon u)(x,t)| = |(g^\varepsilon)'(x,t,(T_\alpha u)(x,t))| \le k_3(x,t) \tag{5.4}$$

for a.e. $(x,t) \in Q$, for all $u \in L^p(Q)$ and for all ε with $0 < \varepsilon < \alpha$. In order to show that (5.4) is true, we see from (H2) (iii) that

$$(g^\varepsilon)'(x,t,(T_\alpha u)(x,t)) \in \frac{1}{\varepsilon} \int_{-\infty}^{+\infty} \partial g(x,t,(T_\alpha u)(x,t) - \zeta)\rho(\frac{\zeta}{\varepsilon})\, d\zeta. \tag{5.5}$$

Here we used Aubin-Clarke Theorem (cf. [13]) whose application is possible due to the inequalities

$$\underline{u}(x,t) - 2\alpha \leq \underline{u}(x,t) - \alpha - \zeta \leq (T_\alpha u)(x,t) - \zeta$$
$$\leq \overline{u}(x,t) + \alpha - \zeta \leq \overline{u}(x,t) + 2\alpha.$$

Using again (H2) (iii) it results that

$$|(g^\varepsilon)'(x,t,(T_\alpha u)(x,t))| \leq \frac{1}{\varepsilon}\int_{-\infty}^{+\infty} k_3(x,t)\rho(\frac{\zeta}{\varepsilon})\, d\zeta = k_3(x,t),$$

i.e. (5.4) is true.

Next we introduce the cut-off function $b : Q \times I\!R \to I\!R$ by

$$b(x,t,s) = \begin{cases} (s - \overline{u}(x,t))^{p-1} & \text{if} \quad s > \overline{u}(x,t) \\ 0 & \text{if} \quad u_0(x,t) \leq s \leq \overline{u}(x,t) \\ -(u_0(x,t) - s)^{p-1} & \text{if} \quad s < u_0(x,t). \end{cases} \tag{5.6}$$

In view of (5.6) it follows that b is a Carathéodory function satisfying the growth condition

$$|b(x,t,s)| \leq k_5(x,t) + c_2|s|^{p-1} \tag{5.7}$$

for a.e. $(x,t) \in Q$ and for all $s \in I\!R$, where $c_2 > 0$ is a constant and $k_5 \in L^q(Q)$. Moreover, one has the following estimate

$$\int_Q b(x,t,u(x,t))\, u(x,t)\, dx\, dt \geq c_3\|u\|^p_{L^p(Q)} - c_4, \quad \forall u \in L^p(Q), \tag{5.8}$$

for some constants $c_3 > 0$ and $c_4 > 0$.

By (5.7), the Nemytski operator $B : L^p(Q) \to L^q(Q)$ defined by

$$Bu(x,t) = b(x,t,u(x,t)) \tag{5.9}$$

is continuous and bounded.

At this moment we are in a position to introduce for every $\varepsilon > 0$ the following regularized truncated problem

$$(P_\varepsilon) \qquad \begin{cases} \dfrac{\partial u}{\partial t} + Au + G^\varepsilon_\alpha u + \lambda Bu & \\ \qquad = F \circ T_0 u + \sum_{i=1}^2 |F \circ T_i u - F \circ T_0 u| + h & \text{in } Q \\ u(\cdot, 0) = 0 & \text{in } \Omega \\ u = 0 & \text{on } \Gamma, \end{cases}$$

where $\lambda > 0$ is any constant satisfying $\lambda > c_1$.

The existence, convergence and comparison properties of problem (P_ε) are studied in the next steps.

Step 2: *Existence of solutions of* (P_ε) $(0 < \varepsilon < \alpha)$. Consider the operator $L = \partial/\partial t : D(L) \subset V_0 \to V_0^*$, with the domain

$$D(L) = \{u \in W_0 \,:\, u(\cdot, 0) = 0 \text{ in } \Omega\},$$

defined by

$$\langle Lu, \varphi \rangle = \int_0^T \langle \frac{\partial u}{\partial t}(t), \varphi(t) \rangle_{W^{-1,q}(\Omega), W_0^{1,p}(\Omega)} \, dt, \quad \forall u \in D(L), \varphi \in V_0.$$

It is well-known that the linear operator L is closed, densely defined and maximal monotone (cf., e.g., [25]).

Fix $0 < \varepsilon < \alpha$. By means of the unbounded operator L, problem (P_ε) can be reformulated as follows

$$u \in D(L), \quad (L + A + G_\alpha^\varepsilon + \lambda B)u = Eu + h \;\text{ in }\; V_0^*, \tag{5.10}$$

where the operator $E : L^p(Q) \to L^q(Q)$ is defined by

$$Eu := F \circ T_0 u + \sum_{i=1}^{2} |F \circ T_i u - F \circ T_0 u|.$$

Using assumption (H3) and the continuity of the truncation operators T_k, $k = 0, 1, 2$, we have that the operator $E : L^p(Q) \to L^q(Q)$ is continuous and uniformly bounded. In addition, since the embedding $W_0 \subset L^p(Q)$ is compact, endowing the domain $D(L) \subset W_0$ of L with the graph norm

$$\|u\|_{D(L)} = \|u\|_{V_0} + \|Lu\|_{V_0^*} = \|u\|_{W_0},$$

we obtain that $E : D(L) \to L^q(Q) \subset V_0^*$ is completely continuous.

Similarly, using now (5.4) and the continuity of the truncation operator T_α, we derive that the operator $G_\alpha^\varepsilon : L^p(Q) \to L^q(Q)$ is continuous and uniformly bounded. Using the compactness of the embedding $W_0 \subset L^p(Q)$ yields that the continuous, bounded operators $G_\alpha^\varepsilon, B : D(L) \to L^q(Q) \subset V_0^*$ are completely continuous on $D(L)$ endowed with the graph norm topology.

The Leray-Lions conditions (A1)-(A3) and the properties of the operators G_α^ε, B, E imply that $A + G_\alpha^\varepsilon + \lambda B - E : D(L) \subset V_0 \to V_0^*$ is continuous, bounded and pseudo-monotone with respect to the graph norm topology of $D(L)$ (see, [8, Theorem E.3.2]). Thus the mapping $L + A + G_\alpha^\varepsilon + \lambda B - E : D(L) \to V_0^*$ is surjective provided that $A + G_\alpha^\varepsilon + \lambda B - E : V_0 \to V_0^*$ is coercive. Therefore, admitting the coerciveness of the nonlinear operator $A + G_\alpha^\varepsilon + \lambda B - E : V_0 \to V_0^*$, there exists at least a solution of problem (5.10), so it solves problem (P_ε).

It remains to show the coerciveness property of $A + G_\alpha^\varepsilon + \lambda B - E : V_0 \to V_0^*$. Towards this, using (5.2), (5.9), (A3), (5.8) as well as the uniformly boundedness of the operators G_α^ε and E, one has

$$\langle (A+G_\alpha^\varepsilon+\lambda B-E)u, u\rangle = \langle Au, u\rangle + \lambda\langle Bu, u\rangle_{L^q(Q),L^p(Q)} + \langle (G_\alpha^\varepsilon - E)u, u\rangle$$

$$\geq \sum_{i=1}^N \int_Q a_i(x,t,u,\nabla u)\frac{\partial u}{\partial x_i}\, dx\, dt + \lambda \int_Q b(x,t,u(x,t))\, u(x,t)\, dx\, dt$$

$$-\|(G_\alpha^\varepsilon - E)u\|_{V_0^*}\, \|u\|_{V_0}$$

$$\geq \nu \int_Q |\nabla u|^p\, dx\, dt - \int_Q k_1(x,t)\, dx\, dt + \lambda c_3 \|u\|_{L^p(Q)}^p - \lambda c_4 - \tilde{c}\|u\|_{V_0}$$

$$\geq \bar{c}\|u\|_{V_0}^p - c, \quad \forall u \in V_0, \tag{5.11}$$

where $\tilde{c}, \bar{c}, c$ are positive constants. Since $p \geq 2$, estimate (5.11) ensures that $A+G_\alpha^\varepsilon+\lambda B-E : V_0 \to V_0^*$ is coercive. Consequently, the existence of a solution of approximate problem (P_ε) is proved.

In the next step we are concerned with the convergence properties of the solution u_ε of problem (P_ε) as $\varepsilon \to 0$.

Step 3: *Convergence of solutions of* (P_ε). Let $\{\varepsilon_n\}$ be a sequence such that $\varepsilon_n \in (0,\alpha)$ and $\varepsilon_n \to 0$ as $n \to \infty$. We know from the previous step that for all n problem (P_{ε_n}) has at least a solution u_{ε_n}. For the sake of simplicity, we denote $u_n = u_{\varepsilon_n}$.

Let us show that the sequence $\{u_n\}$ is bounded in W_0. First, we remark that

$$\langle Lu, u\rangle = \int_0^\tau \langle \frac{\partial u}{\partial t}(t), u(t)\rangle_{W^{-1,q}(\Omega),W_0^{1,p}(\Omega)}\, dt$$

$$= \int_0^\tau \frac{\partial}{\partial t}\left(\frac{1}{2}\|u(t)\|_{L^2(\Omega)}^2\right) dt = \frac{1}{2}\|u(\tau)\|_{L^2(\Omega)}^2 \geq 0, \quad \forall u \in D(L). \tag{5.12}$$

Using that u_n is a solution of (P_{ε_n}), (5.12) and (5.11) with u_n in place of u we can write

$$\|h\|_{V_0^*}\, \|u_n\|_{V_0} \geq \langle h, u_n\rangle$$

$$= \langle Lu_n, u_n\rangle + \langle (A + G_\alpha^\varepsilon + \lambda B - E)u_n, u_n\rangle \geq \bar{c}\|u_n\|_{V_0}^p - c.$$

Since $p \geq 2$, the preceding inequality implies that $\{u_n\}$ is bounded in V_0.

Using again that u_n is a solution of (P_{ε_n}), we have

$$\frac{\partial u_n}{\partial t} = (-A - G_\alpha^\varepsilon - \lambda B + E)u_n + h \ \text{ in } \ V_0^*.$$

The boundedness of the sequence $\{u_n\}$ in V_0 ensures that the right-hand side in the previous equality is bounded in V_0^*. This implies that $(\frac{\partial u_n}{\partial t})$ is bounded in V_0^*, so the sequence $\{u_n\}$ is bounded in W_0.

In the following we justify that there is a subsequence of $\{u_n\}$ having the properties below

(*i*) $u_n \rightharpoonup u$ weakly in W_0, i.e. $u_n \rightharpoonup u$ weakly in V_0 and $\frac{\partial u_n}{\partial t} \rightharpoonup \frac{\partial u}{\partial t}$ weakly in V_0^* as $n \to \infty$,

(*ii*) $u_n \to u$ strongly in $L^p(Q)$ as $n \to \infty$,

(*iii*) $G_\alpha^{\varepsilon_n} u_n \rightharpoonup v$ weakly in $L^q(Q)$ as $n \to \infty$, where we have in addition that $v(x,t) \in \partial g(x,t,(T_\alpha u)(x,t))$ for a.e. $(x,t) \in Q$.

Property (*i*) is a consequence of the boundedness of $\{u_n\}$ in the reflexive Banach space W_0, while condition (*ii*) results from assertion (*i*) and the compactness of the embedding $W_0 \subset L^p(Q)$.

By (5.4) and (H2) (iii), the sequence $\{G_\alpha^{\varepsilon_n} u_n\}$ is bounded in $L^q(Q)$, thus along a relabelled subsequence we may admit that $G_\alpha^{\varepsilon_n} u_n \rightharpoonup v$ weakly in $L^q(Q)$, for some $v \in L^q(Q)$. In order to obtain (*iii*) we have to prove that $v(x,t) \in \partial g(x,t,(T_\alpha u)(x,t))$ for a.e. $(x,t) \in Q$, with $u \in W_0$ entering (*i*), (*ii*) and $v \in L^q(Q)$ in (*iii*).

To this end, let us first establish the following inequality

$$\int_Q \limsup_{n\to\infty} \left(\frac{1}{\varepsilon_n} \int_{-\infty}^{+\infty} g^0(x,t,(T_\alpha u_n)(x,t) - \zeta; w(x,t)) \rho(\frac{\zeta}{\varepsilon_n})\, d\zeta \right) dx\, dt$$

$$\geq \langle v, w\rangle_{L^q(Q),L^p(Q)}, \quad \forall w \in L^p(Q), \tag{5.13}$$

where the notation g^0 stands for the generalized directional derivative (in the sense of Clarke [13]) of g with respect to the third variable. For any $w \in L^p(Q)$, using (5.3), (5.5) and Proposition 2.1.2 in [13], we have that

$$\begin{aligned}
\langle G_\alpha^{\varepsilon_n} u_n, w\rangle_{L^q(Q),L^p(Q)} &= \langle (g^{\varepsilon_n})'(T_\alpha u_n), w\rangle_{L^q(Q),L^p(Q)} \\
&= \int_Q (g^{\varepsilon_n})'(x,t,(T_\alpha u_n)(x,t)) w(x,t)\, dx\, dt \\
&= \int_Q \left(\frac{1}{\varepsilon_n} \int_{-\infty}^{+\infty} z_n(x,t,\zeta) \rho(\frac{\zeta}{\varepsilon_n})\, d\zeta \right) w(x,t)\, dx\, dt \\
&\leq \int_Q \left(\frac{1}{\varepsilon_n} \int_{-\infty}^{+\infty} g^0(x,t,(T_\alpha u_n)(x,t) - \zeta; w(x,t)) \rho(\frac{\zeta}{\varepsilon_n})\, d\zeta \right) dx\, dt,
\end{aligned}$$

with $z_n(x,t,\zeta) \in \partial g(x,t,(T_\alpha u_n)(x,t) - \zeta)$. Passing to the upper limit in the inequality above and using the weak convergence $G_\alpha^{\varepsilon_n} u_n \rightharpoonup v$ in $L^q(Q)$ as well as Fatou's lemma (see, e.g., [2, p.54]) we derive

$$\langle v, w\rangle_{L^q(Q),L^p(Q)} = \lim_{n\to\infty} \langle G_\alpha^{\varepsilon_n} u_n, w\rangle_{L^q(Q),L^p(Q)}$$

$$\leq \limsup_{n\to\infty} \int_Q \left(\frac{1}{\varepsilon_n} \int_{-\infty}^{+\infty} g^0(x,t,(T_\alpha u_n)(x,t) - \zeta; w(x,t)) \rho(\frac{\zeta}{\varepsilon_n})\, d\zeta \right) dx\, dt$$

$$\leq \int_Q \limsup_{n\to\infty} \left(\frac{1}{\varepsilon_n} \int_{-\infty}^{+\infty} g^0(x,t,(T_\alpha u_n)(x,t) - \zeta; w(x,t)) \rho(\frac{\zeta}{\varepsilon_n})\, d\zeta \right) dx\, dt,$$

i.e. (5.13) holds true. The application of Fatou's lemma was possible due to the inequalities

$$\frac{1}{\varepsilon_n} \int_{-\infty}^{+\infty} g^0(x,t,(T_\alpha u_n)(x,t) - \zeta; w(x,t)) \rho(\frac{\zeta}{\varepsilon_n})\, d\zeta$$

$$= \frac{1}{\varepsilon_n} \int_{-\infty}^{+\infty} z_n(x,t,\zeta) w(x,t) \rho(\frac{\zeta}{\varepsilon_n})\, d\zeta$$

$$\leq \frac{1}{\varepsilon_n} \int_{-\infty}^{+\infty} k_3(x,t) w(x,t) \rho(\frac{\zeta}{\varepsilon_n})\, d\zeta = k_3(x,t) w(x,t),$$

with $k_3 w \in L^1(Q)$, and

$$\int_Q \left(\frac{1}{\varepsilon_n} \int_{-\infty}^{+\infty} g^0(x,t,(T_\alpha u_n)(x,t) - \zeta; w(x,t)) \rho(\frac{\zeta}{\varepsilon_n})\, d\zeta \right) dx\, dt$$

$$\geq - \int_Q \left(\frac{1}{\varepsilon_n} \int_{-\infty}^{+\infty} |z_n(x,t,\zeta)|\, |w(x,t)| \rho(\frac{\zeta}{\varepsilon_n})\, d\zeta \right) dx\, dt$$

$$\geq - \int_Q k_3(x,t) |w(x,t)|\, dx\, dt,$$

where $z_n(x,t,\zeta) \in \partial g(x,t,(T_\alpha u_n)(x,t) - \zeta)$ is fixed such that

$$g^0(x,t,(T_\alpha u_n)(x,t) - \zeta; w(x,t)) = z_n(x,t,\zeta) w(x,t).$$

Next we show that

$$\limsup_{n\to\infty} \left(\frac{1}{\varepsilon_n} \int_{-\infty}^{+\infty} g^0(x,t,(T_\alpha u_n)(x,t) - \zeta; w(x,t)) \rho(\frac{\zeta}{\varepsilon_n})\, d\zeta \right)$$

$$\leq g^0(x,t,(T_\alpha u)(x,t); w(x,t)), \quad \text{for a.e. } (x,t) \in Q,\ \forall w \in L^p(Q). \tag{5.14}$$

Towards the proof of (5.14) we note that, by *(ii)* and the continuity of truncation operator T_α, we get that $T_\alpha u_n \to T_\alpha u$ strongly in $L^p(Q)$ as $n \to \infty$. Then passing eventually to a relabelled subsequence it results that

$$(T_\alpha u_n)(x,t) \to (T_\alpha u)(x,t) \ \text{ for a.e. } (x,t) \in Q \ \text{ as } n \to \infty. \tag{5.15}$$

Thus to check (5.14) it is sufficient to show that (5.14) holds for every $w \in L^p(Q)$ and every point $(x,t) \in Q$ satisfying (5.15) (because (5.15) is valid for a.e. $(x,t) \in Q$). To this end, fix $w \in L^p(Q)$ and any point $(x,t) \in Q$ satisfying (5.15). Let us take an arbitrary number $\varepsilon > 0$. The upper semicontinuity of $g^0(x,t,\cdot;w(x,t))$ yields a number $\delta > 0$ such that for all ξ with $|\xi - (T_\alpha u)(x,t)| < \delta$ one has

$$g^0(x,t,\xi;w(x,t)) < g^0(x,t,(T_\alpha u)(x,t);w(x,t)) + \varepsilon. \tag{5.16}$$

On the other hand, the convergence in (5.15) yields a positive integer n_ε (depending on (x,t)) such that

$$|(T_\alpha u_n)(x,t) - \zeta - (T_\alpha u)(x,t)| \leq |(T_\alpha u_n)(x,t) - (T_\alpha u)(x,t)| + |\zeta|$$

$$\leq |(T_\alpha u_n)(x,t) - (T_\alpha u)(x,t)| + \varepsilon_n < \delta, \quad \forall n \geq n_\varepsilon, \ \forall \zeta \in (-\varepsilon_n, \varepsilon_n).$$

This enables us to apply (5.16) with $\xi = (T_\alpha u_n)(x,t) - \zeta$ to get

$$g^0(x,t,(T_\alpha u_n)(x,t) - \zeta; w(x,t)) < g^0(x,t,(T_\alpha u)(x,t);w(x,t)) + \varepsilon$$

for all $n \geq n_\varepsilon$ and all $\zeta \in (-\varepsilon_n, \varepsilon_n)$. Consequently, we may write

$$\frac{1}{\varepsilon_n}\int_{-\infty}^{+\infty} g^0(x,t,(T_\alpha u_n)(x,t) - \zeta; w(x,t))\rho(\frac{\zeta}{\varepsilon_n})\,d\zeta$$

$$= \frac{1}{\varepsilon_n}\int_{-\varepsilon_n}^{\varepsilon_n} g^0(x,t,(T_\alpha u_n)(x,t) - \zeta; w(x,t))\rho(\frac{\zeta}{\varepsilon_n})\,d\zeta$$

$$< g^0(x,t,(T_\alpha u)(x,t);w(x,t)) + \varepsilon.$$

Passing to the upper limit as $n \to \infty$ we derive that

$$\limsup_{n\to\infty}\left(\frac{1}{\varepsilon_n}\int_{-\infty}^{+\infty} g^0(x,t,(T_\alpha u_n)(x,t) - \zeta; w(x,t))\rho(\frac{\zeta}{\varepsilon_n})\,d\zeta\right)$$

$$\leq g^0(x,t,(T_\alpha u)(x,t);w(x,t)) + \varepsilon.$$

As $\varepsilon > 0$ was arbitrary, we conclude that (5.14) holds true.

Combining (5.13) and (5.14) it turns out that

$$\int_Q v(x,t)w(x,t)\,dx\,dt \leq \int_Q g^0(x,t,(T_\alpha u)(x,t);w(x,t))\,dx\,dt \tag{5.17}$$

for all $w \in L^p(Q)$. We use a Lebesgue's point argument in (5.17) for achieving (*iii*).

Specifically, let an arbitrarily fixed number $r \in \mathbb{R}$ and any open ball $B((\bar{x},\bar{t}),\eta)$ in Q centered at some fixed point $(\bar{x},\bar{t})$ and of radius $\eta > 0$. Denote by $\chi_{B((\bar{x},\bar{t}),\eta)}$ the characteristic function of the ball $B((\bar{x},\bar{t}),\eta)$.

Setting as a test function $w = \chi_{B((\bar{x},\bar{t}),\eta)}r$ in equality (5.17), we deduce that

$$\int_Q v(x,t)\chi_{B((\bar{x},\bar{t}),\eta)}(x,t)r\,dx\,dt$$
$$\leq \int_Q g^0(x,t,(T_\alpha u)(x,t);\chi_{B((\bar{x},\bar{t}),\eta)}(x,t)r)\,dx\,dt.$$

This inequality can be equivalently written as

$$\frac{1}{|B((\bar{x},\bar{t}),\eta)|}\int_{B((\bar{x},\bar{t}),\eta)} v(x,t)r\,dx\,dt$$
$$\leq \frac{1}{|B((\bar{x},\bar{t}),\eta)|}\int_{B((\bar{x},\bar{t}),\eta)} g^0(x,t,(T_\alpha u)(x,t);r)\,dx\,dt,$$

where $|B((\bar{x},\bar{t}),\eta)|$ denotes the measure of $B((\bar{x},\bar{t}),\eta)$. Since the functions v and $g^0(\cdot,\cdot,(T_\alpha u)(\cdot,\cdot);r)$ belong to $L^q(Q)$, letting $\eta \to 0$ in the previous inequality, we arrive at

$$v(\bar{x},\bar{t})r \leq g^0(\bar{x},\bar{t},(T_\alpha u)(\bar{x},\bar{t});r), \quad \forall r \in \mathbb{R}.$$

The definition of the generalized gradient leads to the conclusion

$$v(\bar{x},\bar{t}) \in \partial g(\bar{x},\bar{t},(T_\alpha u)(\bar{x},\bar{t})),$$

which completes the proof of assertion *(iii)*.

In order to pass to the weak limit as $n \to \infty$ in (P_{ε_n}) we first show that

$$Au_n \rightharpoonup Au \ \text{ weakly in } V_0^* \ \text{ as } n \to \infty. \tag{5.18}$$

To this end we shall use the pseudo-monotonicity of the operator $A : V_0 \to V_0^*$ with respect to the graph norm topology of the domain $D(L)$ of the linear operator L. Precisely, we claim that

$$\limsup_{n\to\infty}\langle Au_n, u_n - u\rangle \leq 0. \tag{5.19}$$

Taking into account (5.12) we have

$$\langle\frac{\partial u_n}{\partial t}, u_n - u\rangle = \langle\frac{\partial(u_n - u)}{\partial t}, u_n - u\rangle + \langle\frac{\partial u}{\partial t}, u_n - u\rangle \geq \langle\frac{\partial u}{\partial t}, u_n - u\rangle.$$

This inequality combined with the fact that u_n solves problem (P_{ε_n}) implies that

$$\langle\frac{\partial u}{\partial t}, u_n - u\rangle + \langle Au_n, u_n - u\rangle + \langle G_\alpha^\varepsilon u_n, u_n - u\rangle_{L^q(Q),L^p(Q)}$$
$$+\langle(\lambda B - E)u_n, u_n - u\rangle \leq \langle h, u_n - u\rangle.$$

Passing to the upper limit in the relation above and using properties (i)-(iii) as well as the fact that $\lambda B - E : D(L) \subset V_0 \to V_0^*$ is completely continuous with respect to the graph norm, we find that the claim in (5.19) is true.

The pseudo-monotonicity of $A : V_0 \to V_0^*$ with respect to the graph norm topology of the domain $D(L)$ of unbounded linear operator L in conjunction with $u_n \rightharpoonup u$ weakly in W_0 (see assertion (i)) and (5.19) implies (5.18) (cf., e.g., [1]).

Passing to the weak limit (in V_0^*) as $n \to \infty$ in problem (P_{ε_n}) and making use of the convergences (i), (5.18), (iii) as well as of the complete continuity of $\lambda B - E$ from $D(L) \subset W_0$ into V_0^* we conclude that u is a solution of the following problem

$$(P_0) \qquad \begin{cases} \dfrac{\partial u}{\partial t} + Au + v + \lambda Bu = Eu + h & \text{in } V_0^* \\ v \in \partial g(\cdot, \cdot, (T_\alpha u)(\cdot, \cdot)) & \text{a.e. in } Q. \end{cases}$$

Additionally, the operator L being closed, we have that its graph is closed and convex, thus weakly closed. This leads to the conclusion that $u \in D(L)$.

In the next step we show that the solution u of problem (P_0) satisfies the double inequality $u_0 \le u \le \overline{u}$.

Step 4: *Comparison* $u_0 \le u \le \overline{u}$. For proving $u_0 \le u$ we show that $u_k \le u$, $k = 1, 2$. Since $u_k \in \mathcal{S}$ it follows that for $k = 1, 2$, $u_k \in W_0$ and it verifies (5.1), thus

$$\begin{cases} \dfrac{\partial u_k}{\partial t} + Au_k + v_k = Fu_k + h & \text{in } V_0^* \\ v_k \in \partial g(\cdot, \cdot, u_k(\cdot, \cdot)) & \text{a.e. in } Q. \end{cases} \tag{5.20}$$

Substracting the equality in (P_0) from the one in (5.20) it results that

$$\frac{\partial (u_k - u)}{\partial t} + Au_k - Au + v_k - v - \lambda Bu$$

$$= Fu_k - F \circ T_0 u - \sum_{i=1}^{2} |F \circ T_i u - F \circ T_0 u| \text{ in } V_0^*. \tag{5.21}$$

By assumption (A4), for any fixed $\varepsilon > 0$ one can find a number $\delta(\varepsilon) \in (0, \varepsilon)$ such that

$$\int_{\delta(\varepsilon)}^{\varepsilon} \frac{1}{\omega(r)}\, dr = 1.$$

We define the function $\theta_\varepsilon : \mathbb{R} \to \mathbb{R}_+$ by

$$\theta_\varepsilon(s) = \begin{cases} 0 & \text{if} \quad s < \delta(\varepsilon) \\ \displaystyle\int_{\delta(\varepsilon)}^{s} \frac{1}{\omega(r)}\,dr & \text{if} \quad \delta(\varepsilon) \le s \le \varepsilon \\ 1 & \text{if} \quad s > \varepsilon. \end{cases}$$

The function θ_ε is Lipschitz continuous, nondecreasing and satisfies

$$\theta_\varepsilon \to \chi_{\{s>0\}} \quad \text{as} \quad \varepsilon \to 0, \tag{5.22}$$

where $\chi_{\{s>0\}}$ is the characteristic function of the set $\{s > 0\}$. In addition, one has

$$\theta'_\varepsilon(s) = \begin{cases} \dfrac{1}{\omega(s)} & \text{if} \quad \delta(\varepsilon) < s < \varepsilon \\ 0 & \text{if} \quad s \notin [\delta(\varepsilon), \varepsilon]. \end{cases}$$

Taking in the (weak) formulation of (5.21) the test function $\theta_\varepsilon(u_k - u) \in V_0 \cap L^p_+(Q)$ it follows

$$\langle \frac{\partial(u_k - u)}{\partial t}, \theta_\varepsilon(u_k - u)\rangle + \langle Au_k - Au, \theta_\varepsilon(u_k - u)\rangle$$

$$+ \int_Q (v_k - v)\theta_\varepsilon(u_k - u)\,dx\,dt - \lambda \int_Q (Bu)\theta_\varepsilon(u_k - u)\,dx\,dt$$

$$= \int_Q (Fu_k - F \circ T_0 u - \sum_{i=1}^{2} |F \circ T_i u - F \circ T_0 u|)\theta_\varepsilon(u_k - u)\,dx\,dt. \tag{5.23}$$

Let Θ_ε denote the primitive of the function θ_ε defined by

$$\Theta_\varepsilon(s) = \int_0^s \theta_\varepsilon(r)\,dr.$$

We obtain for the first term on the left-hand side of (5.23) (cf., e.g., [11]) that

$$\langle \frac{\partial(u_k - u)}{\partial t}, \theta_\varepsilon(u_k - u)\rangle = \int_\Omega \Theta_\varepsilon(u_k - u)(x, \tau)\,dx \ge 0. \tag{5.24}$$

Using assumptions (A4) and (A2), the second term on the left-hand side of (5.23) can be estimated as follows

$$\langle Au_k - Au, \theta_\varepsilon(u_k - u)\rangle$$

$$= \sum_{i=1}^{N} \int_Q (a_i(x, t, u_k, \nabla u_k) - a_i(x, t, u, \nabla u)) \frac{\partial}{\partial x_i} \theta_\varepsilon(u_k - u)\,dx\,dt$$

$$\geq \sum_{i=1}^{N} \int_Q (a_i(x,t,u_k,\nabla u_k) - a_i(x,t,u_k,\nabla u)) \frac{\partial (u_k - u)}{\partial x_i} \theta'_\varepsilon(u_k - u)\, dx\, dt$$

$$-N \int_Q (k_2 + |u_k|^{p-1} + |u|^{p-1}$$

$$+|\nabla u|^{p-1})\, \omega(|u_k - u|) \theta'_\varepsilon(u_k - u) |\nabla(u_k - u)|\, dx\, dt$$

$$\geq -N \int_{\{\delta(\varepsilon) < u_k - u < \varepsilon\}} \gamma |\nabla (u_k - u)|\, dx\, dt, \tag{5.25}$$

where $\gamma = k_2 + |u_k|^{p-1} + |u|^{p-1} + |\nabla u|^{p-1} \in L^q(Q)$. We also note that the right-hand side of (5.25) tends to zero as $\varepsilon \to 0$.

By (5.22), the application of Lebesgue's dominated convergence theorem implies

$$\lim_{\varepsilon \to 0} \int_Q (v_k - v - \lambda Bu - Fu_k + F \circ T_0 u + \sum_{i=1}^{2} |F \circ T_i u - F \circ T_0 u|) \theta_\varepsilon(u_k - u)\, dx\, dt$$

$$= \int_Q (v_k - v - \lambda Bu - Fu_k + F \circ T_0 u + \sum_{i=1}^{2} |F \circ T_i u - F \circ T_0 u|) \chi_{\{u_k > u\}}\, dx\, dt. \tag{5.26}$$

Using (5.24), (5.25) and passing to the limit as $\varepsilon \to 0$ in (5.23), the convergence in (5.26) and the definition of the truncation operators T_0, T_1, T_2 allow us to deduce

$$-\lambda \int_Q Bu \chi_{\{u_k > u\}}\, dx\, dt$$

$$\leq \int_Q (v - v_k + Fu_k - F \circ T_0 u - \sum_{i=1}^{2} |F \circ T_i u - F \circ T_0 u|) \chi_{\{u_k > u\}}\, dx\, dt$$

$$= \int_{\{u_k > u\}} (v - v_k + Fu_k - F \circ T_0 u - \sum_{i=1}^{2} |F \circ T_i u - F \circ T_0 u|)\, dx\, dt$$

$$\leq \int_{\{u_k > u\}} (v - v_k)\, dx\, dt. \tag{5.27}$$

If the point (x,t) is such that $u(x,t) < u_k(x,t)$, from the definition of T_α we see that $\underline{u}(x,t) - \alpha \leq (T_\alpha u)(x,t) < u_k(x,t) \leq \overline{u}(x,t) + \alpha$. Applying (H2) (ii) we derive

$$v(x,t) - v_k(x,t) \leq c_1 (u_k(x,t) - (T_\alpha u)(x,t))^{p-1},$$

with v in (iii) and v_k in (5.20). Combining the previous inequality with (5.27) and making use of (5.6), (5.9) we obtain that

$$\lambda \int_{\{u_k>u\}} (u_0 - u)^{p-1}\, dx\, dt = -\lambda \int_{\{u_k>u\}} Bu\, dx\, dt$$

$$\leq c_1 \int_{\{u_k>u\}} (u_k - T_\alpha u)^{p-1}\, dx\, dt.$$

Again by the definition of T_α, for (x,t) such that $u(x,t) < u_k(x,t)$, we have $(u_k - T_\alpha u)(x,t) \leq (u_0 - u)(x,t)$, which ensures that

$$(\lambda - c_1) \int_{\{u_k>u\}} (u_0 - u)^{p-1}\, dx\, dt \leq 0.$$

Since $c_1 < \lambda$ (see (H2) (ii)) and $(u_0 - u)(x,t) > 0$ whenever $(u_k - u)(x,t) > 0$, we infer from the previous inequality that the Lebesgue measure of the set $\{u_k > u\}$ is equal to 0. This implies that $u_k \leq u$ a.e. in Q, for $k = 1, 2$, thus $u_0 \leq u$.

In order to prove $u \leq \overline{u}$, we use Definition 5.2 and the fact that u is a solution of problem (P_0) to deduce

$$\langle \frac{\partial(u-\overline{u})}{\partial t}, \theta_\varepsilon(u-\overline{u})\rangle + \langle Au - A\overline{u}, \theta_\varepsilon(u-\overline{u})\rangle$$

$$+ \int_Q (v - \overline{v})\theta_\varepsilon(u-\overline{u})\, dx\, dt + \lambda \int_Q (Bu)\theta_\varepsilon(u-\overline{u})\, dx\, dt$$

$$\leq -\int_Q (F\overline{u} - F\circ T_0 u - \sum_{i=1}^{2} |F\circ T_i u - F\circ T_0 u|)\theta_\varepsilon(u-\overline{u})\, dx\, dt.$$

Using similar arguments as in proving (5.27), on the basis of (5.22) we obtain

$$\lambda \int_Q Bu\chi_{\{u>\overline{u}\}}\, dx\, dt \leq \int_{\{u>\overline{u}\}} (\overline{v} - v)\, dx\, dt.$$

If (x,t) is such that $u(x,t) > \overline{u}(x,t)$, we have that $\underline{u}(x,t) - \alpha \leq \overline{u}(x,t) < T_\alpha u(x,t) \leq \overline{u}(x,t) + \alpha$. Applying assumption (H2) (ii) we find that

$$\overline{v}(x,t) - v(x,t) \leq c_1(T_\alpha u(x,t) - \overline{u}(x,t))^{p-1},$$

with v in (iii) and $\overline{v}$ in Definition 5.2, (ii). Consequently, in view of (5.6), (5.9) we deduce that

$$\lambda \int_{\{u>\overline{u}\}} (u - \overline{u})^{p-1}\, dx\, dt \leq c_1 \int_{\{u>\overline{u}\}} (T_\alpha u - \overline{u})^{p-1}\, dx\, dt.$$

Since $T_\alpha u(x,t) \le u(x,t)$ whenever $u(x,t) > \overline{u}(x,t)$ it follows

$$(\lambda - c_1) \int_{\{u>\overline{u}\}} (u-\overline{u})^{p-1}\, dx\, dt \le 0.$$

In view of $c_1 < \lambda$ (see (H2) (ii)) we obtain that $u \le \overline{u}$ a.e. in Q.

Step 5: *Completion of the proof of Lemma 5.1.* By Step 4 it is known that any solution u of problem (P_0) satisfies $u_0 \le u \le \overline{u}$. Thus $Bu = 0$ and, since $T_i u = u$ for $i = 0, 1, 2$, one has $Eu = Fu$. In addition, we note that $v(x,t) \in \partial g(x,t,u(x,t))$ a.e. $(x,t) \in Q$ because $T_\alpha u = u$. Hence u is a solution of problem (5.1) satisfying $u_0 \le u \le \overline{u}$. The proof is complete. ■

The following lemma plays a basic role in the proof of Theorem 5.1.

Lemma 5.2 The solution set $\mathcal{S}$ is bounded in W_0. Any sequence in $\mathcal{S}$ contains a weakly convergent subsequence in W_0 and its limit belongs to $\mathcal{S}$.

Proof. Since $\mathcal{S} \subset [\underline{u}, \overline{u}]$, from assumption (H2) (iii) we see that the generalized gradient ∂g is bounded in $L^q(Q)$ on $[\underline{u}, \overline{u}]$, while assumption (H3) implies that F is bounded in $L^q(Q)$ on $[\underline{u}, \overline{u}]$.

We claim that $\mathcal{S}$ is bounded in W_0. The coerciveness condition (A3) for $A : V_0 \to V_0^*$ yields

$$\|u\|_{V_0} \le c', \quad \forall u \in \mathcal{S}, \tag{5.28}$$

for some constant $c' > 0$. Indeed, for any $u \in \mathcal{S}$ one has

$$\frac{\partial u}{\partial t} = -Au - v + Fu + h \quad \text{in } V_0^*, \tag{5.29}$$

with $v(x,t) \in \partial g(x,t,u(x,t))$ a.e. $(x,t) \in Q$. Then one obtains

$$\langle \frac{\partial u}{\partial t}, u\rangle + \langle Au, u\rangle = \langle Fu - v, u\rangle_{L^q(Q),L^p(Q)} + \langle h, u\rangle$$

Using (5.12), the boundedness of F, ∂g in $L^q(Q)$ on $[\underline{u}, \overline{u}]$ and (A3) we arrive at

$$M\|u\|_{V_0} \ge \langle Au, u\rangle = \sum_{i=1}^{N} \int_Q a_i(x,t,u,\nabla u)\frac{\partial u}{\partial x_i}\, dx\, dt$$

$$\ge \nu\|\nabla u\|^p_{L^p(Q)} - \|k_1\|_{L^1(Q)} = \nu\|u\|^p_{V_0} - \|k_1\|_{L^1(Q)},$$

for some constant $M > 0$. This proves (5.28).

By (5.28) and the boundedness of $A: V_0 \to V_0^*$ as well as of F, ∂g in $L^q(Q) \subset V_0^*$ we obtain from (5.29) that

$$\|\frac{\partial u}{\partial t}\|_{V_0^*} \leq c'', \quad \forall u \in \mathcal{S}, \tag{5.30}$$

for some constant $c'' > 0$. From (5.28) and (5.30) we infer the boundedness of $\mathcal{S}$ in W_0, which is the first part in Lemma 5.2.

Let a sequence $\{u_n\}$ in $\mathcal{S}$. By the reflexivity of W_0 we find a subsequence of $\{u_n\}$, denoted again by $\{u_n\}$, such that

$$\begin{aligned} &u_n \rightharpoonup u \ \text{ weakly in } W_0, \\ &u_n \to u \ \text{ strongly in } L^p(Q) \text{ and a.e. in } Q \ \text{ as } n \to \infty, \end{aligned} \tag{5.31}$$

for some $u \in W_0$, where the compactness of the embedding $W_0 \subset L^p(Q)$ has been used.

Since L is a closed linear operator, its graph is weakly closed, so the weak convergence $u_n \rightharpoonup u$ in W_0 implies $u \in D(L)$.

From the fact that $\{u_n\} \subset \mathcal{S}$ we have that $u_n \in W_0$ and

$$\frac{\partial u_n}{\partial t} + Au_n + v_n = Fu_n + h, \tag{5.32}$$

with $v_n \in \partial g(\cdot, \cdot, u_n(\cdot, \cdot))$. Hypothesis (H2) (iii) ensures that $\{v_n\}$ is bounded in $L^q(Q)$. Then there exists a subsequence of $\{v_n\}$, denoted by $\{v_n\}$, such that

$$v_n \rightharpoonup v \ \text{ weakly in } L^q(Q) \ \text{ as } n \to \infty, \tag{5.33}$$

for some $v \in L^q(Q)$.

Next we show that

$$v(\cdot, \cdot) \in \partial g(\cdot, \cdot, u(\cdot, \cdot)). \tag{5.34}$$

Using $v_n \rightharpoonup v$ weakly in $L^q(Q)$, $v_n \in \partial g(\cdot, \cdot, u_n(\cdot, \cdot))$, (5.31), Fatou's lemma and the upper semicontinuity of $g^0(x, t, \cdot; w(x, t)) : \mathbb{R} \to \mathbb{R}$, we deduce that

$$\int_Q v(x,t)w(x,t)\,dx\,dt = \lim_{n\to\infty} \int_Q v_n(x,t)w(x,t)\,dx\,dt$$

$$\leq \limsup_{n\to\infty} \int_Q g^0(x, t, u_n(x,t); w(x,t))\,dx\,dt$$

$$\leq \int_Q \limsup_{n\to\infty} g^0(x, t, u_n(x,t); w(x,t))\,dx\,dt$$

$$\leq \int_Q g^0(x,t,u(x,t);w(x,t))\,dx\,dt.$$

In order to use Lebesgue's point argument, fix $r \in I\!R$, $(\bar{x},\bar{t}) \in Q$, $\eta > 0$ and let $w = \chi_{B((\bar{x},\bar{t}),\eta)}r$ in the previous inequality, where $\chi_{B((\bar{x},\bar{t}),\eta)}$ denotes the characteristic function of the open ball $B((\bar{x},\bar{t}),\eta)$. We obtain

$$\frac{1}{|B((\bar{x},\bar{t}),\eta)|}\int_{B((\bar{x},\bar{t}),\eta)} v(x,t)r\,dx\,dt$$

$$\leq \frac{1}{|B((\bar{x},\bar{t}),\eta)|}\int_{B((\bar{x},\bar{t}),\eta)} g^0(x,t,u(x,t);r)\,dx\,dt,$$

where $|B((\bar{x},\bar{t}),\eta)|$ is the measure of $B((\bar{x},\bar{t}),\eta)$. Letting $\eta \to 0$ in the previous inequality we derive

$$v(\bar{x},\bar{t})r \leq g^0(\bar{x},\bar{t},u(\bar{x},\bar{t});r), \quad \forall r \in I\!R.$$

Using the definition of the generalized gradient, we deduce that (5.34) is satisfied.

From (5.32) it results

$$\langle\frac{\partial u_n}{\partial t}, u_n - u\rangle + \langle Au_n, u_n - u\rangle$$

$$= \langle Fu_n, u_n - u\rangle_{L^q(Q),L^p(Q)} - \langle v_n, u_n - u\rangle_{L^q(Q),L^p(Q)} + \langle h, u_n - u\rangle. \quad (5.35)$$

By (5.12) we get

$$\langle\frac{\partial u_n}{\partial t}, u_n - u\rangle = \langle\frac{\partial(u_n - u)}{\partial t}, u_n - u\rangle + \langle\frac{\partial u}{\partial t}, u_n - u\rangle \geq \langle\frac{\partial u}{\partial t}, u_n - u\rangle.$$

Using this inequality in (5.35) and passing to the upper limit as $n \to \infty$, on the basis of (5.31), (5.33) and the boundedness of $F(u_n)$ (see (H3)), we arrive at

$$\limsup_{n\to\infty}\langle Au_n, u_n - u\rangle \leq 0.$$

By the pseudo-monotonicity of A with respect to the graph norm topology of $D(L)$, this inequality and $u_n \rightharpoonup u$ weakly in W_0 imply that $Au_n \rightharpoonup Au$ weakly in V_0^* (cf., e.g., [1]). This allows us to pass to the limit as $n \to \infty$ in (5.32), obtaining

$$\frac{\partial u}{\partial t} + Au + v = Fu + h \text{ in } V_0^*.$$

As v satisfies (5.34) it follows that u is a solution of (5.1).

Combining $u_n \to u$ a.e. in Q (see (5.31)) with $u_n \in [\underline{u},\overline{u}]$ leads to $u \in [\underline{u},\overline{u}]$. Therefore $u \in \mathcal{S}$ and the proof is complete. ■

We proceed now for completing the proof of Theorem 5.1.

Proof of Theorem 5.1. We have to show the existence of the greatest solution of (5.1). Since W_0 is separable we have that $\mathcal{S} \subset W_0$ is separable, so there exists a countable, dense subset $Z = \{z_n : n \in I\!N\}$ of $\mathcal{S}$. By Lemma 5.1, $\mathcal{S}$ is upward directed, so we can construct an increasing sequence $\{u_n\} \subset \mathcal{S}$ as follows. Let $u_1 = z_1$. Select $u_{n+1} \in \mathcal{S}$ such that

$$\max\{z_n, u_n\} \leq u_{n+1} \leq \overline{u}.$$

The existence of u_{n+1} is due to Lemma 5.1. By Lemma 5.2 we find a subsequence of $\{u_n\}$, denoted again $\{u_n\}$, and an element $u \in \mathcal{S}$ such that $u_n \rightharpoonup u$ weakly in W_0, $u_n \to u$ strongly in $L^p(Q)$ and $u_n(x,t) \to u(x,t)$ a.e. $(x,t) \in Q$. This last property of $\{u_n\}$ in conjunction with its increasing monotonicity implies that $u = \sup_n u_n$. By construction, we see that

$$\max\{z_1, z_2, \ldots, z_n\} \leq u_{n+1} \leq u, \quad \forall n,$$

thus $Z \subset [\underline{u}, u]$. Since the interval $[\underline{u}, u]$ is closed in W_0, we infer that

$$\mathcal{S} \subset \overline{Z} \subset \overline{[\underline{u}, u]} = [\underline{u}, u],$$

which combined with $u \in \mathcal{S}$ ensures that u is the greatest solution of problem (5.1).

The existence of the least solution of (5.1) can be established in a similar way using Lemma 5.2 and a corresponding dual formulation of Lemma 5.1. This completes the proof. ∎

An important qualitative property of the solution set $\mathcal{S}$ is pointed out in the result below.

Corollary 5.1 The solution set $\mathcal{S}$ of initial boundary value problem (5.1) is weakly compact in W_0, and compact in V_0.

Proof. The weak compactness in W_0 is the contents of Lemma 5.2. We only need to show that $\mathcal{S}$ is compact in V_0. Let a sequence $\{u_n\} \subset \mathcal{S}$. Then we have to prove that there is a subsequence of $\{u_n\}$ which is strongly convergent in V_0 to some $u \in \mathcal{S}$. The weak compactness of $\mathcal{S}$ in W_0 implies the existence of a subsequence denoted by $\{u_k\}$ which is weakly convergent in W_0 to some $u \in \mathcal{S}$. Hypotheses (A1)-(A3) ensure that the operator A satisfies the (S_+)-property with respect to the graph norm topology of L (see [8, Theorem E.3.2]), which means that whenever (u_k) is weakly convergent to u in W_0 and satisfies $\limsup_{k\to\infty}\langle Au_k, u_k - u\rangle \leq 0$, then (u_k) is strongly convergent in V_0 to u. Since $\limsup_{k\to\infty}\langle Au_k, u_k - u\rangle \leq 0$ has already been shown in the proof of Lemma 5.2, the (S_+)-property of A immediately implies that the weak limit $u \in \mathcal{S}$ in W_0 of the sequence $\{u_k\}$ is its strong limit in V_0. Thus the compactness of the solution set $\mathcal{S}$ in V_0 is justified, which completes the proof. ■

3. Proof of the Existence Result in Proposition 5.1 and an Example

First, we deal with the proof of Proposition 5.1 in Section 1. The argument follows essentially the same steps as those in the proof of Lemma 5.1.

Proof of Proposition 5.1. Step 1: *Preliminaries.*

Consider the following regularized truncated problem

$$(\tilde{P}_\varepsilon) \qquad \begin{cases} \dfrac{\partial u}{\partial t} + Au + G_\alpha^\varepsilon u + \lambda B u = F \circ T u + h & \text{in } Q \\ u(\cdot, 0) = 0 & \text{in } \Omega \\ u = 0 & \text{on } \Gamma, \end{cases}$$

where λ is some constant satisfying $\lambda > c_1$. Here the operator $G_\alpha^\varepsilon : L^p(Q) \to L^q(Q)$ is the one in (5.3) and verifies (5.4) for $0 < \varepsilon < \alpha$. The operator $B : L^p(Q) \to L^q(Q)$ is defined by (5.9) for $b : Q \times I\!R \to I\!R$ given by

$$b(x,t,s) = \begin{cases} (s - \overline{u}(x,t))^{p-1} & \text{if } \; s > \overline{u}(x,t) \\ 0 & \text{if } \; \underline{u}(x,t) \le s \le \overline{u}(x,t) \\ -(\underline{u}(x,t) - s)^{p-1} & \text{if } \; s < \underline{u}(x,t). \end{cases}$$

On sees that b is a Carathéodory function satisfying (5.7) and (5.8). It follows that the operator B is continuous and bounded. Let us now introduce the truncation operator $T : V_0 \to V_0$ defined by

$$(Tu)(x,t) = \begin{cases} \overline{u}(x,t) & \text{if } \; u(x,t) > \overline{u}(x,t) \\ u(x,t) & \text{if } \; \underline{u}(x,t) \le u(x,t) \le \overline{u}(x,t) \\ \underline{u}(x,t) & \text{if } \; u(x,t) < \underline{u}(x,t). \end{cases}$$

It is straightforward to check that the operator T is continuous and bounded.

Step 2: *Existence of solutions to* $(\tilde{P}_\varepsilon)$ $(0 < \varepsilon < \alpha)$. For a fixed ε with $0 < \varepsilon < \alpha$, problem $(\tilde{P}_\varepsilon)$ can be reformulated as follows

$$u \in D(L), \; (L + A + G_\alpha^\varepsilon + \lambda B - F \circ T)u = h \; \text{ in } V_0^*,$$

where $L = \frac{\partial}{\partial t}$ is as in the proof of Lemma 5.1. Following the same arguments as the ones in the proof of Lemma 5.1 we see that the operator

$A + G_\alpha^\varepsilon + \lambda B - F \circ T$ is continuous, bounded, pseudo-monotone with respect to the graph norm of $D(L)$, and coercive, while L is maximal monotone (see [25]). Thus $L + A + G_\alpha^\varepsilon + \lambda B - F \circ T : D(L) \to V_0^*$ is surjective, so problem $(\tilde{P}_\varepsilon)$ has at least a solution.

Step 3: *Convergence of solutions of* $(\tilde{P}_{\varepsilon_n})$. Let a sequence $\{\varepsilon_n\}$ satisfy $\varepsilon_n \in (0, \alpha)$ and $\varepsilon_n \to 0$ as $n \to \infty$. For each n let u_n be a solution of problem $(\tilde{P}_{\varepsilon_n})$. The existence of such a solution is known from the previous step.

Using that u_n is a solution of $(\tilde{P}_{\varepsilon_n})$ in conjunction with (5.12) and (5.11) we obtain that the sequence $\{u_n\}$ is bounded in V_0. This combined with the equation in $(\tilde{P}_{\varepsilon_n})$ implies that $(\frac{\partial u_n}{\partial t})$ is bounded in V_0^*. Hence the sequence $\{u_n\}$ is bounded in W_0.

In the same way as in the proof of Lemma 5.1 we can show that the following properties hold:

(i) $u_n \rightharpoonup u$ weakly in W_0, i.e. $u_n \rightharpoonup u$ weakly in V_0 and $\frac{\partial u_n}{\partial t} \rightharpoonup \frac{\partial u}{\partial t}$ weakly in V_0^* as $n \to \infty$,

(ii) $u_n \to u$ strongly in $L^p(Q)$ as $n \to \infty$,

(iii) $G_\alpha^{\varepsilon_n} u_n \rightharpoonup v$ weakly in $L^q(Q)$ as $n \to \infty$, where $v(x,t) \in \partial g(x,t,(T_\alpha u)(x,t))$ for a.e. $(x,t) \in Q$.

On the basis of problem $(\tilde{P}_{\varepsilon_n})$ and by means of relation (5.12) we have

$$\langle \frac{\partial u}{\partial t}, u_n - u \rangle + \langle A u_n, u_n - u \rangle + \langle G_\alpha^\varepsilon u_n, u_n - u \rangle_{L^q(Q), L^p(Q)}$$

$$+ \langle (\lambda B - F \circ T) u_n, u_n - u \rangle \le \langle h, u_n - u \rangle.$$

Passing here to the upper limit as $n \to \infty$ and using properties (i)-(iii) as well as the fact that $\lambda B - F \circ T : D(L) \subset V_0 \to V_0^*$ is completely continuous with respect to graph norm topology, we obtain

$$\limsup_{n \to \infty} \langle A u_n, u_n - u \rangle \le 0.$$

Taking into account that $u_n \rightharpoonup u$ weakly in W_0, the pseudo-monotonicity of $A : V_0 \to V_0^*$ with respect to the graph norm of $D(L)$ yields

$$A u_n \rightharpoonup A u \text{ weakly in } V_0^* \text{ as } n \to \infty,$$

(see [1]). Letting now $n \to \infty$ in problem $(\tilde{P}_{\varepsilon_n})$ and making use of the above convergence as well as assertions (i), (iii) above and the complete continuity of $\lambda B - F \circ T$ from $D(L) \subset W_0$ into V_0^*, we conclude that u

is a solution of the problem

$$(\tilde{P}_0) \qquad \begin{cases} \dfrac{\partial u}{\partial t} + Au + v + \lambda Bu = F \circ Tu + h & \text{in } V_0^* \\ v \in \partial g(\cdot, \cdot, (T_\alpha u)(\cdot, \cdot)) & \text{a.e. in } Q. \end{cases}$$

In addition, the closedness of the linear operator L yields $u \in D(L)$.

Step 4: *Comparison* $\underline{u} \leq u \leq \overline{u}$. Let us first check that $\underline{u} \leq u$. Using the definition of the lower solution (see Definition 5.2) and the fact that u is a solution of problem $(\tilde{P}_0)$ it results that

$$\langle \frac{\partial(\underline{u} - u)}{\partial t}, \theta_\varepsilon(\underline{u} - u)\rangle + \langle A\underline{u} - Au, \theta_\varepsilon(\underline{u} - u)\rangle$$

$$+ \int_Q (\underline{v} - v)\theta_\varepsilon(\underline{u} - u)\, dx\, dt - \lambda \int_Q (Bu)\theta_\varepsilon(\underline{u} - u)\, dx\, dt$$

$$\leq \int_Q (F\underline{u} - F \circ Tu)\theta_\varepsilon(\underline{u} - u)\, dx\, dt,$$

with θ_ε as it was defined in Step 4 of Lemma 5.1. Proceeding in the same way as in proving (5.27), on the basis of the previous inequality and (5.22) we obtain

$$-\lambda \int_Q Bu\chi_{\{\underline{u}>u\}}\, dx\, dt \leq \int_Q (v - \underline{v} + F\underline{u} - F \circ Tu)\chi_{\{\underline{u}>u\}}\, dx\, dt$$

$$= \int_{\{\underline{u}>u\}} (v - \underline{v} + F\underline{u} - F \circ Tu)\, dx\, dt \leq \int_{\{\underline{u}>u\}} (v - \underline{v})\, dx\, dt. \qquad (5.36)$$

If the point (x,t) is such that $u(x,t) < \underline{u}(x,t)$, then we have the inequality $\underline{u}(x,t) - \alpha \leq T_\alpha u(x,t) < \underline{u}(x,t) \leq \overline{u}(x,t) + \alpha$. Hypothesis (H2) (ii) implies

$$v(x,t) - \underline{v}(x,t) \leq c_1(\underline{u}(x,t) - T_\alpha u(x,t))^{p-1},$$

with v in (*iii*) and $\underline{v} \in \partial g(\cdot, \cdot, \underline{u}(\cdot, \cdot))$. Using (5.6), (5.9), the previous inequality and (5.36), we deduce

$$\lambda \int_{\{\underline{u}>u\}} (\underline{u} - u)^{p-1}\, dx\, dt = -\lambda \int_{\{\underline{u}>u\}} Bu\, dx\, dt$$

$$\leq c_1 \int_{\{\underline{u}>u\}} (\underline{u} - T_\alpha u)^{p-1}\, dx\, dt.$$

For (x,t) such that $u(x,t) < \underline{u}(x,t)$, by the definition of T_α, we have $u(x,t) \leq (T_\alpha u)(x,t)$, thus

$$(\lambda - c_1) \int_{\{\underline{u}>u\}} (\underline{u} - u)^{p-1}\, dx\, dt \leq 0.$$

Since $c_1 < \lambda$ (see (H2) (ii)) it results that the Lebesgue measure of the set $\{\underline{u} > u\}$ is equal to 0. This implies that $\underline{u} \le u$ a.e. in Q.

In order to prove $u \le \overline{u}$, we use Definition 5.2 and that u solves problem $(\tilde{P}_0)$ to deduce

$$\langle \frac{\partial(u-\overline{u})}{\partial t}, \theta_\varepsilon(u-\overline{u})\rangle + \langle Au - A\overline{u}, \theta_\varepsilon(u-\overline{u})\rangle$$

$$+ \int_Q (v-\overline{v})\theta_\varepsilon(u-\overline{u})\, dx\, dt + \lambda \int_Q (Bu)\theta_\varepsilon(u-\overline{u})\, dx\, dt$$

$$\le \int_Q (F\circ Tu - F\overline{u})\theta_\varepsilon(u-\overline{u})\, dx\, dt.$$

Similar arguments as in proving (5.27), based on (5.22), yield

$$\lambda \int_Q Bu\chi_{\{u>\overline{u}\}}\, dx\, dt \le \int_{\{u>\overline{u}\}} (\overline{v} - v + F\circ Tu - F\overline{u})\, dx\, dt$$

$$\le \int_{\{u>\overline{u}\}} (\overline{v}-v)\, dx\, dt.$$

If the point (x,t) is such that $u(x,t) > \overline{u}(x,t)$, then we have the inequality $\underline{u}(x,t) - \alpha \le \overline{u}(x,t) < T_\alpha u(x,t) \le \overline{u}(x,t) + \alpha$. Applying assumption (H2) (ii) we get

$$\overline{v}(x,t) - v(x,t) \le c_1(T_\alpha u(x,t) - \overline{u}(x,t))^{p-1},$$

with v in *(iii)* and $\overline{v}$ in Definition 5.2, (ii). Thus in view of (5.6), (5.9) we obtain

$$\lambda \int_{\{u>\overline{u}\}} (u-\overline{u})^{p-1}\, dx\, dt \le c_1 \int_{\{u>\overline{u}\}} (T_\alpha u - \overline{u})^{p-1}\, dx\, dt.$$

Since $T_\alpha u(x,t) \le u(x,t)$ whenever $u(x,t) > \overline{u}(x,t)$, it results that

$$(\lambda - c_1) \int_{\{u>\overline{u}\}} (u-\overline{u})^{p-1}\, dx\, dt \le 0.$$

Using $c_1 < \lambda$ (see (H2) (ii)) it follows that $u \le \overline{u}$ a.e. in Q.

Step 5: *Completion of the proof.* From Step 4 we know that any solution u of problem $(\tilde{P}_0)$ satisfies the inequality $\underline{u} \le u \le \overline{u}$. It follows that $Bu = 0$ and $Tu = u$. In addition, one has that $v(x,t) \in \partial g(x,t,u(x,t))$ a.e. $(x,t) \in Q$ since $T_\alpha u = u$. We conclude that u is a solution of problem (5.1) satisfying $\underline{u} \le u \le \overline{u}$. The proof of Proposition 5.1 is complete. ∎

In the final part of this Section we consider a significant particular case of initial boundary value problem (5.1), namely when the generalized gradient ∂g becomes a maximal monotone graph β in $\mathbb{R}^2$, i.e. β is the subdifferential of a convex function in the sense of convex analysis. Precisely, we state the following initial boundary value problem

$$\begin{cases} \dfrac{\partial u}{\partial t} + Au + \beta(\cdot,\cdot,u) \ni Fu + h & \text{in } Q \\ u(\cdot,0) = 0 & \text{in } \Omega \\ u = 0 & \text{on } \Gamma. \end{cases} \tag{5.37}$$

Excepting β, all the data appearing in the statement of problem (5.37) have the meaning in problem (5.1). Let us describe now the meaning of the notation β in problem (5.37). Here β stands for a multifunction generated by a function $g : Q \times \mathbb{R} \to \mathbb{R}$ for which we impose, in place of (H2), the following hypothesis:

(H2)$'$ The function $g : Q \times \mathbb{R} \to \mathbb{R}$ satisfies

(i) g is Borel measurable in $Q \times \mathbb{R}$.

(ii) $g(x,t,\cdot) : \mathbb{R} \to \mathbb{R}$ is increasing (possibly discontinuous) for a.e. $(x,t) \in Q$, and it is related with the maximal monotone graph β by

$$\beta(x,t,s) = [g(x,t,s-0), g(x,t,s+0)], \tag{5.38}$$

where the closed interval on the right-hand side of (5.38) is formed by the one-sided limits of g, i.e.,

$$g(x,t,s \pm 0) = \lim_{\varepsilon \downarrow 0} g(x,t,s \pm \varepsilon).$$

(iii) There is a function $k_3 \in L^q_+(Q)$ and a constant $\alpha > 0$ such that

$$|g(x,t,s)| \le k_3(x,t),$$

for a.e. $(x,t) \in Q$ and $s \in [\underline{u}(x,t) - 2\alpha, \overline{u}(x,t) + 2\alpha]$.

We can give now the main result for the existence of extremal solutions to initial boundary value problem (5.37).

Corollary 5.2 (Carl and Motreanu [9]) Under the assumptions of Theorem 5.1 with (H2) replaced by (H2)$'$, problem (5.37) possesses extremal solutions within the order interval $[\underline{u}, \overline{u}]$ formed by the given lower and upper solutions $\underline{u}$ and $\overline{u}$, respectively.

Proof. Let us first point out that problem (5.37) is a particular case of problem (5.1), with the function g replaced by the primitive

$$G(x,t,s) = \int_0^s g(x,t,\tau)\,d\tau, \quad \text{a.e. } (x,t) \in Q, \ \forall s \in \mathbb{R}.$$

Using essentially assumption (H2)$'$, this can be seen as follows. Since $g(x,t,\cdot)$ is an increasing function on $\mathbb{R}$ it follows that $g(x,t,\cdot) \in L^1_{\text{loc}}(\mathbb{R})$, so we can introduce the primitive $G(x,t,s)$ with respect to $s \in \mathbb{R}$. It turns out that $\frac{d}{ds}G(x,t,s) = g(x,t,s)$ for a.e. $s \in \mathbb{R}$. Thus $G(x,t,\cdot)$ is convex. Now we have $\beta = \partial G$, where the notation ∂G means the subdifferential of the convex function G with respect to the third variable. The subdifferential ∂G is equal to the maximal monotone graph β described in relation (5.38). Taking into account that a convex function on $\mathbb{R}$ is locally Lipschitz and its subdifferential coincides with the generalized gradient, the statement of problem (5.37) is a particular case of problem (5.1).

In order to complete the proof we need only to justify that assumption (H2) is more general than (H2)$'$. Specifically, assertion (H2)$'$ (ii) implies property (H2) (ii) with $c_1 = 0$ because the graph $\beta = \partial G$ is monotone. Applying Theorem 5.1 one obtains the conclusion of Corollary 5.2. The proof is thus complete. ■

In the final part of this Section we present an example of application of Corollary 5.2 in studying problem (5.37).

Example 5.1. The existence of nonnegative bounded solutions of initial boundary value problem (5.37) can, in particular, be ensured under the following assumptions:

(i) the Leray-Lions type (i.e. $(A1)$-$(A4)$ hold) operator $A : V \to V^*$ satisfying $Ar = 0$, $\forall r \in \mathbb{R}$;

(ii) $h = 0 \in V_0^*$;

(iii) $g : Q \times \mathbb{R} \to \mathbb{R}$ Borel measurable, $g(x,t,\cdot)$ increasing for a.e. $(x,t) \in Q$, $g(\cdot,\cdot,0), g(\cdot,\cdot,1) \in L^q(\Omega)$;

(iv) $f : Q \times \mathbb{R} \to \mathbb{R}$ Carathéodory function, $f(\cdot,\cdot,0), f(\cdot,\cdot,1) \in L^q(\Omega)$ such that

$$g(x,t,0) \le f(x,t,0), \quad \text{for a.e. } (x,t) \in Q; \tag{5.39}$$

$$f(x,t,1) \le g(x,t,1), \quad \text{for a.e. } (x,t) \in Q; \tag{5.40}$$

$$|g(x,t,s)| \le k_3(x,t), \quad \text{for a.e. } (x,t) \in Q,\ \forall s \in [-2\alpha, 2\alpha+1], \tag{5.41}$$

for some $\alpha > 0$ and $k_3 \in L^q(Q)$;

$$|f(x,t,s)| \le k_4(x,t,s), \quad \text{for a.e. } (x,t) \in Q,\ \forall s \in [0,1]. \tag{5.42}$$

By (5.39) we see that

$$\underline{u} = 0 \text{ is a lower solution of problem (5.37)}, \tag{5.43}$$

while by (5.40) we note that

$$\overline{u} = 1 \text{ is a upper solution of problem (5.37).} \tag{5.44}$$

In checking assertions (5.43), (5.44) we use that $g(x,t,s) \in \beta(x,t,s)$ (see (5.38)). Therefore, assumption $(H1)$ is verified. In view of (5.41), assumption (H2) is fulfilled. Finally, $(H3)$ follows from (5.42). Corollary 5.2 can be applied. Since we get the existence of extremal solutions in the interval $[0,1]$, we conclude that there exist nonnegative bounded solutions of problem (5.37) as required.

Concrete choices for satisfying assumptions (iii), (iv) and relations (5.39)-(5.42) are for instance $g(x,t,s) = g(s)$, with an increasing function $g : I\!R \to I\!R$, and $f(x,t,s) = f(s)$, with a continuous function $f : I\!R \to I\!R$, such that $g(0) \leq f(0)$ and $g(1) \leq f(1)$.

References

[1] J. Berkovits and V. Mustonen, Monotonicity methods for nonlinear evolution equations, *Nonlinear Anal.* **27** (1996), 1397-1405.

[2] H. Brézis, *Analyse Fonctionelle. Théorie et Applications*, Masson, Paris, 1983.

[3] S. Carl, Leray-Lions operators perturbed by state-dependent subdifferentials, *Nonlinear World* **3** (1996), 505-518.

[4] S. Carl, Existence and comparison results for quasilinear parabolic inclusions with state-dependent subdifferentials, *Optimization* **49** (2000), 51-66.

[5] S. Carl, Extremal solutions of parabolic hemivariational inequalities, *Nonlinear Analysis* **47** (2001), 5077-5088.

[6] S. Carl, *A survey of recent results on the enclosure and extremality of solutions for quasilinear hemivariational inequalities,* in: From Convexity to Nonconvexity (Eds. R.P. Gilbert, P.D. Panagiotopoulos and P.M. Pardalos), pp. 15-28, Kluwer Academic Publishers, Dordrecht/Boston/London, 2001.

[7] S. Carl, Existence of extremal solutions of boundary hemivariational inequalities, *J. Differ. Equations* **171** (2001), 370-396.

[8] S. Carl and S. Heikkilä, *Nonlinear Differential Equations in Ordered Spaces*, Chapman & Hall/CRC, Boca Raton, 2001.

[9] S. Carl and D. Motreanu, Extremal solutions of quasilinear parabolic subdifferential inclusions, *Diff. Int. Eqns*, to appear.

[10] S. Carl and D. Motreanu, Extremal solutions of quasilinear parabolic inclusions with generalized Clarke's gradient, submitted.

[11] M. Chipot and J.F. Rodrigues, Comparison and stability of solutions to a class of quasilinear parabolic problems, *Proc. Royal Soc. Edinburgh* **110 A** (1988), 275-285.

[12] Ph. Ciarlet, *Mathematical Elasticity*, Vol. 3: Theory of shells, Studies in Mathematics and its Applications **29**, North-Holland, Amsterdam, 2000.

[13] F.H. Clarke, *Optimization and Nonsmooth Analysis*, John Wiley & Sons, Inc., New York, 1983.

[14] G. Dincă and G. Pop, *Existence results for variational-hemivariational inequalities: A F. E. Browder technique*, in: From Convexity and Nonconvexity (R. P. Gilbert at al. (eds.)), Kluwer Academic Publishers, Dordrecht/Boston/London, 2001, pp. 233-241.

[15] D. Y. Gao, *Duality, Principles in Nonconvex Systems. Theory, Methods and Applications*, Nonconvex Optimization and its Applications **39**, Kluwer Academic Publishers, Dordrecht/Boston/London, 2000.

[16] L. Gasiński and M. Smolka, Existence of Solutions for Wave-Type Hemivariational Inequalities with Noncoercive Viscosity Damping, Jagiellonian University, Preprint no. 2001/002.

[17] D. Goeleven, M. Miettinen and P. D. Panagiotopoulos, Dynamic Hemivariational Inequalities and Their Applications, *J. Optimization Theory Appl.* **103** (1999), 567-601.

[18] M. Miettinen, *Hemivariational inequalities and hysteresis*, in: From Convexity and Nonconvexity (R. P. Gilbert at al. (eds.)), Kluwer Academic Publishers, Dordrecht/Boston/London, 2001, pp. 193-206.

[19] M. Miettinen and P. D. Panagiotopoulos, Hysteresis and hemivariational inequalities: semilinear case, *J. Glob. Optim.* **13** (1998), 269-298.

[20] E. S. Mistakidis and G. E. Stavroulakis, *Nonconvex Optimization in Mechanics*, Nonconvex Optimization and Its Applications **21**, Kluwer Academic Publishers, Dordrecht/Boston/London, 1998.

[21] D. Motreanu and P. D. Panagiotopoulos, *Minimax Theorems and Qualitative Properties of the Solutions of Hemivariational Inequalities*, Nonconvex Optimization and its Applications **29**, Kluwer Academic Publishers, Dordrecht/Boston/London, 1998.

[22] Z. Naniewicz and P. D. Panagiotopoulos, *Mathematical Theory of Hemivariational Inequalities and Applications*, Marcel Dekker, New York, 1995.

[23] P.D. Panagiotopoulos, *Hemivariational Inequalities and Applications in Mechanics and Engineering*, Springer-Verlag, New York, 1993.

[24] P. D. Panagiotopoulos and G. Pop, On a type of hyperbolic variational-hemivariational inequalities, *J. Appl. Anal.* 5 (1999), 95-112.

[25] E. Zeidler, *Nonlinear Functional Analysis and its Applications*, Vols. II A/B, Springer-Verlag, Berlin, 1990.

Chapter 6

VARIATIONAL, HEMIVARIATIONAL AND VARIATIONAL-HEMIVARIATIONAL INEQUALITIES: EXISTENCE RESULTS

The celebrated Hartman-Stampacchia theorem (see [6], Lemma 3.1, or [9], Theorem I.3.1) asserts that if V is a finite dimensional Banach space, $K \subset V$ is non-empty, compact and convex, $A : K \to V^*$ is continuous, then there exists $u \in K$ such that, for every $v \in K$,

$$\langle Au, v - u \rangle \geq 0 . \tag{6.1}$$

The simplest proof of this result (which does not coincide with the original one) is due to H. Brézis. Assuming $V = I\!R^N$ and $K \subset I\!R^N$ is compact and convex, proving (6.1) is equivalent to show that there exists $u \in K$ such that $(u, v - u) \geq (u - \pi Au, v - u)$, $\forall v \in K$, where $(\cdot,\cdot)$ denotes the scalar product in $I\!R^N$ and $\pi : (I\!R^N)^* \to I\!R^N$ is the canonical identification. By Brouwer's fixed point theorem, the mapping $P_K(id - \pi A) : K \to K$, with P_K the projection onto K, admits a fixed point $u \in K$, $u = P_K(id - \pi A)u$. Consequently, by the characterization of the projection on closed and convex sets in Hilbert spaces (Theorem V.2 in [1]), we see that (6.1) holds.

If we weaken the hypotheses and consider the case where K is a nonempty, closed and convex subset of the finite dimensional space V, Hartman and Stampacchia proved (see [9], Theorem I.4.2) that a necessary and sufficient condition which ensures the existence of a solution to Problem (6.1) is that there is some $R > 0$ such that a solution u of (6.1) with $\|u\| \leq R$ satisfies $\|u\| < R$.

We shall develop in this Chapter a similar theory in the framework of hemivariational inequalities and we will make several connections with other theories, such as the KKM principle of Knaster, Kuratowski and Mazurkiewicz. The abstract results are applied in the last Section of this Chapter for solving several concrete problems arising in Nonsmooth Mechanics and Engineering.

1. Hartman-Stampacchia Type Results for Hemivariational Inequalities

Let V be a real Banach space and let $T : V \to L^p(\Omega; \mathbb{R}^k)$ be a linear continuous operator, where $1 \le p < \infty$, $k \ge 1$, and Ω is a bounded open set in $\mathbb{R}^N$. Throughout this Section, K is a subset of V, $A : K \to V^*$ an operator and $j = j(x, y) : \Omega \times \mathbb{R}^k \to \mathbb{R}$ is a Carathéodory function which is locally Lipschitz with respect to the second variable $y \in \mathbb{R}^k$ and satisfies the following assumption

(j) there exists $h_1 \in L^{\frac{p}{p-1}}(\Omega; \mathbb{R})$ and $h_2 \in L^\infty(\Omega; \mathbb{R})$ such that

$$|z| \le h_1(x) + h_2(x)|y|^{p-1},$$

for a.e. $x \in \Omega$, every $y \in \mathbb{R}^k$ and $z \in \partial j(x, y)$. Denoting by $Tu = \hat{u}$, $u \in V$, our aim is to study the problem

(P) Find $u \in K$ such that, for every $v \in K$,

$$\langle Au, v - u \rangle + \int_\Omega j^0(x, \hat{u}(x); \hat{v}(x) - \hat{u}(x))dx \ge 0.$$

Recall that $j^0(x, y; h)$ denotes the Clarke's generalized directional derivative of the locally Lipschitz mapping $j(x, \cdot)$ at the point $y \in \mathbb{R}^k$ with respect to the direction $h \in \mathbb{R}^k$, where $x \in \Omega$, while $\partial j(x, y)$ is the Clarke's generalized gradient of this mapping at $y \in \mathbb{R}^k$, that is

$$j^0(x, y; h) = \limsup_{\substack{y' \to y \\ t \downarrow 0}} \frac{j(x, y' + th) - j(x, y')}{t},$$

$$\partial j(x, y) = \{z \in \mathbb{R}^k : \langle z, h \rangle \le j^0(x, y; h), \text{ for all } h \in \mathbb{R}^k\}$$

(see Definitions 1.1 and 1.2).

The euclidean norm in $\mathbb{R}^k$, $k \ge 1$, resp. the duality pairing between a Banach space and its dual will be denoted by $|\cdot|$, respectively $\langle \cdot, \cdot \rangle$. We also denote by $\|\cdot\|_p$ the norm in the space $L^p(\Omega; \mathbb{R}^k)$ defined by

$$\|\hat{u}\|_p = \left(\int_\Omega |\hat{u}(x)|^p dx\right)^{\frac{1}{p}}, \quad 1 \le p < \infty.$$

Definition 6.1 The operator $A : K \to V^*$ is w^*-demicontinuous if for any sequence $\{u_n\} \subset K$ converging to u, the sequence $\{Au_n\}$ converges to Au for the w^*-topology in V^*.

Definition 6.2 The operator $A : K \to V^*$ is continuous on finite dimensional subspaces of K if for any finite dimensional space $F \subset V$, which intersects K, the operator $A|_{K\cap F}$ is demicontinuous, that is $\{Au_n\}$ converges weakly to Au in V^* for each sequence $\{u_n\} \subset K \cap F$ which converges to u.

Remark 6.1 In reflexive Banach spaces the following hold:
a) the w^*-demicontinuity and demicontinuity are the same.
b) a demicontinuous operator $A : K \to V^*$ is continuous on finite dimensional subspaces of K.

Theorem 6.1 Let K be a compact and convex subset of the infinite dimensional Banach space V and let j satisfy condition (j). If the operator $A : K \to V^*$ is w^*-demicontinuous, then problem (P) admits a solution.

The condition of w^*-demicontinuity on the operator $A : K \to V^*$ in Theorem 6.1 may be replaced equivalently by the assumption:
(A_1) the mapping $K \ni u \to \langle Au, v\rangle$ is weakly upper semicontinuous, for each $v \in V$.

Indeed, since on the compact set K the weak-topology is in fact the normed topology, we can replace equivalently the weak upper semicontinuity by upper semicontinuity. So we have to prove that the w^*-demicontinuity of A follows from the assumption (A_1); but for any sequence $\{u_n\} \subset K$ converging to u one finds (by (A_1)):

$$\limsup_{n\to\infty}\langle Au_n, v\rangle \leq \langle Au, v\rangle$$

and

$$\limsup_{n\to\infty}\langle Au_n, -v\rangle \leq \langle Au, -v\rangle \Longleftrightarrow \liminf_{n\to\infty}\langle Au_n, v\rangle \geq \langle Au, v\rangle,$$

for each fixed point $v \in V$. Thus, there exists $\lim_{n\to\infty}\langle Au_n, v\rangle$, and $\lim_{n\to\infty}\langle Au_n, v\rangle = \langle Au, v\rangle$, for every $v \in V$. Consequently, the sequence $\{Au_n\}$ converges to Au for the w^*-topology in V^*.

We also point out that if A is w^*-demicontinuous, $\{u_n\} \subset K$ and $u_n \to u$, then $\lim_{n\to\infty}\langle Au_n, u_n\rangle = \langle Au, u\rangle$. This follows from the w^*-boundedness of $\{Au_n\}$ in V^* (as a w^*-convergent sequence) and from the fact that in real dual Banach spaces each w^*-bounded set is a (strongly) bounded set (see [20], Proposition IV.5.2). Thus, in this case, one can write $\lim_{n\to\infty}\langle Au_n, v - u_n\rangle = \langle Au, v - u\rangle$, for each $v \in V$.

In finite dimensional Banach spaces Theorem 6.1 has the following equivalent form.

Corollary 6.1 Let V be a finite dimensional Banach space and let K be a compact and convex subset of V. If assumption (j) is fulfilled and if $A : K \to V^*$ is a continuous operator, then problem (P) has at least a solution.

The proof of Theorem 6.1 is based on Corollary 6.1. That is why we provide an independent proof of Corollary 6.1. For this purpose we need the following auxiliary result.

Lemma 6.1 (a) If assumption (j) is satisfied and V_1, V_2 are nonempty subsets of V, then the mapping $V_1 \times V_2 \to I\!R$ defined by

$$(u, v) \to \int_\Omega j^0(x, \hat{u}(x); \hat{v}(x))dx \tag{6.2}$$

is upper semicontinuous.

(b) Moreover, if $T : V \to L^p(\Omega; I\!R^k)$ is a linear compact operator, then the above mapping is weakly upper semicontinuous.

Proof. a) Let $\{(u_m, v_m)\}_{m\in I\!N} \subset V_1 \times V_2$ be a sequence converging to $(u, v) \in V_1 \times V_2$, as $m \to \infty$. Since $T : V \to L^p(\Omega; I\!R^k)$ is continuous, it follows that

$$\hat{u}_m \to \hat{u}, \quad \hat{v}_m \to \hat{v} \text{ in } L^p(\Omega; I\!R^k), \quad \text{as } m \to \infty.$$

There exists a subsequence $\{(\hat{u}_n, \hat{v}_n)\}$ of the sequence $\{(\hat{u}_m, \hat{v}_m)\}$ such that

$$\limsup_{m\to\infty} \int_\Omega j^0(x, \hat{u}_m(x); \hat{v}_m(x))dx = \lim_{n\to\infty} \int_\Omega j^0(x, \hat{u}_n(x); \hat{v}_n(x))dx.$$

By Proposition 4.11 in [8], one may suppose the existence of two functions $\hat{u}_0, \hat{v}_0 \in L^p(\Omega; I\!R^+)$, and of two subsequences of $\{\hat{u}_n\}$ and $\{\hat{v}_n\}$ denoted again by the same symbols and such that

$$|\hat{u}_n(x)| \leq \hat{u}_0(x), \qquad |\hat{v}_n(x)| \leq \hat{v}_0(x),$$

$$\hat{u}_n(x) \to \hat{u}(x), \qquad \hat{v}_n(x) \to \hat{v}(x), \quad \text{as } n \to \infty$$

for a.e. $x \in \Omega$. On the other hand, for each x where holds true condition (j) and for each $y, h \in I\!R^k$, let $z \in \partial j(x, y)$ be such that

$$j^0(x, y; h) = \langle z, h\rangle = \max\{\langle w, h\rangle : \ w \in \partial j(x, y)\}.$$

Now, by (j),

$$|j^0(x, y; h)| \leq |z| \ |h| \leq (h_1(x) + h_2(x)|y|^{p-1})|h|.$$

Consequently, denoting $F(x) = (h_1(x) + h_2(x)|\hat{u}_0(x)|^{p-1})|\hat{v}_0(x)|$, we find that

$$|j^0(x, \hat{u}_n(x); \hat{v}_n(x))| \le F(x),$$

for all $n \in \mathbb{N}$ and for a.e. $x \in \Omega$.

From Hölder's inequality and from condition (j) for the functions h_1 and h_2 it follows that $F \in L^1(\Omega; \mathbb{R})$. Fatou's lemma yields

$$\lim_{n\to\infty} \int_\Omega j^0(x, \hat{u}_n(x); \hat{v}_n(x))dx \le \int_\Omega \limsup_{n\to\infty} j^0(x, \hat{u}_n(x); \hat{v}_n(x))dx.$$

Next, the upper-semicontinuity of the mapping $j^0(x, \cdot; \cdot)$ yields

$$\limsup_{n\to\infty} j^0(x, \hat{u}_n(x); \hat{v}_n(x)) \le j^0(x, \hat{u}(x); \hat{v}(x)),$$

for a.e. $x \in \Omega$, because

$$\hat{u}_n(x) \to \hat{u}(x) \quad \text{and} \quad \hat{v}_n(x) \to \hat{v}(x), \quad \text{as } n \to \infty$$

for a.e. $x \in \Omega$. Hence

$$\limsup_{m\to\infty} \int_\Omega j^0(x, \hat{u}_m(x); \hat{v}_m(x))dx \le \int_\Omega j^0(x, \hat{u}(x); \hat{v}(x))dx,$$

which proves the upper semicontinuity of the mapping defined by (6.2).

b) Let $\{(u_m, v_m)\} \subset V_1 \times V_2$ be now a sequence weakly converging to $\{u, v\} \in V_1 \times V_2$, as $m \to \infty$. Thus $u_m \rightharpoonup u$, $v_m \rightharpoonup v$ weakly as $m \to \infty$. Since $T : V \to L^p(\Omega; \mathbb{R}^k)$ is a linear compact operator, it follows that

$$\hat{u}_m \to \hat{u}, \quad \hat{v}_m \to \hat{v} \text{ in } L^p(\Omega; \mathbb{R}^k).$$

From now on the proof follows the same argument as in the case a). ■

Proof of Corollary 6.1. Arguing by contradiction, for every $u \in K$, there exists $v = v_u \in K$ such that

$$\langle Au, v - u\rangle + \int_\Omega j^0(x, \hat{u}(x); \hat{v}(x) - \hat{u}(x))dx < 0.$$

For every $v \in K$, set

$$N(v) = \{u \in K : \langle Au, v - u\rangle + \int_\Omega j^0(x, \hat{u}(x); \hat{v}(x) - \hat{u}(x))dx < 0\}.$$

For any fixed $v \in K$ the mapping $K \to \mathbb{R}$ defined by

$$u \mapsto \langle Au, v - u\rangle + \int_\Omega j^0(x, \hat{u}(x); \hat{v}(x) - \hat{u}(x))dx$$

is upper semicontinuous, by Lemma 6.1 and the continuity of A. Thus, by the definition of the upper semicontinuity, $N(v)$ is an open set. Our initial assumption implies that $\{N(v); v \in K\}$ is a covering of K. Hence, by the compactness of K, there exist $v_1, \cdots, v_n \in K$ such that

$$K \subset \bigcup_{j=1}^{n} N(v_j).$$

Let $\rho_j(u)$ be the distance from u to $K \setminus N(v_j)$. Then ρ_j is a Lipschitz map which vanishes outside $N(v_j)$ and the functionals

$$\psi_j(u) = \frac{\rho_j(u)}{\sum_{i=1}^{n} \rho_i(u)}$$

define a partition of the unity subordinated to the covering $\{\rho_1, \cdots, \rho_n\}$. Moreover, the mapping $p(u) = \sum_{j=1}^{n} \psi_j(u) v_j$ is continuous and maps K into itself, because of the convexity of K. Thus, by Brouwer's fixed point theorem, there exists u_0 in the convex closed hull of $\{v_1, \cdots, v_n\}$ such that $p(u_0) = u_0$. Define

$$q(u) = \langle Au, p(u) - u \rangle + \int_{\Omega} j^0(x, \hat{u}(x); \widehat{p(u)}(x) - \hat{u}(x))dx.$$

The convexity of the map $j^0(\hat{u}; \cdot)$ and the fact that $\sum_{j=1}^{n} \psi_j(u) = 1$ on K imply

$$q(u) \leq \sum_{j=1}^{n} \psi_j(u) \langle Au, v_j - u \rangle + \sum_{j=1}^{n} \psi_j(u) \int_{\Omega} j^0(x, \hat{u}(x); \hat{v}_j(x) - \hat{u}(x))dx.$$

For arbitrary $u \in K$, there are only two possibilities: if $u \notin N(v_i)$, then $\psi_i(u) = 0$. On the other hand, for all $1 \leq j \leq n$ (there exists at least such an index) such that $u \in N(v_j)$, we have $\psi_j(u) > 0$. Thus, by the definition of $N(v_j)$,

$$q(u) < 0, \quad \text{for every } u \in K.$$

But $q(u_0) = 0$, which gives a contradiction. ■

For the proof of Theorem 6.1 we need Lemma 6.2 below. Let F be an arbitrary finite dimensional subspace of V which intersects K. Let $i_{K \cap F}$ be the canonical injection of $K \cap F$ into K and i_F^* be the adjoint of the canonical injection i_F of F into V.

Lemma 6.2 The operator

$$B : K \cap F \to F^*, \quad B = i_F^* A i_{K \cap F}$$

is continuous.

Proof. We have to prove that the sequence $\{Bu_n\}$ converges to Bu in F^* for any sequence $\{u_n\} \subset K \cap F$ converging to u in $K \cap F$ (or in V). In order to do this, we prove that the sequence $\{Bu_n\}$ is weakly ($= w^*$) convergent to Bu, because F^* is a finite dimensional Banach space. Let v be an arbitrary point of F. Then by the w^*-demicontinuity of the operator $A : K \to V^*$ it follows that

$$\langle Bu_n, v\rangle = \langle i_F^* A i_{K\cap F} u_n, v\rangle = \langle i_F^* A u_n, v\rangle$$

$$= \langle Au_n, i_F v\rangle = \langle Au_n, v\rangle \underset{n\to\infty}{\rightarrow} \langle Au, v\rangle = \langle Bu, v\rangle.$$

Therefore $\{Bu_n\}$ converges weakly to Bu. ■

Remark 6.2 The above lemma also holds true if the operator A is continuous on finite dimensional subspaces of K.

Proof of Theorem 6.1. For any $v \in K$, set

$$S(v) = \{u \in K \ : \ \langle Au, v-u\rangle + \int_\Omega j^0(x, \hat{u}(x); \hat{v}(x) - \hat{u}(x))dx \geq 0\}.$$

Step 1. $S(v)$ is closed set.

We first observe that $S(v) \neq \emptyset$, since $v \in S(v)$. Let $\{u_n\} \subset S(v)$ be an arbitrary sequence which converges to u as $n \to \infty$. We have to prove that $u \in S(v)$. By $u_n \in S(v)$ and by Lemma 6.1 (a),

$$0 \leq \limsup_{n\to\infty}[\langle Au_n, v-u_n\rangle + \int_\Omega j^0(x, \hat{u}_n(x); \hat{v}(x) - \hat{u}_n(x))]dx$$

$$= \lim_{n\to\infty}\langle Au_n, v-u_n\rangle + \limsup_{n\to\infty}\int_\Omega j^0(x, \hat{u}_n(x); \hat{v}(x) - \hat{u}_n(x))dx$$

$$\leq \langle Au, v-u\rangle + \int_\Omega j^0(x, \hat{u}(x); \hat{v}(x) - \hat{u}(x))dx.$$

This is equivalent to $u \in S(v)$.

Step 2. The family $\{S(v); v \in K\}$ has the finite intersection property.

Let $\{v_1, \cdots, v_n\}$ be an arbitrary finite subset of K and let F be the linear space spanned by this family. Applying Corollary 6.1 to the operator B defined in Lemma 6.2, we find $u \in K \cap F$ such that $u \in \cap_{j=1}^n S(v_j)$, which means that the family of closed sets $\{S(v); v \in K\}$ has the finite intersection property. But the set K is compact. Hence

$$\bigcap_{v\in K} S(v) \neq \emptyset,$$

which means that the problem (P) has at least one solution. ■

Weakening more the hypotheses on K by assuming that K is a closed, bounded and convex subset of the Banach space V, we need something more about the operators A and T. We first recall that an operator $A : K \to V^*$ is said to be monotone if, for every $u, v \in K$,

$$\langle Au - Av, u - v \rangle \geq 0.$$

The following result generalizes Theorem 1.1 in [6].

Theorem 6.2 Let V be an (infinite dimensional) reflexive Banach space and let $T : V \to L^p(\Omega; I\!R^k)$ be a linear and compact operator. Assume K is a closed, bounded and convex subset of V and $A : K \to V^*$ is monotone and continuous on finite dimensional subspaces of K. If j satisfies condition (j) then the problem (P) has at least one solution.

Proof. Let F be an arbitrary finite dimensional subspace of V, which intersects K. Consider the canonical injections $i_{K\cap F} : K \cap F \to K$ and $i_F : F \to V$ and let $i_F^* : V^* \to F^*$ be the adjoint of i_F. Applying Corollary 6.1 to the continuous operator $B = i_F^* A i_{K\cap F}$ (see Remark 6.2) we find u_F in the compact set $K \cap F$ such that, for every $v \in K \cap F$,

$$\langle i_F^* A i_{K\cap F} u_F, v - u_F \rangle + \int_\Omega j^0(x, \hat{u}_F(x); \hat{v}(x) - \hat{u}_F(x))dx \geq 0. \quad (6.3)$$

But

$$0 \leq \langle Av - Au_F, v - u_F \rangle = \langle Av, v - u_F \rangle - \langle Au_F, v - u_F \rangle. \quad (6.4)$$

Hence, by (6.3), (6.4) and the observation that $\langle i_F^* A i_{K\cap F} u_F, v - u_F \rangle = \langle Au_F, v - u_F \rangle$, we have

$$\langle Av, v - u_F \rangle + \int_\Omega j^0(x, \hat{u}_F(x); \hat{v}(x) - \hat{u}_F(x))dx \geq 0, \quad (6.5)$$

for any $v \in K \cap F$. The set K is weakly closed as a closed convex set. Thus it is weakly compact because it is bounded and V is a reflexive Banach-space.

Now, for every $v \in K$ define

$$M(v) = \{u \in K \, : \, \langle Av, v - u \rangle + \int_\Omega j^0(x, \hat{u}(x); \hat{v}(x) - \hat{u}(x))dx \geq 0\}.$$

The set $M(v)$ is weakly closed by part (b) of Lemma 6.1 and the fact that this set is weakly sequentially dense (see, e.g., [7], pp. 145-149 or [18],

p. 3). We now show that the set $M = \cap_{v \in K} M(v) \subset K$ is non-empty. To prove this, it suffices to prove that

$$\bigcap_{j=1}^{n} M(v_j) \neq \emptyset, \tag{6.6}$$

for any $v_1, \cdots, v_n \in K$. Let F be the finite dimensional linear space spanned by $\{v_1, \cdots, v_n\}$. Hence, by (6.5), there exists $u_F \in F$ such that, for every $v \in K \cap F$,

$$\langle Av, v - u_F \rangle + \int_\Omega j^0(x, \hat{u}_F(x); \hat{v}(x) - \hat{u}_F(x))dx \geq 0.$$

This means that $u_F \in M(v_j)$, for every $1 \leq j \leq n$, which implies (6.6). Consequently, it follows that $M \neq \emptyset$. Therefore there is some $u \in K$ such that, for every $v \in K$,

$$\langle Av, v - u \rangle + \int_\Omega j^0(x, \hat{u}(x); \hat{v}(x) - \hat{u}(x))dx \geq 0. \tag{6.7}$$

We shall prove that from (6.7) we can conclude that u is a solution of problem (P). Fix $w \in K$ and $\lambda \in]0, 1[$. Putting $v = (1 - \lambda)u + \lambda w \in K$ in (6.7) we find

$$\langle A((1-\lambda)u + \lambda w), \lambda(w-u)\rangle + \int_\Omega j^0(x, \hat{u}(x); \lambda(\hat{w} - \hat{u})(x))dx \geq 0. \tag{6.8}$$

But $j^0(x, \hat{u}; \lambda\hat{v}) = \lambda j^0(x, \hat{u}; \hat{v})$, for any $\lambda > 0$. Therefore (6.8) may be written, equivalently,

$$\langle A((1 - \lambda)u + \lambda w), w - u\rangle + \int_\Omega j^0(x, \hat{u}(x); (\hat{w} - \hat{u})(x))dx \geq 0. \tag{6.9}$$

Let F be the vector space spanned by u and w. Taking into account the hemicontinuity of the operator $A|_{K\cap F}$ and passing to the limit in (6.9) as $\lambda \to 0$, we obtain that u is a solution of problem (P). ■

We also give a generalization of Theorem III.1.7. in [2].

Theorem 6.3 Assume that the same hypotheses as in Theorem 6.2 hold without the assumption of boundedness of K. Then a necessary and sufficient condition for the hemivariational inequality (P) to have a solution is that there exists $R > 0$ with the property that at least one solution of the problem

$$\begin{cases} u_R \in K \cap \{u \in V; \|u\| \leq R\}; \\ \langle Au_R, v - u_R\rangle + \int_\Omega j^0(x, \hat{u}_R(x); \hat{v}(x) - \hat{u}_R(x))dx \geq 0, \\ \text{for every } v \in K \text{ with } \|v\| \leq R, \end{cases} \tag{6.10}$$

satisfies the inequality $\|u_R\| < R$.

Remark 6.3 As the set $K \cap \{x \in V : \|u\| \leq R\}$ is a closed, bounded and convex set in V, it follows from Theorem 6.2 that the problem (6.10) has at least one solution for any fixed $R > 0$.

Proof of Theorem 6.3. The necessity is evident.

Let us now suppose that there exists a solution u_R of (6.10) with $\|u_R\| < R$. We prove that u_R is solution of (P). For any fixed $v \in K$, we choose $\varepsilon > 0$ small enough so that $w = u_R + \varepsilon(v - u_R)$ satisfies $\|w\| < R$. Hence, by (6.10),

$$\langle Au_R, \varepsilon(v - u_R)\rangle + \int_\Omega j^0(x, \hat{u}_R(x); \varepsilon(\hat{v} - \hat{u}_R)(x))dx \geq 0$$

and, using again the positive homogeneity of the map $v \mapsto j^0(u; v)$, the conclusion follows. ■

2. Variational-Hemivariational Inequality Problems with Lack of Convexity

Throughout this Section, X will denote a real reflexive Banach space, (T, μ) will be a measure space of positive and finite measure and $A : X \to X^*$ will stand for a nonlinear operator. We also assume that there are given $m \in I\!N$, $p \geq 1$ and a compact mapping $\gamma : X \to L^p(T; I\!R^m)$. We shall denote by p' the conjugated exponent of p. If $\varphi : X \to I\!R$ is a locally Lipschitz functional then $\varphi^0(u; v)$ will stand for the Clarke derivative of φ at $u \in X$ with respect to the direction $v \in X$, while $\partial\varphi(u)$ will denote the Clarke generalized gradient of φ at u (see Definitions 1.1 and 1.2).

Let $j : T \times I\!R^m \to I\!R$ be a function such that the mapping

$$j(\cdot, y) : T \to I\!R \quad \text{is measurable, for every } y \in I\!R^m . \tag{6.11}$$

We assume that at least one of the following conditions hold: either there exists $k \in L^{p'}(T; I\!R)$ such that

$$|j(x, y_1) - j(x, y_2)| \leq k(x)\,|y_1 - y_2|\,, \quad \forall x \in T\,, \forall y_1, y_2 \in I\!R^m\,, \tag{6.12}$$

or

$$\text{the mapping } j(x, \cdot) \text{ is locally Lipschitz, } \forall x \in T, \tag{6.13}$$

and there exists $C > 0$ such that

$$|z| \leq C(1 + |y|^{p-1})\,, \quad \forall x \in T\,, \forall y_1, y_2 \in I\!R^m\,, \forall z \in \partial_y j(x, y)\,. \tag{6.14}$$

Let K be a nonempty closed, convex subset of X, $f \in X^*$ and $\Phi : X \to I\!\!R \cup \{+\infty\}$ a convex, lower semicontinuous functional such that

$$D(\Phi) \cap K \neq \emptyset . \tag{6.15}$$

Consider the problem: Find $u \in K$ such that

$$\langle Au - f, v - u\rangle + \Phi(v) - \Phi(u)$$
$$+ \int_T j^0(x, \gamma(u(x)); \gamma(v(x) - u(x)))d\mu \geq 0, \quad \forall v \in K. \tag{6.16}$$

The following two situations are of particular interest in applications: (i) $T = \Omega$, $\mu = dx$, $X = W^{1,q}(\Omega; I\!\!R^m)$ and $\gamma : X \to L^p(\Omega; I\!\!R^m)$, with $p < q^*$, is the Sobolev embedding operator;
(ii) $T = \partial\Omega$, $\mu = d\sigma$, $X = W^{1,p}(\Omega; I\!\!R^m)$ and $\gamma = i \circ \eta$, where $\eta : X \to W^{1-\frac{1}{p},p}(\partial\Omega; I\!\!R^m)$ is the trace operator and $i : W^{1-\frac{1}{p},p}(\partial\Omega; I\!\!R^m) \to L^p(\partial\Omega; I\!\!R^m)$ is the embedding operator.

Lemma 6.3 Let K be a nonempty, bounded, closed, convex subset of X, $\Phi : X \to I\!\!R \cup \{+\infty\}$ a convex, lower semicontinuous functional such that (6.15) holds. Consider a Banach space Y such that there exists a linear and compact mapping $L : X \to Y$ and let $J : Y \to I\!\!R$ be an arbitrary locally Lipschitz function. Suppose in addition that the mapping $K \ni v \mapsto \langle Av, v - u\rangle$ is weakly lower semicontinuous, for every $u \in K$.

Then, for every $f \in X^*$, there exists $u \in K$ such that

$$\langle Au - f, v - u\rangle + \Phi(v) - \Phi(u) + J^0\left(L(u); L(v - u)\right) \geq 0\,, \forall v \in K\,. \tag{6.17}$$

The proof of Lemma 6.3 relies on the celebrated Knaster-Kuratowski-Mazurkiewicz (KKM, in short) principle (see [10] or [4]) that we prove in what follows. We first recall some basic definitions. Let E be a vector space. A subset A of E is said to be *finitely closed* if its intersection with any finite-dimensional linear variety $L \subset E$ is closed in the Euclidean topology of L. Let X be an arbitrary subspace of E. A function $G : X \to 2^E$ is called a *KKM-mapping* if

$$\operatorname{conv}\{x_1, \ldots, x_n\} \subset \bigcup_{i=1}^{n} G(x_i)$$

for any finite set $\{x_1, \ldots, x_n\} \subset X$.

The KKM Principle Let E be a vector space, X an arbitrary subspace of E, and $G : X \to 2^E$ a KKM-mapping such that $G(x)$ is finitely closed for any $x \in X$. Then the family $\{G(x)\}_{x\in X}$ has the finite intersection property.

Proof. Arguing by contradiction, let $x_1, \dots, x_n \in X$ be such that $\bigcap_{i=1}^n G(x_i) = \emptyset$. Let L be the linear manifold spanned by $\{x_1, \dots, x_n\}$. Hence

$$\mathrm{conv}\,\{x_1, \dots, x_n\} \subset L\,.$$

Let d be the Euclidean metric in L. Since $L \cap G(x_i)$ is closed in L, it follows that $d(x, L \cap G(x_i)) = 0$ if and only if $x \in L \cap G(x_i)$.

Define $\lambda : \mathrm{conv}\,\{x_1, \dots, x_n\} \to I\!R$ by

$$\lambda(u) = \sum_{i=1}^n d(u, L \cap G(x_i)), \qquad \forall u \in \mathrm{conv}\,\{x_1, \dots, x_n\}\,.$$

Our assumption by contradiction ensures that

$$\bigcap_{i=1}^n (L \cap G(x_i)) = \emptyset\,.$$

Hence $\lambda(u) \neq 0$, for any $u \in \mathrm{conv}\,\{x_1, \dots, x_n\}$. Thus we may define a continuous function

$$f : \mathrm{conv}\,\{x_1, \dots, x_n\} \to \mathrm{conv}\,\{x_1, \dots, x_n\}$$

by setting

$$f(u) = \frac{1}{\lambda(u)} \sum_{i=1}^n d\,(u, L \cap G(x_i))\; x_i\,.$$

The Brouwer fixed point theorem ensures the existence of a fixed point $u_0 \in \mathrm{conv}\,\{x_1, \dots, x_n\}$ of f. Set

$$I = \{i \; : \; d\,(u_0, L \cap G(x_i)) \neq 0\}\,.$$

Then u_0 cannot belong to $\bigcup_{i\in I} G(x_i)$. On the other hand

$$u_0 = f(u_0) \in \mathrm{conv}\,\{x_i \; : \; i \in I\} \subset \bigcup_{i\in I} G(x_i)\,.$$

This contradiction concludes the proof. ■

Proof of Lemma 6.3. Let us first define the set-valued mapping $G : K \cap D(\Phi) \to 2^X$ by

$$G(x) = \{v \in K \cap D(\Phi) \; : \; \langle Av - f, v - x\rangle$$
$$- J^0(L(v); L(x) - L(v)) + \Phi(v) - \Phi(x) \leq 0\}.$$

We claim that the set $G(x)$ is weakly closed. Indeed, if $G(x) \ni v_n \rightharpoonup v$ then, by our hypotheses,

$$\langle Av, v - x\rangle \leq \liminf_{n\to\infty} \langle Av_n, v_n - x\rangle$$

and

$$\Phi(v) \leq \liminf_{n\to\infty} \Phi(v_n)\,.$$

Moreover, $L(v_n) \to L(v)$ and thus, by the upper semicontinuity of J^0, we also obtain

$$\limsup_{n\to\infty} J^0\left(L(v_n); L(x - v_n)\right) \leq J^0\left(L(v); L(x - v)\right)\,.$$

Therefore

$$-J^0\left(L(v); L(x - v)\right) \leq \liminf_{n\to\infty} \left(-J^0\left(L(v_n); L(x - v_n)\right)\right)\,.$$

So, if $v_n \in G(x)$ and $v_n \rightharpoonup v$ then

$$\begin{aligned}\langle Av - f, v - x\rangle - J^0\left(L(v); L(x - v)\right) + \Phi(v) - \Phi(x)\\ \leq \liminf\{\langle Av_n - f, v_n - x\rangle\\ -J^0\left(L(v_n); L(x - v_n)\right) + \Phi(v_n) - \Phi(x)\} \leq 0\,,\end{aligned}$$

which shows that $v \in G(x)$. Since K is bounded, it follows that $G(x)$ is weakly compact. This implies that

$$\bigcap_{x\in K\cap D(\Phi)} G(x) \neq \emptyset\,,$$

provided that the family $\{G(x)\ :\ x \in K \cap D(\Phi)\}$ has the finite intersection property. We may conclude by using the KKM principle after showing that G is a KKM-mapping. Suppose by contradiction that there exist $x_1, \cdots, x_n \in K \cap D(\Phi)$ and $y_0 \in \text{conv}\,\{x_1, \cdots, x_n\}$ such that $y_0 \notin \bigcup_{i=1}^n G(x_i)$. Then

$$\langle Ay_0 - f, y_0 - x_i\rangle + \Phi(y_0) - \Phi(x_i) - J^0\left(L(y_0); L(x_i - y_0)\right) > 0\,,$$

for all $i = 1, \cdots, n$. Therefore $x_i \in \Lambda$, for all $i \in \{1, \cdots, n\}$, where

$$\Lambda := \{x \in X; \langle Ay_0 - f, y_0 - x\rangle + \Phi(y_0) - \Phi(x) - J^0\left(L(y_0); L(x - y_0)\right) > 0\}\,.$$

The set Λ is convex and thus $y_0 \in \Lambda$, leading to an obvious contradiction. So,

$$\bigcap_{x\in K\cap D(\Phi)} G(x) \neq \emptyset\,.$$

This yields an element $u \in K \cap D(\Phi)$ such that, for any $v \in K \cap D(\Phi)$,

$$\langle Au - f, v - u \rangle + \Phi(v) - \Phi(u) + J^0 \left(L(u); L(v-u) \right) \geq 0 \, .$$

This inequality is trivially satisfied if $v \notin D(\Phi)$ and the conclusion follows. ■

We may now derive a result applicable to the inequality problem (6.16). Specifically, suppose that the above hypotheses are satisfied and set $Y = L^p(T; I\!R^m)$. Let $J : Y \to I\!R$ be the function defined by

$$J(u) = \int_T j(x, u(x)) d\mu \, . \tag{6.18}$$

Conditions (6.12) or (6.13)-(6.14) on j ensure that J is locally Lipschitz on Y and

$$\int_T j^0(x, u(x); v(x)) d\mu \geq J^0(u; v) \, , \quad \forall u, v \in L^p(T; I\!R^m) \, .$$

It follows that

$$\int_T j^0(x, \gamma(u(x)); \gamma(v(x))) d\mu \geq J^0(\gamma(u); \gamma(v)) \, , \quad \forall u, v \in X \, . \tag{6.19}$$

It results that if $u \in K$ is a solution of (6.17) then u solves inequality problem (6.16), too. ■

Thus, the result below has been proven.

Theorem 6.4 (Motreanu and Rădulescu [13]) Assume that the hypotheses of Lemma 6.3 are fulfilled for $Y = L^p(T; I\!R^m)$ and $L = \gamma$. Then problem (6.16) has at least a solution.

In order to establish a variant of Lemma 6.3 for monotone and hemicontinuous operators we need the following result which is due to Mosco (see [11]).

Mosco's Theorem Let K be a nonempty convex and compact subset of a topological vector space X. Let $\Phi : X \to I\!R \cup \{+\infty\}$ be a proper, convex and lower semicontinuous function such that $D(\Phi) \cap K \neq \emptyset$. Let $f, g : X \times X \to I\!R$ be two functions such that
(i) $g(x, y) \leq f(x, y)$, for every $x, y \in X$;
(ii) the mapping $f(\cdot, y)$ is concave, for any $y \in X$;
(iii) the mapping $g(x, \cdot)$ is lower semicontinuous, for every $x \in X$.
Let λ be an arbitrary real number. Then the following alternative holds: either there exists $y_0 \in D(\Phi) \cap K$ such that $g(x, y_0) + \Phi(y_0) - \Phi(x) \leq \lambda$, for any $x \in X$, or there exists $x_0 \in X$ such that $f(x_0, x_0) > \lambda$.

We notice that two particular cases of interest for the above result are if $\lambda = 0$ or $f(x,x) \leq 0$, for every $x \in X$.

Lemma 6.4 Let K be a nonempty, bounded, closed subset of the real reflexive Banach space X, and $\Phi : X \to I\!R \cup \{+\infty\}$ a convex and lower semicontinuous function such that (6.15) holds. Consider a linear subspace Y of X^* such that there exists a linear and compact mapping $L : X \to Y$. Let $J : Y \to I\!R$ be a locally Lipschitz function. Suppose in addition that the operator $A : X \to X^*$ is monotone and hemicontinuous. Then for each $f \in X^*$, the inequality problem (6.17) has at least a solution.

Proof. Set

$$g(x,y) = \langle Ax - f, y - x \rangle - J^0(L(y); L(x) - L(y))$$

and

$$f(x,y) = \langle Ay - f, y - x \rangle - J^0(L(y); L(x) - L(y)).$$

The monotonicity of A implies that

$$g(x,y) \leq f(x,y), \quad \forall x, y \in X.$$

The mapping $x \mapsto f(x,y)$ is concave while the mapping $y \mapsto g(x,y)$ is weakly lower semicontinuous. Applying Mosco's Theorem with $\lambda = 0$, we obtain the existence of $u \in K \cap D(\Phi)$ satisfying

$$g(w,u) + \Phi(u) - \Phi(w) \leq 0, \quad \forall w \in K,$$

that is

$$\langle Aw - f, w - u \rangle + \Phi(w) - \Phi(u) + J^0(L(u); L(w-u)) \geq 0, \quad \forall w \in K. \quad (6.20)$$

Fix $v \in K$ and set $w = u + \lambda(v - u) \in K$, for $\lambda \in [0,1[$. So, by (6.20), we get

$$\lambda \langle A(u + \lambda(v-u)) - f, v - u \rangle + \Phi(\lambda v + (1-\lambda)u)) - \Phi(u)$$
$$+ J^0(L(u); \lambda L(v-u)) \geq 0.$$

Using the convexity of Φ, the fact that $J^0(u;\cdot)$ is positive homogeneous and dividing then by $\lambda > 0$ we find

$$\langle A(\lambda v + (1-\lambda)u) - f, v - u \rangle + \Phi(v) - \Phi(u) + J^0(L(u); L(v-u)) \geq 0.$$

Now, letting $\lambda \to 0$ and using the hemicontinuity of A we find that u solves (6.17). ■

In particular, we obtain the analogue of Theorem 6.4 for monotone and hemicontinuous operators.

Theorem 6.5 Assume that the hypotheses of Lemma 6.4 are fulfilled for $Y = L^p(T; \mathbb{R}^m)$ and $L = \gamma$. Then the inequality problem (6.16) admits at least a solution.

We observe that if j satisfies conditions (6.11) and (6.12) then, by the Cauchy-Schwarz inequality,

$$|\int_T j^0(x, \gamma(u(x)); \gamma(v(x)))d\mu| \leq \int_T k(x)|\gamma(v(x))|d\mu$$

$$\leq |k|_{p'} \cdot |\gamma(v)|_p \leq C\,|k|_{p'}\,\|v\|\,, \tag{6.21}$$

where $|\cdot|_p$ denotes the norm in the space $L^p(T; \mathbb{R}^m)$ and $\|\cdot\|$ stands for the norm in X. On the other hand, if j satisfies conditions (6.11), (6.13) and (6.14) then

$$|j^0(x, \gamma(u(x)); \gamma(v(x)))| \leq C\,(1 + |\gamma(u(x))|^{p-1})\,|\gamma(v(x))|$$

and thus

$$\left|\int_T j^0(x, \gamma(u(x)); \gamma(v(x)))d\mu\right| \leq C\,(|\gamma(v)|_1 + |\gamma(u)|_p^{p-1}|\gamma(v)|_p)$$

$$\leq C_1\,\|v\| + C_2\,\|u\|^{p-1}\|v\|\,, \tag{6.22}$$

for some suitable constants $C_1, C_2 > 0$. We discuss in this framework the solvability of coercive variational-hemivariational inequalities.

Theorem 6.6 (Motreanu and Rădulescu [13]) Let K be a nonempty closed convex subset of X, $\Phi : X \to \mathbb{R} \cup \{+\infty\}$ a proper, convex and lower semicontinuous function such that $K \cap D(\Phi) \neq \emptyset$ and $A : X \to X^*$ an operator such that the mapping $v \mapsto \langle Av, v - x\rangle$ is weakly lower semicontinuous, for all $x \in K$. The following hold

(i) If j satisfies conditions (6.11) and (6.12), and if there exists $x_0 \in K \cap D(\Phi)$ such that

$$\frac{\langle Aw, w - x_0\rangle + \Phi(w)}{\|w\|} \to +\infty\,, \quad \text{as } \|w\| \to +\infty \tag{6.23}$$

then for each $f \in X^*$, there exists $u \in K$ such that

$$\langle Au - f, v - u\rangle + \Phi(v) - \Phi(u) + \int_T j^0(x, \gamma(u(x)); \gamma(v(x)) - \gamma(u(x)))d\mu \geq 0\,, \tag{6.24}$$

for all $v \in K$.

(ii) If j satisfies conditions (6.11), (6.13) and (6.14) and if there exist $x_0 \in K \cap D(\Phi)$ and $\theta \geq p$ such that

$$\frac{\langle Aw, w - x_0 \rangle}{\|w\|^\theta} \to +\infty, \quad \text{as } \|w\| \to +\infty \tag{6.25}$$

then for each $f \in X^*$, there exists $u \in K$ satisfying (6.24).

Proof. There exists a positive integer n_0 such that

$$x_0 \in K_n := \{x \in K : \|x\| \leq n\}, \quad \forall n \geq n_0 .$$

Applying Lemma 6.3 with J as defined in (6.18) we find some $u_n \in K_n$ such that, for every $n \geq n_0$ and any $v \in K_n$,

$$\langle Au_n - f, v - u_n \rangle + \Phi(v) - \Phi(u_n) + J^0(\gamma(u_n); \gamma(v) - \gamma(u_n)) \geq 0 . \tag{6.26}$$

We claim that the sequence $\{u_n\}$ is bounded. Suppose by contradiction that $\|u_n\| \to +\infty$. Then, passing eventually to a subsequence, we may assume that

$$v_n := \frac{u_n}{\|u_n\|} \rightharpoonup v .$$

Setting $v = x_0$ in (6.26) and using (6.19), we obtain

$$\langle Au_n, u_n - x_0 \rangle + \Phi(u_n) \leq \Phi(x_0) + \langle f, u_n - x_0 \rangle + J^0(\gamma(u_n); \gamma(x_0 - u_n))$$

$$\leq \Phi(x_0) + \langle f, u_n - x_0 \rangle + \left| \int_T j^0(x, \gamma(u_n); \gamma(x_0 - u_n)) d\mu \right| . \tag{6.27}$$

Case (i). Using (6.21) we obtain

$$\langle Au_n, u_n - x_0 \rangle + \Phi(u_n) \leq \Phi(x_0) + \langle f, u_n - x_0 \rangle + c|k|_{p'} \|u_n - x_0\|$$

and thus

$$\frac{\langle Au_n, u_n - x_0 \rangle + \Phi(u_n)}{\|u_n\|}$$

$$\leq \frac{\Phi(x_0)}{\|u_n\|} + \langle f, v_n - x_0 \|u_n\|^{-1} \rangle + c|k|_{p'} \|v_n - x_0 \|u_n\|^{-1}\| . \tag{6.28}$$

Passing to the limit as $n \to \infty$ we observe that the left-hand side term in (6.28) tends to $+\infty$ while the right-hand term remains bounded which yields a contradiction.

Case (ii). The function Φ being convex and lower semicontinuous, we may apply the Hahn-Banach separation theorem to find that

$$\Phi(x) \geq \langle \alpha, x \rangle + \beta , \quad \forall x \in X ,$$

for some $\alpha \in X^*$ and $\beta \in \mathbb{R}$. This implies that

$$\Phi(x) \geq -\|\alpha\|_* \|x\| + \beta, \quad \forall x \in X.$$

From (6.27) and (6.22) we deduce that

$$\langle Au_n, u_n - x_0 \rangle \leq \Phi(x_0) + \|\alpha\|_* \|u_n\| - \beta$$

$$+\langle f, u_n - x_0 \rangle + C_1 \|u_n - x_0\| + C_2 \|u_n\|^{p-1} \|u_n - x_0\|.$$

Thus

$$\frac{\langle Au_n, u_n - x_0 \rangle}{\|u_n\|^\theta} \leq \|\alpha\|_* \|u_n\|^{1-\theta} + (\Phi(x_0) - \beta)\|u_n\|^{-\theta}$$

$$+\langle f, v_n \|u_n\|^{1-\theta} - x_0 \|u_n\|^{-\theta} \rangle + C_1 \|v_n \|u_n\|^{1-\theta} - x_0 \|u_n\|^{-\theta}\|$$
$$+C_2 \|v_n - x_0 \|u_n\|^{-1}\| \cdot \|u_n\|^{p-\theta}$$

and taking the limit as $n \to \infty$ we obtain a contradiction, since we supposed that $\theta \geq p \geq 1$.

Thus in both cases (i) and (ii), the sequence $\{u_n\}$ is bounded. This implies that, up to a subsequence, $u_n \rightharpoonup u \in K$. Let $v \in K$ be given. For all n large enough we have $v \in K_n$ and thus by (6.26),

$$\langle Au_n - f, u_n - v \rangle + \Phi(u_n) - \Phi(v) - J^0(\gamma(u_n); \gamma(v) - \gamma(u_n)) \leq 0. \quad (6.29)$$

Passing to the limit as $n \to \infty$ we obtain

$$\langle Au - f, u - v \rangle \leq \liminf_{n \to \infty} \langle Au_n - f, u_n - v \rangle,$$

$$\Phi(u) \leq \liminf_{n \to \infty} \Phi(u_n),$$

$$\gamma(u) = \lim_{n \to \infty} \gamma(u_n),$$

and

$$-J^0(\gamma(u); \gamma(v) - \gamma(u)) \leq \liminf_{n \to \infty} \left(-J^0(\gamma(u_n); \gamma(v) - \gamma(u_n)) \right).$$

Taking the inferior limit in (6.29) we obtain

$$\langle Au - f, u - v \rangle + \Phi(u) - \Phi(v) - J^0(\gamma(u); \gamma(v) - \gamma(u)) \leq 0.$$

Since v has been chosen arbitrarily we obtain

$$\langle Au - f, v - u \rangle + \Phi(v) - \Phi(u) + J^0(\gamma(u); \gamma(v) - \gamma(u)) \geq 0, \quad \forall v \in K.$$

Using now again (6.19) we conclude that u solves (6.24). ■

The following result gives a corresponding variant for monotone hemicontinuous operators.

Theorem 6.7 Let K be a nonempty closed convex subset of X, $\Phi : X \to I\!R \cup \{+\infty\}$ a proper convex and lower semicontinuous function such that $D(\Phi) \cap K \neq \emptyset$. Let $A : X \to X^*$ be a monotone and hemicontinuous operator. Assume (6.23) or (6.25) as in Theorem 6.6. Then the conclusion of Theorem 6.6 holds true.

Proof. Using Lemma 6.4 we find a sequence $u_n \in K_n$ such that

$$\langle Au_n - f, v - u_n \rangle + \Phi(v) - \Phi(u_n) + J^0(\gamma(u_n); \gamma(v) - \gamma(u_n)) \geq 0, \quad (6.30)$$

for all $v \in K_n$.

As in the proof of Theorem 6.6 we justify that $\{u_n\}$ is bounded and thus, up to a subsequence, we may assume that $u_n \rightharpoonup u$. By (6.30) and the monotonicity of A we deduce that

$$\langle Av - f, v - u_n \rangle + \Phi(v) - \Phi(u_n) + J^0(\gamma(u_n); \gamma(v) - \gamma(u_n)) \geq 0.$$

Let $v \in K$ be given. For n large enough we obtain

$$\langle Av - f, u_n - v \rangle + \Phi(u_n) - \Phi(v) - J^0(\gamma(u_n); \gamma(v) - \gamma(u_n)) \leq 0$$

and taking the inferior limit we obtain

$$\langle Av - f, u - v \rangle + \Phi(u) - \Phi(v) - J^0(\gamma(u); \gamma(v) - \gamma(u)) \leq 0.$$

Since v has been chosen arbitrarily it follows that

$$\langle Av - f, v - u \rangle + \Phi(v) - \Phi(u) + J^0(\gamma(u); \gamma(v) - \gamma(u)) \geq 0, \quad \forall v \in K.$$

Using now the same argument as in the proof of Lemma 6.4 we obtain that

$$\langle Au - f, v - u \rangle + \Phi(v) - \Phi(u) + J^0(\gamma(u); \gamma(v) - \gamma(u)) \geq 0, \quad \forall v \in K$$

and the conclusion follows now by (6.19). ∎

In order to treat noncoercive cases we shall apply a minimax approach for studying the inequality problem (6.17) (in particular, (6.16)).

First, we describe the abstract functional framework of our variational approach in studying the inequality problem (6.17) without the assumptions of boundedness for set K or of coerciveness as in Theorem 6.6. Let X and Y be Banach spaces, with X reflexive, and let $L : X \to Y$ be a linear compact operator. Consider the functionals $E \in C^1(X; I\!R)$

(in (6.17) we will take $A := E' : X \to X^*$), $\Phi : X \to \mathbb{R}$ convex, lower semicontinuous, Gâteaux differentiable and $J : Y \to \mathbb{R}$ locally Lipschitz. Given a closed convex cone K of X, with $0 \in K$, let I_K denote the indicator function of K. We apply the nonsmooth version of the Mountain Pass Theorem in Corollary 3.1.

The following result follows readily from Definition 2.2.

Lemma 6.5 Every critical point $u \in X$ of the functional $I = E + J \circ L$ in the sense of Definition 2.2 is a solution to problem (6.17) with $A = E'$.

The following statement provides a sufficient condition for the Palais-Smale condition of functional I in Lemma 6.5.

Lemma 6.6 Assume in addition that the following hypotheses are satisfied:
(H1) There exist positive constants a_0, a_1, α with $\alpha < a_0$ such that

$$E(v) + \Phi(v) + J(Lv) - \alpha(\langle E'(v) + \Phi'(v), v\rangle + J^0(Lv; Lv))$$

$$\geq a_0\|v\| - a_1, \quad \forall v \in K;$$

(H2) If $\{u_n\}$ is a sequence in K satisfying $u_n \rightharpoonup u$ in X and

$$\limsup_{n\to\infty}\langle E'(u_n), u_n - u\rangle \leq 0$$

for some $u \in X$, then $\{u_n\}$ contains a subsequence denoted again by $\{u_n\}$ with $u_n \to u$ in X.
Then the functional I satisfies the Palais-Smale condition in the sense of Definition 2.3.

Proof. Let $\{u_n\}$ be a sequence in X with the properties required in Definition 2.3. In particular, we know that $\{u_n\} \subset K$ and there exist a constant $M > 0$ and a sequence $\{\varepsilon_n\} \subset \mathbb{R}^+$ with $\varepsilon_n \to 0$ such that

$$|I(u_n)| \leq M, \quad \forall n \geq 1,$$

and

$$\langle E'(u_n), v - u_n\rangle + J^0(Lu_n; Lv - Lu_n) + \Phi(v) - \Phi(u_n) \geq -\varepsilon_n\|v - u_n\|,$$

for all $v \in K$. Using the convexity and the Gâteaux differentiability of Φ, setting $v = (1 + t)u_n$, with $t > 0$, in the inequality above and then letting $t \to 0$ one obtains that

$$\langle E'(u_n) + \Phi'(u_n), u_n\rangle + J^0(Lu_n; Lu_n) \geq -\varepsilon_n\|u_n\|, \quad \forall n \geq 1.$$

The inequalities above ensure that for n sufficiently large (so that $\varepsilon_n \leq 1$) one has

$$M + \alpha\|u_n\|$$

$$\geq E(u_n) + \Phi(u_n) + J(Lu_n) - \alpha[\langle E'(u_n) + \Phi'(u_n), u_n\rangle + J^0(Lu_n; Lu_n)].$$

Here α denotes the positive constant entering assumption $(H1)$. Then on the basis of condition $(H1)$ we deduce that the sequence $\{u_n\}$ is bounded in X.

Consequently, the sequence $\{u_n\}$ contains a subsequence again denoted by $\{u_n\}$ such that $u_n \rightharpoonup u$ in X and $Lu_n \to Lu$ in Y for some $u \in K$. On the other hand setting $v = u$, we derive that

$$\langle E'(u_n), u - u_n\rangle + J^0(Lu_n; Lu - Lu_n) + \Phi(u) - \Phi(u_n) \geq -\varepsilon_n\|u - u_n\|.$$

Since J^0 is upper semicontinuous and Φ is lower semicontinuous, this yields that

$$\limsup_{n\to\infty}\langle E'(u_n), u_n - u\rangle \leq 0.$$

Assumption $(H2)$ completes the proof. ■

The main result of this Section is stated below.

Theorem 6.8 Assume $(H1)$, $(H2)$ in Lemma 6.6 together with

$(H3)$ There exist an element $\overline{u} \in K \setminus \{0\}$ satisfying $\|\overline{u}\| > a_1/a_0$, for the constants a_0, a_1 in $(H1)$, and $E(\overline{u}) + \Phi(\overline{u}) + J(L\overline{u}) < 0$;

$(H4)$ There exist a constant $\rho > 0$ such that

$$\inf_{\|v\|=\rho} (E(v) + \Phi(v) + J(Lv)) > E(0) + \Phi(0) + J(0).$$

Then problem (6.17) with $A = E'$ admits at least a solution $u \in K \setminus \{0\}$.

Proof. Let us apply the nonsmooth version of Mountain Pass Theorem in Corollary 3.1 to our functional I. Lemma 6.6 establishes that I satisfies the Palais-Smale condition in the sense of Definition 2.3.

By the properties of generalized gradients we have

$$\partial_t(t^{-\frac{1}{\alpha}}(E + \Phi)(tu) + t^{-\frac{1}{\alpha}}J(tLu))$$

$$\subset -\frac{1}{\alpha}t^{-\frac{1}{\alpha}-1}(E + \Phi)(tu) + t^{-\frac{1}{\alpha}}\langle(E' + \Phi')(tu), u\rangle$$

$$-\frac{1}{\alpha}t^{-\frac{1}{\alpha}-1}J(tLu) + t^{-\frac{1}{\alpha}}\partial J(tLu)Lu, \ \forall t > 0, \ \forall u \in X,$$

where the notation ∂_t stands for the generalized gradient with respect to t. Lebourg's mean value theorem allows to find some $\tau = \tau(u) \in (1, t)$ such that

$$t^{-\frac{1}{\alpha}}(E(tu) + \Phi(tu) + J(tLu)) - (E(u) + \Phi(u) + J(Lu))$$

$$\in \frac{1}{\alpha}\tau^{-\frac{1}{\alpha}-1}[\alpha(\langle E'(\tau u) + \Phi'(\tau u), \tau u\rangle + \partial J(\tau Lu)\tau Lu)$$

$$-(E(\tau u) + \Phi(\tau u) + J(\tau Lu))](t-1), \quad \forall t > 1, \ \forall u \in X.$$

Combining with assumption $(H1)$ it follows that

$$t^{-\frac{1}{\alpha}}(E(tu) + \Phi(tu) + J(tLu)) - (E(u) + \Phi(u) + J(Lu))$$

$$\leq \frac{1}{\alpha}\tau^{-\frac{1}{\alpha}-1}(-a_0\tau\|u\| + a_1)(t-1), \quad \forall t > 1, \ \forall u \in K.$$

It is then clear from assumption $(H3)$ that one can write

$$I(t\overline{u}) = E(t\overline{u}) + \Phi(t\overline{u}) + J(tL\overline{u}) \leq t^{\frac{1}{\alpha}}[E(\overline{u}) + \Phi(\overline{u}) + J(L\overline{u})], \quad \forall t > 1.$$

This fact in conjunction with assumption $(H3)$ leads to the conclusion that

$$\lim_{t \to +\infty} I(t\overline{u}) = -\infty.$$

Then assumption $(H4)$ enables us to apply the nonsmooth version of Mountain Pass Theorem (see Corollary 3.1) for $e = t\overline{u}$, with a sufficiently large positive number t. According to Mountain Pass Theorem the functional I possesses a nontrivial critical point $u \in X$ in the sense of Definition 2.2. Finally, Lemma 6.5 shows that u is a (nontrivial) solution of problem (6.17) with $A = E'$. The proof of Theorem 6.8 is thus complete. ∎

We give in what follows an example of application of Theorem 6.8 in the case of variational-hemivariational inequality (6.16). For the sake of simplicity we consider a uniformly convex Banach space X, a convex closed cone K in X with $0 \in K$, $f = 0$, $\Phi = 0$ and a self-adjoint linear continuous operator $A : X \to X^*$ satisfying $\langle Av, v\rangle \geq c_0\|v\|^2$, for all $v \in X$, with a constant $c_0 > 0$.

Assume that the function $j : T \times \mathbb{R}^m \to \mathbb{R}$ verifies the conditions (6.11), (6.13), (6.14) with $p > 2$, as well as the following assumptions of Ambrosetti-Rabinowitz type:

(AR_1) there exist constants $0 < \alpha < 1/2$ and $c \in \mathbb{R}$ such that

$$j(x, y) \geq \alpha j_y^0(x, y; y) + c, \quad \text{for a.e. } x \in T, \ \forall y \in \mathbb{R}^m;$$

(AR_2) $\liminf_{y\to 0} \frac{1}{|y|^2} j(x,y) \geq 0$ uniformly with respect to $x \in T$, and $j(x,0) = 0$ a.e. $x \in T$;

(AR_3) there exists an element $u_0 \in K \setminus \{0\}$ such that

$$\liminf_{t\to\infty} \left[\frac{1}{2} \langle Au_0, u_0 \rangle t^2 + \int_T j(x, tu_0(x))dx \right] < 0.$$

Let us apply Theorem 6.8 for the functional J given by (6.18) and $E(v) = (1/2)\langle Av, v\rangle$, $\forall v \in X$. We see that hypotheses (AR_1) and (AR_2) imply $(H1)$ and $(H4)$, respectively. Taking $\overline{u} = tu_0$ for $t > 0$ sufficiently large, we get $(H3)$ from (AR_3). It is straightforward to check that condition $(H2)$ holds true. Therefore Theorem 6.8 yields a nontrivial solution of variational-hemivariational inequality (6.16) in our setting.

3. Double Eigenvalue Hemivariational Inequalities with Non-locally Lipschitz Energy Functional

In this Section we prove an existence result for a new type of hemivariational inequalities that are called "double eigenvalue problems" and which has been introduced by D. Motreanu and P. D. Panagiotopoulos in [12].

Let V be a Hilbert space and let $\Omega \subset \mathbb{R}^m$ be an open bounded subset of $\mathbb{R}^m, m \geq 1$, with $\partial\Omega$ sufficiently smooth. We suppose that V is compactly embedded into $L^p(\Omega; \mathbb{R}^N), N \geq 1$, for some $p \in (1, +\infty)$. In particular, the continuity of this embedding implies the existence of a constant $C_p(\Omega) > 0$ such that

$$(*) \qquad \|u\|_{L^p} \leq C_p(\Omega) \cdot \|u\|_V, \text{ for all } u \in V,$$

where by $\|\cdot\|_{L^p}$ and $\|\cdot\|_V$ we have denoted the norms in $L^p(\Omega; \mathbb{R}^N)$ and V respectively. We suppose that $V \cap L^\infty(\Omega; \mathbb{R}^N)$ is dense in V. Let $a_1, a_2 : V \times V \to \mathbb{R}$ be two bilinear and continuous forms on V which are coercive in the sense that there exist two real-valued functions $c_1, c_2 : \mathbb{R}_+ \to \mathbb{R}_+$, with $\lim_{r\to\infty} c_i(r) = +\infty$, such that for all $v \in V$

$$a_i(v,v) \geq c_i(\|v\|_V) \cdot \|v\|_V, \ i = 1, 2.$$

Denote by $A_1, A_2 : V \to V$ the operators associated to the forms considered above, defined by

$$(A_i u, v)_V = a_i(u,v), \ i = 1, 2.$$

The operators A_1 and A_2 are linear, continuous and coercive in the sense that for each $i = 1, 2$ we have

$$(A_i u, u)_V \geq c_i(\|u\|_V) \cdot \|u\|_V, \text{ for all } u \in V.$$

In addition, we suppose that the operators A_1 and A_2 are weakly continuous, i.e., if $u_n \rightharpoonup u$, weakly in V then $A_i u_n \rightharpoonup A_i u$, also weakly in V, for each $i = 1, 2$. Consider two bounded selfadjoint linear and weakly continuous operators $B_1, B_2 : V \to V$. Let $j : \Omega \times \mathbb{R}^N \to \mathbb{R}$ be a Carathéodory function which is locally Lipschitz in the second variable for a.e. $x \in \Omega$.

In order to ensure the integrability of $j(\cdot, u(\cdot))$ and $j^0(\cdot, u(\cdot); v(\cdot))$ for any $u, v \in V \cap L^\infty(\Omega; \mathbb{R}^N)$ we admit the existence of a function $\beta : \Omega \times \mathbb{R}_+ \to \mathbb{R}$ fulfilling the conditions

(β_1) $\beta(\cdot, r) \in L^1(\Omega)$, for each $r \geq 0$;
(β_2) if $r_1 \leq r_2$ then $\beta(x, r_1) \leq \beta(x, r_2)$, for almost all $x \in \Omega$, and such that

$$|j(x,\xi) - j(x,\eta)| \leq \beta(x,r) \cdot |\xi - \eta|, \ \forall \xi, \eta \in B(0,r), \ r \geq 0, \quad (6.31)$$

where $B(0, r) = \{\xi \in \mathbb{R}^N : |\xi| \leq r\}$, the symbol $|\cdot|$ denoting the norm in $\mathbb{R}^N$.

Let $1 \leq s < p$ and let $k : \Omega \to \mathbb{R}_+$ and $\alpha : \Omega \times \mathbb{R}_+ \to \mathbb{R}$ be two functions satisfying the assumptions:

$$k(\cdot) \in L^q(\Omega), \quad \text{where } \frac{1}{p} + \frac{1}{q} = 1, \quad (6.32)$$

$$\alpha(\cdot, r) \in L^{q'}(\Omega), \text{ for each } r > 0, \text{ where } q' = \frac{p}{p - s} \quad (6.33)$$

and

$$\text{if } 0 < r_1 \leq r_2 \text{ then } \alpha(x, r_1) \leq \alpha(x, r_2), \text{ for almost all } x \in \Omega. \quad (6.34)$$

We also impose the following directional growth conditions:

$$j^0(x, \xi; -\xi) \leq k(x) \cdot |\xi|, \text{ for all } \xi \in \mathbb{R}^N \text{ and a.e. } x \in \Omega; \quad (6.35)$$

$$j^0(x, \xi; \eta - \xi) \leq \alpha(x, r)\left(1 + |\xi|^s\right), \text{ for all } \xi, \eta \in \mathbb{R}^N, \quad (6.36)$$

with $\eta \in B(0, r)$, $r > 0$, and a.e. $x \in \Omega$.

Remark 6.4 We point out that the growth conditions (6.35) and (6.36) do not ensure the finite integrability of $j(\cdot, u(\cdot))$ and $j^0(\cdot, u(\cdot); v(\cdot))$ in Ω for any $u, v \in V$. We can also remark that they do not guarantee that the functional $J : V \to \mathbb{R}$ given by

$$J(v) = \int_\Omega j(x, v(x)) dx,$$

is locally Lipschitz on V. In fact, (6.35) and (6.36) do not allow us to conclude even that the effective domain of J coincides with the whole space V. For more details on this type of conditions we refer to Naniewicz [14] and Naniewicz and Panagiotopoulos [15].

Remark 6.5 Notice that we do not impose any coerciveness assumption on the operators B_i $(i = 1, 2)$, as done in [12], Section 4, for the case of a double eigenvalue problem on a sphere. We suppose however that these operators satisfy the additional hypothesis of weak continuity.

Let us now consider two nonlinear monotone and demicontinuous operators $C_1, C_2 : V \to V$.

We study the following double eigenvalue problem:

(P) Find $u_1, u_2 \in V$ and $\lambda_1, \lambda_2 \in I\!R$ such that

$$a_1(u_1, v_1) + a_2(u_2, v_2) + (C_1(u_1), v_1)_V + (C_2(u_2), v_2)_V$$

$$+ \int_\Omega j^0(x, (u_1 - u_2)(x); (v_1 - v_2)(x))dx$$

$$\geq \lambda_1 (B_1 u_1, v_1)_V + \lambda_2 (B_2 u_2, v_2)_V, \quad \forall v_1, v_2 \in V.$$

By Remark 6.4 we derive that in order to find a solution for the problem (P) we cannot follow the classical technique of Clarke [3]. For this reason, our problem (P) is a nonstandard one.

Definition 6.3 We say that $(u_1, u_2, \lambda_1, \lambda_2) \in V \times V \times I\!R \times I\!R$ is a solution of (P) if there exists $\chi \in L^1(\Omega; I\!R^N) \cap V$ such that

$$a_1(u_1, v_1) + a_2(u_2, v_2) + (C_1(u_1), v_1)_V + (C_2(u_2), v_2)_V$$

$$+ \int_\Omega \chi(x) \cdot (v_1 - v_2)(x)dx = \lambda_1 (B_1 u_1, v_1)_V + \lambda_2 (B_2 u_2, v_2)_V, \quad (6.37)$$

for any $v_1, v_2 \in V \cap L^\infty(\Omega; I\!R^N)$, and

$$\chi(x) \in \partial j(x, (u_1 - u_2)(x)), \text{ for a.e. } x \in \Omega. \quad (6.38)$$

We shall prove the following existence result.

Theorem 6.9 Assume that the above hypotheses are fulfilled. Then the double eigenvalue problem (P) has at least one solution.

The difficulties mentioned in the Remark 6.4 will be surmounted by employing the Galerkin approximation method combined with the finite intersection property. For the treatment of finite dimensional problem we shall use Kakutani's fixed point theorem for multivalued mappings. This technique has been introduced in Naniewicz and Panagiotopoulos [15].

Let Λ be the family of all finite dimensional subspaces F of $V \cap L^\infty(\Omega; I\!R^N)$, ordered by inclusion. For any $F \in \Lambda$ we formulate the following finite dimensional problem

(P_F) Find $u_{1F}, u_{2F} \in F, \lambda_1, \lambda_2 \in I\!R$ and $\chi_F \in L^1(\Omega; I\!R^N)$ such that

$$a_1(u_{1F}, v_1) + a_2(u_{2F}, v_2) + (C_1(u_{1F}), v_1)_V + (C_2(u_{2F}), v_2)_V$$

$$+ \int_\Omega \chi_F(x) \cdot (v_1 - v_2)(x) dx = \lambda_1 (B_1 u_{1F}, v_1)_V + \lambda_2 (B_2 u_{2F}, v_2)_V, \quad (6.39)$$

for any $v_1, v_2 \in F$, and

$$\chi_F(x) \in \partial j(x, (u_{1F} - u_{2F})(x)), \text{ for a.e. } x \in \Omega. \quad (6.40)$$

Let $\Gamma_F : F \to 2^{L^1(\Omega; I\!R^N)}$ defined as follows: for any $v_F \in F$, $\Gamma_F(v_F)$ is the set of all $\Psi \in L^1(\Omega; I\!R^N)$ such that

$$\int_\Omega \Psi w dx \leq \int_\Omega j^0(x, v_F(x); w(x)) dx,$$

for all $w \in L^\infty(\Omega; I\!R^N)$.

It is immediate that if $\Psi \in \Gamma_F(v_F)$ then we have $\Psi(x) \in \partial j(x, v_F(x))$, for a.e. $x \in \Omega$. Let $v_F \in F$ for some $F \in \Lambda$. By Lemma 3.1 in [14], $\Gamma(v_F)$ is a nonempty convex and weakly compact subset of $L^1(\Omega; I\!R^N)$. For $F \in \Lambda$, we denote by $i_F : F \to V$ and by $i_F^* : V^* \to F^*$ the inclusion and the dual projection mappings respectively. Let us define $\gamma_F : L^1(\Omega; I\!R^N) \to F^*$, by

$$\langle \gamma_F \Psi, v \rangle_F = \int_\Omega \Psi \cdot v dx, \ \forall v \in F.$$

Consider the map $T_F : F \to 2^{F^*}$ defined by

$$T_F(v_F) = \gamma_F \Gamma_F(v_F).$$

The main properties of T_F are pointed out by the following result which has been established in [14].

Lemma 6.7 For each $v_F \in F$, $T_F(v_F)$ is a nonempty bounded closed convex subset of F^*. Moreover, T_F is upper semicontinuous as a map from F into 2^{F^*}.

We are ready to formulate the existence result for the finite dimensional problem (P_F).

Theorem 6.10 Assume the above hypotheses are fulfilled. Then, for each $F \in \Lambda$, there exist $u_{1F}, u_{2F} \in F, \lambda_1, \lambda_2 \in \mathbb{R}$ and $\chi_F \in L^1(\Omega; \mathbb{R}^N)$ which solve the problem (P_F). Moreover, there exists a positive constant M, independent by F such that

$$\|u_{1F}\|_V + \|u_{2F}\|_V \leq M. \tag{6.41}$$

Proof. In what follows we shall be able to find a solution of the problem (P_F) by restraining the searching area for λ_i, $i \in \{1,2\}$ on the class of all those numbers $\lambda_1, \lambda_2 \in \mathbb{R}$ which satisfy the relation

$$\delta := \inf_{\substack{w_1 \in V \cap L^\infty(\Omega;\mathbb{R}^N) \\ w_2 \in V \cap L^\infty(\Omega;\mathbb{R}^N)}} \frac{\sum_{i=1}^{2} [(C_i(w_i), w_i)_V - \lambda_i \|B_i\| \|w_i\|_V^2]}{\|w_1\|_V + \|w_2\|_V} > -\infty. \tag{6.42}$$

Define $A_{1F} = i_F^* A_1 i_F, A_{2F} = i_F^* A_2 i_F$, and let $\overline{G} : V \times V \to V$ be the map given by

$$\overline{G}(v_1, v_2) = v_1 - v_2.$$

Fix $F \in \Lambda$. We denote by G the map $\overline{G}$ restricted to $F \times F$. Let us consider the multivalued mapping $\Delta : F \times F \to 2^{F^* \times F^*}$ defined by

$$\Delta(u_1, u_2) = (U_1, U_2),$$

where

$$U_1 = A_{1F} u_1 + (C_1(u_1), \cdot)_V - \lambda_1 (B_1 u_1, \cdot)$$

and

$$U_2 = A_{2F} u_2 + (C_2(u_2), \cdot)_V - \lambda_2 (B_2 u_2, \cdot)_V + (G^* \circ T_F \circ G)(u_1, u_2).$$

By $(G^* \circ T_F \circ G)(u_1, u_2)$ we mean the set

$$\{G^*(f) \; : \; f \in T_F(u_1 - u_2)\} \subset F^* \times F^*.$$

The first step consists in proving the upper semicontinuity of $G^* \circ T_F \circ G$. For this aim, let us consider $u_n^1 \to u_1, u_n^2 \to u_2$, strongly in F and $\Psi_n \in G^*(T_F(u_n^1 - u_n^2))$ converging strongly to $\Psi \in F^* \times F^*$. It must be proven that $\Psi \in G^*(T_F(u_1 - u_2))$. We first observe that G fulfills the conditions which permit to apply Theorem II.19 in [1]. From there we draw the conclusion that $\Re(G^*) = \{G^*\theta \; : \theta \in F^*\}$ is closed. Since $\Psi_n \in \Re(G^*)$, for all $n \geq 1$ and $\Psi_n \to \Psi$ in $F^* \times F^*$, it follows that $\Psi \in \Re(G^*)$. Thus we obtain the existence of $\xi^* \in F^*$ such that $\Psi = G^*(\xi^*)$. We have

$$\langle G^*(\gamma_F \chi_n), (v, w)\rangle_{F\times F} \to \langle \Psi, (v, w)\rangle_{F\times F}, \text{ for all } v, w \in F,$$

which implies that $\langle \gamma_F \chi_n, v - w\rangle_F$ tends to $\langle \xi^*, v - w\rangle_F, \forall v, w \in F$ and thus, due to the fact that $\dim F < +\infty$, we get the strong convergence of $\gamma_F \chi_n$ to ξ^* in F^*. Since T_F is upper semicontinuous (see Lemma 6.7), we obtain that there exists $\chi \in \Gamma_F(u_1 - u_2)$ such that $\xi^* = \gamma_F \chi$. Hence, $\Psi = G^*(\gamma_F \chi)$, which means that $\Psi \in (G^* \circ T_F)(u_1 - u_2)$. This ends the proof of the upper semicontinuity of $G^* \circ T_F \circ G$.

On the other hand, the weak continuity of A_1 and A_2 implies the continuity of A_{1F} and A_{2F} from F into F^*. The hypotheses on B_i and $C_i (i = 1, 2)$ and the above considerations lead us to the upper semicontinuity of Δ from $F \times F$ to $2^{F^* \times F^*}$. Applying again Lemma 6.7 and the hypotheses imposed on B_i, C_i and A_i, we can directly derive that for each $(u_1, u_2) \in F \times F, \Delta(u_1, u_2)$ is a nonempty, bounded, closed and convex subset of $F^* \times F^*$. Moreover, from the coercivity of a_1 and a_2 and from the definition of T_F we have

$$\langle \Delta(u_1, u_2), (u_1, u_2)\rangle_{F\times F}$$

$$\geq c_1(\|u_1\|_V)\|u_1\|_V + c_2(\|u_2\|_V)\|u_2\|_V + (C_1(u_1), u_1)_V + (C_2(u_2), u_2)_V$$

$$-\lambda_1 \|B_1\| \cdot \|u_1\|_V^2 - \lambda_2 \|B_2\| \cdot \|u_2\|_V^2 + \int_\Omega \Psi(u_1 - u_2) dx,$$

where $\Psi \in \Gamma_F(u_1 - u_2)$. By (*) and (6.35) we obtain

$$\langle \Delta(u_1, u_2), (u_1, u_2)\rangle_{F\times F}$$

$$\geq c_1(\|u_1\|_V)\|u_1\|_V + c_2(\|u_2\|_V)\|u_2\|_V + (C_1(u_1), u_1)_V + (C_2(u_2), u_2)_V$$

$$-\lambda_1 \|B_1\| \cdot \|u_1\|_V^2 - \lambda_2 \|B_2\| \cdot \|u_2\|_V^2 - \int_\Omega j^0(x, (u_1 - u_2)(x); -(u_1 - u_2)(x)) dx$$

$$\geq c_1(\|u_1\|_V)\|u_1\|_V + c_2(\|u_2\|_V)\|u_2\|_V + (C_1(u_1), u_1)_V + (C_2(u_2), u_2)_V$$

$$-\lambda_1 \|B_1\| \cdot \|u_1\|_V^2 - \lambda_2 \|B_2\| \cdot \|u_2\|_V^2 - C_p(\Omega)\|k\|_{L^q} (\|u_1\|_V + \|u_2\|_V).$$

Taking into account (6.42) we obtain the coercivity of Δ. Thus, Δ fulfills the conditions which allow us to apply Kakutani's fixed point theorem (see [2], Proposition 10, p. 270). Thus $\Re(\Delta) = F^* \times F^*$, which implies the existence of $u_{1F}, u_{2F} \in F$ such that $0 \in \Delta(u_{1F}, u_{2F})$. From the definition of Δ we obtain some $\chi_F \in L^1(\Omega; \mathbb{R}^N)$ such that (6.39) and (6.40) hold. In order to prove the final part of Theorem 6.10 we use the estimates:

$$\lambda_1\|B_1\|\,\|u_{1F}\|_V^2+\lambda_2\|B_2\|\,\|u_{2F}\|_V^2 \geq \lambda_1\left(B_1u_{1F},u_{1F}\right)_V+\lambda_2\left(B_2u_{2F},u_{2F}\right)_V$$

$$= a_1(u_{1F},u_{1F})+a_2(u_{2F},u_{2F})$$

$$+\left(C_1(u_{1F}),u_{1F}\right)_V+\left(C_2(u_{2F}),u_{2F}\right)_V+\int_\Omega \chi_F(u_{1F}-u_{2F})(x)dx$$

$$\geq c_1(\|u_{1F}\|_V)\|u_{1F}\|_V+c_2(\|u_{2F}\|_V)\|u_{2F}\|_V$$

$$+\left(C_1(u_{1F}),u_{1F}\right)_V+\left(C_2(u_{2F}),u_{2F}\right)_V$$

$$-\int_\Omega j^0(x,(u_{1F}-u_{2F})(x);-(u_{1F}-u_{2F})(x))dx.$$

Taking into account relations (6.35) and (6.42) we get

$$\frac{c_1(\|u_{1F}\|_V)\|u_{1F}\|_V+c_2(\|u_{2F}\|_V)\|u_{2F}\|_V}{\|u_{1F}\|_V+\|u_{2F}\|_V} \leq C_p(\Omega)\|k\|_{L^q}-\delta,$$

which by the properties of c_1 and c_2 implies the existence of a positive constant M such that (6.41) holds. ■

Lemma 6.8 For every $F\in\Lambda$, let $u_{1F},u_{2F}\in F$, $\lambda_1,\lambda_2\in I\!R$ and $\chi_F\in L^1(\Omega;I\!R^N)$ solve the problem (P_F). Then the set $\{\chi_F:\ F\in\Lambda\}$ is weakly precompact in $L^1(\Omega;I\!R^N)$.

Proof. The proof is based on the Dunford-Pettis theorem. We have to prove that for each $\epsilon>0$, there exists $\delta_\epsilon>0$ such that for any $\omega\subset\Omega$ with $|\omega|<\delta_\epsilon$,

$$\int_\omega |\chi_F|dx<\epsilon,\ \ F\in\Lambda.$$

Fix $r>0$ and let $\eta\in I\!R^N$ be such that $|\eta|\leq r$. From $\chi_F\in\partial j(x,(u_{1F}-u_{2F})(x))$, for a.e. $x\in\Omega$ we derive that

$$\chi_F(x)\left(\eta-(u_{1F}-u_{2F})(x)\right)\leq j^0(x,(u_{1F}-u_{2F})(x);\eta-(u_{1F}-u_{2F})(x)).$$

Taking into account (6.36) it follows that

$$\chi_F(x)\cdot\eta\leq\chi_F(x)\cdot(u_{1F}-u_{2F})(x)$$

$$+\alpha(x,r)\left(1+|u_{1F}(x)-u_{2F}(x)|^s\right), \tag{6.43}$$

for a.e. $x\in\Omega$.

Denote by $\chi_{Fi}(x), i = 1, 2, \cdots, N$ the components of $\chi_F(x)$ and set

$$\eta(x) = \frac{r}{\sqrt{N}} \left(sgn\chi_{F1}(x), \cdots, sgn\chi_{Fn}(x) \right).$$

It follows that $|\eta(x)| \leq r$ a.e. $x \in \Omega$ and

$$\chi_F(x) \cdot \eta(x) \geq \frac{r}{\sqrt{N}} \cdot |\chi_F(x)|.$$

From (6.43) we obtain

$$\frac{r}{\sqrt{N}} \cdot |\chi_F(x)| \leq \chi_F(x) \cdot (u_{1F} - u_{2F})(x)$$

$$+\alpha(x, r) \left(1 + |u_{1F}(x) - u_{2F}(x)|^s\right)$$

Integrating over $\omega \subset \Omega$ the above inequality yields

$$\int_\omega |\chi_F(x)| dx \leq \frac{\sqrt{N}}{r} \int_\omega \chi_F(x) \cdot (u_{1F} - u_{2F})(x) dx$$

$$+\frac{\sqrt{N}}{r} \|\alpha(\cdot, r)\|_{L^{q'}(\omega)} \cdot |\omega|^{\frac{s}{p}}$$

$$+\frac{\sqrt{N}}{r} \|\alpha(\cdot, r)\|_{L^{q'}(\omega)} \cdot \|u_{1F} - u_{2F}\|^s_{L^p(\omega)}.$$

Thus, from (*) and (6.41) we get

$$\int_\omega |\chi_F(x)| dx \leq \frac{\sqrt{N}}{r} \int_\omega \chi_F(x) \cdot (u_{1F} - u_{2F})(x) dx$$

$$+\frac{\sqrt{N}}{r} \|\alpha(\cdot, r)\|_{L^{q'}(\Omega)} \cdot |\omega|^{\frac{s}{p}}$$

$$+\frac{\sqrt{N}}{r} \|\alpha(\cdot, r)\|_{L^{q'}(\omega)} \cdot (C_p(\Omega))^s \cdot \|u_{1F} - u_{2F}\|^s_V$$

$$\leq \frac{\sqrt{N}}{r} \int_\omega \chi_F(x) \cdot (u_{1F} - u_{2F})(x) dx + \frac{\sqrt{N}}{r} \|\alpha(\cdot, r)\|_{L^{q'}(\Omega)} \cdot |\omega|^{\frac{s}{p}}$$

$$+\frac{\sqrt{N}}{r} \|\alpha(\cdot, r)\|_{L^{q'}(\omega)} \cdot (C_p(\Omega))^s \cdot M^s. \tag{6.44}$$

Observe that (6.35) implies

$$\chi_F(x) \cdot (u_{1F}(x) - u_{2F}(x)) + k(x) \cdot (1 + |u_{1F}(x) - u_{2F}(x)|) \geq 0,$$

for a.e. $x \in \Omega$. Therefore

$$\int_{\omega} (\chi_F(x) \cdot (u_{1F} - u_{2F})(x) + k(x)(1 + |u_{1F}(x) - u_{2F}(x)|))\, dx$$

$$\leq \int_{\Omega} (\chi_F(x) \cdot (u_{1F} - u_{2F})(x) + k(x)(1 + |u_{1F}(x) - u_{2F}(x)|))\, dx$$

and we derive that

$$\int_{\omega} \chi_F(x) \cdot (u_{1F} - u_{2F})(x) dx$$

$$\leq \int_{\Omega} \chi_F(x) \cdot (u_{1F} - u_{2F})(x) dx + \|k\|_{L^q(\Omega)} \cdot C_p(\Omega) \cdot \|u_{1F} - u_{2F}\|_V$$

$$+ \|k\|_{L^q(\Omega)} \cdot |\Omega|^{\frac{1}{p}} \leq \int_{\Omega} \chi_F(x) \cdot (u_{1F} - u_{2F})(x) dx$$

$$+ \|k\|_{L^q(\Omega)} \cdot |\Omega|^{\frac{1}{p}} + \|k\|_{L^q(\Omega)} \cdot C_p(\Omega) \cdot M.$$

We have

$$\int_{\Omega} \chi_F(u_{1F} - u_{2F}) dx = -(A_1 u_{1F}, u_{1F})_V - (A_2 u_{2F}, u_{2F})_V$$

$$-(C_1(u_{1F}), u_{1F})_V - (C_2(u_{2F}), u_{2F})_V$$

$$+ \lambda_1 (B_1 u_{1F}, u_{1F})_V + \lambda_2 (B_2 u_{2F}, u_{2F})_V .$$

Taking into account that C_i are monotone operators and that A_i, being weakly continuous, maps bounded sets into bounded sets, the relation

$$\int_{\Omega} \chi_F(u_{1F} - u_{2F}) dx$$

$$\leq \sum_{i=1}^{2} \{\|A_i\| \|u_{iF}\|_V^2 + \lambda_i \|B_i\| \|u_{iF}\|_V^2 - (C_i(u_{iF}), u_{iF})_V\},$$

implies that there exists a positive constant $\tilde{C}$ such that

$$\int_{\Omega} \chi_F(u_{1F} - u_{2F}) dx \leq \tilde{C}. \tag{6.45}$$

Now, from (6.44) and (6.45) we obtain

$$\int_{\omega} |\chi_F(x)| dx \leq \frac{\sqrt{N}}{r} \cdot C + \frac{\sqrt{N}}{r} \cdot \|\alpha(\cdot, r)\|_{L^{q'}(\Omega)} \cdot |\omega|^{\frac{s}{p}}$$

$$+\frac{\sqrt{N}}{r}\cdot\|\alpha(\cdot,r)\|_{L^{q'}(\omega)}\cdot(C_p(\Omega))^s\cdot M^s, \tag{6.46}$$

where

$$C:=\tilde{C}+\|k\|_{L^q(\Omega)}\cdot|\Omega|^{\frac{1}{p}}+\|k\|_{L^q(\Omega)}\cdot C_p(\Omega)\cdot M.$$

Let $\epsilon>0$. We choose $r>0$ such that $\frac{\sqrt{N}}{r}\cdot C<\frac{\epsilon}{2}$. Since $\alpha(\cdot,r)\in L^{q'}(\Omega)$ we can determine $\delta_\epsilon>0$ small enough so that if $|\omega|<\delta_\epsilon$, we have

$$\frac{\sqrt{N}}{r}\|\alpha(\cdot,r)\|_{L^{q'}(\Omega)}\cdot|\omega|^{\frac{s}{p}}+\frac{\sqrt{N}}{r}\|\alpha(\cdot,r)\|_{L^{q'}(\omega)}\cdot(C_p(\Omega))^s\cdot M^s<\frac{\epsilon}{2}.$$

By (6.46) it follows that

$$\int_\omega|\chi_F(x)|dx\leq\epsilon,$$

for any $\omega\subset\Omega$ with $|\omega|<\delta_\epsilon$. The weak precompactness of $\{\chi_F\ :\ F\in\Lambda\}$ in $L^1(\Omega;\mathbb{R}^N)$ is established. ■

Proof of Theorem 6.9. For every $F\in\Lambda$ let

$$W_F=\bigcup_{\substack{F'\in\Lambda\\ F'\supset F}}\{(u_{1F'},u_{2F'},\chi_{F'})\}\subset V\times V\times L^1(\Omega;\mathbb{R}^N),$$

with $(u_{1F'},u_{2F'},\chi_{F'})$ being a solution of $(P_{F'})$. Moreover, let

$$Z=\bigcup_{F\in\Lambda}\{\chi_F\}\subset L^1(\Omega;\mathbb{R}^N).$$

Denoting by $weakcl(W_F)$ the weak closure of W_F in $V\times V\times L^1(\Omega;\mathbb{R}^N)$ and by $weakcl(Z)$ the weak closure of Z in $L^1(\Omega;\mathbb{R}^N)$ we obtain, taking into account relation (6.42), that

$$weakcl(W_F)\subset B_V(0,M)\times B_V(0,M)\times weakcl(Z),\quad\forall F\in\Lambda.$$

Since V is reflexive it follows that $B_V(0,M)$ is weakly compact in V. Using Lemma 6.8 we get that the family $\{weakcl(W_F)\ :\ F\in\Lambda\}$ is contained in a weakly compact set of $V\times V\times L^1(\Omega;\mathbb{R}^N)$. It follows that this family has the finite intersection property, so we may infer that

$$\bigcap_{F\in\Lambda}weakcl(W_F)\neq\emptyset.$$

We choose some (u_1,u_2,χ) belonging to the nonempty set above. We prove that this is the searched solution for problem (P).

Let $v_1, v_2 \in L^\infty(\Omega; I\!R^N)$ and let F be an element of Λ such that $(v_1, v_2) \in F \times F$. We note that such an F exists, for example we can take $F = span\{v_1, v_2\}$. Since $(u_1, u_2, \chi) \in \bigcap_{F\in\Lambda} weakcl(W_F)$ it follows that there exists a sequence $\{(u_{1F_n}, u_{2F_n}, \chi_{F_n})\}$ in W_F, simply denoted by (u_{1n}, u_{2n}, χ_n) converging weakly to (u_1, u_2, χ) in $V \times V \times L^1(\Omega; I\!R^N)$. We have $u_{in} \rightharpoonup u_i$ weakly in $V (i = 1, 2)$ and $\chi_n \rightharpoonup \chi$ weakly in $L^1(\Omega; I\!R^N)$. Since (u_{1n}, u_{2n}, χ_n) is a solution of (P_F) we get

$$(A_1 u_{1n}, v_1)_V + (A_2 u_{2n}, v_2)_V + (C_1(u_{1n}), v_1)_V + (C_2(u_{2n}), v_2)_V$$

$$+ \int_\Omega \chi_n(v_1 - v_2)dx = \lambda_1 (B_1 u_{1n}, v_1)_V + \lambda_2 (B_2 u_{2n}, v_2)_V$$

The hypotheses on $A_i, B_i, C_i (i = 1, 2)$ and the above convergences imply the equality

$$\sum_{i=1}^{2} \{(A_i u_i, v_i)_V + (C_i(u_i), v_i)_V - \lambda_i (B_i u_i, v_i)_V\} + \int_\Omega \chi(v_1 - v_2)dx = 0,$$

which is satisfied for any $v_1, v_2 \in V \cap L^\infty(\Omega; I\!R^N)$. By the density of $V \cap L^\infty(\Omega; I\!R^N)$ in V we draw the conclusion that the relation (6.37) is valid for any $v_1, v_2 \in V$.

In what follows we prove relation (6.38). Due to the compact embedding $V \subset L^p(\Omega; I\!R^N)$ it results from the weak convergences $u_{in} \rightharpoonup u_i$ in V that

$$u_{in} \to u_i \text{ strongly in } L^p(\Omega; I\!R^N), \text{ for each } i = 1, 2.$$

So, passing eventually to a subsequence,

$$u_{in} \to u_i \text{ a.e. in } \Omega.$$

By the Egoroff theorem we obtain that for any $\epsilon > 0$ there exists a subset $\omega \subset \Omega$ with $|\omega| < \epsilon$ and such that for each $i \in \{1, 2\}$

$$u_{in} \to u_i \text{ uniformly on } \Omega \setminus \omega,$$

with $u_i \in L^\infty(\Omega \setminus \omega; I\!R^N)$ for every $i \in \{1, 2\}$. Let $v \in L^\infty(\Omega \setminus \omega; I\!R^N)$ be arbitrarily chosen. Fatou's lemma implies that for any $\mu > 0$ there exist $\delta_\mu > 0$ and a positive integer N_μ such that

$$\int_{\Omega\setminus\omega} \frac{1}{\lambda}(j(x, (u_{1n} - u_{2n})(x) - \theta + \lambda v(x)) - j(x, (u_{1n} - u_{2n})(x) - \theta))dx$$

$$\leq \int_{\Omega \setminus \omega} j^0(x, (u_1 - u_2)(x); v(x))dx + \mu, \tag{6.47}$$

for every $n \geq N_\mu, |\theta| < \delta_\mu$ and $\lambda \in (0, \delta_\mu)$. Taking into account that $\chi_n \in \partial j(x, (u_{1n} - u_{2n})(x))$ for a.e. $x \in \Omega$ we have

$$\int_{\Omega \setminus \omega} \chi_n(x) \cdot v(x)dx \leq \int_{\Omega \setminus \omega} j^0(x, (u_{1n} - u_{2n})(x); v(x))dx. \tag{6.48}$$

Passing to the limit as $\lambda \to 0$ in (6.47) and employing relation (6.48) it follows that

$$\int_{\Omega \setminus \omega} \chi_n(x) \cdot v(x)dx \leq \int_{\Omega \setminus \omega} j^0(x, (u_1 - u_2)(x); v(x))dx + \mu.$$

From the relation above and the weak convergence of χ_n to χ in the space $L^1(\Omega; I\!R^N)$ we derive that

$$\int_{\Omega \setminus \omega} \chi(x) \cdot v(x)dx \leq \int_{\Omega \setminus \omega} j^0(x, (u_1 - u_2)(x); v(x))dx + \mu.$$

Since $\mu > 0$ was chosen arbitrarily, we see that

$$\int_{\Omega \setminus \omega} \chi(x) \cdot v(x)dx \leq \int_{\Omega \setminus \omega} j^0(x, (u_1 - u_2)(x); v(x))dx, \ \forall v \in L^\infty(\Omega \setminus \omega; I\!R^N).$$

The last inequality implies that

$$\chi(x) \in \partial j(x, (u_1 - u_2)(x)), \text{ for a.e. } x \in \Omega \setminus \omega,$$

where $|\omega| < \epsilon$. Since $\epsilon > 0$ was chosen arbitrarily we have the inclusion

$$\chi(x) \in \partial j(x, (u_1 - u_2)(x)), \text{ for a.e. } x \in \Omega,$$

which means that relation (6.38) holds. The proof of Theorem 6.9 is now complete. ∎

4. Applications

Noncoercive Hemivariational Inequalities. We consider noncoercive forms of the coercive and semicoercive hemivariational problems treated in [15], pp. 65-77. The results are more general from the point of view of the absence of coercivity or semicoercivity assumption, but less general from the point of view of the boundedness of the set K. For this purpose, let us assume that V is a real Hilbert space and that the continuous injections

$$V \subset [L^2(\Omega; \mathbb{R}^k)]^N \subset V^*$$

hold, where V^* denotes the dual space of V. Moreover let $T : V \to L^2(\Omega; \mathbb{R}^k)$, $T(u) = \hat{u}$, $\hat{u}(x) \in \mathbb{R}^k$ be a linear and continuous mapping. Consider the operator A appearing in our abstract framework as $Au = A_1 u + f$, where $f \in V^*$ is a prescribed element, while A_1 satisfies, respectively, the assumptions of Theorems 6.1, 6.2 or 6.3. Then Theorem 6.1 is applicable for the problem

(P_1) Find $u \in K$ such that, for every $v \in K$,

$$\langle Au, v - u \rangle + \int_\Omega j^0(x, \hat{u}(x); \hat{v}(x) - \hat{u}(x))dx \geq 0.$$

Moreover, if T is a linear compact operator, then Theorems 6.2 and 6.3 apply for the above problem.

Suppose further that Γ is the Lipschitz boundary of Ω and that the linear mapping $T : V \to L^2(\Gamma; \mathbb{R}^k)$ is continuous. Then the conclusion of Theorem 6.1 holds for the problem

(P_2) Find $u \in K$ such that, for every $v \in K$,

$$\langle Au, v - u \rangle + \int_\Gamma j^0(x, \hat{u}(x); \hat{v}(x) - \hat{u}(x))dx \geq 0.$$

Furthermore, if T is compact, then the conclusion of Theorems 6.2 and 6.3 remain valid for (P_2).

Nonmonotone Laws in Networks with Convex Constraints. We give now an application in Economics concerning a network flow problem. We follow the basic ideas of W. Prager [19], [18] and, for the consideration of nonlinearities, we combine them with the notion of nonconvex superpotential.

The nonlinearity generally nonmonotone is caused by the law relating the two branch variables of the network, the "flow intensity" and the "price differential" which here can also be vectors. The problem is formulated as a hemivariational inequality and the existence of its solution is discussed further. We consider networks with directed branches. The nodes are denoted by Latin and the branches by Greek letters. We suppose that we have m nodes and ν branches. We take as branch variables the "flow intensity" s_γ and the "price differential" e_γ. As node variables the "amount of flow" p_k and the "shadow price" u_k are considered. Moreover, each branch may have an "initial price differential" vector e^0_γ. The above given quantities are assembled in vectors e, e^0, u, s, p. The node-branch incidence matrix G is denoted by G, where the lines of G are linearly independent. Upper index T denotes the transpose of a matrix or a vector. The network law is a relation between the "flow intensity" s_γ and the "price differential" e_γ. We accept that s_γ is a nonmonotone function of the e_γ expressed by the relation

$$e_\gamma - e^0_\gamma \in \partial j_\gamma(s_\gamma) + \frac{1}{2}\partial s^T_\gamma C_\gamma s_\gamma \,, \tag{6.49}$$

where k_γ is a positive definite symmetric matrix and ∂ is the generalized gradient. The graph of the $s_\gamma - e_\gamma$ law is called γ-characteristic.

The problem to be solved consists in the determination for the whole network of the vectors s, e, u, with given vectors p and e_0.

Further let $C = \operatorname{diag}[C_1, \cdots, C_\gamma, \cdots]$ and let the summation $\sum_\gamma$ be extended over all branches. Now we consider the graph which corresponds to the network and a corresponding tree. The tree results from the initial graph by cutting all the branches creating the closed loops. Let us denote by s_T (resp. s_M) the part of the vector s corresponding to the tree branches (resp. to the cut branches giving rise to closed loops). Then we may write instead of $Gs = p$ the relation

$$G_T s_T + G_M s_M = p\,.$$

Here G_T is nonsingular and thus we may write that

$$s = \begin{bmatrix} s_T \\ s_M \end{bmatrix} = \begin{bmatrix} G_T^{-1} \\ 0 \end{bmatrix} p + \begin{bmatrix} -G_T^{-1}G_M \\ I \end{bmatrix} s_M = s_0 + Bs_M\,, \tag{6.50}$$

where I denotes the unit matrix. Using (6.49) and (6.50) we obtain (cf. [15]) a hemivariational inequality with respect to s_M which reads as follows: find $s_M \in \mathbb{R}^{n_1}$ (n_1 is the dimension of s_M) such that

$$\sum_\gamma j^0_\gamma((s_0 + Bs_M)_\gamma; (Bs^*_M - Bs_M)_\gamma) + s^T_M B^T CB(s^*_M - s_M)$$

$$+ s^T_0 CB(s^*_M - s_M) + e^{0T} B(s^*_M - s_M) \geq 0, \quad \forall s^*_M \in \mathbb{R}^{n_1}\,. \tag{6.51}$$

Let us now assume that the flow intensities s_M are constrained to belong to a bounded and closed convex subset $K \subset \mathbb{R}^{n_1}$ (box constraints are very common). In this way the problem takes the form: find $s_M \in K$ which satisfies (6.51), for every $s^*_M \in K$.

Since the rank of B is equal to the number of its columns and C is symmetric and positive definite the same happens for B^TCB. In the finite dimensional case treated here, one can easily verify that Corollary 6.1 holds if $j_\gamma(\cdot,\cdot)$ satisfies condition (j). Thus (6.51) has at least one solution.

On the Nonconvex Semipermeability Problem. We consider an open, bounded, connected subset Ω of $\mathbb{R}^3$ referred to a fixed Cartesian coordinate system $0x_1x_2x_3$ and we formulate the equation

$$-\Delta u = f \quad \text{in } \Omega \tag{6.52}$$

for stationary problems.

Here u represents the temperature in the case of heat conduction problems, whereas in problems of hydraulics and electrostatics the pressure and the electric potential are represented, respectively. We denote by Γ the boundary of Ω and we assume that Γ is sufficiently smooth ($C^{1,1}$-boundary is sufficient). If $n = \{n_i\}$ denotes the outward unit normal to Γ then $\partial u/\partial n$ is the flux of heat, fluid or electricity through Γ for the aforementioned classes of problems.

We may consider the interior and the boundary semipermeability problems.

In the first class of problems the classical boundary conditions

$$u = 0 \quad \text{on } \Gamma \tag{6.53}$$

are assumed to hold, whereas in the second class the boundary conditions are defined as a relation between $\partial u/\partial n$ and u. In the first class the semipermeability conditions are obtained by assuming that $f = \bar{f} + \bar{\bar{f}}$ where $\bar{\bar{f}}$ is prescribed and $\bar{f}$ is a known function of u. Here, we consider (6.53) for the sake of simplicity. All these problems may be put in the following general framework. For the first class we seek a function u such as to satisfy (6.52), (6.53) with

$$f = \bar{f} + \bar{\bar{f}}, \quad -\bar{f} \in \partial j_1(x,u) \text{ in } \Omega\,. \tag{6.54}$$

For the second class we seek a function u such that (6.52) is satisfied together with the boundary condition

$$-\frac{\partial u}{\partial n} \in \partial j_2(x,u) \quad \text{on } \Gamma_1 \subset \Gamma \quad \text{and} \quad u = 0 \text{ on } \Gamma \setminus \Gamma_1\,. \tag{6.55}$$

Both $j_1(x,\cdot)$ and $j_2(x,\cdot)$ are locally Lipschitz functions and ∂ denotes the generalized gradient. Note, that if $q = \{q_i\}$ denotes the heat flux vector and $k > 0$ is the coefficient of thermal conductivity of the material we may write by Fourier's law that $q_i n_i = -k\partial u/\partial n$.

Let us introduce the notations

$$a(u,v) = \int_\Omega \nabla u \cdot \nabla v dx$$

and

$$(f,u) = \int_\Omega f u dx\,.$$

We may ask in addition that u is constrained to belong to a convex bounded closed set $K \subset V$ due to some technical reasons, e.g. constraints for the temperature or the pressure of the fluid etc.

The hemivariational inequalities correspond to the two classes of problems. For the first class consider $V = H_0^1(\Omega)$ and $\bar{\bar{f}} \in L^2(\Omega)$; for the second class, $V = \{v : v \in H^1(\Omega),\, v = 0 \text{ on } \Gamma \setminus \Gamma_1\}$ and $f \in L^2(\Omega)$. Then from the Green-Gauss theorem applied to (6.52), with (6.54) and (6.55) we are led to the following two hemivariational inequalities for the first and for the second class of semipermeability problems, respectively,

(i) Find $u \in K$ such that

$$a(u, v-u) + \int_\Omega j_1^0(x, u(x); v(x) - u(x))dx \geq (\bar{\bar{f}}, v-u)\,,$$

for all $v \in K$.

(ii) Find $u \in K$ such that

$$a(u, v-u) + \int_{\Gamma_1} j_2^0(x, u(x); v(x) - u(x))d\Gamma \geq (f, v-u)\,,$$

for all $v \in K$.

Since $a(\cdot,\cdot)$ is (strongly) monotone on V both in (i) and (ii), and the embeddings $V \subset L^2(\Omega)$ and $V \subset L^2(\Gamma_1)$ are compact, we can prove the existence of solutions of (i) and of (ii) by applying Theorem 6.2 if j_1 and j_2 satisfy condition (j).

Adhesively Supported Elastic Plate between two Rigid Supports. Let us consider a Kirchoff plate. The elastic plate is referred to a right-handed orthogonal Cartesian coordinate system $Ox_1x_2x_3$. The plate has constant thickness h_1, and the middle surface of the plate coincides with the Ox_1x_2-plane. Let Ω be an open, bounded and connected subset of $I\!R^2$ and suppose that the boundary Γ is Lipschitzian ($C^{0,1}$-boundary). The domain Ω is occupied by the plate in its undeformed state. On $\Omega' \subset \Omega$ (Ω' is such that $\overline{\Omega}' \cap \Gamma = \emptyset$) the plate is bonded to a support through an adhesive material. We denote by $\zeta(x)$ the deflection of the point $x = (x_1, x_2, x_3)$ and by $f = (0, 0, f_3)$, $f_3 = f_3(x)$ (hereafter called f for simplicity) the distributed load of the considered plate per unit area of the middle surface. Concerning the laws for adhesive forces and the formulation of the problems we refer to [16]. Here we make the additional assumption that the displacements of the plate are prevented by some rigid supports. Thus we may put as an additional assumption the following one:

$$z \in K, \tag{6.56}$$

where K is a convex closed bounded subset of the displacement space. One could have e.g. that $a_0 \leq z \leq b_0$.

We assume that any type of boundary conditions may hold on Γ. Here we admit that the plate boundary is free. There is no need to guarantee that the strain energy of the plate is coercive. Thus the whole space $H^2(\Omega)$ is the kinematically admissible set of the plate. If one takes now into account relation (6.56), then $z \in K \subset H^2(\Omega)$, where K is a closed convex bounded subset of $H^2(\Omega)$ and the problem has the following form:

Find $\zeta \in K$ such as to satisfy

$$a(\zeta, z - \zeta) + \int_{\Omega'} j^0(\zeta; z - \zeta) d\Omega \geq (f, z - \zeta) \quad \forall z \in K. \tag{6.57}$$

Here $a(\cdot, \cdot)$ is the elastic energy of the Kirchoff plate, i.e.

$$a(\zeta, z) = k \int_\Omega [(1 - \nu)\, \zeta_{,\alpha\beta}\, z_{,\alpha\beta} + \nu\, \Delta\zeta\, \Delta z] d\Omega \quad \alpha, \beta = 1, 2, \tag{6.58}$$

where $k = Eh^3/12(1 - \nu^2)$ is the bending rigidity of the plate with E and ν the modulus of elasticity and the Poisson ratio, respectively, and h is its thickness. Moreover, j is the binding energy of the adhesive which is a locally Lipschitz function on $H^2(\Omega)$ and $f \in L^2(\Omega)$ denotes the external forces. Furthermore, if j fulfills the growth condition (j) then, taking into consideration that $a(\cdot, \cdot)$ appearing in (6.58) is continuous and monotone, we deduce, by Theorem 6.2, the existence of a solution of problem (6.57).

The Multiple Loading Buckling. We consider two elastic beams (linear elasticity) of length l measured along the axis Ox of the coordinate system yOx, and with the same cross-section. The beams, numbered here by $i = 1, 2$, are simply supported at their ends $x = 0$ and $x = l$. On the interval (l_1, l_2), $l_1 < l_2 < l$, they are connected with an adhesive material of negligible thickness. The displacements of the i-th beam are denoted by $x \to u_i(x)$, $i = 1, 2$, and the behavior of the adhesive material is described by a nonmonotone possibly multivalued law between $-f(x)$ and $[u(x)]$, where $x \to f(x)$ denotes the reaction force per unit length vertical to the Ox axis, due to the adhesive material (cf. [17], p. 87 and [15], p. 110) and $[u] = u_1 - u_2$ is the relative deflection of the two beams. Recall that u_i is referred to the middle line of the beam i and that each beam has constant thickness which remains the same after the deformation. The adhesive material can sustain a small tensile force before rupture (debonding). The beams are assumed to have the same moduli of elasticity E and let I be the moment of inertia of them. The sandwich beam is subjected to the compressive forces P_1 and P_2 and we want to determine the buckling loading of it. This problem is yet an open problem in Engineering. From the large deflection theory of beams we may write the following relations which describe the behavior of the i-th beam:

$$u_i''''(x) + \frac{1}{a_i^2} u_i''(x) = f_i(x) \quad \text{on } (0,l); \tag{6.59}$$

$$u_i(0) = u_i(l) = 0, \quad u_i''(0) = u_i''(l) = 0 \quad i = 1, 2. \tag{6.60}$$

Here $a_i^2 := IE/P_i$. We assume that the $(-f, [u])$ graph results from a non locally Lipschitz function $j : \mathbb{R} \to \overline{\mathbb{R}}$ such that

$$-f(x) \in \partial j([u(x)]), \quad \forall x \in (l_1, l_2), \tag{6.61}$$

where ∂ denotes the generalized gradient of Clarke. Set

$$V := H^2(\Omega) \cap H_0^1(\Omega) \quad \Omega = (0, l).$$

Then V is a Hilbert space with the inner product (see [5], p. 216, Lemma 4.2)

$$a(u, v) := \int_0^l u''(x) v''(x) dx.$$

Let $L : V \to V^*$ be the linear operator defined by

$$\langle Lu, v \rangle := \int_0^l u'(x) v'(x) dx, \quad \forall u, v \in V.$$

We observe that L is bounded, weak continuous and satisfies

$$\langle Lu, v \rangle = \langle Lv, u \rangle, \quad \text{for all } u, v \in V.$$

The superpotential law (6.61) implies

$$j^0([u(x)]; y) \geq -f(x) y, \quad \forall x \in (l_1, l_2), \forall y \in \mathbb{R}.$$

Multiplying (6.59) by $v_i(x) - u_i(x)$, integrating over $(0, l)$ and adding the resulting relations for $i = 1, 2$, implies by taking into account the boundary condition (6.60), the hemivariational inequality: find $u = \{u_1, u_2\} \in$

$V \times V$ such that

$$\sum_{i=1}^{2} \int_0^l u_i''(x)[v_i''(x) - u_i''(x)]dx - \sum_{i=1}^{2} \frac{1}{a_i^2} \int_0^l u_i'(x)[v_i'(x) - u_i'(x)]dx$$

$$+ \int_{l_1}^{l_2} j^0([u(x)]; [v(x)] - [u(x)])dx \geq 0,$$

for all $v = \{v_1, v_2\} \in V \times V$.

Thus buckling of the beam occurs if $\lambda_i := 1/a_i^2$ $(i = 1, 2)$ is an eigenvalue for the following hemivariational inequality

$$\sum_{i=1}^{2} a_i(u_i, v_i - u_i) - \sum_{i=1}^{2} \lambda_i \langle u_i, v_i - u_i \rangle$$

$$+ \int_{l_1}^{l_2} j^0([u(x)]; [v(x)] - [u(x)])dx \geq 0, \qquad (6.62)$$

for all $v = \{v_1, v_2\} \in V \times V$. According to Theorem 6.9, Problem (6.62) admits at least one solution $\{u_1, u_2, \lambda_1, \lambda_2\}$, provided that j fulfills the growth assumptions (6.31), (6.35) and (6.36).

References

[1] H. Brézis, *Analyse fonctionnelle. Théorie et applications,* Masson, 1992.

[2] F. Browder and P. Hess, Nonlinear mappings of monotone type in Banach spaces, *J. Funct. Anal.* **11** (1972), 251-294.

[3] F. Clarke, *Optimization and Nonsmooth Analysis*, John Wiley & Sons, New York, 1983.

[4] J. Dugundji and A. Granas, KKM-maps and variational inequalities, *Ann. Scuola Norm. Sup. Pisa* **5** (1978), 679-682.

[5] G. Duvaut and J.-L. Lions, *Les Inéquations en Mécanique et en Physique*, Dunod, Paris, 1972.

[6] G. J. Hartman and G. Stampacchia, On some nonlinear elliptic differential equations, *Acta Math.* **15** (1966), 271-310.

[7] R. B. Holmes, *Geometric Functional Analysis and its Applications*, Springer-Verlag-New York, 1975.

[8] O. Kavian, *Introduction à la théorie des points critiques et applications aux problémes elliptiques*, Springer-Verlag, Paris, Berlin, Heidelberg, New York, London, Tokyo, Hong Kong, Barcelona, Budapest, 1993.

[9] D. Kinderlehrer and G. Stampacchia, *An Introduction to Variational Inequalities*, Academic Press, New York, 1980.

[10] B. Knaster, K. Kuratowski and S. Mazurkiewicz, Ein Beweis des Fixpunktsatzes für n-dimensionale Simplexe, *Fund. Mat.* **14** (1929), 132-137.

[11] U. Mosco, Implicit variational problems and quasi-variational inequalities, in *Nonlinear Operators and the Calculus of Variations* (J.P. Gossez, E.J. Lami Dozo, J. Mawhin and L. Waelbroeck, Eds.), Lecture Notes in Mathematics 543, Springer-Verlag, Berlin, 1976, pp. 83-156.

[12] D. Motreanu and P. D. Panagiotopoulos, Double eigenvalue problems for hemivariational inequalities, *Arch. Rat. Mech. Analysis* **140** (1997), 225-251.

[13] D. Motreanu and V. Rădulescu, Existence results for inequality problems with lack of convexity, *Numer. Funct. Anal. Optimiz.* **21** (2000), 869-884.

[14] Z. Naniewicz, Hemivariational Inequalities with functionals which are not locally Lipschitz, *Nonlinear Analysis, T.M.A.*, **25** (1995), No. 12, pp. 1307-1320.

[15] Z. Naniewicz and P. D. Panagiotopoulos, *Mathematical Theory of Hemivariational Inequalities and Applications*, Marcel Dekker, New York, 1995.

[16] P. D. Panagiotopoulos, *Hemivariational Inequalities: Applications to Mechanics and Engineering,* Springer-Verlag, New-York/Boston/Berlin, 1993.

[17] P. D. Panagiotopoulos, *Inequality Problems in Mechanics and Applications. Convex and Nonconvex Energy Functionals*, Birkhäuser-Verlag, Basel, 1985.

[18] D. Pascali and S. Sburlan, *Nonlinear Mappings of Monotone Type*, Sijthoff and Noordhoff International Publishers, The Netherlands, 1978.

[19] W. Prager, Problems of network flow, *Zeitschrift für Angewandte Mathematik und Physik (Z.A.M.P.)* **16** (1965), 185-193.

[20] H. H. Schaefer, *Topological Vector Spaces*, Macmillan Series in Advances Mathematics and Theoretical Physics, New York, 1966.

Chapter 7

EIGENVALUE PROBLEMS WITH SYMMETRIES

In this Chapter we consider several classes of inequality problems involving hemivariational inequalities with various kinds of symmetry and possibly with constraints. We establish multiplicity results, including cases of infinitely many solutions. The proofs use powerful tools of nonsmooth critical point theory combined with arguments from Algebraic Topology. Results in this direction in the framework of elliptic equations have been initially established by Ambrosetti and Rabinowitz (see [1], [16]), while pioneering results in the study of multiple solutions for periodic problems can be found in Fournier and Willem [6], and Mawhin and Willem [9].

In the first part of the Chapter we obtain an equivariant deformation lemma in order to be able to consider nonsmooth functions satisfying some invariant properties. By adapting some classical ideas in critical point theory, we prove several abstract results concerning the existence of multiple critical points for periodic functionals. Next, we apply our theoretical results to several classes of eigenvalue problems for hemivariational inequalities. In the last part of this Chapter we establish a multiplicity result for a new class of hemivariational inequalities introduced in Motreanu and Panagiotopoulos [11].

1. Orbits of Critical Points

In this Section we discuss the existence of critical orbits for locally Lipschitz functionals invariant with respect to a discrete group. Our approach is based on the ideas developed in the smooth framework by Mawhin and Willem [8].

Let G be the discrete subgroup of a Banach space X and let $\pi : X \to X/G$ be the canonical surjection. A subset C of X is said to be *G-invariant* if $\pi^{-1}(\pi(C)) = C$. A function $f : X \to I\!R$ is called G-invariant if

$$f(u+g) = f(u), \quad \forall\, u \in X,\, \forall\, g \in G.$$

Lemma 7.1 Assume that the function $f : X \to I\!R$ is G-invariant and locally Lipschitz. Then

$$\partial f(x) = \partial f(x+g), \quad \forall x \in X,\, \forall\, g \in G.$$

In particular, the set K of critical points of f, i.e.,

$$K = \{x \in X : \; 0 \in \partial f(x)\}$$

is G-invariant.

Proof. The result is derived from the equivalences below:

$$w \in \partial f(x) \Longleftrightarrow f^0(x; v) \geq \langle w, v\rangle, \quad \forall v \in X$$

$$\Longleftrightarrow f^0(x+g; v) \geq \langle w, v\rangle, \quad \forall v \in X,$$

which, in turn, is equivalent to say that $w \in \partial f(x+g)$. ■

For a locally Lipschitz function $f : X \to I\!R$ and any $c \in I\!R$ we denote

$$f_c := \{x \in X : f(x) \leq c\},$$

$$K_c := \{x \in X : \; 0 \in \partial f(x) \text{ and } f(x) = c\} = K \cap f_c.$$

Any element $x \in K$ is called a *critical point* of f . If the function f is G-invariant, Lemma 7.1 reveals that, for each $x \in K$, $x+G = \pi^{-1}(\pi(x))$ is a set of critical points of f which is called a *critical orbit* of f. The mapping λ defined by

$$X \ni x \mapsto \lambda(x) := \min_{w \in \partial f(x)} \|w\|_*$$

is lower semicontinuous (see Proposition 1.2). Moreover, Lemma 7.1 gives rise to the following.

Lemma 7.2 Under the assumption of Lemma 7.1, the function λ is G-invariant.

We shall use a compactness condition of Palais-Smale type. A G-invariant, locally Lipschitz function $f : X \to I\!R$ is said to satisfy the $(PS)_{c,G}$-condition if, for every sequence $\{u_k\}$ in X such that $f(u_k) \to c$ and $\lambda(u_k) \to 0$, there exists $v_k \in \pi^{-1}(\pi(u_k))$ such that $\{v_k\}$ contains a convergent subsequence in X. If the $(PS)_{c,G}$-condition is satisfied for all

$c \in I\!R$, we simply say that f verifies the $(PS)_G$-condition. The usual Palais-Smale condition is obtained when $G = \{0\}$.

For a later use we review the notion of Ljusternik- Schnirelman category. More details can be found in Ambrosetti [1], Browder [3], Mawhin and Willem [8] (see also Chapter 4). A subset A of a topological space Y is *contractible* in Y if there exists a homotopy $h \in C([0,1] \times A, Y)$ and a point $y \in Y$ such that $h(0,u) = u$ and $h(1,u) = y$ for all $u \in A$. We say that a subset A of a topological space Y has the category k in Y if k is the smallest positive integer such that A can be covered by k closed and contractible sets in Y. If no such k exists we say that A is of category $+\infty$. The category of A in Y is denoted by $\mathrm{Cat}_Y(A)$. For convenience we set $\mathrm{Cat}_Y(\emptyset) = 0$. We recall in what follows some basic properties of this notion (see [8] for proofs).

Proposition 7.1 Let A and B be subsets of Y. Then
(i) $A \subset B \Longrightarrow \mathrm{Cat}_Y(A) \leq \mathrm{Cat}_Y(B)$
(ii) $\mathrm{Cat}_Y(A \cup B) \leq \mathrm{Cat}_Y(A) + \mathrm{Cat}_Y(B)$
(iii) if A is closed and $B = \eta(1, A)$, where $\eta \in C([0,1] \times A, Y)$ is such that $\eta(0,u) = u$, $\forall\, u \in A$, then $\mathrm{Cat}_Y(A) \leq \mathrm{Cat}_Y(B)$.

We also precise some additional notations. By $B_\delta(x)$ we denote the closed ball in the Banach space X with center $x \in X$ and radius $\delta > 0$. For a subset C of X the notation $N_\delta(C)$ means the open δ-neighborhood of C in X.

Lemma 7.3 Suppose that $f : X \to I\!R$ is a locally Lipschitz and G-invariant functional on the Banach space X satisfying the $(PS)_{c,G}$-condition with a given $c \in I\!R$. Then, for every $\delta > 0$, there exist constants $\bar{\varepsilon} > 0$, $b > 0$ and a locally Lipschitz X-valued mapping v on

$$M(f, \bar{\varepsilon}, \delta, c) := f^{-1}([c - \bar{\varepsilon}, c + \bar{\varepsilon}]) \setminus N_\delta(K_c)$$

satisfying the conditions
(i) $\|v(x)\| \leq 1$;
(ii) $\langle x^*, v(x) \rangle > b/2$;
(iii) $v(x + g) = v(x)$,
for all $x \in M(f, \bar{\varepsilon}, \delta, c)$, $x^* \in \partial f(x)$ and $g \in G$.

Proof. Fix $\delta > 0$. The $(PS)_{c,G}$-condition implies that there exist $b > 0$ and $\bar{\varepsilon} > 0$ such that

$$\lambda(x) \geq b, \quad \forall\, x \in M(f, \bar{\varepsilon}, \delta, c). \tag{7.1}$$

In order to check (7.1), suppose the contrary. Then there exists a sequence $\{x_n\}$ in $X \setminus N_\delta(K_c)$ such that $f(x_n) \to c$ and $\lambda(x_n) \to 0$ as

$n \to \infty$. By $(PS)_{c,G}$-condition we can assume, going if necessary to a subsequence, that there exist a renamed sequence $\{y_n\}$ in X and an element $x \in X$ such that

$$\pi(y_n) = \pi(x_n), \quad \forall n \geq 1, \tag{7.2}$$

and

$$y_n \to x \quad \text{in } X \text{ as } n \to \infty. \tag{7.3}$$

But (7.2) reads as

$$y_n = x_n + g_n \quad \text{for some } g_n \in G, \; \forall\, n \geq 1.$$

Since $x_n \notin N_\delta(K_c)$, Lemma 7.1 yields

$$\text{dist}\,(y_n, K_c) = \text{dist}\,(x_n + g_n, K_c) = \text{dist}\,(x_n, K_c) \geq \delta.$$

Taking into account (7.3) we deduce that $x \notin N_\delta(K_c)$. The lower semicontinuity of λ and Lemma 7.2 show that

$$\lambda(x) \leq \liminf_{n\to\infty} \lambda(y_n) = \liminf_{n\to\infty} \lambda(x_n) = 0,$$

so $0 \in \partial f(x)$. In addition, we obtain from (7.3) that $f(x) = c$. It follows that $x \in K_c$. The contradiction establishes the claim in (7.1). Given $x_0 \in M(f, \bar{\varepsilon}, \delta, c)$ there exists $w_0 \in \partial f(x_0)$ such that $\|w_0\|_* = \lambda(x_0)$. The set $\partial f(x_0)$ is nonempty, w^*-compact and convex. Since the open ball $\overset{o}{B}_*(0, \|w_0\|_*)$ in X^* is nonempty, convex and

$$\partial f(x_0) \cap \overset{o}{B}_*(0, \|w_0\|_*) = \emptyset,$$

it follows by the Eberlein Separation Theorem that there exists $h_0 \in X$ such that $\|h_0\| = 1$ and

$$\langle x^*, h_0\rangle \geq \langle w, h_0\rangle, \quad \forall w \in B_*(0, \|w_0\|_*), \quad \forall x^* \in \partial f(x_0). \tag{7.4}$$

The Hahn-Banach Theorem implies

$$\max_{w \in B_*(0, \|w_0\|_*)} \langle w, h_0\rangle = \|w_0\|_* \, \|h_0\| = \|w_0\|_*.$$

Then, by (7.1) and (7.4),

$$\langle x^*, h_0\rangle \geq \|w_0\|_{X^*} = \lambda(x_0) \geq b, \quad \forall x^* \in \partial f(x_0).$$

The weak upper semicontinuity of ∂f at x_0 ensures that for any $\varepsilon > 0$ there exists $\eta_0 = \eta(x_0) > 0$ such that to each $x^* \in \partial f(x)$ with $\|x - x_0\| < \eta_0$ it corresponds $z_0 \in \partial f(x_0)$ satisfying

$$|\langle x^* - z_0, h_0\rangle| < \varepsilon.$$

It turns out

$$\langle x^*, h_0 \rangle = \langle x^* - z_0, h_0 \rangle + \langle z_0, h_0 \rangle \geq b - \varepsilon .$$

Thus, taking a small ε,

$$\langle x^*, h_0 \rangle > \frac{b}{2}, \quad \forall x^* \in \partial f(x), \ \forall\, x \in \overset{o}{B} (x_0, \eta_0). \tag{7.5}$$

Letting x_0 run in $M(f, \bar{\varepsilon}, \delta, c)$, we consider the open covering

$$\mathcal{N} = \{\overset{o}{B} (x_0 \,, \eta(x_0)) \ : \ x_0 \in M(f, \bar{\varepsilon}, \delta, c)\}$$

of $M(f, \bar{\varepsilon}, \delta, c)$. The assumption upon G to be discrete assures that for each $u \in X$ we can find an open neighbourhood A_u of u in X with the property that each orbit $\pi(x) = x + G$ intersects A_u at finitely many points whose number is locally constant with respect to x. We form the open covering

$$\mathcal{A} = \{A_u \ : \ u \in M(f, \bar{\varepsilon}, \delta, c)\}$$

of $M(f, \bar{\varepsilon}, \delta, c)$. Then there exists an open covering $\{U_j\}_{j \in J}$ of the set $M(f, \bar{\varepsilon}, \delta, c)$ which is locally finite and finer than $\mathcal{N}$ and $\mathcal{A}$. We assign to each $j \in J$ an $x_j \in M(f, \bar{\varepsilon}, \delta, c)$ with

$$U_j \subset \overset{o}{B} (x_j, \eta(x_j)) .$$

Set

$$\varphi_j(x) := \frac{\sum_{g \in G} \operatorname{dist}(x + g, X \setminus U_j)}{\sum_{i \in J} \sum_{g \in G} \operatorname{dist}(x + g, X \setminus U_i)} \quad \text{if } x \in \bigcup_{i \in J} U_i$$

and

$$v(x) := \sum_{j \in J} \varphi_j(x) h_j \,, \quad \forall x \in \bigcup_{j \in J} U_j \,,$$

where h_j plays the same role for x_j as h_0 for x_0 previously. It follows that v is well-defined and locally Lipschitz. We see that

$$\|v(x)\| \leq \sum_{j \in J} \varphi_j(x) \|h_j\| = \sum_{j \in J} \varphi_j(x) = 1$$

and, by (7.5),

$$\langle x^*, v(x) \rangle = \sum_{j \in J} \varphi_j(x) \langle x^*, h_j \rangle > \frac{b}{2} \,,$$

for all $x \in \bigcup_{j \in J} U_j$ and for any $x^* \in \partial f(x)$. Additionally one has

$$v(x + g) = \sum_{j \in J} \varphi_j(x + g) h_j = \sum_{j \in J} \varphi_j(x) h_j \,,$$

for all $x \in M(f, \bar{\varepsilon}, \delta, c)$ and $g \in G$, after observing that $x + g$ belongs to $M(f, \bar{\varepsilon}, \delta, c)$. Assertions (i)-(iii) are thus established. ■

We proceed to construct a suitable equivariant deformation following the argument in Chang [4]. Given $\varepsilon \in (0, \bar{\varepsilon})$ choose locally Lipschitz functions $\varphi : X \to [0,1]$ and $\psi : X \to [0,1]$ satisfying

$$\begin{array}{lll} \varphi(x) = 1 & \text{if} & x \in f^{-1}([c - \varepsilon, c + \varepsilon]), \\ \varphi(x) = 0 & \text{if} & x \notin f^{-1}([c - \bar{\varepsilon}, c + \bar{\varepsilon}]) \end{array}$$

and

$$\begin{array}{lll} \psi(x) = 1 & \text{if} & x \notin N_{4\delta}(K_c), \\ \psi(x) = 0 & \text{if} & x \in N_{2\delta}(K_c). \end{array}$$

Then we introduce $V : X \to X$ by

$$V(x) = \varphi(x)\psi(x)v(x), \quad x \in X,$$

with $v(x)$ entering Lemma 7.3. In view of the choice of functions φ and ψ, V is well defined, locally Lipschitz and bounded. Consequently, the Cauchy problem

$$\begin{cases} \frac{d}{dt}\eta(t,x) = -V(\eta(t,x)) \\ \eta(0,x) = x \end{cases} \tag{7.6}$$

determines a unique solution $\eta : [0,1] \times X \to X$.

Lemma 7.4 The global flow $\eta : [0,1] \times X \to X$ of (7.6) fulfills the requirements

(i) $\|\eta(t,x) - x\| \leq t$, for any $(t,x) \in [0,1] \times X$;
(ii) $f(x) - f(\eta(t,x)) > \frac{b}{2}t$ if $\eta(t,x) \in M(f, \bar{\varepsilon}, \delta, c)$;
(iii) $\eta(t, x + g) = \eta(t,x) + g$, for all $(t,x) \in [0,1] \times X$ and $g \in G$.

Proof. Property (i) follows directly from (7.6) and the boundedness of V. To justify (ii) we note that Lebourg's mean value theorem and (ii) of Lemma 7.3 imply

$$f(x) - f(\eta(t,x)) = -\int_0^t \frac{d}{ds} f(\eta(s,x))ds > \frac{b}{2}t,$$

for t and x as required in (ii). The equivariant property (iii) holds because we infer from (iii) of Lemma 7.3 that

$$\frac{d}{dt}(\eta(t,x) + g) = \frac{d}{dt}\eta(t,x) = -V(\eta(t,x)) = -V(\eta(t,x) + g)$$

and

$$\eta(0,x) + g = x + g,$$

whenever $g \in G$. ■

We are now in position to state the needed deformation result.

Theorem 7.1 (Equivariant Deformation Lemma). Let f be a locally Lipschitz function on a real Banach space X which is invariant with respect to a discrete subgroup G of X and satisfies the $(PS)_{c,G}$-condition for a fixed $c \in I\!R$. Then, for any neighborhood $\mathcal{N}$ of K_c and $\varepsilon_0 > 0$ there exist $\varepsilon \in (0, \varepsilon_0)$ and a homeomorphism $\eta_0 : X \to X$ such that

(i) $\eta_0(x) = x$, for all $x \notin f^{-1}([c - \varepsilon_0, c + \varepsilon_0])$;

(ii) $\eta_0(f_{c+\varepsilon} \setminus \mathcal{N}) \subset f_{c-\varepsilon}$;

(iii) if $K_c = \emptyset$, then $\eta_0(f_{c+\varepsilon}) \subset f_{c-\varepsilon}$;

(iv) $\eta_0(x + g) = \eta_0(x) + g$, for all $x \in X$, and $g \in G$.

Proof. By $(PS)_{c,G}$-condition , K_c is a compact set. Hence we can choose $\delta > 0$ such that $N_{6\delta}(K_c) \subset \mathcal{N}$. Lemma 7.3 provides numbers $\bar{\varepsilon} > 0$ and $b > 0$ and a locally Lipschitz vector field v on $M(f, \bar{\varepsilon}, \delta, c)$ such that the assertions (i)-(iii) of Lemma 7.3 are verified. Moreover, it is clear that if $0 < \varepsilon_1 \leq \varepsilon_2$ then $M(f, \varepsilon_1, \delta, c) \subset M(f, \varepsilon_2, \delta, c)$, and this allows to impose $\bar{\varepsilon} \leq \min\{\varepsilon_0, b\delta/4\}$. Fixing $\varepsilon \in (0, \bar{\varepsilon})$ let $\eta \in C([0,1] \times X, X)$ be defined by (7.6). Putting

$$t_0 = \frac{4\bar{\varepsilon}}{b}$$

we introduce $\eta_0 : X \to X$ by

$$\eta_0(x) = \eta(t_0, x), \quad \forall x \in X.$$

If $x \notin f^{-1}([c - \varepsilon_0, c + \varepsilon_0])$, then we get $x \notin f^{-1}([c - \bar{\varepsilon}, c + \bar{\varepsilon}])$ and $V(x) = 0$ which implies (i).

Concerning (ii), since f is non-increasing along $\eta(.,x)$, we may restrict to $x \in M(f, \varepsilon, 6\delta, c)$. Arguing by contradiction, suppose there exists $x \in M(f, \varepsilon, 6\delta, c)$ such that $\eta_0(x) \notin f_{c-\varepsilon}$. It is seen that $\eta(s, x) \in f_{c+\varepsilon} \setminus f_{c-\varepsilon}$ for all $s \in [0, t_0]$. We claim that

$$\eta(s, x) \notin N_{4\delta}(K_c), \quad \forall s \in [0, t_0].$$

If not, there exists $s_0 \in [0, t_0]$ such that $\eta(s, x) \in N_{4\delta}(K_c)$ for all $s \in [0, s_0)$, and $\eta(s_0, x) \notin N_{4\delta}(K_c)$. Since then

$$\eta(s, x) \in M(f, \varepsilon, 4\delta, c), \quad \forall s \in [0, s_0),$$

we derive from (i), (ii) of Lemma 7.4 where t, $\bar{\varepsilon}$, δ are replaced by s, ε, 4δ, respectively,

$$\|\eta(s_0, x) - x\| \leq s_0 \leq \frac{2}{b}(f(x) - f(\eta(s_0, x)))$$

$$\leq \frac{2}{b}(c+\varepsilon-(c-\varepsilon)) = \frac{4\varepsilon}{b} < \delta.$$

We arrive at a contradiction with $x \notin N_{6\delta}(K_c)$. We thus established that

$$\eta_0(x) \in M(f, c, \varepsilon, 4\delta)\,.$$

Making use once more of (ii) of Lemma 7.4 we get

$$2\varepsilon = c+\varepsilon-(c-\varepsilon) \geq f(x) - f(\eta_0(x)) > \frac{b}{2}t_0 = 2\bar{\varepsilon}\,.$$

This contradicts the choice of ε. The statement (ii) is valid, thereby (iii). Property (iv) follows directly from (iii) of Lemma 7.4. The proof is complete. ■

The main result of this Section is the following multiplicity theorem which is in the same spirit as Theorem 4.1 (see also [10], [17]).

Theorem 7.2 (Goeleven, Motreanu and Panagiotopoulos [7]) Let $f : X \to \mathbb{R}$ be a G-invariant and locally Lipschitz functional which satisfies the $(PS)_G$-condition. If f is bounded from below and if the dimension N of the linear subspace of X spanned by G is finite, then f has at least $N+1$ critical orbits.

Proof. For each $1 \leq j \leq N+1$ we set

$$\mathcal{A}_j = \{A \subset X : A \text{ is compact and } \mathrm{Cat}_{\pi(X)}\pi(A) \geq j\}\,.$$

Due to the fact that $\dim(\mathrm{span}\,G) = N$ we have (see [8])

$$\mathrm{Cat}_{\pi(X)}(\pi([0,1]^N \times \{0\})) = N+1,$$

where the identification of $\mathbb{R}^N$ with an N-dimensional subspace of X has been used. Therefore every set $\mathcal{A}_j$ is nonempty. We introduce for every $1 \leq j \leq N+1$ the minimax value

$$c_j = \inf_{A\in\mathcal{A}_j} \max_{x\in\mathcal{A}} f(x).$$

Since $\mathcal{A}_{j+1} \subset \mathcal{A}_j$ it follows that

$$-\infty < \inf_X f = c_1 \leq c_2 \leq \ldots \leq c_{N+1} < +\infty\,.$$

We now prove that if $c := c_j = c_k$ for some $1 \leq j \leq k \leq N+1$ then the set of critical points K_c contains at least $k-j+1$ critical orbits. Suppose the contrary. Thus K_c contains $0 \leq n \leq k-j$ distinct critical orbits $\pi(u_1), \ldots, \pi(u_n)$. Let $\delta > 0$ be such that π restricted to $B(u_i, \delta)$ is a homeomorphism, for any $1 \leq i \leq n$. Then we have that

$$\mathcal{N} = \bigcup_{i=1}^{n} \bigcup_{g \in G} \overset{o}{B}(u_i + g, \delta)$$

is an open neighbourhood of K_c (in the case where $K_c = \emptyset$ one takes $\mathcal{N} = \emptyset$). Corresponding to $\mathcal{N}$, Theorem 7.1 supplies some $\varepsilon > 0$ and the homeomorphism η_0 of X satisfying (i)-(iv) of Theorem 7.1. The definition of $c = c_k$ shows the existence of $A \in \mathcal{A}_k$ such that $A \subset f_{c+\varepsilon}$. Let $B = A \setminus \mathcal{N}$. By Proposition 7.1 we infer that

$$k \leq \mathrm{Cat}_{\pi(X)}\pi(A) \leq \mathrm{Cat}_{\pi(X)}(\pi(B) \cup \pi(\mathcal{N}))$$

$$\leq \mathrm{Cat}_{\pi(X)}\pi(B) + n \leq \mathrm{Cat}_{\pi(X)}\pi(B) + k - j.$$

It follows that $B \in \mathcal{A}_j$ and $B \subset f_{c+\varepsilon} \setminus \mathcal{N}$. From (ii) of Theorem 7.1 we deduce that $\eta_0(B) \subset f_{c-\varepsilon}$. The equivariance of the homeomorphism η_0 ensures that

$$\mathrm{Cat}_{\pi(X)}\pi(\eta_0(B)) = \mathrm{Cat}_{\pi(X)}\pi(B) \geq j\,,$$

so $\eta_0(B) \in \mathcal{A}_j$. Consequently, we find

$$c \leq \max_{\eta_0(B)} f \leq c - \varepsilon\,.$$

This contradiction completes the proof. ■

2. Multiple Eigensolutions for Symmetric Functionals

Throughout this Section, X stands for an infinite dimensional Hilbert space with the scalar product $(\cdot,\cdot)$ and its associated norm $\|\cdot\|$. Let G represent a discrete subgroup of the group of linear isometries of X. For every fixed $r > 0$ we denote by S_r the sphere in X centered at 0 and of radius r, that is,

$$S_r = \{x \in X :\ \|x\| = r\}\,.$$

We are concerned with a functional $f : X \to I\!R$ which is G-invariant, that is,

$$f(gx) = f(x), \quad \forall x \in X,\ \forall g \in G.$$

We denote the restriction of f to S_r by $\tilde{f}$, thus $f = \tilde{f}\,|_{Sr}$: $S_r \to I\!R$.

Assuming that $f : X \to I\!R$ is locally Lipschitz, the generalized gradient $\partial\tilde{f}$ of $\tilde{f}$ is determined by ∂f at an arbitrary $x \in S_r$ as follows

$$\partial\tilde{f}(x) = \{z - r^{-2}\langle z, x\rangle \Lambda x : z \in \partial f(x)\}, \tag{7.7}$$

where $\Lambda : X \to X^*$ designates the duality map of X, namely, $\langle \Lambda x, y \rangle = (x, y)$, for any x, $y \in X$.

Lemma 7.5 If $f : X \to I\!R$ is G-invariant and locally Lipschitz, then the generalized gradient $\partial \tilde{f}$ of $\tilde{f} = f_{|S_r}$ satisfies

$$\partial \tilde{f}(x) = g^* \partial \tilde{f}(gx), \quad \forall x \in S_r, \ \forall g \in G.$$

In particular, the set $\tilde{K}$ of critical points of $\tilde{f}$, that is,

$$\tilde{K} = \{x \in S_r : \ 0 \in \partial \tilde{f}(x)\}$$

is G-invariant.

Proof. Notice that

$$gS_r = S_r, \quad \forall g \in G,$$

because G consists of isometries of X.

Let us check the formula

$$\partial f(x) = g^* \partial f(gx), \quad \forall x \in X, \ \forall g \in G. \tag{7.8}$$

The fact that $w \in \partial f(x)$ is equivalent to

$$\langle w, v \rangle \leq \limsup_{\substack{u \to x \\ t \downarrow 0}} \frac{1}{t} \left(f(u + tv) - f(u) \right)$$

$$= \limsup_{\substack{u \to x \\ t \downarrow 0}} \frac{1}{t} \left(f(gu + tgv) - f(gu) \right) = f^0(gx; gv), \quad \forall v \in X, \ \forall g \in G.$$

Equality (7.8) follows. Combining (7.7) and (7.8) we see that the conclusion of Lemma 7.5 is true. ■

Lemma 7.6 Under the assumptions of Lemma 7.5, the function $\tilde{\lambda}$ defined by

$$\tilde{\lambda}(x) = \min_{w \in \partial \tilde{f}(x)} \|w\|_*, \quad x \in S_r,$$

is G-invariant.

Proof. Applying Lemma 7.5 we obtain

$$\tilde{\lambda}(gx) = \min_{w \in \partial \tilde{f}(gx)} \|w\|_* = \min_{z \in \partial \tilde{f}(x)} \|g^{*-1} z\|_* = \min_{z \in \partial \tilde{f}(x)} \|z\|_* = \tilde{\lambda}(x),$$

for all $x \in S_r$ and $g \in G$. ■

Let $\pi : X \to X/G$ denote the quotient map $\pi(x) = Gx$, $x \in X$.

We introduce a compactness condition of Palais-Smale type on the sphere S_r. Given $r > 0$ and $c \in I\!R$, a G-invariant, locally Lipschitz function $f : X \to I\!R$ is said to satisfy the $(PS)_{r,c,G}$-condition if for every sequence $\{u_k\} \subset S_r$ with $f(u_k) \to c$ and $\tilde{\lambda}(u_k) \to 0$, where $\tilde{\lambda}$ enters Lemma 7.6, there exist a subsequence $\{u_{kn}\}$ and a sequence $\{x_n\} \subset S_r$ with $\pi(u_{kn}) = \pi(x_n)$ such that $\{x_n\}$ converges. We say that the $(PS)_{r,G}$-condition is satisfied if the preceding holds for every $c \in I\!R$.

Set

$$\tilde{K}_c := \tilde{K} \cap f^{-1}(c) = \{x \in S_r : 0 \in \partial\tilde{f}(x) \text{ and } f(x) = c\}.$$

A result analogue to Lemma 7.3 is stated below.

Lemma 7.7 Assume that $f : X \to I\!R$ is locally Lipschitz, G-invariant and satisfies the $(PS)_{r,c,G}$-condition . Then, for each $\delta > 0$, there exist positive constants $\bar{\varepsilon}$, b and a locally Lipschitz map $v : \tilde{M}(f, \bar{\varepsilon}, \delta, c) \to X$, where

$$\tilde{M}(f, \bar{\varepsilon}, \delta, c) := \tilde{f}^{-1}([c - \bar{\varepsilon}, c + \bar{\varepsilon}]) \setminus N_\delta(\tilde{K}_c),$$

such that

(i) $(v(x), x) = 0$,
(ii) $\|v(x)\| \leq 1$,
(iii) $\langle x^*, v(x)\rangle > \frac{b}{2}$,
(iv) $v(gx) = gv(x)$,

for all $x \in \tilde{M}(f, \bar{\varepsilon}, \delta, c)$, $x^* \in \partial\tilde{f}(x)$ and $g \in G$.

Proof. For a fixed $\delta > 0$ there exist constants $b > 0$ and $\bar{\varepsilon}$ such that

$$\tilde{\lambda}(x) \geq b, \quad \forall\, x \in \tilde{M}(f, \bar{\varepsilon}, \delta, c). \tag{7.9}$$

The proof of (7.9) relies on the $(PS)_{r,c,G}$ condition, the G-invariance in Lemma 7.6 and the lower semicontinuity of $\tilde{\lambda}$. Since it follows the same lines as for verifying (7.1) we omit the details.

Take now a fixed $x_0 \in \tilde{M}(f, \bar{\varepsilon}, \delta, c)$. Arguing in the tangent space of the sphere S_r at x_0, i.e.,

$$T_{x_0}S_r = \{x \in X :\ (x, x_0) = 0\},$$

as it was proceeded in the proof of Lemma 7.3 in the space X with $\partial\tilde{f}(x_0)$ in place of $\partial f(x_0)$, we find $h_0 = h(x_0) \in T_{x_0}S_r$ satisfying $\|h_0\| = 1$ and

$$\langle x^*, h_0\rangle \geq b, \quad \forall x^* \in \partial\tilde{f}(x_0).$$

In addition, we can suppose

$$h(gx_0) = gh(x_0), \quad \forall g \in G,$$

because, by Lemma 7.1,

$$\langle \partial\tilde{f}(gx_0), gh(x_0)\rangle = \langle g^*\partial\tilde{f}(gx_0), h(x_0)\rangle = \langle \partial\tilde{f}(x_0), h(x_0)\rangle\,,$$

for all $g \in G$, whose minimum is at least b. The upper semicontinuity of ∂f in the weak sense (see Chang [4]) and the fact that h_0 belongs to $T_{x_0}S_r$ assure that for every $\varepsilon > 0$ there exists $\eta_0 = \eta(x_0) > 0$ with the property that for each $x^* \in \partial\tilde{f}(x)$ with $\|x - x_0\| < \eta_0$ and $x \in S_r$ there is some $z_0 \in \partial\tilde{f}(x_0)$ satisfying

$$|\langle x^* - z_0, h_0\rangle| < \varepsilon\,.$$

Choosing $\varepsilon > 0$ small enough we conclude

$$\langle x^*, h_0\rangle > \frac{b}{2}, \quad \forall x^* \in \partial\tilde{f}(x)\,, \tag{7.10}$$

whenever $\|x - x_0\| < \eta_0$ and $x \in S_r$.

Corresponding to the open covering

$$\mathcal{N} = \{\overset{o}{B}(x_0, \eta_0)\;:\; x_0 \in \tilde{M}(f, \bar{\varepsilon}, \delta, c)\}$$

of $\tilde{M}(f, \bar{\varepsilon}, \delta, c)$ there exists a locally finite open covering $\{U_j\}_{j\in J}$ of $\tilde{M}(f, \bar{\varepsilon}, \delta, c)$ which is a refinement of $\mathcal{N}$ and has the property that any orbit $\pi(u) = Gu$ meets an arbitrary U_j at a finite number of points, which is locally constant with respect to $u \in X$. We assign to each $j \in J$ a point $x_j \in \tilde{M}(f, \bar{\varepsilon}, \delta, c)$ such that $U_j \subset \overset{o}{B}(x_j, \eta_j)$. For every $j \in J$ we define

$$\varphi_j(x) = \frac{\sum_{g\in G} \operatorname{dist}(gx, S_r \setminus U_j)}{\sum_{i\in J}\sum_{g\in G} \operatorname{dist}(gx, S_r \setminus U_i)}\,.$$

Hence

$$v(x) = \sum_{i\in J} \varphi_i(x)\left(h_i - r^{-2}(h_i, x)x\right)\,,$$

for all $x \in \bigcup_{i\in J} U_i \cap S_r$, where $h_i = h(x_i)$, for any $i \in J$.

Then v is well defined and locally Lipschitz. A direct computation yields (i) and (ii). From (7.7) and (7.10) we infer that (iii) holds. A preceding remark yields

$$\begin{aligned} v(gx) = &\sum_{j\in J} \varphi_j(gx)(h_j - r^{-2}(h_j, gx)gx) \\ &= \sum_{j\in J} \varphi_j(gx)(gh_j - r^{-2}(gh_j, gx)gx) \\ &= g\sum_{j\in J} \varphi_j(x)(h_j - r^{-2}(h_j, x)x) = gv(x)\,, \end{aligned}$$

whenever $x \in \tilde{M}(f, \bar{\varepsilon}, \delta, c)$ and $g \in G$. This means that (iv) is fulfilled. ∎

Choosing $\varepsilon \in (0, \bar{\varepsilon})$ we consider locally Lipschitz functions $\varphi : S_r \to [0,1]$ and $\psi : S_r \to [0,1]$ having the same properties as in the previous Section where we replace f and K_c by $\tilde{f}$ and $\tilde{K}_c$, respectively. We define $\tilde{V} : S_r \to X$ by means of mapping v of Lemma 7.7 as follows

$$\tilde{V}(x) = \varphi(x)\psi(x)v(x), \quad \forall x \in S_r .$$

In view of Lemma 7.7 and the properties of the functions φ and ψ, $\tilde{V}$ is well defined, locally Lipschitz and tangent to S_r, i.e., $\tilde{V}(x) \in T_x S_r$, for any $x \in S_r$. The tangency to S_r and the boundedness of $\tilde{V}$ ensure the existence of the global flow $\tilde{\eta} : [0,1] \times S_r \to S_r$ described by the initial value problem

$$\begin{cases} \dfrac{d}{dt}\tilde{\eta}(t,x) = -\tilde{V}(\tilde{\eta}(t,x)) \\ \tilde{\eta}(0,x) = x \in S_r . \end{cases}$$

Lemma 7.8 The flow $\tilde{\eta} \in C([0,1] \times S_r, S_r)$ has the properties
(i) $\|\tilde{\eta}(t,x) - x\| \le t$, for any $x \in S_r$,
(ii) $f(x) - f(\tilde{\eta}(t,x)) > \frac{b}{2}t$, for any $t \in [0,1]$ and $x \in \tilde{M}(f, \bar{\varepsilon}, \delta, c)$,
(iii) $\tilde{\eta}(t, gx) = g\tilde{\eta}(t,x)$, for any $t \in [0,1]$, $x \in S_r$ and $g \in G$.

Proof. The argument is the same as in the proof of Lemma 7.4 involving the vector field $\tilde{V}$ on the sphere S_r in place of V on X. ∎

By means of the flow $\tilde{\eta}$ one obtains the equivariant deformation on the sphere S_r.

Theorem 7.3 Assume that $f : X \to \mathbb{R}$ satisfies the hypotheses of Lemma 7.7 corresponding to a discrete group G of linear isometries of X and with fixed numbers $c \in \mathbb{R}$ and $r > 0$. Given a neighbourhood $\mathcal{N}$ of $\tilde{K}_c$ and a number $\varepsilon_0 > 0$, there exist $\varepsilon \in (0, \varepsilon_0)$ and a homeomorphism $\tilde{\eta}_0 : S_r \to S_r$ such that
(i) $\tilde{\eta}_0(x) = x$, for all $x \notin \tilde{f}^{-1}([c - \varepsilon_0, c + \varepsilon_0])$;
(ii) $\tilde{\eta}_0(\tilde{f}_{c+\varepsilon} \setminus \mathcal{N}) \subset \tilde{f}_{c-\varepsilon}$;
(iii) if $\tilde{K}_c = \emptyset$, then $\tilde{\eta}_0(\tilde{f}_{c+\varepsilon}) \subset \tilde{f}_{c-\varepsilon}$;
(iv) $\tilde{\eta}_0(gx) = g\tilde{\eta}_0(x)$, for all $x \in S_r$ and $g \in G$.

Proof. We can use the pattern of proof for Theorem 7.1 with the difference in employing Lemmas 7.7 and 7.8 instead of Lemmas 7.3 and 7.4. It is necessary to replace in the proof of Theorem 7.1, the data K_c and $\eta \in C([0,1] \times X, X)$ by $\tilde{K}_c$ and $\tilde{\eta} \in C([0,1] \times S_r, S_r)$, respectively. Comparing with Theorem 7.1 we must also take into account the different way of acting (on S_r) of the group G. ∎

Following an idea in Browder [3] we discuss the category of $\pi(S_r)$ over compact sets, that is,

$$\operatorname{Cat}\pi(S_r) = \sup\{\operatorname{Cat}_{\pi(S_r)}\pi(A) : A \text{ is a compact subset of } S_r\}$$

where $\pi : X \to X/G$ is the quotient projection.

Lemma 7.9 For the sphere S_r in X we have

$$\operatorname{Cat}\pi(S_r) = +\infty\,,$$

if X is infinite dimensional and G is cyclic of prime order.

Proof. This follows from Proposition 8.3 of Browder [3]. ∎

The next result establishes the existence of infinitely many orbits of solutions to the abstract eigenvalue problem formulated for a symmetric functional f.

Theorem 7.4 (Goeleven, Motreanu and Panagiotopoulos [7]) Let G be a discrete subgroup of linear isometries of a Hilbert space X with the usual action on X and let $f : X \to I\!R$ be a G-invariant and locally Lipschitz functional which is bounded from below on S_r and satisfies the $(PS)_{r,G}$-condition for some $r > 0$. Then there exist $\operatorname{Cat}\pi(S_r)$ many distinct orbits $Gu_j \subset S_r$ of eigenelements of f on S_r, $j \geq 1$, in the sense that for each $j \geq 1$ there exist $\lambda_j \in I\!R$ and $u_j \in S_r$ such that

$$\lambda_j \Lambda x_j \in \partial f(x_j)\,,$$

for all $x_j \in Gu_j$. Under the additional assumptions of Lemma 7.9, there exist infinitely many orbits Gu_j.

Proof. With the family of sets

$$\tilde{\mathcal{A}}_j := \{A \subset S_r : A \text{ is compact and } \operatorname{Cat}_{\pi(S_r)}\pi(A) \geq j\},$$

for $1 \leq j \leq \operatorname{Cat}\pi(S_r)$, we construct the sequence of real numbers

$$\tilde{c}_j = \inf_{A \in \tilde{\mathcal{A}}_j} \max_{x \in A} f, \quad 1 \leq j \leq \operatorname{Cat}\pi(S_r)\,.$$

Since $1 \leq j \leq \operatorname{Cat}\pi(S_r)$, every collection $\tilde{\mathcal{A}}_j$ is nonempty, so each $\tilde{c}_j$ is finite. Our goal is to show that if $c := \tilde{c}_k = \tilde{c}_j$ for $j \leq k$ then $\tilde{K}_c$ contains at least $k-j+1$ critical orbits of $\tilde{f} = f\,|_{S_r}$. Indeed, if not, there exist at most $n = k-j$ distinct critical orbits $\pi(u_1), ..., \pi(u_n)$ with representatives $u_1, ..., u_n \in S_r$. In view of the fact that G is discrete we may choose mutually disjoint, open and contractible neighbourhoods $U_1, ..., U_n$ of

$u_1, ..., u_n$ in S_r, respectively, such that π is a homeomorphism on each U_i, $i = 1, ..., n$. Then

$$\mathcal{N} = \bigcup_{j=1}^{n} \bigcup_{g \in G} gU_j$$

is an open neighbourhood of $\tilde{K}_c$ in S_r. Corresponding to $\mathcal{N}$, Theorem 7.3 supplies an $\varepsilon > 0$ and a homeomorphism η_0 if S_r with the properties there stated. Let us take $A \in \tilde{\mathcal{A}}_k$ such that $A \subset \tilde{f}_{c+\varepsilon}$ and set $B = A \backslash \mathcal{N}$. It turns out that

$$\mathrm{Cat}_{\pi(S_r)} \pi(B) \geq k - \mathrm{Cat}_{\pi(S_r)} \pi(\mathcal{N}) = k - n \geq j\,,$$

so $B \in \tilde{\mathcal{A}}_j$. Theorem 7.3 implies further that $\tilde{\eta}_0(B) \subset \tilde{f}_{c-\varepsilon}$. The equivariance property (iv) in Theorem 7.3 for $\tilde{\eta}_0$ and (iii) in Proposition 7.1 enable us to write

$$\mathrm{Cat}_{\pi(S_r)} \pi(\tilde{\eta}_0(B)) \geq \mathrm{Cat}_{\pi(S_r)} \pi(B) \geq j\,.$$

Hence $\tilde{\eta}_0(B) \in \tilde{\mathcal{A}}_j$. This leads to the contradiction

$$c \leq \max_{\tilde{\eta}_0(B)} f \leq c - \varepsilon\,.$$

Therefore the existence of at least $\mathrm{Cat}\, \pi(S_r)$ critical orbits for $\tilde{f}$ is justified. Finally, we note that $u \in \tilde{K}$ if and only if there exists $x^* \in \partial f(u)$ such that

$$x^* = r^{-2} \langle x^*, u \rangle \Lambda u\,.$$

This remark completes the proof. ∎

3. Periodic Solutions of Hemivariational Inequalities. Multiple Eigensolutions

For a given number $T > 0$ we assume that

(H_1) $M(t, x)$ is a symmetric matrix of order n continuously differentiable on $[0, T] \times \mathbb{R}^N$ such that

$$\langle M(t, x)y, y \rangle \;\geq\; \alpha |y|^2, \quad \forall (t, x, y) \in [0, T] \times \mathbb{R}^N \times \mathbb{R}^N\,,$$

for some constant $\alpha > 0$;

(H_2) $j(t, x)$ is a real-valued function which is measurable in t, for every $x \in \mathbb{R}^N$, and for which there exists $K \in L^2([0, T])$ such that

$$|j(t, x) - j(t, y)| \leq K(t)|x - y|, \quad \forall x, y \in \mathbb{R}^N, \forall t \in [0, T]\,;$$

(H_3) there exists $h \in L^1([0,T])$ such that

$$|j(t,x)| \leq h(t), \quad \forall x \in \mathbb{R}^N \text{ and a.e. } t \in [0,T];$$

(H_4) $M(t,x)$ and $j(t,x)$ are T_i-periodic in x_i for prescribed $T_i > 0$, $i = 1, ..., N$, where $x = (x_1, ..., x_N)$;

(H_5) $g(t)$ is a $\mathbb{R}^N$-valued function on $[0,T]$ with $g \in L^1(0,T;\mathbb{R}^N)$ and $\int_0^T g(t)dt = 0$.

Using the data above we formulate the following problem in the form of a hemivariational inequality: find $u \in H_T^1$ such that

$$\int_0^T \left(\frac{1}{2}\langle M(t,u)\dot{u},\dot{v}\rangle + \frac{1}{2}\sum_{i=1}^{N}\langle D_{x_i}M(t,u)\dot{u},\dot{u}\rangle v \right) dt + \int_0^T j^0(t,u;v)dt \geq \int_0^T \langle g(t),v(t)\rangle dt, \quad \forall v \in H_T^1.$$

Here H_T^1 stands for the Sobolev space obtained as the completion of the set of smooth $\mathbb{R}^N$-valued T-periodic functions with respect to the H^1-norm.

The general variational background for periodic problems regarding ordinary differential equations is presented in Mawhin and Willem [8].

To fit the problem in our abstract setting we consider the discrete subgroup G of H_T^1 defined by

$$G = \left\{ \sum_{i=1}^{N} k_i T_i e_i \, : \, k_i \in \mathbb{Z}, \ 1 \leq i \leq N \right\},$$

where $\{e_i\}_{1 \leq i \leq N}$ denotes the canonical basis of $\mathbb{R}^N$.

We state the following existence result.

Theorem 7.5 (Goeleven, Motreanu and Panagiotopoulos [7]) Under hypotheses (H_1)-(H_5) the foregoing hemivariational inequality possesses at least $(N+1)$ T-periodic solutions which are distinct.

Proof. In view of (H_2) the functional

$$J(u) := \int_0^T j(t,u(t))dt$$

is well defined and Lipschitz continuous on $L^2(0,T;\mathbb{R}^N)$. We introduce the functional $f : H_T^1 \to \mathbb{R}$ by

$$f(u) := \int_0^T \left(\langle \frac{1}{2}M(t,u)\dot{u},\dot{u}\rangle - \langle g(t),u(t)\rangle \right) dt + J(u).$$

It is clear that f is locally Lipschitz. A straightforward computation based on (H_4) and (H_5) yields

$$f(u+\sum_{i=1}^{N} k_i T_i e_i) = f(u), \quad \forall u \in H^1_T, \forall k_i \in \mathbb{Z}, i=1,...,N.$$

Therefore f is G-invariant.

For every $u \in L^1(0,T;\mathbb{R}^N)$ we write $u = \bar{u} + \tilde{u}$, where

$$\bar{u} = \frac{1}{T}\int_0^T u(t)dt.$$

By $(H_1),(H_3),(H_5)$ and the usual embedding theorems the estimate below holds

$$\begin{aligned} f(u) &\geq \alpha\|\dot{u}\|^2_{L^2} - \int_0^T h(t)dt + \int_0^T \langle g(t), u(t)\rangle dt \\ &\geq \alpha\|\tilde{u}\|^2_{L^2} - c_1 - \|g\|_{L^1}\|\tilde{u}\|_{L^\infty} \\ &\geq \tilde{\alpha}\|\tilde{u}\|^2_{H^1_T} - c_1 - c_2\|\tilde{u}\|_{H^1_T}, \quad \forall u \in H^1_T, \end{aligned} \tag{7.11}$$

with positive constants $\tilde{\alpha}$, c_1, c_2. It follows that f is bounded from below.

Our next aim is to justify that f satisfies the $(PS)_G$-condition (see Section 1). Towards this we need the following remark: if $u \in H^1_T$ has the above decomposition $u = \bar{u} + \tilde{u}$, then there exist uniquely $k_i \in \mathbb{Z}$ such that

$$(\bar{u}, e_i) - k_i T_i \in [0, T_i), \quad \forall 1 \leq i \leq N.$$

Set

$$\bar{u}^0 = ((\bar{u}, e_1) - k_1 T_1, ..., (\bar{u}, e_N) - k_N T_N).$$

Then $v = \bar{u}^0 + \tilde{u}$ is a representative of $\pi(u) \in H^1_T/G$.

In order to check the $(PS)_G$-condition, let $\{u_n\}$ be a sequence in H^1_T such that $f(u_n)$ is bounded and $\lambda(u_n) \to 0$. Firstly, we prove that

$$v_n := \bar{u}^0_n + \tilde{u}_n$$

is bounded. Since

$$\langle \bar{u}^0_n, e_i\rangle \leq T_i, \quad 1 \leq i \leq N, \ \forall n \geq 1,$$

the sequence $\bar{u}^0_n$ is bounded. Taking into account the boundedness of $f(u_n)$, it follows from (7.11) that $\{\tilde{u}_n\}$ is bounded in H^1_T. Hence $\{v_n\}$ is bounded in H^1_T .

There exists a subsequence of $\{v_n\}$, denoted again by $\{v_n\}$, such that $v_n \rightharpoonup v$ weakly in H^1_T, $v_n \to v$ strongly in $L^2(0,T;\mathbb{R}^N)$ and $v_n \to v$

strongly in $C([0,T];\mathbb{R}^N)$. By Lemma 7.2 we know that $\lambda(v_n) \to 0$. Thus there exists $w_n \in \partial f(v_n)$ with $w_n \to 0$. Putting $\varepsilon_n := \|w_n\|$ we get

$$\int_0^T (\langle M(t,v_n)\dot{v}_n, \dot{z}\rangle + \frac{1}{2}\sum_{i=1}^N \langle D_{x_i} M(t,v_n)\dot{v}_n, \dot{v}_n\rangle z_i)dt$$

$$+ \int_0^T j^0(t, v_n; z)dt \geq -\varepsilon_n \|z\|_{H_T^1} + \int_0^T \langle g(t), z\rangle dt\,, \tag{7.12}$$

for all $z = (z_1, ..., z_N) \in H_T^1$.

If we take $z := v - v_n$ in (7.12), by employing (H_1) and standard inequalities we obtain

$$\varepsilon_n \|v - v_n\|_{H_T^1} + \left| \int_0^T j^0(t, v_n; v - v_n)dt \right|$$

$$- \int_0^T \langle M(t,v_n)\dot{v}_n, \dot{v}_n - \dot{v}\rangle dt + \int_0^T \langle f(t), v_n - v\rangle dt$$

$$\geq \alpha \|\dot{v}_n - \dot{v}\|_{L^2}^2 - \frac{1}{2}\sum_{i=1}^N \|\langle D_{xi} M(t,v_n)\dot{v}_n, \dot{v}_n\rangle\|_{L^1} \|v_{n,i} - v_i\|_{L^\infty}, \tag{7.13}$$

where we used the element $v = (v_1, ..., v_N) \in H_T^1$. The boundedness of

$$\|v - v_n\|_{H_T^1}$$

implies

$$\varepsilon_n \|v - v_n\|_{H_T^1} \to 0\,.$$

Hypothesis (H_2) ensures

$$\left| \int_0^T j^0(t, v_n; v - v_n)dt \right| \leq \|K\|_{L^2} \|v - v_n\|_{L^2}\,,$$

so it is seen that

$$\int_0^T j^0(t, v_n; v - v_n)dt \to 0 \quad \text{as } n \to \infty.$$

The weak convergence of $\{v_n\}$ to v in H_T^1 shows

$$\int_0^T \langle f(t), v_n - v\rangle dt \to 0 \quad \text{as } n \to \infty\,.$$

The uniform convergence of $\{v_n\}$ to v and the weak convergence of $\{\dot{v}_n\}$ to $\dot{v}$ in $L^2(0,T;\mathbb{R}^N)$ imply

$$\int_0^T \langle M(t,v_n)\dot{v}_n, \dot{v}_n - \dot{v}\rangle dt \to 0 \quad \text{as } n \to \infty\,.$$

Since the partial derivatives $D_{xi}M$ are continuous, the sequence $\{v_n\}$ is uniformly convergent and the sequence $\{\dot{v}_n\}$ is bounded in $L^2(0,T;\mathbb{R}^N)$, it turns out that $\langle D_{xi}M(t,v_n)\dot{v}_n,\dot{v}_n\rangle$ is bounded in $L^1(0,T;\mathbb{R}^N)$. Hence

$$\frac{1}{2}\sum_{i=1}^{N}\|\langle D_{xi}M(t,v_n)\dot{v}_n,\dot{v}_n\rangle\|_{L^1}\,\|v_{n,i}-v_i\|_{L^\infty}\to 0 \quad \text{as } n\to\infty\,.$$

Then, from (7.13), we deduce the strong convergence of $\{\dot{v}_n\}$ to $\dot{v}$ in $L^2(0,T;\mathbb{R}^N)$. Combining with $v_n \to v$ in $L^2(0,T;\mathbb{R}^N)$ we conclude that $v_n \to v$ in H^1_T. Recalling the equality $\pi(u_n)=\pi(v_n)$, for any $n\geq 1$, the $(PS)_G$-condition is verified.

We also note that $\dim\{\mathrm{span}\,(G)\}=N$. Therefore, for the functional f on H^1_T and the group G, all the assumptions of Theorem 7.2 are satisfied. Then Theorem 7.2 insures that at least $N+1$ critical orbits of f exist. It suffices now to observe that each critical point of f is a solution of the stated hemivariational inequality. The essential point here is that

$$\partial J(u)\subset\int_0^T\partial_u j(t,u)dt\,,\quad \forall u\in L^2(0,T;\mathbb{R}^N)\,,$$

because (H_2) is just Clarke's Hypothesis A with $q=2$ (see Clarke [5], p. 83, or Section 1 in Chapter 1). This completes the proof. ■

In the second part of the present Section we consider the following eigenvalue problem: find $u\in X$ with $\|u\|=r$ and $\lambda\in\mathbb{R}$ such that

$$a(u,v)+\int_\Omega j^0(x,u;v)dx\geq\lambda(u,v),\quad \forall v\in X\,. \qquad (P_r)$$

In the statement of problem (P_r), X denotes an infinite dimensional real Hilbert space, with the scalar product $(\cdot,\cdot)$ and the associated norm $\|\cdot\|$, which is densely and compactly imbedded in $L^p(\Omega;\mathbb{R}^N)$, for a bounded domain Ω in $\mathbb{R}^m$, $m\geq 1$, where $p\geq 2$ and $N\geq 1$ is an integer. Problem (P_r) is stated for a prescribed number $r>0$. The data in (P_r) have the following meaning: $a: X\times X\to\mathbb{R}$ is a continuous, symmetric bilinear form whose corresponding self-adjoint operator is denoted by $A: X\to X$, i.e.,

$$(Au,v)=a(u,v),\quad \forall u,v\in X\,,$$

and $j:\Omega\times\mathbb{R}^N\to\mathbb{R}$ is a Carathéodory function for which we impose assumption (A_1):

(i) $j(\cdot,y):\Omega\to\mathbb{R}$ is measurable for all $y\in\mathbb{R}^N$;

(ii) $j(x,\cdot):\mathbb{R}^N\to\mathbb{R}$ is locally Lipschitz for all $x\in\Omega$;

(iii) $j(\cdot, 0) : \Omega \to I\!R$ is essentially bounded.

To simplify the notation we utilize the symbol $j^0(x, \xi; \eta)$ to denote the generalized directional derivative of j with respect to the second variable and, accordingly, the generalized gradient of $j(x, \cdot)$ is denoted by ∂j (see Definitions 1.1 and 1.2).

Let G be a discrete subgroup of the group of linear isometries of X with the properties in Lemma 7.9. We ask for three further assumptions to be verified:

(A_2) $|z| \leq c(1 + |y|^{p-1})$, for all $(x, y) \in \Omega \times I\!R^N$ and $z \in \partial j(x, y)$, with a constant $c > 0$, and there exists $v \in X$ with $\int_\Omega j^0(x, 0; v(x))dx < 0$.

(A_3) Whenever a sequence $\{v_n\} \subset X$ with $\|v_n\| = r$, a number $\alpha \in [-r^2\|A\|, r^2\|A\|]$ and a measurable map $z : \Omega \to I\!R^N$ are such that $v_n \to v$ in $L^p(\Omega; I\!R^N)$, for some $v \in X \setminus \{0\}$, $a(v_n, v_n) \to \alpha$ and

$$z(x) \in \partial j(x, v(x)) \quad \text{for a.e. } x \in \Omega\,,$$

then

$$\inf_{\|w\|=1} \{a(w, w)\} - \frac{1}{r^2}(\alpha + \int_\Omega \langle z(x), v(x)\rangle dx) > 0\,.$$

(A_4) a and j are G-invariant in the sense that

$$a(gu, gv) = a(u, v), \quad \forall u, v \in X, \ \forall g \in G\,,$$

and

$$j(x, (gu)(x)) = j(x, u(x)), \quad \forall u \in X, \forall g \in G \text{ and for a.e. } x \in \Omega\,.$$

Our existence and multiplicity result for the solution to problem (P_r) is now formulated.

Theorem 7.6 (Goeleven, Motreanu and Panagiotopoulos [7]) Assume that conditions $(A_1) - (A_4)$ are satisfied. Then the eigenvalue problem (P_r) admits infinitely many pairs $(\lambda_n, Gu_n), n \geq 1$, such that $\lambda = \lambda_n$ and every $u \in Gu_n$ solves (P_r). In addition, the sequence $I(u_n)$ is bounded.

Proof. Hypotheses (A_1) and (A_2) imply that the functional I defined by

$$I(v) = \int_\Omega j(x, v(x))dx\,, \quad \forall v \in L^p(\Omega; I\!R^N)$$

is well defined and locally Lipschitz. Then the same properties are valid for the functional $f : X \to I\!R$ defined as

$$f(v) = \frac{1}{2}\,a(v, v) + I(v), \quad \forall v \in X\,.$$

The growth condition (A_2), Lebourg's mean value theorem and (iii) of (A_1) guarantee the estimate

$$|j(x, y)| \leq c_1 + c_2|y|^p, \quad \forall\, (x, y) \in \Omega \times I\!R^N,$$

with positive constants c_1, c_2. Then using the continuous embedding $X \subset L^p(\Omega; I\!R^N)$ it follows that

$$|f(v)| \leq \frac{1}{2}\|A\|\,\|v\|^2 + c_1 + c_2\|v\|^p, \quad \forall v \in X,$$

for possibly new positive constants c_1 and c_2. We conclude that the functional f is bounded on the sphere S_r

$$|f(v)| \leq M, \quad \forall v \in S_r. \tag{7.14}$$

Our assumption (A_4) shows that f is G-invariant, that is

$$f(gv) = f(v), \quad \forall v \in X, \ \forall g \in G. \tag{7.15}$$

We claim that the functional f satisfies the $(PS)_{r,G}$-condition. Indeed, let $\{v_n\} \subset S_r$ be such that $\tilde{\lambda}(v_n) \to 0$ (due to (7.14), the condition concerning the convergence of $f(v_n)$ is not relevant). Denoting by $\Lambda : X \to X^*$ the duality mapping

$$\langle \Lambda u, v\rangle = (u, v), \quad \forall u, v \in X,$$

we find a sequence $\{z_n\} \subset X^*$ such that

$$z_n \in \partial I(v_n), \quad \forall n \geq 1$$

and

$$\Lambda A v_n + z_n - \frac{1}{r^2}\langle \Lambda A v_n + z_n, v_n\rangle \Lambda v_n \to 0 \quad \text{in } X^* \text{ as } n \to \infty.$$

The boundedness of $\{v_n\}$ and the compactness of the embedding $X \subset L^p(\Omega; I\!R^N)$ allow to assume that a renamed subsequence $\{v_n\}$ can be extracted to have

$$a(v_n, v_n) \to \alpha \quad \text{in } I\!R,$$

$$v_n \rightharpoonup v \quad \text{weakly in } X$$

and

$$v_n \to v \quad \text{strongly in } L^p(\Omega; I\!R^N),$$

for a number α. Since the function I is locally Lipschitz and the embedding $L^p(\Omega; I\!R^N)^* \subset X^*$ is compact, it follows that

$$z_n \rightharpoonup z \quad \text{weakly in } L^p(\Omega; I\!R^N)^*$$

and

$$z_n \to z \quad \text{strongly in } X^*.$$

It turns out that

$$z \in \partial I(v) \tag{7.16}$$

and

$$Av_n - \frac{1}{r^2}(\alpha + \langle z, v\rangle)\, v_n \quad \text{converges in } X. \tag{7.17}$$

Relation (7.16) can be interpreted as the condition required for z in (A_3) since $v \neq 0$. (If $v = 0$ we derive from (7.17) and the convergence property in the choice of z_n that $z = 0 \in \partial J(0)$ which contradicts the last part of assumption (A_2)). This expresses that we can invoke assumption (A_3). On the other hand

$$\left(\inf_{\|w\|=1} a(w,w) - \frac{1}{r^2}\left(\alpha + \int_\Omega \langle z(x), v(x)\rangle dx\right)\right) \|v_n - v_k\|^2$$

$$\leq \left\| A(v_n - v_k) - \frac{1}{r^2}\left(\alpha + \int_\Omega \langle z(x), v(x)\rangle dx\right)(v_n - v_k)\right\| \|v_n - v_k\|,$$

for all $n, k \geq 1$.

From (A_3) and (7.17) it follows that $\{v_n\}$ contains a strongly convergent subsequence. Consequently, the Palais-Smale condition for $\tilde{f} = f|_{S_r}$ holds. In particular, it follows that the $(PS)_{r,G}$-condition is true.

Then, by Theorem 7.2, there exist infinitely many pairs (λ_n, Gu_n), $n \geq 1$, such that

$$\Lambda(\lambda_n x_n - Ax_n) \in \partial I(x_n), \quad \forall x_n \in Gu_n,\ \forall n \geq 1. \tag{7.18}$$

The final step in the proof is to make use of relation

$$\partial I(u) \subset \int_\Omega \partial j(x, u)dx, \quad \forall u \in L^p(\Omega; I\!R^N). \tag{7.19}$$

The inclusion above is derived from (A_2) and Section 1 of Chapter 1. Combining (7.18) and (7.19) we see that $\lambda = \lambda_n$ and $u = x_n$ for $x_n \in Gu_n, n \geq 1$, form an eigensolution to problem (P_r). The last assertion of theorem follows from property (7.14). The proof is thus complete. ■

Remark 7.1 Theorem 7.6 extends Theorem 2 in Motreanu and Panagiotopoulos [12] that can be obtained when $G = \{id_X, -id_X\}$. An advantage of Theorem 7.6 is that it points out the relationship between a group of symmetries and the multiple solutions. The class of eigenvalue problems (P_r) where groups of symmetries are present is very large. We refer to Panagiotopoulos [15] where specific problems in Mechanics are discussed in the framework of Problem (P_r) and assumptions (A_1)-(A_4).

4. Multiple Solutions for a Double Eigenvalue Hemivariational Inequality with Constraints

In this Section we prove the existence of infinitely many solutions for a symmetric double eigenvalue hemivariational inequality. The solutions are searched on a sphere-like type manifold and it is obtained a result similar to that established in the preceding Section.

Let V be a real Hilbert space, with the scalar product and the associated norm denoted by $(\cdot,\cdot)_V$ and $\|\cdot\|_V$, respectively. We shall suppose that V is densely and compactly embedded in $L^p(\Omega;\mathbb{R}^N)$ for some $p \geq 2$, an integer $N \geq 1$ and a smooth, bounded domain $\Omega \subset \mathbb{R}^m, m \geq 1$. We denote by $\langle\cdot,\cdot\rangle_V$, $\langle\cdot,\cdot\rangle_{V\times V}$ and $\langle\cdot,\cdot\rangle$, the duality pairings on V, $V\times V$ and $\mathbb{R}^N$, respectively. Let us denote by $C_p(\Omega)$ the constant of the continuous embedding $V \subset L^p(\Omega;\mathbb{R}^N)$ which means that

$$\|v\|_{L^p} \leq C_p(\Omega)\cdot\|v\|_V, \quad \text{for all } v \in V\,.$$

Let $a_1, a_2 : V \times V \to \mathbb{R}$ be two continuous symmetric bilinear forms on V and let $B_1, B_2 : V \to V$ be two bounded self-adjoint linear operators which are coercive in the sense that

$$(B_i v, v)_V \geq b_i \cdot \|v\|_V^2, \quad \text{for all } v \in V,\ i = 1,2,$$

for some constants $b_1, b_2 > 0$. For fixed positive numbers a, b, r we consider the submanifold $S_r^{a,b}$ of $V \times V$ described as follows

$$S_r^{a,b} = \left\{(v_1,v_2) \in V\times V:\ a(B_1v_1,v_1)_V + b(B_2v_2,v_2)_V = r^2\right\}.$$

For a later use we denote by π_i, $i = 1,2$, the projection maps of $V \times V$ onto V, namely $\pi_i(x_1,x_2) = x_i$.

Let $j : \Omega \times \mathbb{R}^N \to \mathbb{R}$ satisfy the following assumptions

(i) $j(\cdot,y)$ is measurable in Ω for each $y \in \mathbb{R}^N$ and $j(\cdot,0)$ is essentially bounded on Ω;

(ii) $j(x,\cdot)$ is locally Lipschitz in $\mathbb{R}^N$ for a.e. $x \in \Omega$.

We also suppose

(H_1) There exist $\theta \in L^{\frac{p}{p-1}}(\Omega)$ and $\rho \in \mathbb{R}$ such that

$$|z| \leq \theta(x) + \rho|y|^{p-1}\,, \tag{7.20}$$

for all $(x,y) \in \Omega\times\mathbb{R}^N$ and each $z \in \partial_y j(x,y)$, for a.e. $x \in \Omega$.

Let $C : S_r^{a,b} \times V \times V \to \mathbb{R}$ be a real function to which we impose no continuity assumption. Consider the following double eigenvalue problem : Find $u_1, u_2 \in V$ and $\lambda_1, \lambda_2 \in \mathbb{R}$ such that

$$(P_{r,a,b}) \begin{cases} a_1(u_1,v_1) + a_2(u_2,v_2) + C((u_1,u_2),v_1,v_2) \\ + \int\limits_{\Omega} j_y^0(x,(u_1-u_2)(x);(v_1-v_2)(x))dx \\ \geq \lambda_1(B_1u_1,v_1)_V + \lambda_2(B_2u_2,v_2)_V, \ \forall v_1, v_2 \in V, \\ a(B_1u_1,u_1)_V + b(B_2u_2,u_2)_V = r^2 . \end{cases}$$

Three additional assumptions are needed for our approach. The first one is the following

(H_2) There exist two locally Lipschitz maps $f_i : V \to \mathbb{R}$, bounded on $\pi_i(S_r^{a,b})$, $(i = 1,2)$ respectively, and such that the following inequality holds

$$C((u_1,u_2),v_1,v_2) \geq f_1^0(u_1;v_1) + f_2^0(u_2;v_2), \tag{7.21}$$

for all $(u_1,u_2) \in S_r^{a,b}$ and for all $(v_1,v_2) \in T_{(u_1,u_2)}S_r^{a,b}$, where the notation $T_{(u_1,u_2)}S_r^{a,b}$ stands for the tangent space

$$\{(v_1,v_2) \in V \times V;\ a(B_1u_1,v_1)_V + b(B_2u_2,v_2)_V = 0\}.$$

In addition, we suppose that the sets

$$\{z \in V^* :\ z \in \partial f_i(u_i), u_i \in \pi_i(S_r^{a,b})\} \tag{7.22}$$

are relatively compact in V^*, $i = 1,2$.

Let us define the map $(A_1,A_2) : V \times V \to V^* \times V^*$ by

$$\langle (A_1,A_2)(u_1,u_2),(v_1,v_2)\rangle_{V\times V} = a_1(u_1,v_1) + a_2(u_2,v_2) \tag{7.23}$$

and the duality map $\Lambda : V \times V \to V^* \times V^*$ expressed by the formula

$$\langle \Lambda(u_1,u_2),(v_1,v_2)\rangle_{V\times V} = a(B_1u_1,v_1)_V + b(B_2u_2,v_2)_V, \tag{7.24}$$

for all u_1, u_2, v_1, $v_2 \in V$.

We also assume

(H_3) For every sequence $\{(u_n^1,u_n^2)\} \subset S_r^{a,b}$ such that $u_n^i \rightharpoonup u_i$ weakly in V, for any $z_n^i \in \partial f_i(u_n^i)$ with

$$a_i(u_n^i,u_n^i) + \langle z_n^i,u_n^i\rangle_V \to \alpha_i \in \mathbb{R}, \tag{7.25}$$

$i = 1,2$ and for all $w \in L^{\frac{p}{p-1}}(\Omega;\mathbb{R}^N)$ satisfying

$$w(x) \in \partial_y j(x,(u_1-u_2)(x)) \quad \text{for a.e. } x \in \Omega, \tag{7.26}$$

for which

$$[(A_1, A_2) - \lambda_0 \cdot \Lambda]\,(u_n^1, u_n^2)$$

converges in $V^* \times V^*$, where

$$\lambda_0 = r^{-2}(\alpha_1 + \alpha_2 + \int_\Omega \langle w(x), (u_1 - u_2)(x)\rangle dx), \tag{7.27}$$

there exists a convergent subsequence of $\{(u_n^1, u_n^2)\}$ in $V \times V$ (thus in $S_r^{a,b}$).

The last assumption is a symmetry hypothesis:

(H_4) j is even with respect to the second variable $y \in I\!R^N$, i.e.,

$$j(x, -y) = j(x, y), \text{ for a.e. } x \in \Omega \text{ and every } y \in I\!R^N,$$

and f_i is even on $\pi_i(S_r^{a,b})$, i.e.,

$$f_i(-u_i) = f_i(u_i), \text{ for all } (u_1, u_2) \in S_r^{a,b}, i = 1, 2.$$

We are now in position to formulate the main result of this Section which asserts the existence of infinitely many solutions to the above problem. We point out that other results of this type are proved in Motreanu and Panagiotopoulos [13], [14].

Theorem 7.7 (Bocea, Motreanu and Panagiotopoulos [2]) Assume that the hypotheses (H_1), (H_2), (H_3) and (H_4) are fulfilled. Then the double eigenvalue problem $(P_{r,a,b})$ admits infinitely many pairs of solutions $\{\pm(u_n^1, u_n^2), (\lambda_n^1, \lambda_n^2)\} \subset S_r^{a,b} \times I\!R^2$ with $\lambda_n^1 = a \cdot \lambda_n$ and $\lambda_n^2 = b \cdot \lambda_n$, where

$$\lambda_n = r^{-2}\left(a_1(u_n^1, u_n^1) + a_2(u_n^2, u_n^2) + \langle z_n^1, \pm u_n^1\rangle_V\right)$$

$$+ r^{-2}\left(\langle z_n^2, \pm u_n^2\rangle_V + \int_\Omega \langle w_n(x), \pm(u_n^1 - u_n^2)(x)\rangle dx\right), \tag{7.28}$$

for some $z_n^i \in V^*$ and $w_n \in L^{\frac{p}{p-1}}(\Omega; I\!R^N)$ satisfying

$$z_n^i \in \partial f_i(\pm u_n^i), \quad i = 1, 2$$

and

$$w_n(x) \in \partial_y j(x, \pm(u_n^1 - u_n^2)(x)) \quad \text{a.e. } x \in \Omega,$$

for every $n \geq 1$.

Proof. Consider the locally Lipschitz functional $I : V \times V \to I\!R$ given by

$$I(u_1, u_2) = \frac{1}{2} \cdot [a_1(u_1, u_1) + a_2(u_2, u_2)] + f_1(u_1)$$
$$+f_2(u_2) + J(u_1 - u_2)\,, \tag{7.29}$$

where $J : L^p(\Omega; I\!R^N) \to I\!R$ is defined by

$$J(u) = \int_\Omega j(x, u(x))dx, \quad \forall u \in L^p(\Omega; I\!R^N)\,.$$

By (H_1), J is locally Lipschitz. From (H_4) it follows that I is even on $S_r^{a,b}$, i.e.,

$$I(-u_1, -u_2) = I(u_1, u_2), \quad \text{for all } (u_1, u_2) \in S_r^{a,b}\,. \tag{7.30}$$

We claim that I is bounded from below on $S_r^{a,b}$. Notice first that, for a.e. $(x, y) \in \Omega \times I\!R^N$, by (7.20), we have

$$|j(x, y)| \le |j(x, 0)| + |j(x, y) - j(x, 0)|$$
$$\le |j(x, 0)| + \sup\{|z| : \ z \in \partial_y j(x, Y),\ Y \in [0, y]\} \cdot |y|$$
$$\le |j(x, 0)| + \theta(x)|y| + \rho|y|^p\,.$$

Therefore

$$|J(u)| \le \|j(\cdot, 0)\|_{L^1} + \|\theta\|_{L^{\frac{p}{p-1}}} \cdot \|u\|_{L^p} + \rho\|u\|_{L^p}^p.$$

Hence $I|_{S_r^{a,b}}$ satisfies the estimate

$$(I|_{S_r^{a,b}})(v_1, v_2) \ge -\frac{1}{2}(\|a_1\| \cdot \|v_1\|_V^2 + \|a_2\| \cdot \|v_2\|_V^2) + f_1(v_1) + f_2(v_2)$$
$$-\|j(\cdot, 0)\|_{L^\infty} \cdot |\Omega| - \|\theta\|_{L^{\frac{p}{p-1}}} \cdot C_p(\Omega) \cdot r \cdot \left(\frac{1}{\sqrt{ab_1}} + \frac{1}{\sqrt{bb_2}}\right)$$
$$-\rho \cdot C_p^p(\Omega) \cdot r^p \left(\frac{1}{\sqrt{ab_1}} + \frac{1}{\sqrt{bb_2}}\right)^p\,. \tag{7.31}$$

Taking into account that f_i are bounded on $\pi_i(S_r^{a,b})(i = 1, 2)$, as stated in hypothesis (H_2), we conclude that I is bounded from below on $S_r^{a,b}$ as claimed.

The expression of the generalized gradient $\partial(I|_{S_r^{a,b}})$ at $(u_1, u_2) \in S_r^{a,b}$ is given by the formula

$$\partial(I|_{S_r^{a,b}})(u_1, u_2) = \{\xi - r^{-2}\langle\xi, (u_1, u_2)\rangle_{V\times V} \cdot \Lambda(u_1, u_2);\ \xi \in \partial I(u_1, u_2)\},$$

where $\Lambda : V \times V \to V^* \times V^*$ is the duality map in (7.24). Here, the duality pairing $\langle \cdot, \cdot \rangle_{V\times V}$ is taken for the norm on $V \times V$ defined by

$$\|(u_1, u_2)\|_{V\times V} := \sqrt{a(B_1 u_1, u_1)_V + b(B_2 u_2, u_2)_V}\,, \qquad (7.32)$$

for all u_1, $u_2 \in V$.

The next step is to prove that I fulfills the Palais-Smale condition (in the sense of Chang [4]) on $S_r^{a,b}$. Accordingly, let us consider a sequence $\{(u_n^1, u_n^2)\} \subset S_r^{a,b}$ such that

$$\sup_n |(I|_{S_r^{a,b}})(u_n^1, u_n^2)| < +\infty$$

and there exists a sequence $\{J_n\} \subset V^* \times V^*$ fulfilling the conditions

$$J_n \in \partial I(u_n^1, u_n^2)\,, \quad n \geq 1 \qquad (7.33)$$

and

$$J_n - r^{-2}\langle J_n, (u_n^1, u_n^2)\rangle_{V\times V} \cdot \Lambda(u_n^1, u_n^2) \to 0\,, \qquad (7.34)$$

strongly in $V^* \times V^*$. We have to prove that $\{(u_n^1, u_n^2)\}$ contains a convergent subsequence in $V \times V$. Under hypothesis (H_1), the functional J is Lipschitz continuous on bounded sets in $L^p(\Omega; \mathbb{R}^N)$ and its generalized gradient satisfies (cf. Theorem 1.3)

$$\partial J(v) \subset \int_\Omega \partial_y j(x, v(x))dx, \quad \forall v \in L^p(\Omega; \mathbb{R}^N)\,.$$

The density of V into $L^p(\Omega; \mathbb{R}^N)$ allows us to apply Corollary 1.2. Thus

$$\partial(J|_V)(v) \subset \partial J(v), \quad \forall v \in V.$$

From $J_n \in \partial I(u_n^1, u_n^2)$ we derive that there exist $z_n^i \in \partial f_i(u_n^i)$, $i = 1, 2$, and $w_n \in \partial(J|_V)(u_n^1 - u_n^2)$ such that

$$J_n = (a_1(u_n^1, \cdot) + z_n^1, a_2(u_n^2, \cdot) + z_n^2) + G^* w_n\,,$$

where $G : V \times V \to V$ is the map given by

$$G(v_1, v_2) = v_1 - v_2\,.$$

The above considerations allow us to write

$$w_n(x) \in \partial_y j(x, (u_n^1 - u_n^2)(x)) \quad \text{for a.e. } x \in \Omega.$$

Taking into account relation (7.34), we get

$$\left(a_1(u_n^1, \cdot) + z_n^1, a_2(u_n^2, \cdot) + z_n^2\right) + G^* w_n$$

$$-r^{-2}\langle[(a_1(u_n^1,\cdot)+z_n^1, a_2(u_n^2,\cdot)+z_n^2)+G^*w_n],(u_n^1,u_n^2)\rangle_{V\times V}\cdot\Lambda(u_n^1,u_n^2)$$

which converges strongly to 0 in $V^*\times V^*$.

Due to the fact that the sequence $\{(u_n^1,u_n^2)\}$ is contained in $S_r^{a,b}$ and by the coercivity property of B_1 and B_2 it follows easily that each sequence $\{u_n^1\}$ and $\{u_n^2\}$ is bounded in V. So, up to a subsequence, we may conclude that

$$u_n^i \rightharpoonup u_i \text{ weakly in } V, \quad \text{for some } u_i\in V \ (i=1,2).$$

The compactness assumptions in hypothesis (H_2) imply that, again up to a subsequence,

$$z_n^i \to z_i \ \text{ strongly in } V^*, \text{ for some } z_i\in V^* \ (i=1,2).$$

Also we have

$$w_n\in\partial(J|_V)(u_n^1-u_n^2)\subset\partial J(u_n^1-u_n^2). \tag{7.35}$$

The compactness of the embedding $V\subset L^p(\Omega;\mathbb{R}^N)$ implies that, passing again to a subsequence, we can assume

$$u_n^i\to u_i \quad \text{strongly in } L^p(\Omega;\mathbb{R}^N), \ (i=1,2). \tag{7.36}$$

Since J is locally Lipschitz on $L^p(\Omega;\mathbb{R}^N)$, the above property ensures that $\{w_n\}$ is bounded in $L^{\frac{p}{p-1}}(\Omega;\mathbb{R}^N)$. By the reflexivity of $L^{\frac{p}{p-1}}(\Omega;\mathbb{R}^N)$ and the compactness of the embedding $L^{\frac{p}{p-1}}(\Omega;\mathbb{R}^N)\subset V^*$, there exists $w\in L^{\frac{p}{p-1}}(\Omega;\mathbb{R}^N)$ such that

$$w_n\to w \quad \text{strongly in } V^* \text{ and weakly in } L^{\frac{p}{p-1}}(\Omega;\mathbb{R}^N).$$

The properties of the generalized gradient and relations (7.35), (7.36) yield

$$w\in\partial J(u_1-u_2). \tag{7.37}$$

At this moment we may assume that

$$a_i(u_n^i,u_n^i) \ \text{ converges in } \mathbb{R},$$

and

$$\langle\left[(z_n^1,z_n^2)+G^*w_n\right],(u_n^1,u_n^2)\rangle_{V\times V}$$

possesses a convergent subsequence in $\mathbb{R}$. From (7.34), taking into account the convergences stated above, we derive that

$$\left(a_1(u_n^1,\cdot),a_2(u_n^2,\cdot)\right)-\lambda_0\cdot\Lambda(u_n^1,u_n^2)$$

converges strongly in $V^* \times V^*$, where λ_0 is the number required in (H_3). Now we apply hypothesis (H_3) and it follows that $\{(u_n^1, u_n^2)\}$ has a convergent subsequence in $V \times V$, so in $S_r^{a,b}$. Thus the Palais-Smale condition for the functional I on $S_r^{a,b}$ is satisfied, as claimed.

Let us denote by Υ the family of closed and symmetric (with respect to the origin $0_{V\times V}$), subsets of $S_r^{a,b}$. Let us denote by $\gamma(S)$ the Krasnoselski genus of the set $S \in \Upsilon$, that is, the smallest integer $k \in I\!N \cup \{+\infty\}$ for which there exists an odd continuous mapping from S into $I\!R^k \setminus \{0\}$. For every $n \geq 1$, set

$$\Gamma_n = \{S \subset S_r^{a,b};\ S \in \Upsilon, \gamma(S) \geq n\}.$$

Let us define the corresponding minimax values of I over Γ_n, i.e.,

$$\beta_n = \inf_{S \subset \Gamma_n} \sup_{(u_1,u_2)\in S} I(u_1, u_2).$$

Each class Γ_n contains compact sets, for instance $S_r^{a,b} \cap F_{n+1}$, where F_{n+1} is a $(n+1)$-dimensional linear subspace of $V \times V$. Since I is bounded from below, it is clear that each β_n is a real number. Since the submanifold $S_r^{a,b}$ becomes the sphere S_r in the Hilbert space $V \times V$ endowed with the new norm defined in (7.32), we may apply now Theorem 3.2 of Chang [4]. This implies that $\beta_n \in I\!R$ are critical values of I on $S_r^{a,b}$. Hence there exists a critical point (u_n^1, u_n^2) (in fact, $\pm(u_n^1, u_n^2)$, since I is an even function on $S_r^{a,b}$), which means

$$0_{V\times V} \in \left(\partial I|_{S_r^{a,b}}\right)(\pm(u_n^1, u_n^2)) \tag{7.38}$$

with

$$I(\pm(u_n^1, u_n^2)) = \beta_n, \tag{7.39}$$

for any $n \geq 1$. From (7.38) it follows that there exists $\xi_n \in \partial I(\pm(u_n^1, u_n^2))$ such that

$$\xi_n - r^{-2}\langle \xi_n, \pm(u_n^1, u_n^2)\rangle_{V\times V} \cdot \Lambda(u_n^1, u_n^2) = 0. \tag{7.40}$$

Clarke's calculus with generalized gradients implies

$$\langle \xi_n, (v_1, v_2)\rangle_{V\times V} \in \partial I(\pm(u_n^1, u_n^2))(v_1, v_2) \subset a_1(\pm u_n^1, v_1) + a_2(\pm u_n^2, v_2)$$

$$+\partial f_1(u_n^1)v_1 + \partial f_2(u_n^2)v_2 + \int_\Omega \partial_y j(x, \pm(u_n^1 - u_n^2)(x))(v_1 - v_2)(x)dx.$$

So, there exist $z_n^i \in \partial f_i(\pm u_n^i)$ $(i = 1, 2)$ and $w_n \in L^{\frac{p}{p-1}}(\Omega; I\!R^N)$ with

$$w_n(x) \in \partial_y j(x, \pm(u_n^1 - u_n^2)(x)) \quad \text{for a.e. } x \in \Omega,$$

such that

$$\langle \xi_n, (v_1, v_2) \rangle_{V \times V} = a_1(\pm u_n^1, v_1) + a_2(\pm u_n^2, v_2)$$
$$+\langle z_n^1, v_1 \rangle_V + \langle z_n^2, v_2 \rangle_V + \int_\Omega \langle w_n(x), (v_1 - v_2)(x) \rangle dx .$$

From (7.40) it is seen that

$$a_1(\pm u_n^1, v_1) + a_2(\pm u_n^2, v_2) + \langle z_n^1, v_1 \rangle_V + \langle z_n^2, v_2 \rangle_V$$
$$+ \int_\Omega \langle w_n(x), (v_1 - v_2)(x) \rangle dx - r^{-2}\{a_1(u_n^1, u_n^1) + a_2(u_n^2, u_n^2)$$
$$+\langle z_n^1, \pm u_n^1 \rangle_V + \langle z_n^2, \pm u_n^2 \rangle_V + \int_\Omega \langle w_n(x), \pm(u_n^1 - u_n^2)(x) \rangle dx\} \cdot$$
$$\cdot \Big(a(B_1(\pm u_n^1), v_1)_V + b(B_2(\pm u_n^2), v_2)_V \Big) = 0, \ \forall v_1, v_2 \in V .$$

Using the notations introduced in the formulation of Theorem 7.7 we may write

$$a_1(\pm u_n^1, v_1) + a_2(\pm u_n^2, v_2) + \langle z_n^1, v_1 \rangle_V$$
$$+\langle z_n^2, v_2 \rangle_V + \int_\Omega \langle w_n(x), (v_1 - v_2)(x) \rangle dx$$
$$= \lambda_n^1 (B_1(\pm u_n^1), v_1)_V + \lambda_n^2 (B_2(\pm u_n^2), v_2)_V, \quad \forall v_1, v_2 \in V .$$

Taking into account the expressions of z_n^1, z_n^2, w_n as well as assumption (H_2) we obtain

$$\begin{cases} a_1(\pm u_n^1, v_1) + a_2(\pm u_n^2, v_2) + C(\pm(u_1, u_2), v_1, v_2) \\ + \int_\Omega j_y^0(x, \pm(u_n^1 - u_n^2)(x); (v_1 - v_2)(x)) dx \\ \geq \lambda_n^1 (B_1(\pm u_n^1), v_1)_V + \lambda_n^2 (B_2(\pm u_n^2), v_2)_V, \quad \forall v_1, v_2 \in V , \\ a(B_1(\pm u_n^1), \pm u_n^1)_V + b(B_2(\pm u_n^2), \pm u_n^2)_V = r^2 . \end{cases}$$

The above relations complete the proof of Theorem 7.7. ■

We are now briefly concerned with a variant of eigenvalue problem $(P_{r,a,b})$. Let V be a Hilbert space and let be given a bounded linear operator $L : V \to L^p(\Gamma; I\!R^N)$, with $\Gamma = \partial\Omega$. We consider the following problem: find $u_1, u_2 \in V$ and $\lambda_1, \lambda_2 \in I\!R$ such that

$$(P'_{r,a,b}) \begin{cases} a_1(u_1, v_1) + a_2(u_2, v_2) + C((u_1, u_2), v_1, v_2) \\ + \int_\Gamma j_y^0(\omega, (\hat{u}_1 - \hat{u}_2)(\omega); (v_1 - v_2)(\omega)) d\omega \\ \geq \lambda_1 (B_1 u_1, v_1)_V + \lambda_2 (B_2 u_2, v_2)_V, \quad \forall v_1, v_2 \in V , \\ a(B_1 u_1, u_1)_V + b(B_2 u_2, u_2)_V = r^2 . \end{cases}$$

In the formulation of the above problem we have used the notation $\hat{u} := Lu$, for all $u \in V$. All the other notations have the same meaning as before, except j, which is now a mapping $j : \Gamma \times I\!R^N \to I\!R$ that satisfies (i) and (ii) for $x \in \Gamma$. Hypotheses (H_i), $i = 1, 4$ are also assumed to hold with the obvious change that now, $x \in \Gamma$. The following condition replaces (H_3).

$(H_3^{'})$ For every sequence $\{(u_n^1, u_n^2)\} \subset S_r^{a,b}$ with $u_n^i \rightharpoonup u_i$ weakly in V, for any $z_n^i \in \partial f_i(u_n^i)$, with

$$a_i(u_n^i, u_n^i) + \langle z_n^i, u_n^i \rangle \to \alpha_i \in I\!R, \tag{7.41}$$

$i = 1, 2$ and for all $w \in L^{\frac{p}{p-1}}(\Gamma; I\!R^N)$ which satisfy the relation

$$w(\omega) \in \partial_y j(\omega, (\hat{u}_1 - \hat{u}_2)(\omega)) \text{ for a.e. } \omega \in \Gamma\,, \tag{7.42}$$

such that

$$[(A_1, A_2) - \lambda_0 \cdot \Lambda]\,(u_n^1, u_n^2)$$

converges in $V^* \times V^*$, where

$$\lambda_0 = r^{-2}(\alpha_1 + \alpha_2 + \int_\Gamma \langle w(\omega), (\hat{u}_1 - \hat{u}_2)(\omega) \rangle d\Gamma)\,, \tag{7.43}$$

there exists a convergent subsequence of $\{(u_n^1, u_n^2)\}$ in $V \times V$ (thus in $S_r^{a,b}$).

Our multiplicity result concerning problem $(P_{r,a,b}^{'})$ is the following

Theorem 7.8 (Bocea, Motreanu and Panagiotopoulos [2]) Suppose that (H_1), (H_2), $(H_3^{'})$, (H_4) hold, for some positive numbers a, b, r. Then problem $(P_{r,a,b}^{'})$ admits infinitely many pairs of solutions $\{\pm(u_n^1, u_n^2), (\lambda_n^1, \lambda_n^2)\} \subset S_r^{a,b} \times I\!R^2$ with $\lambda_n^1 = a \cdot \lambda_n$ and $\lambda_n^2 = b \cdot \lambda_n$, where

$$\lambda_n = r^{-2}\left\{a_1(u_n^1, u_n^1) + a_2(u_n^2, u_n^2) + \langle z_n^1, \pm u_n^1 \rangle_V\right\}$$

$$+ r^{-2}\left\{\langle z_n^2, \pm u_n^2 \rangle_V + \int_\Gamma \langle w_n(x), \pm(\hat{u}_n^1 - \hat{u}_n^2)(x) \rangle d\Gamma\right\}\,, \tag{7.44}$$

for some $z_n^i \in V^*$ and $w_n \in L^{\frac{p}{p-1}}(\Gamma; I\!R^N)$ satisfying

$$z_n^i \in \partial f_i(\pm u_n^i),\ i = 1, 2$$

and

$$w_n(\omega) \in \partial_y j(\omega, \pm(\hat{u}_n^1 - \hat{u}_n^2)(\omega)),\ \text{for a.e. } \omega \in \Gamma,$$

for every $n \geq 1$.

Proof. The argument is the same as in the proof of Theorem 7.7. Only the functional J will be replaced in this case with the corresponding functional $\hat{J} : L^p(\Gamma; \mathbb{R}^N) \to \mathbb{R}$ given by

$$\hat{J}(u) = \int_\Gamma j(\omega, u(\omega))d\omega, \quad \forall u \in L^p(\Gamma; \mathbb{R}^N).$$

■

Remark 7.2 The proofs of Theorems 7.7 and 7.8 also hold if we replace $C((u_1, u_2), v_1, v_2)$ in the formulation of the problem $(P_{r,a,b})$ by the sum $(C_1(u_1), v_1)_V + (C_2(u_2), v_2)_V$, where $C_i : V \to V$, $i = 1, 2$, are two operators satisfying condition (H_2), in which we make the same replacement.

References

[1] A. Ambrosetti, Critical points and nonlinear variational problems, *Suppl. Bull. Math. France* **120** (1992), fasc. 2, Mémoire No. 49.

[2] M. Bocea, D. Motreanu and P. D. Panagiotopoulos, Multiple solutions for a double eigenvalue hemivariational inequality on a sphere-like type manifold, *Nonlinear Anal., TMA* **42A** (2000), 737-749.

[3] F. Browder, *Nonlinear eigenvalue problems and group invariance.* In *Functional Analysis and Related Fields*, Springer-Verlag, Berlin, 1970.

[4] K. C. Chang, Variational methods for non-differentiable functionals and their applications to partial differential equations, *J. Math. Anal. Appl.* **80** (1981), 102-129.

[5] F.H. Clarke, *Optimization and Nonsmooth Analysis*, Willey, New York, 1983.

[6] G. Fournier and M. Willem, Multiple solutions of the forced pendulum equation, *Ann. Inst. H. Poincaré, Analyse Non Linéaire* **6** (1989), 259-282.

[7] D. Goeleven, D. Motreanu and P. D. Panagiotopoulos, Multiple solutions for a class of hemivariational inequalities involving periodic energy functionals, *Math. Meth. Appl. Sciences* **20** (1997), 548-568.

[8] J. Mawhin and M. Willem, *Critical Point Theory and Hamiltonian Systems*, Springer-Verlag, New-York, 1989.

[9] J. Mawhin and M. Willem, Multiple solutions for the periodic boundary value problem for some forced pendulum type equations, *J. Diff. Equations* **52** (1984), 264-287.

[10] P. Mironescu and V. Rădulescu, A multiplicity theorem for locally Lipschitz periodic functionals, *J. Math. Anal. Appl.* **195** (1995), 621-637.

[11] D. Motreanu and P. D. Panagiotopoulos, Double eigenvalue problems for hemivariational inequalities, *Arch. Rat. Mech. Anal.* **140** (1997), 225-251.

[12] D. Motreanu and P. D. Panagiotopoulos, On the eigenvalue problem for hemivariational inequalities: existence and multiplicity of solutions, *J. Math. Anal. Appl.* **197** (1996), 75-89.

[13] D. Motreanu and P. D. Panagiotopoulos, An eigenvalue problem for a hemivariational inequality involving a nonlinear compact operator, *Set-Valued Anal.* **3** (1995), 157-166.

[14] D. Motreanu and P. D. Panagiotopoulos, Nonconvex energy functions, related eigenvalue hemivariational inequalities on the sphere and applications, *J. Global Optimiz.* **6** (1995), 163-177.

[15] P. D. Panagiotopoulos, *Hemivariational Inequalities.: Applications in Mechanics and Engineering*, Springer-Verlag, New-York/Boston/Berlin, 1993.

[16] P. Rabinowitz, *Minimax Methods in Critical Point Theory with Applications to Differential Equations*, CBMS Reg. Conf. Ser. Math. 65, Amer. Math. Soc., Providence, RI, 1986.

[17] V. Rădulescu, Mountain Pass theorems for non-differentiable functions and applications, *Proc. Japan Acad.* **69A** (1993), 193-198.

Chapter 8

NON-SYMMETRIC PERTURBATION OF SYMMETRIC EIGENVALUE PROBLEMS

In this Chapter we establish the influence of an arbitrary small perturbation for several classes of symmetric hemivariational eigenvalue inequalities with constraints. If the symmetric problem has infinitely many solutions we show that the number of solutions of the perturbed problem tends to infinity if the perturbation approaches zero with respect to an appropriate topology. This is a very natural phenomenon that occurs often in concrete situations. We illustrate it with the following elementary example: consider on the real axis the equation $\sin x = 1/2$. This is a "symmetric" problem (due to the periodicity) with infinitely many solutions. Let us now consider an arbitrary non-symmetric "small" perturbation of the above equation. For instance, the equation $\sin x = 1/2 + \varepsilon x^2$ has finitely many solutions, for any $\varepsilon \neq 0$. However, the number of solutions of the perturbed equation becomes greater and greater if the perturbation (that is, $|\varepsilon|$) is smaller and smaller. In contrast with this elementary example, our proofs rely on powerful tools such as topological methods in nonsmooth critical point theory. For different perturbation results and their applications we refer to [1], [15], [20] (see also [9] for a nonsmooth setting) in the case of elliptic equations, [8] for variational inequalities and [3], [5], [6], [14], [16], [17], [18] for various perturbations of hemivariational inequalities. This abstract developments are motivated by important appications in Mechanics (see [12], [13]).

1. Non-symmetric Perturbations of Eigenvalue Problems for Periodic Hemivariational Inequalities with Constraints

Throughout this Section V denotes a real Hilbert space which is densely and compactly embedded in $L^p(\Omega; \mathbb{R}^N)$, for some $1 < p < \infty$ and $N \geq 1$, where Ω is a bounded domain in $\mathbb{R}^m$, $m \geq 1$. Denote by $\|\cdot\|$ the norm on V and by $(\cdot, \cdot)$ the corresponding inner product. Let $a : V \times V \to \mathbb{R}$ be a continuous, symmetric and bilinear form, not necessarily coercive. We denote by $A : V \to V$ the self-adjoint bounded linear operator corresponding to a, i.e.

$$(Au, v) = a(u, v) \quad \text{for all } u, v \in V.$$

Denote by $|\cdot|$ the Euclidean norm on $\mathbb{R}^N$, while the duality pairing between V^* and V (resp., between $(\mathbb{R}^N)^*$ and $\mathbb{R}^N$) is denoted by $\langle \cdot, \cdot \rangle_V$ (resp., $\langle \cdot, \cdot \rangle$). For $r > 0$, let S_r denote the sphere of radius r in V centered at the origin, i.e.,

$$S_r = \{u \in V \;:\; \|u\| = r\}.$$

Let $j : \Omega \times \mathbb{R}^N \to \mathbb{R}$ be a Carathéodory function which is locally Lipschitz with respect to the second variable and such that $j(\cdot, 0) \in L^1(\Omega)$. The generalized directional derivative of $j(x, \cdot)$ is denoted by $j^0(x, \xi; \eta)$ for $\xi, \eta \in \mathbb{R}^N$, while the generalized gradient of $j(x, \cdot)$ is designated by $\partial_y j(x, y)$ for any $(x, y) \in \Omega \times \mathbb{R}^N$ (see Definitions 1.1 and 1.2).

Let G be a finite subgroup of the group of linear isometries of V. Assume further that the following conditions are satisfied

($\mathbf{A}_1$) a and j are G-invariant in the sense that

$$a(gu, gv) = a(u, v), \ \forall u, v \in V, \ \forall g \in G,$$

and

$$j(x, (gu)(x)) = j(x, u(x)), \ \forall u \in V, \ \forall g \in G \text{ and for a.e. } x \in \Omega;$$

($\mathbf{A}_2$) there exist $a_1 \in L^{p/(p-1)}(\Omega)$ and $b \in \mathbb{R}_+$ such that

$$|w| \leq a_1(x) + b|y|^{p-1}, \text{ for a.e. } (x, y) \in \Omega \times \mathbb{R}^N \text{ and all } w \in \partial_y j(x, y),$$

and there exists $v \in L^p(\Omega; \mathbb{R}^N)$ with

$$\int_\Omega j^0(x, 0; v(x))\, dx \; < 0.$$

Consider $\Lambda : V \to V^*$ the duality isomorphism

$$\langle \Lambda u, v \rangle_V = (u, v), \quad \text{for all } u, v \in V.$$

Suppose also that the following "compactness" assumption holds

($\mathbf{A}_3$) For every sequence $\{u_n\} \subset V$ with $\|u_n\| = r$, for every number $\alpha \in [-r^2\|A\|, r^2\|A\|]$ and for every measurable map $w : \Omega \to (\mathbb{R}^N)^*$ such that $u_n \to u$ strongly in $L^p(\Omega; \mathbb{R}^N)$ for some $u \in V \setminus \{0\}$, $w(x) \in \partial_y j(x, u(x))$ for a.e. $x \in \Omega$ and $a(u_n, u_n) \to \alpha$, we have that

$$\inf_{\|\tau\|=1} \{a(\tau, \tau)\} - r^{-2} \left(\alpha + \int_\Omega \langle w(x), u(x) \rangle \, dx \right) > 0.$$

Consider the following eigenvalue problem

$$(\mathbf{P}_1) \begin{cases} (u, \lambda) \in V \times \mathbb{R} \\ a(u, v) + \int_\Omega j^0(x, u(x); v(x)) \, dx \geq \lambda(u, v), \quad \text{for all } v \in V \\ \|u\| = r. \end{cases}$$

Under assumptions ($\mathbf{A}_1$) – ($\mathbf{A}_3$), we have proved in Theorem 7.6 that problem ($\mathbf{P}_1$) admits infinitely many distinct pairs of solutions $(Gu_n, \lambda_n)_{n \geq 1}$ in $S_r \times \mathbb{R}$ (see also [10]). We also remark that in Theorem 7.6 it is assumed $a_1 =$ const. in ($\mathbf{A}_2$), so the statement therein is formulated under a slightly less general hypothesis. We observe that in order to show that the arguments of Theorem 7.6 hold in our case, it is sufficient to verify that the energy functional

$$F(u) = \frac{1}{2} a(u, u) + J(u), \quad u \in V \tag{8.1}$$

is bounded from below on S_r, where the locally Lipschitz function $J : L^p(\Omega; \mathbb{R}^N) \to \mathbb{R}$ is defined by

$$J(u) = \int_\Omega j(x, u(x)) \, dx \,.$$

Indeed, by Lebourg's mean value theorem,

$$|j(x, y)| \leq |j(x, 0)| + |j(x, y) - j(x, 0)|$$

$$\leq |j(x, 0)| + \sup\{|w| \, : \, w \in \partial_y j(x, Y), \; Y \in [0, y]\} \cdot |y|$$

$$\leq |j(x, 0)| + a_1(x)|y| + b|y|^p. \tag{8.2}$$

Therefore

$$|J(u)| \leq \|j(\cdot, 0)\|_{L^1} + \|a_1\|_{L^{p/(p-1)}} \|u\|_{L^p} + b\|u\|_{L^p}^p. \tag{8.3}$$

The continuity of the embedding $V \subset L^p(\Omega; \mathbb{R}^N)$ ensures the existence of a positive constant $C_p(\Omega)$ such that

$$\|u\|_{L^p} \le C_p(\Omega)\|u\|_V, \quad \text{for all } u \in V.$$

From (8.1) and (8.3) it follows that

$$F|_{S_r}(u) \ge -\frac{1}{2}\|A\|r^2 - \|j(\cdot,0)\|_{L^1} - C_p(\Omega)\, r\, \|a_1\|_{L^{p/(p-1)}} - b\, C_p^p(\Omega)\, r^p,$$

for any $u \in S_r$. From now on the proof follows in the same way as in Theorem 7.6.

Let us now consider the following non-symmetric perturbed hemivariational inequality:

$$(\mathbf{P}_2)\begin{cases} (u,\lambda) \in V \times \mathbb{R} \\ a(u,v) + \int_\Omega (j^0(x,u(x);v(x)) + h^0(x,u(x);v(x)))\, dx + \langle \varphi, v\rangle_V \\ \ge \lambda(u,v), \quad \forall\, v \in V \\ \|u\| = r, \end{cases}$$

where $\varphi \in V^*$ and $h : \Omega \times \mathbb{R}^N \to \mathbb{R}$ is a Carathéodory function which is locally Lipschitz with respect to the second variable and such that $h(\cdot,0) \in L^1(\Omega)$. We point out that we do not make any symmetry assumption on h. We require only the natural growth condition

($\mathbf{A}_4$) $|z| \le a_2(x) + c|y|^{p-1}$, for a.e. $(x,y) \in \Omega \times \mathbb{R}^N$ and for all $z \in \partial_y h(x,y)$, where $a_2 \in L^{p/(p-1)}(\Omega)$ and $c > 0$, and if $\varphi \in L^{p/(p-1)}(\Omega; \mathbb{R}^N)$ there exists $v \in L^p(\Omega; \mathbb{R}^N)$ with

$$\int_\Omega (\varphi(x)v(x) + j^0(x,0;v(x)) + h^0(x,0;v(x)))\, dx < 0.$$

The corresponding variant of compactness condition ($\mathbf{A}_3$) is

($\mathbf{A}_5$) For every sequence $\{u_n\} \subset V$ with $\|u_n\| = r$, for every number $\alpha \in [-r^2\|A\|, r^2\|A\|]$, and for every measurable maps $z, w : \Omega \to (\mathbb{R}^N)^*$ such that $u_n \to u$ strongly in $L^p(\Omega; \mathbb{R}^N)$ with some $u \in V \setminus \{0\}$, $w(x) \in \partial_y j(x,u(x))$, $z(x) \in \partial_y h(x,u(x))$ for a.e. $x \in \Omega$ and $a(u_n,u_n) \to \alpha$, we have that

$$\inf_{\|\tau\|=1}\{a(\tau,\tau)\} - r^{-2}(\alpha + \langle \varphi, u\rangle_V + \int_\Omega \langle w(x)+z(x), u(x)\rangle\, dx) > 0. \quad (8.4)$$

This Section deals with the study of (possibly non-symmetric) perturbed hemivariational inequality ($\mathbf{P}_2$). Our main result asserts that the number of solutions of $(\mathbf{P}_2)$ goes to infinity as the perturbation becomes smaller and smaller.

Theorem 8.1 (Ciulcu, Motreanu and Rădulescu [5]) Suppose that assumptions $(\mathbf{A}_1)-(\mathbf{A}_5)$ hold. Then, for every $n \geq 1$, there exists $\delta_n > 0$ such that problem $(\mathbf{P}_2)$ admits at least n distinct solutions, provided that $\|h(\cdot,0)\|_{L^1} \leq \delta_n$, $\|a_2\|_{L^{p/(p-1)}} \leq \delta_n$, $c \leq \delta_n$ and $\|\varphi\|_{V^*} \leq \delta_n$.

We start the proof with some auxiliary results. Let us first define the energy functional $W : V \to I\!R$ associated to the hemivariational problem $(\mathbf{P}_2)$ by

$$W(u) = \frac{1}{2}a(u,u) + J(u) + H(u) + \langle \varphi, u \rangle_V, \quad \forall u \in V,$$

where $H(u) = \int_\Omega h(x,u(x))\,dx$. We first prove that W can be viewed as a small perturbation of the functional F in (8.1) whenever the data h and φ are sufficiently small in a suitable sense.

Lemma 8.1 For every number $\varepsilon > 0$ there exists $\delta_\varepsilon > 0$ such that

$$\sup_{u \in S_r} |F(u) - W(u)| < \varepsilon,$$

provided $\|h(\cdot,0)\|_{L^1} \leq \delta_\varepsilon$, $\|a_2\|_{L^{p/(p-1)}} \leq \delta_\varepsilon$, $c \leq \delta_\varepsilon$ and $\|\varphi\|_{V^*} \leq \delta_\varepsilon$.

Proof. Proceeding as for proving (8.2) we obtain

$$|h(x,y)| \leq |h(x,0)| + a_2(x)|y| + c|y|^p.$$

Thus, for all $u \in S_r$ we have

$$|F(u) - W(u)| \leq |H(u)| + |\langle \varphi, u \rangle_V| \leq |H(u)| + r\|\varphi\|_{V^*}$$

$$\leq \|h(\cdot,0)\|_{L^1} + \|a_2\|_{L^{p/(p-1)}} C_p(\Omega) r + c C_p^p(\Omega) r^p + r\|\varphi\|_{V^*} < \varepsilon$$

for small $h(\cdot,0)$, a_2, c and φ. ∎

Our next result shows that $W|_{S_r}$ satisfies the Palais-Smale condition in the sense of Chang [4].

Lemma 8.2 The functional W satisfies the Palais-Smale condition (in short, (PS) condition) on S_r.

Proof. Let $\{u_n\}$ be a sequence in S_r such that

$$\sup_n |W(u_n)| < \infty \tag{8.5}$$

and

$$\lambda_{W|_{S_r}}(u_n) \to 0 \quad \text{as } n \to \infty, \tag{8.6}$$

where $\lambda_{W|_{S_r}}(u) = \min\{\|\theta\| : \theta \in \partial(W|_{S_r})(u)\}$. We already know (see Chapter 1) that the functional $\lambda_{W|_{S_r}}$ is well defined and lower semicontinuous. The expression of the generalized gradient of W on S_r is given by

$$\partial(W|_{S_r})(u) = \{\xi - r^{-2}\langle \xi, u\rangle_V \Lambda u : \xi \in \partial W(u)\}. \tag{8.7}$$

Notice that (8.5) is automatically fulfilled due to the growth conditions in $(\mathbf{A}_2)$ and $(\mathbf{A}_4)$. From (8.6) and (8.7) we deduce the existence of a sequence $\{\xi_n\} \subset V^*$ such that

$$\xi_n \in \partial W(u_n) \tag{8.8}$$

and

$$\xi_n - r^{-2}\langle \xi_n, u_n\rangle_V \Lambda u_n \to 0 \quad \text{strongly in } V^*. \tag{8.9}$$

For every $u \in V$, the generalized gradient $\partial W(u) \subset V^*$ satisfies

$$\partial W(u) \subset \Lambda A u + \partial(J|_V)(u) + \partial(H|_V)(u) + \varphi. \tag{8.10}$$

From (8.8), (8.9) and (8.10) it follows that there exist

$$w_n \in \partial(J|_V)(u_n) \quad \text{and} \quad z_n \in \partial(H|_V)(u_n)$$

such that

$$\Lambda A u_n + w_n + z_n + \varphi - r^{-2}\langle \Lambda A u_n + w_n + z_n + \varphi, u_n\rangle_V \Lambda u_n \to 0 \tag{8.11}$$

strongly in V^*. The density of V in $L^p(\Omega; \mathbb{R}^N)$ implies (see Corollary 1.2)

$$\partial(J|_V)(u) \subset \partial J(u) \quad \text{and} \quad \partial(H|_V)(u) \subset \partial H(u), \quad u \in V. \tag{8.12}$$

So, by (8.12), we see that

$$w_n \in \partial J(u_n) \quad \text{and} \quad z_n \in \partial H(u_n). \tag{8.13}$$

Since V is a reflexive space and $\|u_n\| = r$, we can extract a subsequence, denoted again by $\{u_n\}$, such that

$$u_n \rightharpoonup u \quad \text{weakly in } V \text{ as } n \to \infty. \tag{8.14}$$

The compactness of the embedding $V \subset L^p(\Omega; \mathbb{R}^N)$ implies that, up to a subsequence,

$$u_n \to u \quad \text{strongly in } L^p(\Omega; \mathbb{R}^N) \text{ as } n \to \infty. \tag{8.15}$$

Using (8.13), (8.15) and the fact that the functionals J and H are locally Lipschitz on $L^p(\Omega; \mathbb{R}^N)$ we deduce that the sequences $\{w_n\}$ and $\{z_n\}$ are bounded in $L^{p/(p-1)}(\Omega; \mathbb{R}^N)$. Thus, passing eventually to subsequences, we have

$$w_n \rightharpoonup w \quad \text{weakly in } L^{p/(p-1)}(\Omega; \mathbb{R}^N) \text{ as } n \to \infty, \tag{8.16}$$

$$z_n \rightharpoonup z \quad \text{weakly in } L^{p/(p-1)}(\Omega; \mathbb{R}^N) \text{ as } n \to \infty. \tag{8.17}$$

Since the embedding $L^{p/(p-1)}(\Omega; \mathbb{R}^N) \subset V^*$ is compact, relations (8.16) and (8.17) imply (up to subsequences)

$$w_n \to w \quad \text{strongly in } V^* \text{ as } n \to \infty, \tag{8.18}$$

$$z_n \to z \quad \text{strongly in } V^* \text{ as } n \to \infty. \tag{8.19}$$

Combining (8.14), (8.18) and (8.19) we obtain

$$\langle w_n + z_n, u_n \rangle_V \to \langle w + z, u \rangle_V \quad \text{as } n \to \infty. \tag{8.20}$$

In virtue of the boundedness of the sequence $\{u_n\}$ in V and the continuity of the bilinear form a we may suppose that, along a subsequence, we have

$$a(u_n, u_n) \to \alpha \quad \text{as } n \to \infty \text{ for some } \alpha \in [-r^2\|A\|, r^2\|A\|].$$

Taking into account (8.18)-(8.20) we see that (8.11) implies that

$$Au_n - r^{-2}(\alpha + \langle \varphi, u \rangle_V + \langle w + z, u \rangle_V)u_n \quad \text{converges strongly in } V \tag{8.21}$$

as $n \to \infty$. Using (8.13), (8.15), (8.16), (8.17) and the fact that the Clarke generalized gradient is a weak*-closed multifunction (see Proposition 1.1) we deduce

$$w \in \partial J(u) \tag{8.22}$$

and

$$z \in \partial H(u). \tag{8.23}$$

Hypotheses ($\mathbf{A}_2$) and ($\mathbf{A}_4$) allow to apply Theorem 2.7.5 in Clarke [7], and from relations (8.22) and (8.23) we get the existence of two measurable mappings $w, z : \Omega \to (\mathbb{R}^N)^*$ such that

$$w(x) \in \partial_y j(x, u(x)) \quad \text{for a.e. } x \in \Omega \tag{8.24}$$

$$z(x) \in \partial_y h(x, u(x)) \quad \text{for a.e. } x \in \Omega \tag{8.25}$$

$$\langle w, u \rangle_V = \langle w, u \rangle_{L^p(\Omega; \mathbb{R}^N)} = \int_\Omega \langle w(x), u(x) \rangle \, dx \tag{8.26}$$

$$\langle z, u \rangle_V = \langle z, u \rangle_{L^p(\Omega; I\!R^N)} = \int_\Omega \langle z(x), u(x) \rangle \, dx. \tag{8.27}$$

Remark that, due to the first part of $(\mathbf{A}_2)$, (8.24) and $u \in L^p(\Omega; I\!R^N)$, we have that $\langle w(\cdot), u(\cdot) \rangle \in L^1(\Omega; I\!R)$ since

$$\int_\Omega |\langle w(x), u(x) \rangle| \, dx \leq \int_\Omega (a_1(x) + b|u(x)|^{p-1})|u(x)| \, dx$$

$$\leq \|a_1\|_{L^{p/(p-1)}} \|u\|_{L^p} + b\|u\|^p_{L^p}.$$

In the same way, using the first part of $(\mathbf{A}_4)$, (8.25) and $u \in L^p(\Omega; I\!R^N)$, we obtain that $\langle z(\cdot), u(\cdot) \rangle \in L^1(\Omega; I\!R)$. Replacing (8.26) and (8.27) in (8.21) we obtain that the sequence

$$Au_n - r^{-2}(\alpha + \langle \varphi, u \rangle_V + \int_\Omega \langle w(x) + z(x), u(x) \rangle \, dx) u_n \tag{8.28}$$

converges in V, with w and z satisfying (8.24) and (8.25), respectively.

We note that $u \neq 0$. Indeed, if $u = 0$, then (8.11) yields that $w_n + z_n \rightharpoonup -\varphi$ weakly in $L^{p/(p-1)}(\Omega; I\!R^N)$. Hence $-\varphi \in \partial J(0) + \partial H(0)$ which contradicts the final part of assumption $(\mathbf{A}_4)$. Consequently, in view of (8.15), (8.24), (8.25), we are in a position to use assumption $(\mathbf{A}_5)$ and therefore inequality (8.4) is valid. For all n, k we have

$$\begin{aligned}
&\left(\inf_{\|\tau\|=1} \{a(\tau, \tau)\} - r^{-2}(\alpha + \langle \varphi, u \rangle_V + \int_\Omega \langle (w+z)(x), u(x) \rangle \, dx) \right) \cdot \\
&\cdot \|u_n - u_k\|^2 \leq a(u_n - u_k, u_n - u_k) - r^{-2}((\alpha + \langle \varphi, u \rangle_V \\
&+ \int_\Omega \langle (w+z)(x), u(x) \rangle \, dx)(u_n - u_k), u_n - u_k) \\
&= (A(u_n - u_k) - r^{-2}(\alpha + \langle \varphi, u \rangle_V \\
&+ \int_\Omega \langle (w+z)(x), u(x) \rangle \, dx)(u_n - u_k), u_n - u_k) \\
&\leq \|A(u_n - u_k) - r^{-2}(\alpha + \langle \varphi, u \rangle_V \\
&+ \int_\Omega \langle (w+z)(x), u(x) \rangle \, dx)(u_n - u_k)\| \, \|u_n - u_k\|.
\end{aligned}$$

The convergence of the sequence in (8.28), the above estimate and (8.4) show that $\{u_n\}$ contains a Cauchy subsequence in V. Hence $\{u_n\}$ converges strongly along a subsequence in V to u. This completes the proof of lemma. ∎

The next result shows that W plays indeed the role of energy functional for the perturbed problem $(\mathbf{P}_2)$.

Lemma 8.3 If $u \in S_r$ is a critical point of $W|_{S_r}$ then there exists $\lambda \in I\!R$ such that (u, λ) is a solution of problem $(\mathbf{P}_2)$.

Proof. Since $0 \in \partial(W|_{S_r})(u)$ it follows from (8.7), (8.10), (8.12) that there exist

$$w \in \partial(J|_V)(u) \subset \partial J(u) \quad \text{and} \quad z \in \partial(H|_V)(u) \subset \partial H(u) \tag{8.29}$$

such that u is a solution of

$$\Lambda Au + w + z + \varphi = r^{-2}\langle \Lambda Au + w + z + \varphi, u\rangle_V \Lambda u. \tag{8.30}$$

By Theorem 1.3 we know that for every $u \in L^p(\Omega; I\!R^N)$ we have

$$\partial J(u) \subset \int_\Omega \partial_y j(x, u(x))\, dx \quad \text{and} \quad \partial H(u) \subset \int_\Omega \partial_y h(x, u(x))\, dx.$$

Thus, by (8.29), the mappings $w, z : \Omega \to (I\!R^N)^*$ satisfy

$$w(x) \in \partial_y j(x, u(x)) \quad \text{for a.e. } x \in \Omega, \tag{8.31}$$

$$z(x) \in \partial_y h(x, u(x)) \quad \text{for a.e. } x \in \Omega, \tag{8.32}$$

and, for all $v \in V$,

$$\langle w, v\rangle_V = \int_\Omega \langle w(x), v(x)\rangle\, dx, \tag{8.33}$$

$$\langle z, v\rangle_V = \int_\Omega \langle z(x), v(x)\rangle\, dx. \tag{8.34}$$

Denote

$$\lambda = r^{-2}\left(\langle \Lambda Au + \varphi, u\rangle_V + \int_\Omega \langle w(x) + z(x), u(x)\rangle\, dx\right). \tag{8.35}$$

From (8.30)-(8.35) it follows that, for every $v \in V$, we have

$$\begin{aligned}
&\lambda(u, v) - a(u, v) - \langle \varphi, v\rangle_V = \int_\Omega \langle w(x) + z(x), v(x)\rangle\, dx \\
&\le \int_\Omega \max\{\langle \mu_1, v(x)\rangle \;:\; \mu_1 \in \partial_y j(x, u(x))\}\, dx \\
&+ \int_\Omega \max\{\langle \mu_2, v(x)\rangle \;:\; \mu_2 \in \partial_y h(x, u(x))\}\, dx \\
&= \int_\Omega j^0(x, u(x); v(x))\, dx + \int_\Omega h^0(x, u(x); v(x))\, dx.
\end{aligned}$$

The last equality is obtained by applying Proposition 1.1. The proof of lemma is complete. ■

We now recall some basic definitions and properties related to trivial pairs and essential values of continuous functionals. We refer to [8] and [19] for further properties and complete proofs. In the next definitions X

denotes a metric space, A is a subset of X and i stands for the inclusion map of A into X.

Definition 8.1 A map $r : X \to A$ is said to be a retraction if it is continuous and $r|_A = id_A$.

Definition 8.2 A retraction r is called a strong deformation retraction provided that there exists a homotopy $\zeta : X \times [0,1] \to X$ of $i \circ r$ and id_X which satisfies $\zeta(x,t) = \zeta(x,0)$ for all $(x,t) \in A \times [0,1]$.

Definition 8.3 The metric space X is said to be weakly locally contractible if for every $u \in X$ there exists a neighborhood U of u contractible in X.

Given a continuous function $f : X \to I\!R$, for every $a \in I\!R$ we denote

$$f^a = \{u \in X : f(u) \leq a\}.$$

Definition 8.4 Let $a, b \in I\!R$ with $a \leq b$. The pair (f^b, f^a) is said to be trivial provided that, for all neighborhoods $[a', a'']$ of a and $[b', b'']$ of b, there exist closed sets A and B such that $f^{a'} \subseteq A \subseteq f^{a''}$, $f^{b'} \subseteq B \subseteq f^{b''}$ and there is a strong deformation retraction of B onto A.

Definition 8.5 A real number c is called an essential value of f if for every $\varepsilon > 0$ there exist $a, b \in (c - \varepsilon, c + \varepsilon)$ with $a < b$ such that the pair (f^b, f^a) is not trivial.

The following property of essential values is due to Degiovanni and Lancelotti ([8], Theorem 2.6).

Proposition 8.1 Let c be an essential value of f. Then for every $\varepsilon > 0$ there exists $\delta > 0$ such that each continuous function $g : X \to I\!R$ with

$$\sup\{|g(u) - f(u)| : u \in X\} < \delta$$

admits an essential value in $(c - \varepsilon, c + \varepsilon)$.

We turn now to the use of the notions above in the setting of problem $(\mathbf{P}_1)$. For every $n \geq 1$, we introduce the class of sets

$$\mathcal{A}_n = \{A \subset S_r : A \text{ is compact and } \mathrm{Cat}_{\pi(S_r)}\pi(A) \geq n\},$$

where $\pi : V \to V/G$ is the canonical projection. We recall that $\mathrm{Cat}_Y A$ is the smallest $k \in I\!N \cup \{+\infty\}$ such that A can be covered by k closed and contractible sets in the topological space Y. In Theorem 7.2 we have proved that the corresponding minimax values of F in (8.1) over $\mathcal{A}_n$

$$c_n = \inf_{A \in \mathcal{A}_n} \max_{u \in A} F(u), \quad n \geq 1,$$

are critical values of $F|_{S_r}$.

The result below is useful in proving the main result (Theorem 8.1 below).

Proposition 8.2 Under assumptions $(\mathbf{A}_1) - (\mathbf{A}_3)$ there exists an increasing sequence $\{b_n\}$ of essential values of $F|_{S_r}$ converging to $\sup_{j\geq 1} c_j$.

Proof. The proof is inspired from an argument in Degiovanni and Lancelotti [8] and some constructions in Rabinowitz [15]. We follow the steps:

Step 1. $c_1 = \inf_{u \in S_r} F(u)$.

Step 2. $c_n < \bar{c} := \sup_{j\geq 1} c_j$, $\forall n \geq 1$, and $c_n \to \bar{c}$ as $n \to +\infty$.

Step 3. There exists the sequence $\{b_n\}$ as required in the statement of Proposition 8.2.

Proof of Step 1. We have

$$c_1 = \inf_{A \in \mathcal{A}_1} \max_{u \in A} F(u),$$

where

$$\mathcal{A}_1 = \{A \subset S_r \ : \ A \text{ compact and } \mathrm{Cat}_{\pi(S_r)} \pi(A) \geq 1\},$$

with $\pi : S_r \to S_r/G$, $\pi(x) = Gx$. Consider any $x \in S_r$ and let $A_0 := \{x\}$ which is a compact set. Since S_r is weakly locally contractible, there exists a neighborhood U_x of x (in S_r) contractible in S_r. We set $T_x := \bigcup_{g \in G} gU_x = \pi(U_x)$. The contractibility of U_x in S_r implies that there exists a homotopy $\mathcal{H}_x : \bar{U}_x \times [0,1] \to S_r$ and a point $z_x \in S_r$ such that $\mathcal{H}_x(y,0) = y$ and $\mathcal{H}_x(y,1) = z_x$ for all $y \in \bar{U}_x$. We can also suppose that π is a homeomorphism on $\bar{U}_x$. Let us define $\mathcal{K}_x : \bar{T}_x \times [0,1] \to \pi(S_r)$ by $\mathcal{K}_x(y,t) = (\pi \circ \mathcal{H}_x)(\pi^{-1}(y), t)$. We have

$$\mathcal{K}_x(y,0) = (\pi \circ \pi^{-1})(y) = y \quad \text{and} \quad \mathcal{K}_x(y,1) = \pi(z_x) \quad \forall y \in \bar{T}_x = \pi(\bar{U}_x).$$

Therefore $\bar{T}_x$ is contractible in $\pi(S_r)$ and $\mathrm{Cat}_{\pi(S_r)}\pi(A_0) = 1$. This is a consequence of the fact that $\pi(\bar{U}_x)$ is a closed and contractible subset of $\pi(S_r)$ which contains $\pi(A_0)$. Hence

$$c_1 = \inf_{A \in \mathcal{A}_1} \max_{u \in A} F(u) \leq \max_{u \in A_0} F(u) = F(x)$$

and therefore $c_1 \leq \inf_{u \in S_r} F(u)$. The converse inequality is obvious.

Proof of Step 2. Since $\mathcal{A}_1 \supseteq \mathcal{A}_2 \supseteq \ldots \supseteq \mathcal{A}_n \supseteq \ldots$ it follows that $c_1 \le c_2 \le \ldots \le c_n \le \ldots$. Taking into account that $c_n \le \sup_{u \in S_r} F(u)$ for all n and that F is bounded on S_r (cf. $(\mathbf{A}_2)$), we deduce that the sequence $\{c_n\}$ converges to $\bar{c}$. This establishes the second part of Step 2.

For proving the first part of Step 2 we argue by contradiction. Let us admit that there exists j with $c_j = \bar{c}$. By the monotonicity of the sequence $\{c_n\}$, one has necessarily that $c_n = \bar{c}$ for all $n \ge j$. As shown in the proof of Theorem 7.4 we have that $c := c_j = c_{j+1} = \ldots = c_{j+p}$ yields $\mathrm{Cat}_{\pi(S_r)}\pi(K_c) \ge p+1$ for every p, where K_c stands for the set of critical points of $F|_{S_r}$ at level c. This ensures that $\mathrm{Cat}_{\pi(S_r)}\pi(K_{\bar{c}}) = +\infty$. Since F satisfies the (PS) condition (see the proof of Theorem 7.6), the set $K_{\bar{c}}$ is compact, which in turn implies that $\mathrm{Cat}_{\pi(S_r)}\pi(K_{\bar{c}}) < +\infty$. The obtained contradiction proves the claim, so the first part of Step 2.

Proof of Step 3. Proceeding inductively, first we construct the essential value b_1. Let us assume by contradiction that there are no essential values in the open interval $(c_1, \bar{c})$. By Theorem 2.5 in [8] the pair $(F^{\bar{c}}, F^{c_1})$ is trivial. Choose $\alpha', \alpha'' \in I\!R$ and the least positive integer m such that $\alpha' < c_1 < \alpha'' < c_m$. This is possible because Step 2 holds. Then we fix $\beta', \beta'' \in I\!R$ such that $c_m < \beta' < \bar{c} < \beta''$. Since the pair $(F^{\bar{c}}, F^{c_1})$ is trivial, we can find two closed subsets A, B of S_r and a strong deformation retraction $r : B \to A$ such that $A \subseteq F^{\alpha''}$, $F^{\beta'} \subseteq B$ and, with a homotopy $\eta : B \times [0,1] \to B$,

$$\begin{aligned} &\eta(x,0) = x, \quad \forall x \in B \\ &\eta(x,1) = r(x), \quad \forall x \in B \\ &\eta(x,0) = \eta(x,t), \quad \forall (x,t) \in A \times [0,1]. \end{aligned}$$

The inequality $c_m < \beta'$ ensures that there exists $C \in \mathcal{A}_m$ such that $C \subseteq F^{\beta'}$, while the inequality $c_m > \alpha''$ enables us to deduce that for every set $D \in \mathcal{A}_m$ there exists a point $u \in D$ satisfying $\alpha'' < F(u)$.

The inclusions $C \subseteq F^{\beta'} \subseteq B$ insure that $\eta(C,1) \subseteq \eta(B,1) = r(B) = A \subseteq F^{\alpha''}$. We show that $\eta(C,1) \in \mathcal{A}_m$. To this end we observe that for the set C one can find a subset $\tilde{C} \subseteq C$ such that $\pi(\tilde{C}) = \pi(C)$ and π is a homeomorphism on $\tilde{C}$. We note that a homotopy $\tilde{\eta} : \pi(\tilde{C}) \times [0,1] \to \pi(S_r)$ can be defined by the relation $\tilde{\eta}(\pi(x), t) = \pi(\eta(x,t))$, $\forall x \in \tilde{C}$, $\forall t \in [0,1]$. From Proposition 7.1 we derive that

$$\begin{aligned} \mathrm{Cat}_{\pi(S_r)}\pi(\eta(C,1)) &\ge \mathrm{Cat}_{\pi(S_r)}\pi(\eta(\tilde{C},1)) = \mathrm{Cat}_{\pi(S_r)}\tilde{\eta}(\pi(\tilde{C}),1) \\ &\ge \mathrm{Cat}_{\pi(S_r)}\pi(\tilde{C}) = \mathrm{Cat}_{\pi(S_r)}\pi(C) \ge m, \end{aligned}$$

which expresses that $\eta(C,1) \in \mathcal{A}_m$. This leads to a contradiction between $\eta(C,1) \subseteq F^{\alpha''}$ and the property of $D = \eta(C,1)$ to contain a point $u \in D$ with $\alpha'' < F(u)$. The achieved contradiction allows to conclude that there exists an essential value b_1 of $F|_{S_r}$ satisfying $c_1 < b_1 < \bar{c}$.

Suppose now inductively the existence of essential values $b_1, ..., b_{n-1} \in \mathbb{R}$ with $b_1 < ... < b_{n-1} < \bar{c}$. Step 2 guarantees that there exists some c_p, with p depending on n, which satisfies $b_{n-1} < c_p < \bar{c}$. Repeating the reasoning used for constructing b_1, with c_1 replaced by c_p, we find an essential value b_n belonging to the open interval $(c_p, \bar{c})$. This completes the inductive process. In view of Step 2 we obtain that the sequence $\{b_n\}$ converges to $\bar{c}$. The proof of Proposition 8.2 is thus complete. ■

Proof of Theorem 8.1. Fix any $n \geq 1$. We observe from Lemmas 8.1 and 8.3 that it is sufficient to establish the existence of some $\delta_n > 0$ such that the functional $W|_{S_r}$ has at least n distinct critical values, provided that $\|h(\cdot,0)\|_{L^1} \leq \delta_n$, $\|a_2\|_{L^{p/(p-1)}} \leq \delta_n$, $c \leq \delta_n$ and $\|\varphi\|_{V^*} \leq \delta_n$. By Proposition 8.2 we can find an increasing sequence $\{b_k\}$ of essential values of $F|_{S_r}$ which converges to $\bar{c} = \sup c_k$. Let $\varepsilon_0 > 0$ be chosen such that $\varepsilon_0 < 1/2 \min_{2 \leq i \leq n}(b_i - b_{i-1})$. Apply Proposition 8.1 to $F|_{S_r}$ and $W|_{S_r}$, for every $1 \leq j \leq n$. There exists $\eta_j > 0$ such that if

$$\sup_{u \in S_r} |F(u) - W(u)| < \eta_j$$

then there is an essential value e_j of $W|_{S_r}$ in $(b_j - \varepsilon_0, b_j + \varepsilon_0)$. Consequently, from Lemma 8.1 for $\eta = \min\{\eta_1, \ldots, \eta_n\}$, we get the existence of some $\delta_n > 0$ such that

$$\sup_{u \in S_r} |F(u) - W(u)| < \eta,$$

provided $\|h(\cdot,0)\|_{L^1} \leq \delta_n$, $\|a_2\|_{L^{p/(p-1)}} \leq \delta_n$, $c \leq \delta_n$ and $\|\varphi\|_{V^*} \leq \delta_n$. Therefore in this situation the functional $W|_{S_r}$ has at least n distinct essential values $e_1, e_2, \ldots, e_n$ in $(b_1 - \varepsilon_0, b_n + \varepsilon_0)$.

For completing the proof of Theorem 8.1 it suffices to show that $e_1, e_2, \ldots, e_n$ are critical values of $W|_{S_r}$. Assuming the contrary, there exists $j \in \{1, 2, \ldots, n\}$ such that e_j is not a critical value of $W|_{S_r}$. In what follows we are going to prove that this fact implies the properties

(A) There exists $\bar{\varepsilon} > 0$ so that $W|_{S_r}$ has no critical values in $(e_j - \bar{\varepsilon}, e_j + \bar{\varepsilon})$;

(B) For every $a, b \in (e_j - \bar{\varepsilon}, e_j + \bar{\varepsilon})$ with $a < b$, the pair $((W|_{S_r})^b, (W|_{S_r})^a)$ is trivial.

Suppose that (A) is not true. Then we get the existence of a sequence $\{d_k\}$ of critical values of $W|_{S_r}$ with $d_k \to e_j$ as $k \to \infty$. Since d_k is a critical value it follows that there exists $u_k \in S_r$ such that

$$W(u_k) = d_k \quad \text{and} \quad \lambda_{W|_{S_r}}(u_k) = 0.$$

Using the condition (PS) (see Lemma 8.2) we can suppose that, up to a subsequence, $\{u_k\}$ converges to some $u \in S_r$ as $k \to \infty$. Taking into account the continuity of W and the lower semicontinuity of $\lambda_{W|_{S_r}}$ we obtain

$$W(u) = e_j \quad \text{and} \quad \lambda_{W|_{S_r}}(u) = 0,$$

which contradicts the assumption that e_j is not a critical value.

To justify (B), we notice that on the basis of (A) we can apply the Noncritical Interval Theorem in the theory of Degiovanni and Marzocchi on every interval $[a, b]$ as described in (B). It implies that there exists a continuous map $\chi : S_r \times [0, 1] \to S_r$ such that

$$\begin{array}{ll} \chi(u,0) = u, \quad W(\chi(u,t)) \le W(u), & \forall (u,t) \in S_r \times [0,1], \\ W(u) \le b \Rightarrow W(\chi(u,1)) \le a, & W(u) \le a \Rightarrow \chi(u,t) = u. \end{array} \tag{8.36}$$

Define the map $\rho : (W|_{S_r})^b \to (W|_{S_r})^a$ by $\rho(u) = \chi(u, 1)$. From (8.36) we have that ρ is well defined and it is a retraction. Set

$$\mathcal{H} : (W|_{S_r})^b \times [0,1] \to (W|_{S_r})^b, \quad \mathcal{H}(u,t) = \chi(u,t).$$

Again from (8.36) we see that, for every $u \in (W|_{S_r})^b$,

$$\mathcal{H}(u,0) = u, \quad \mathcal{H}(u,1) = \rho(u), \tag{8.37}$$

and for each $(u,t) \in (W|_{S_r})^a \times [0,1]$,

$$\mathcal{H}(u,t) = u. \tag{8.38}$$

From (8.37) and (8.38) it follows that the pair $((W|_{S_r})^b, (W|_{S_r})^a)$ is trivial. Combining (A), (B) and Definition 8.5 it follows that e_j is not an essential value of $W|_{S_r}$. The achieved contradiction completes the proof. ■

2. Perturbations of Double Eigenvalue Problems for General Hemivariational Inequalities with Constraints

In this Section we study the effect of a non-symmetric small perturbation for a double eigenvalue hemivariational inequality considered in Theorem 7.7. Precisely, the perturbation is made in the nonsmooth and non-convex term of the involved problem with constraint on a sphere-like type manifold.

Let V be a real Hilbert space, with the scalar product and the associated norm denoted by $(\cdot,\cdot)_V$ and $\|\cdot\|_V$, respectively. We suppose that V is densely and compactly embedded in $L^p(\Omega;\mathbb{R}^N)$ for some $p\geq 2$, where $N\geq 1$ and $\Omega\subset\mathbb{R}^m$, $m\geq 1$, is a smooth, bounded domain. Denote by $\langle\cdot,\cdot\rangle_V$ and $\langle\cdot,\cdot\rangle$ the duality products on V and $\mathbb{R}^N$, respectively. Let $C_p(\Omega)$ be the constant of the continuous embedding $V\subset L^p(\Omega;\mathbb{R}^N)$, which means that

$$\|v\|_{L^p}\leq C_p(\Omega)\cdot\|v\|_V,\quad \text{for all } v\in V.$$

Let $a_1,a_2: V\times V\to\mathbb{R}$ be two continuous symmetric bilinear forms on V and let $B_1,B_2: V\to V$ be two bounded self-adjoint linear operators which are coercive in the sense that

$$(B_i v,v)_V\geq b_i\cdot\|v\|_V^2,\quad \text{for all } v\in V,\ i=1,2,$$

for some constants $b_1,b_2>0$. For fixed positive numbers a,b,r we consider the submanifold $S_r^{a,b}$ of $V\times V$ defined by

$$S_r^{a,b}:=\{(v_1,v_2)\in V\times V\ :\ a(B_1v_1,v_1)_V+b(B_2v_2,v_2)_V=r^2\}.$$

Consider the tangent space associated to the manifold introduced above, that is

$$T_{(u_1,u_2)}S_r^{a,b}:=\{(v_1,v_2)\in V\times V\ :\ a(B_1u_1,v_1)_V+b(B_2u_2,v_2)_V=0\},$$

and the projections $\pi_i: V\times V\to V$, $\pi_i(x_1,x_2)=x_i$, $i=1,2$. Let $j:\Omega\times\mathbb{R}^N\to\mathbb{R}$ satisfy the following assumptions

(*i*) $j(\cdot,y)$ is measurable in Ω for each $y\in\mathbb{R}^N$ and $j(\cdot,0)$ is essentially bounded in Ω;

(*ii*) $j(x,\cdot)$ is locally Lipschitz on $\mathbb{R}^N$ for a.e. $x\in\Omega$.

We also assume

(H_1) There exist $\theta\in L^{\frac{p}{(p-1)}}(\Omega)$ and $\rho\in\mathbb{R}$ such that

$$|z|\leq\theta(x)+\rho|y|^{p-1}, \tag{8.39}$$

for a.e. $(x,y)\in\Omega\times\mathbb{R}^N$ and each $z\in\partial_y j(x,y)$.

Let us consider a real function $C: S_r^{a,b}\times V\times V\to\mathbb{R}$ to which we impose no continuity assumption.

We are now prepared to consider the following double eigenvalue problem:

Find $u_1, u_2 \in V$ and $\lambda_1, \lambda_2 \in \mathbb{R}$ such that

$$(P^1_{r,a,b}) \begin{cases} a_1(u_1, v_1) + a_2(u_2, v_2) + C((u_1, u_2), v_1, v_2) \\ + \int_\Omega j^0_y(x, (u_1 - u_2)(x); (v_1 - v_2)(x))dx \\ \geq \lambda_1 (B_1 u_1, v_1)_V + \lambda_2 (B_2 u_2, v_2)_V, \ \forall v_1, v_2 \in V, \\ a(B_1 u_1, u_1)_V + b(B_2 u_2, u_2)_V = r^2. \end{cases}$$

We impose the following hypothesis

(H_2) There exist two locally Lipschitz maps $f_i : V \to \mathbb{R}$, bounded on $\pi_i(S^{a,b}_r)$, $(i = 1, 2)$ respectively, and such that the following inequality holds

$$\begin{aligned} &C((u_1, u_2), v_1, v_2) \geq f^0_1(u_1; v_1) + f^0_2(u_2; v_2), \\ &\forall (u_1, u_2) \in S^{a,b}_r, \ \forall (v_1, v_2) \in T_{(u_1,u_2)} S^{a,b}_r . \end{aligned} \tag{8.40}$$

In addition we suppose that the sets

$$\{z \in V^* \ : \ z \in \partial f_i(u_i), u_i \in \pi_i(S^{a,b}_r)\}$$

are relatively compact in V^*, for $i = 1, 2$.

Set $(A_1, A_2) : V \times V \to V^* \times V^*$ defined by

$$\langle (A_1, A_2)(u_1, u_2), (v_1, v_2) \rangle_{V \times V} := a_1(u_1, v_1) + a_2(u_2, v_2) . \tag{8.41}$$

The duality map $\Lambda : V \times V \to V^* \times V^*$ is given by

$$\langle \Lambda(u_1, u_2), (v_1, v_2) \rangle_{V \times V} := a(B_1 u_1, v_1)_V + b(B_2 u_2, v_2)_V . \tag{8.42}$$

We also assume the compactness hypothesis

(H_3) For every sequence $\{(u^1_n, u^2_n)\} \subset S^{a,b}_r$ with $u^i_n \rightharpoonup u_i$ weakly in V, for any $z^i_n \in \partial f_i(u^i_n)$, with

$$a_i(u^i_n, u^i_n) + \langle z^i_n, u^i_n \rangle_V \to \alpha_i \in \mathbb{R}, \tag{8.43}$$

$i = 1, 2$, and for all $w \in L^{\frac{p}{p-1}}(\Omega; \mathbb{R}^N)$ which satisfies the relation

$$w(x) \in \partial_y j(x, (u_1 - u_2)(x)) \text{ for a.e. } x \in \Omega, \tag{8.44}$$

such that

$$[(A_1, A_2) - \lambda_0 \cdot \Lambda] (u^1_n, u^2_n)$$

converges in $V^* \times V^*$, where

$$\lambda_0 = r^{-2}(\alpha_1 + \alpha_2 + \int_\Omega \langle w(x), (u_1 - u_2)(x)\rangle \, dx), \tag{8.45}$$

there exists a convergent subsequence of $\{(u_n^1, u_n^2)\}$ in $V \times V$ (thus, in $S_r^{a,b}$).

(H_4) j is even with respect to the second variable, i.e.,

$$j(x, -y) = j(x, y), \text{ for a.e. } x \in \Omega, \text{ and any } y \in \mathbb{R}^N,$$

and f_i is even on $\pi_i(S_r^{a,b})$, i.e.,

$$f_i(-u_i) = f_i(u_i), \text{ for all } (u_1, u_2) \in S_r^{a,b}, i = 1, 2.$$

Under assumptions (H_1), (H_2), (H_3) and (H_4), Theorem 7.7 asserts that the double eigenvalue problem $(P^1_{r,a,b})$ admits infinitely many pairs of solutions $\{\pm(u_n^1, u_n^2), (\lambda_n^1, \lambda_n^2)\}$ in $S_r^{a,b} \times \mathbb{R}^2$.

Let us now consider an arbitrary element φ in V^* and $g : \Omega \times \mathbb{R}^N \to \mathbb{R}$ a Carathéodory function which is locally Lipschitz with respect to the second variable and such that $g(\cdot, 0) \in L^1(\Omega)$. We are concerned with the following non-symmetric perturbed double eigenvalue problem: find $(u_1, u_2) \in V \times V$ and $(\lambda_1, \lambda_2) \in \mathbb{R}^2$ such that

$$(P^2_{r,a,b}) \begin{cases} a_1(u_1, v_1) + a_2(u_2, v_2) + C((u_1, u_2), v_1, v_2) \\ + \int_\Omega \{j_y^0(x, (u_1 - u_2)(x); (v_1 - v_2)(x)) \\ + g_y^0(x, (u_1 - u_2)(x); (v_1 - v_2)(x))\} dx \\ + \langle \varphi, v_1 \rangle_V + \langle \varphi, v_2 \rangle_V \\ \geq \lambda_1 (B_1 u_1, v_1)_V + \lambda_2 (B_2 u_2, v_2)_V, \quad \forall v_1, v_2 \in V, \\ a(B_1 u_1, u_1)_V + b(B_2 u_2, u_2)_V = r^2. \end{cases}$$

Fix $\delta > 0$. We impose to g the growth condition

(H_5) There exists $\theta_1 \in L^{\frac{p}{(p-1)}}(\Omega)$ such that

$$|z| \leq \theta_1(x) + \delta |y|^{p-1}, \tag{8.46}$$

for a.e. $(x, y) \in \Omega \times \mathbb{R}^N$ and each $z \in \partial_y g(x, y)$.

Let us denote by J and G the (locally Lipschitz, by hypotheses (H_1) and (H_5)) functionals from $L^p(\Omega; \mathbb{R}^N)$ into $\mathbb{R}$, defined by

$$J(u) = \int_\Omega j(x, u(x)) dx \quad \text{and} \quad G(u) = \int_\Omega g(x, u(x)) dx.$$

We associate to the problems $(P^1_{r,a,b})$ and $(P^2_{r,a,b})$ the energy functions $I_1, I_2 : V \times V \to \mathbb{R}$, defined by

$$I_1(u_1,u_2) = \frac{1}{2} \cdot [a_1(u_1,u_1) + a_2(u_2,u_2)]$$
$$+f_1(u_1) + f_2(u_2) + J(u_1 - u_2) \tag{8.47}$$

and

$$I_2(u_1,u_2) = I_1(u_1,u_2) + G(u_1 - u_2) + \langle \varphi, u_1 \rangle_V + \langle \varphi, u_2 \rangle_V, \tag{8.48}$$

for all $u_1, u_2 \in V$.

We denote by Υ the family of all subsets of $S^{a,b}_r$ that are closed and symmetric with respect to the origin $0_{V\times V}$. Let us denote, as usually, by $\gamma(S)$ the Krasnoselski genus of the set $S \in \Upsilon$, that is, the smallest integer $k \in \mathbb{N} \cup \{+\infty\}$ for which there exists an odd continuous mapping from S into $\mathbb{R}^k \setminus \{0\}$. For every $n \geq 1$, set

$$\Gamma_n = \{S \subset S^{a,b}_r \ : \ S \in \Upsilon, \gamma(S) \geq n\}.$$

Recall (see Theorem 7.7) that the corresponding minimax values of I_1 over Γ_n

$$\beta_n = \inf_{S\in\Gamma_n} \sup_{(u_1,u_2)\in S} \{I_1(u_1,u_2)\},$$

are critical values of I_1 on $S^{a,b}_r$.

Lemma 8.4 Let $s := \sup\limits_{(u_1,u_2)\in S^{a,b}_r} \{I_1(u_1,u_2)\}$. Then the supremum is not achieved and $\lim\limits_{n\to\infty} \beta_n = s$. Moreover, there exists a sequence $\{b_n\}$ of essential values of the restriction of I_1 to $S^{a,b}_r$, strictly increasing to s.

Proof. We follow some ideas developed in Degiovanni and Lancelotti [8] (see also [2], [11] and [14]). We point out only the main steps of the proof:

Step 1. The functional $I_1|_{S^{a,b}_r}$ satisfies the Palais-Smale condition (see the proof of Theorem 7.7). So, if there exist $u_0 = (u_{01}, u_{02}) \in S^{a,b}_r$ and $m < n$ such that $\beta_m = \beta_n \leq I_1(u_0)$, then $\gamma(K_{\beta_m}) \geq n - m + 1$, where

$$K_{\beta_m} := \{u = (u_1,u_2) \in S^{a,b}_r \ : \ I_1(u) = \beta_m \quad \text{and} \quad \lambda_{I_1}(u) = 0\}.$$

Step 2. If the sequence $\{\beta_n\}$ is stationary and if there exists $u_0 \in S^{a,b}_r$ such that Step 1 holds, then $\gamma(K_{\beta_m}) = +\infty$, for some $m \geq 1$. This is not possible, since $S_r^{a,b}$ is a weakly locally contractible space and K_{β_m} is a compact set, which implies $\gamma(K_{\beta_m}) < +\infty$.

Step 3. It follows by the previous steps, the definition of Krasnoselski's genus and the fact that $I_1 \not\equiv$const. on $S_r^{a,b}$, that $\sup_{u\in S_r^{a,b}} I_1(u)$ is not achieved and

$$\lim_{n\to\infty} \beta_n = \sup_{u\in S_r^{a,b}} I_1(u)\,.$$

Without loss of generality, we may assume that $\sup_{u\in S_r^{a,b}} I_1(u) = +\infty$. Let us define

$$\overline{\Gamma}_n = \{\varphi(S^{n-1}) \ : \ \varphi : S^{n-1} \to S_r^{a,b} \text{ is continuous and odd}\},$$

and

$$\overline{\beta}_n = \inf_{C\in\overline{\Gamma}_n} \sup_{u\in C} I_1(u)\,.$$

Of course, $\overline{\beta}_n \geq \beta_n$, so that $\lim_{n\to\infty} \overline{\beta}_n = \sup_{u\in S_r^{a,b}} I_1(u) = +\infty$. By Theorem 2.12 of [8] it follows that there exists a sequence $\{b_n\}$ of essential values of $I_1|_{S_r^{a,b}}$ strictly increasing to $\sup_{u\in S_r^{a,b}} I_1(u)$. ■

For continuing, we need two additional hypotheses

(H_6) Assume that

$$\|\theta_1\|_{L^{\frac{p}{p-1}}} \leq \delta, \ \|g(\cdot,0)\|_{L^1} \leq \delta \text{ and } \|\varphi\|_{V^*} \leq \delta. \tag{8.49}$$

The next assumption is actually a variant of the compactness hypothesis (H_3).

(H_7) For every sequence $\{(u_n^1, u_n^2)\} \subset S_r^{a,b}$ with $u_n^i \rightharpoonup u_i$ weakly in V, for any $z_n^i \in \partial f_i(u_n^i)$, with

$$a_i(u_n^i, u_n^i) + \langle z_n^i, u_n^i\rangle_V + \langle \varphi, u_n^i\rangle_V \to \alpha_i \in I\!R, \tag{8.50}$$

$i = 1, 2$ and for all $w, z \in L^{\frac{p}{p-1}}(\Omega; I\!R^N)$ satisfying

$$\begin{aligned} &w(x) \in \partial_y j(x, (u_1 - u_2)(x)), \\ &z(x) \in \partial_y g(x, (u_1 - u_2)(x)), \text{ for a.e. } x \in \Omega, \end{aligned} \tag{8.51}$$

such that

$$[(A_1, A_2) - \lambda_0 \cdot \Lambda]\,(u_n^1, u_n^2)$$

converges in $V^* \times V^*$, where

$$\lambda_0 = r^{-2}(\alpha_1 + \alpha_2 + \int_\Omega \langle w(x) + z(x), (u_1 - u_2)(x)\rangle\, dx), \tag{8.52}$$

there exists a convergent subsequence of $\{(u_n^1, u_n^2)\}$ in $V \times V$.

The next result proves that if $\delta > 0$ is sufficiently small in hypotheses (H_5) and (H_6), then I_2 is a small perturbation of I_1 on $S_r^{a,b}$.

Lemma 8.5 For every $\varepsilon > 0$, there exists $\delta_0 > 0$ such that for all $\delta \leq \delta_0$ we have

$$\sup_{(u_1,u_2)\in S_r^{a,b}} |I_1(u_1, u_2) - I_2(u_1, u_2)| < \varepsilon.$$

Proof. By Lebourg's mean value theorem for locally Lipschitz functionals and hypothesis (H_5) we find

$$|G(u)| \leq \|g(\cdot, 0)\|_{L^1} + \|\theta_1\|_{L^{\frac{p}{p-1}}} \cdot \|u\|_{L^p} + \delta \|u\|_{L^p}^p.$$

Taking into account hypothesis (H_6) and the fact that $(u_1, u_2) \in S_r^{a,b}$, we derive that

$$|I_1(u_1, u_2) - I_2(u_1, u_2)| = |G(u_1 - u_2) + \langle \varphi, u_1 \rangle_V + \langle \varphi, u_2 \rangle_V|$$

$$\leq \|g(\cdot, 0)\|_{L^1} + \|\theta_1\|_{L^{\frac{p}{p-1}}} \cdot C_p(\Omega) \cdot r \cdot \left(\frac{1}{\sqrt{ab_1}} + \frac{1}{\sqrt{bb_2}} \right)$$

$$+\delta \cdot C_p^p(\Omega) \cdot r^p \left(\frac{1}{\sqrt{ab_1}} + \frac{1}{\sqrt{bb_2}} \right)^p + \delta \cdot r \cdot \left(\frac{1}{\sqrt{ab_1}} + \frac{1}{\sqrt{bb_2}} \right) < \varepsilon,$$

for $\delta > 0$ small enough. ■

Lemma 8.6 The functional I_2 satisfies the Palais-Smale condition on $S_r^{a,b}$.

Proof. For the beginning it is important to remark that the expression of the generalized gradient $\partial(I|_{S_r^{a,b}})$ at the point $(u_1, u_2) \in S_r^{a,b}$ is given by

$$\partial(I|_{S_r^{a,b}})(u_1, u_2) = \{\xi - r^{-2} \langle \xi, (u_1, u_2) \rangle_{V \times V} \Lambda(u_1, u_2) : \xi \in \partial I(u_1, u_2)\},$$

where $\Lambda : V \times V \to V^* \times V^*$ is the appropriate duality map given in (8.42). Here, the duality $\langle \cdot, \cdot \rangle_{V \times V}$ is taken for the norm

$$\|(u_1, u_2)\|_{V \times V} := \sqrt{a(B_1 u_1, u_1)_V + b(B_2 u_2, u_2)_V}, \quad \forall u_1, u_2 \in V.$$

Let us consider a sequence $\{(u_n^1, u_n^2)\} \subset S_r^{a,b}$ such that

$$\sup_n \left| (I_2|_{S_r^{a,b}})(u_n^1, u_n^2) \right| < +\infty$$

and such that there exists some sequence $\{J_n\} \subset V^* \times V^*$ fulfilling the conditions

$$J_n \in \partial I_2(u_n^1, u_n^2), \quad n \geq 1$$

and

$$J_n - r^{-2}\langle J_n, (u_n^1, u_n^2)\rangle_{V\times V} \cdot \Lambda(u_n^1, u_n^2) \to 0, \tag{8.53}$$

strongly in $V^* \times V^*$. For concluding it suffices to prove that $\{(u_n^1, u_n^2)\}$ contains a convergent subsequence in $V \times V$. Under hypotheses (H_1) and (H_5) the functionals J and G are Lipschitz continuous on bounded sets in $L^p(\Omega; I\!R^N)$ and their generalized gradients satisfy (cf. Theorem 1.3)

$$\partial J(v) \subset \int_\Omega \partial_y j(x, v(x))dx$$

and

$$\partial G(v) \subset \int_\Omega \partial_y g(x, v(x))dx, \quad \forall\, v \in L^p(\Omega; I\!R^N).$$

The density of V into $L^p(\Omega; I\!R^N)$ allows us to apply Corollary 1.2. Thus

$$\partial(J|_V)(v) \subset \partial J(v)$$

and

$$\partial(G|_V)(v) \subset \partial G(v), \quad \forall v \in V.$$

From $J_n \in \partial I_2(u_n^1, u_n^2)$ we derive that there exists $z_n^i \in \partial f_i(u_n^i)$ $(i = 1, 2)$, $w_n \in \partial(J|_V)(u_n^1 - u_n^2)$ and $z_n \in \partial(G|_V)(u_n^1 - u_n^2)$ such that

$$J_n = (a_1(u_n^1, \cdot) + z_n^1 + \varphi, a_2(u_n^2, \cdot) + z_n^2 + \varphi) + K^*(w_n) + K^*(z_n),$$

where $K : V \times V \to V$ is the map given by

$$K(v_1, v_2) = v_1 - v_2.$$

By the above considerations we have

$$w_n(x) \in \partial_y j(x, (u_n^1 - u_n^2)(x))$$

and

$$z_n(x) \in \partial_y g(x, (u_n^1 - u_n^2)(x)), \text{ for a.e. } x \in \Omega.$$

By (8.53) we get

$$\Big(a_1(u_n^1, \cdot) + z_n^1 + \varphi, a_2(u_n^2, \cdot) + z_n^2 + \varphi\Big) + K^*(w_n) + K^*(z_n)$$

$$-r^{-2}\langle[(a_1(u_n^1, \cdot) + z_n^1 + \varphi, a_2(u_n^2, \cdot) + z_n^2 + \varphi)$$

$$+K^*(w_n) + K^*(z_n)], (u_n^1, u_n^2)\rangle_{V\times V} \cdot \Lambda(u_n^1, u_n^2) \to 0$$

strongly in $V^* \times V^*$. Since the sequence $\{(u_n^1, u_n^2)\}$ is contained in $S_r^{a,b}$, it follows by the coercivity property of B_1 and B_2 that $\{u_n^1\}$ and $\{u_n^2\}$ are bounded in V. So, up to a subsequence,

$$u_n^i \rightharpoonup u_i \text{ weakly in } V, \text{ for some } u_i \in V \ (i = 1, 2).$$

The compactness assumptions in hypothesis (H_2) implies that, again up to a subsequence,

$$z_n^i \to z_i \ \text{ strongly in } V^*, \ \text{for some } z_i \in V^* \ (i = 1, 2).$$

Also we have

$$\begin{aligned} w_n &\in \partial(J|_V)(u_n^1 - u_n^2) \subset \partial J(u_n^1 - u_n^2), \\ z_n &\in \partial(G|_V)(u_n^1 - u_n^2) \subset \partial G(u_n^1 - u_n^2). \end{aligned} \tag{8.54}$$

The compactness of the embedding $V \subset L^p(\Omega; I\!R^N)$ provides the convergence

$$u_n^i \to u_i, \ \text{ strongly in } L^p(\Omega; I\!R^N) \ (i = 1, 2). \tag{8.55}$$

Since J and G are locally Lipschitz on $L^p(\Omega; I\!R^N)$, the above property ensures that $\{w_n\}$ and $\{z_n\}$ are bounded in $L^{\frac{p}{p-1}}(\Omega; I\!R^N)$. By the reflexivity of the space $L^{\frac{p}{p-1}}(\Omega; I\!R^N)$ and the compactness of the embedding $L^{\frac{p}{p-1}}(\Omega; I\!R^N) \subset V^*$, there exist $w, z \in L^{\frac{p}{p-1}}(\Omega; I\!R^N)$ such that, up to a subsequence,

$$w_n \to w \ \text{ strongly in } V^* \ \text{ and weakly in } L^{\frac{p}{p-1}}(\Omega; I\!R^N)$$

and

$$z_n \to z \ \text{ strongly in } V^* \ \text{ and weakly in } L^{\frac{p}{p-1}}(\Omega; I\!R^N).$$

By (8.54), (8.55) and Proposition 2.1.5 in [7] we obtain

$$w \in \partial J(u_1 - u_2) \ \text{ and } \ z \in \partial G(u_1 - u_2). \tag{8.56}$$

With the above remarks we may suppose that

$$a_i(u_n^i, u_n^i) \ \text{ converges in } I\!R, \ i = 1, 2,$$

and

$$\langle \Big[(z_n^1 + \varphi, z_n^2 + \varphi) + K^*(w_n) + K^*(z_n)\Big], (u_n^1, u_n^2)\rangle_{V\times V}$$

possesses a convergent subsequence in $I\!R$. From (8.53) and taking into account the convergences stated above we derive that

$$\Big(a_1(u_n^1, \cdot), a_2(u_n^2, \cdot)\Big) - \lambda_0 \cdot \Lambda(u_n^1, u_n^2),$$

converges strongly in $V^* \times V^*$, where λ_0 is the one required in (H_7). So, hypothesis (H_7) allows us to conclude that $\{(u_n^1, u_n^2)\}$ has a convergent subsequence in $V \times V$, so in $S_r^{a,b}$. Thus the Palais-Smale condition for the functional I_2 on $S_r^{a,b}$ is satisfied and the proof is now complete. ■

Lemma 8.7 If $u = (u_1, u_2)$ is a critical point of $I_2|_{S_r^{a,b}}$ then there exists a pair $(\lambda_1, \lambda_2) \in \mathbb{R}^2$ such that $((u_1, u_2), (\lambda_1, \lambda_2))$ is a solution of the problem $(P^2_{r,a,b})$.

Proof. Since u is a critical point for $I_2|_{S_r^{a,b}}$, it follows that

$$0_{V\times V} \in \left(\partial I_2|_{S_r^{a,b}}\right)(u_1, u_2). \tag{8.57}$$

Taking into account the expression of the generalized gradient of the restriction of I_2 to $S_r^{a,b}$, we may conclude the existence of $\xi \in \partial I_2(u_1, u_2)$ such that

$$\xi - r^{-2} \langle \xi, (u_1, u_2) \rangle_{V\times V} \cdot \Lambda(u_1, u_2) = 0. \tag{8.58}$$

By Clarke's calculus and the inclusions stated in the proof of Lemma 8.6 we derive

$$\partial I_2(u_1, u_2)(v_1, v_2) \subset a_1(u_1, v_1) + a_2(u_2, v_2)$$

$$+\partial f_1(u_1)v_1 + \partial f_2(u_2)v_2 + \int_\Omega \partial_y j(x, (u_1 - u_2)(x))(v_1 - v_2)(x)dx$$

$$+ \int_\Omega \partial_y g(x, (u_1 - u_2)(x))(v_1 - v_2)(x)dx + \langle \varphi, v_1 \rangle_V + \langle \varphi, v_2 \rangle_V,$$

for all $v_1, v_2 \in V$. So, there exist elements $z_i \in \partial f_i(u_i)$ $(i = 1, 2)$ and $w, z \in L^{\frac{p}{p-1}}(\Omega; \mathbb{R}^N)$ with

$$w(x) \in \partial_y j(x, (u_1 - u_2)(x)) \quad \text{for a.e. } x \in \Omega,$$

and

$$z(x) \in \partial_y g(x, (u_1 - u_2)(x)) \quad \text{for a.e. } x \in \Omega,$$

such that

$$\langle \xi, (v_1, v_2) \rangle_{V\times V} = a_1(u_1, v_1) + a_2(u_2, v_2) + \langle z_1, v_1 \rangle_V + \langle z_2, v_2 \rangle_V$$

$$+ \int_\Omega \langle w(x), (v_1 - v_2)(x) \rangle dx + \int_\Omega \langle z(x), (v_1 - v_2)(x) \rangle dx$$

$$+ \langle \varphi, v_1 \rangle_V + \langle \varphi, v_2 \rangle_V.$$

From (8.58) it follows that

$$a_1(u_1, v_1) + a_2(u_2, v_2)$$

$$+\langle z_1, v_1\rangle_V + \langle z_2, v_2\rangle_V + \int_\Omega \langle w(x), (v_1 - v_2)(x)\rangle dx$$

$$+ \int_\Omega \langle z(x), (v_1 - v_2)(x)\rangle dx + \langle \varphi, v_1\rangle_V + \langle \varphi, v_2\rangle_V$$

$$-r^{-2}[a_1(u_1, u_1) + a_2(u_2, u_2) + \langle z_1, u_1\rangle_V + \langle z_2, u_2\rangle_V$$

$$+ \int_\Omega \langle w(x), (u_1 - u_2)(x)\rangle dx + \int_\Omega \langle z(x), (u_1 - u_2)(x)\rangle dx$$

$$+\langle \varphi, u_1\rangle_V + \langle \varphi, u_2\rangle_V] \cdot (a(B_1 u_1, v_1)_V + b(B_2 u_2, v_2)_V) = 0,$$

for all $v_1, v_2 \in V$. Set

$$\lambda = r^{-2}[a_1(u_1, u_1) + a_2(u_2, u_2) + \langle z_1, u_1\rangle_V + \langle z_2, u_2\rangle_V$$

$$+ \int_\Omega \langle (w + z)(x), (u_1 - u_2)(x)\rangle dx + \langle \varphi, u_1\rangle_V + \langle \varphi, u_2\rangle_V].$$

Let us now observe that we have

$$\int_\Omega \langle (w + z)(x), (v_1 - v_2)(x)\rangle dx$$

$$\leq \int_\Omega \max\{\langle \mu_1, (v_1 - v_2)(x)\rangle; \mu_1 \in \partial_y j(x, (u_1 - u_2)(x))\}$$

$$+ \int_\Omega \max\{\langle \mu_2, (v_1 - v_2)(x)\rangle; \mu_2 \in \partial_y g(x, (u_1 - u_2)(x))\}$$

$$= \int_\Omega j_y^0(x, (u_1 - u_2)(x); (v_1 - v_2)(x))dx$$

$$+ \int_\Omega g_y^0(x, (u_1 - u_2)(x); (v_1 - v_2)(x))dx\,.$$

In the above relation, the last equality holds because of Proposition 1.1. Taking into account the choice of $z_i (i = 1, 2), z$ and w, it is easy to observe that if we denote $\lambda_1 = \lambda a$ and $\lambda_2 = \lambda b$, our hypothesis (H_2) and some simple calculation lead us to the conclusion claimed in the formulation of Lemma 8.7. ■

With the preliminary results stated above we are now ready to formulate our perturbation result.

Theorem 8.2 (Bocea, Panagiotopoulos and Rădulescu [3]) Assume that the hypotheses $(H_1) - (H_7)$ are fulfilled. Then, for every $n \geq 1$, there exists $\delta_n > 0$ such that, for each $\delta \leq \delta_n$, problem $(P^2_{r,a,b})$ admits at least n distinct solutions.

Proof. Fix $n \geq 1$. By Lemma 8.7 it suffices to prove the existence of some $\delta_n > 0$ such that, for every $\delta \leq \delta_n$, the functional $I_2|_{S_r^{a,b}}$ has at least n distinct critical values. We may use now the conclusion of Lemma 8.4 and this implies that it is possible to consider a sequence $\{b_n\}$ of essential values of $I_1|_{S_r^{a,b}}$, strictly increasing to s. Choose arbitrarily $\varepsilon_0 < \frac{1}{2} \min_{1 \leq i \leq n-1} (b_{i+1} - b_i)$. We now apply Theorem 2.6 from [8] to the functionals $I_1|_{S_r^{a,b}}$ and $I_2|_{S_r^{a,b}}$. Thus, for every $1 \leq i \leq n-1$, there exists $\eta_i > 0$ such that the relation

$$\sup_{(u_1,u_2) \in S_r^{a,b}} |I_1(u_1,u_2) - I_2(u_1,u_2)| < \eta_i$$

implies the existence of an essential value c_i of $I_2|_{S_r^{a,b}}$ in the interval $(b_i - \varepsilon_0, b_i + \varepsilon_0)$. By taking $\varepsilon = \min\{\varepsilon_0, \eta_1, \cdots, \eta_{n-1}\}$ in Lemma 8.5, we derive the existence of a $\delta_n > 0$ such that

$$\sup_{(u_1,u_2) \in S_r^{a,b}} |I_1(u_1,u_2) - I_2(u_1,u_2)| < \varepsilon,$$

provided $\delta \leq \delta_n$ in (H_5) and (H_6). So, the functional $I_2|_{S_r^{a,b}}$ has at least n distinct essential values $c_1, c_2, \cdots, c_n$ in the interval $(-\infty, b_n + \varepsilon)$. For concluding our proof it suffices to show that $c_1 \cdots, c_n$ are critical values of $I_2|_{S_r^{a,b}}$. The first step is to prove that there exists $\varepsilon > 0$ such that $I_2|_{S_r^{a,b}}$ has no critical value in $(c_i - \varepsilon, c_i + \varepsilon)$. Indeed, if this is not the case, there exists a sequence $\{d_n\}$ of critical values of $I_2|_{S_r^{a,b}}$ with $d_n \to c_i$ as $n \to \infty$. The fact that d_n are critical values for the restriction of I_2 to $S_r^{a,b}$ implies that for every $n \geq 1$, there exists $(u_n^1, u_n^2) \in S_r^{a,b}$ such that

$$I_2(u_n^1, u_n^2) = d_n \text{ and } \lambda^*(u_n^1, u_n^2) = 0,$$

where λ^* is the lower semicontinuous functional defined by

$$\lambda^*(u_1, u_2) := \min\{\|(\xi_1, \xi_2)\|_{V^* \times V^*} \ : \ (\xi_1, \xi_2) \in \partial I_2(u_1, u_2)\}.$$

Thus, passing eventually to a relabelled subsequence, we may admit that $(u_n^1, u_n^2) \to (u_1, u_2) \in S_r^{a,b}$, strongly in $V \times V$. The continuity of I_2 and the lower semicontinuity of λ^* implies that

$$I_2(u_1, u_2) = c_i \quad \text{and} \quad \lambda^*(u_1, u_2) = 0,$$

which contradicts the initial conditions on c_i. Let us fix $c_i - \varepsilon < a < b < c_i + \varepsilon$. By Lemma 8.6, I_2 satisfies the Palais-Smale condition on $S_r^{a,b}$. So, for every point $e \in [a, b]$, the Palais-Smale condition $(PS)_e$ holds. We have fulfilled the set of conditions which allow us to apply the Noncritical Interval Theorem in the theory of Degiovanni and Marzocchi, on the complete metric space $\left(S_r^{a,b}, d(\cdot,\cdot)\right)$. We have denoted by $d(\cdot,\cdot)$ the geodesic distance on $S_r^{a,b}$, that is, for every points $x, y \in S_r^{a,b}, d(x, y)$ is equal to the infimum of the smooth lengths of all paths on $S_r^{a,b}$ joining x and y. We obtain that there exists a continuous map $\eta : S_r^{a,b} \times [0,1] \to S_r^{a,b}$ such that, for each $(u = (u_1, u_2), t) \in S_r^{a,b} \times [0, 1]$, are satisfied the conditions

(a) $\eta(u, 0) = u$,

(b) $I_2(\eta(u, t)) \leq I_2(u)$,

(c) $I_2(u) \leq b \Longrightarrow I_2(\eta(u, 1)) \leq a$,

(d) $I_2(u) \leq a \Longrightarrow \eta(u, t) = u$.

By the above conditions, it follows that the map

$$[I_2|_{S_r^{a,b}} \leq b] \ni u \mapsto \eta(u, 1) \in [I_2|_{S_r^{a,b}} \leq b]$$

is a retraction. Define the map

$$\Psi : [I_2|_{S_r^{a,b}} \leq b] \times [0,1] \to [I_2|_{S_r^{a,b}} \leq b], \quad \text{by} \quad \Psi(u, t) = \eta(u, t).$$

Since for every $u \in [I_2|_{S_r^{a,b}} \leq b]$, we have

$$\Psi(u, 0) = u, \;\; \Psi(u, 1) = \eta(u, 1),$$

and for each $(u, t) \in [I_2|_{S_r^{a,b}} \leq a] \times [0, 1]$, the equality $\Psi(u, t) = \Psi(u, 0)$ holds, it follows that Ψ is $[I_2|_{S_r^{a,b}} \leq a]$-homotopic to the identity of $[I_2|_{S_r^{a,b}} \leq b]$. Thus, Ψ is a strong deformation retraction which implies that the pair

$$\left([I_2|_{S_r^{a,b}} \leq b], [I_2|_{S_r^{a,b}} \leq a]\right)$$

is trivial. With this argument, we get that c_i is not an essential value of the restriction of I_2 at $S_r^{a,b}$. This contradiction concludes our proof. ■

References

[1] A. Bahri and H. Berestycki, A perturbation method in critical point theory and applications, *Trans. Amer. Math. Soc.* **267** (1981), 1-32.

[2] M. Bocea, D. Motreanu and P. D. Panagiotopoulos, Multiple solutions for a double eigenvalue hemivariational inequality on a sphere-like type manifold, *Nonlinear Analysis, T.M.A.* **42A** (2000), 737-749.

[3] M. Bocea, P. D. Panagiotopoulos and V. Rădulescu, A perturbation result for a double eigenvalue hemivariational inequality with constraints and applications, *J. Global Optimiz.* **14** (1999), 137-156.

[4] K. C. Chang, Variational methods for non-differentiable functionals and their applications to partial differential equations, *J. Math. Anal. Appl.* **80** (1981), 102-129.

[5] C. Ciulcu, D. Motreanu and V. Rădulescu, Multiplicity of solutions for a class of non-symmetric eigenvalue hemivariational inequalities, *Math. Methods Appl. Sciences*, in press.

[6] F. Cîrstea and V. Rădulescu, Multiplicity of solutions for a class of non-symmetric eigenvalue hemivariational inequalities, *J. Global Optimiz.* **17** (1/4) (2000), 43-54.

[7] F. H. Clarke, *Optimization and Nonsmooth Analysis,* Willey, New York, 1983.

[8] M. Degiovanni and S. Lancelotti, Perturbations of even non-smooth functionals, *Differential Integral Equations* **8** (1995), 981-992.

[9] M. Degiovanni and V. Rădulescu, Perturbations of non-smooth symmetric non-linear eigenvalue problems, *C.R. Acad. Sci. Paris* **329** (1999), 281-286.

[10] D. Goeleven, D. Motreanu and P. D. Panagiotopoulos, Multiple solutions for a class of hemivariational inequalities involving periodic energy functionals, *Math. Methods Appl. Sciences*, **20** (1997), 548-568.

[11] D. Motreanu and P. D. Panagiotopoulos, Double eigenvalue problems for hemi-variational inequalities, *Arch. Rat. Mech. Anal.* **140** (1997), 225-251.

[12] P. D. Panagiotopoulos, *Inequality Problems in Mechanics and Applications. Convex and Nonconvex Energy Functionals*, Birkhäuser-Verlag, Boston, Basel, 1985.

[13] P. D. Panagiotopoulos, *Hemivariational Inequalities: Applications to Mechanics and Engineering,* Springer-Verlag, New York, Boston, Berlin, 1993.

[14] P. D. Panagiotopoulos and V. Rădulescu, Perturbations of hemivariational in-equalities with constraints and applications, *J. Global Optimiz.*, **12**, 285-297 (1998).

[15] P. Rabinowitz, *Minimax Methods in Critical Point Theory with Applications to Differential Equations*, CBMS Reg. Conf. Ser. Math. 65, Amer. Math. Soc., Providence, R.I., 1986.

[16] V. Rădulescu, Perturbations of hemivariational inequalities with constraints, *Revue Roumaine Math. Pures Appl.* **44** (1999), 455-461.

[17] V. Rădulescu, Perturbations of eigenvalue problems with constraints for hemivariational inequalities, *From Convexity to Nonconvexity, volume dedicated to the memory of Prof. G. Fichera*, Nonconvex Optim. Appl., 55, Kluwer Acad. Publ., Dordrecht, 2001 (Gilbert, Pardalos, Eds.), 243-253.

[18] V. Rădulescu, Perturbations of symmetric hemivariational inequalities, in *Nonsmooth/Nonconvex Mechanics with Applications in Engineering*, Editions Ziti, Thessaloniki, 2002 (C. Baniotopoulos, Ed.), 61-72.

[19] E. H. Spanier, *Algebraic Topology*, McGraw-Hill, New York, 1966.

[20] M. Struwe, *Variational Methods*, Springer-Verlag, Berlin, Heidelberg, 1990.

Chapter 9

LOCATION OF SOLUTIONS FOR GENERAL NONSMOOTH PROBLEMS

The aim of the present Chapter is to study from a qualitative point of view a general eigenvalue problem associated to a variational-hemivariational inequality with a constraint for the eigenvalue. The basic feature of our approach is that we are mainly concerned with the location of eigensolution (u, λ), where u and λ stand for the eigenfunction and the eigenvalue, respectively. This is done in Section 2, where the location of eigensolutions is achieved by means of the graph of the derivative of a C^1 function. Section 1 presents a general existence result for variational-hemivariational inequalities with assumptions of Ambrosetti and Rabinowitz type. Section 2 deals with the exposition of our abstract location results. In Section 3 we discuss the location of solutions to variational-hemivariational inequalities by applying the abstract results. The case of nonlinear Dirichlet boundary value problems is contained.

1. Existence of Solutions by Minimax Methods for Variational-Hemivariational Inequalities

In this Section we give an existence result based on nonsmooth critical point theory in Section 2 of Chapter 2 for studying the variational-hemivariational inequalities in the sense of P. D. Panagiotopoulos ([22], [23], [21], [9]-[11], [14], [19]). The present approach enables us to obtain additional multiplicity information for the set of solutions as well as other qualitative properties. The contents of this Section is taken from [18] and [12] (see also [16]).

Let X be a real reflexive Banach space, endowed with the norm $\|\cdot\|$, which is compactly embedded in $L^p(\Omega)$, where $p > 2$ and Ω is a bounded open subset of $\mathbb{R}^N$ with a sufficiently smooth boundary. Let $\alpha : X \to$

$\mathbb{R}\cup\{+\infty\}$ be a convex, lower semicontinuous (in short, l.s.c.) and proper (i.e. $\alpha \not\equiv +\infty$) function with the domain $\mathrm{dom}(\alpha) = \{x \in X : \alpha(x) < +\infty\}$. We suppose that $\mathrm{dom}(\alpha)$ is a nonempty, closed, convex cone in the Banach space X.

Let $a : X \times X \to \mathbb{R}$ be a continuous, bilinear, symmetric form which is coercive, that is

$$a(v,v) \geq \alpha_0 \|v\|^2, \ \forall v \in X, \tag{9.1}$$

for a constant $\alpha_0 > 0$. Consider a function $j : \Omega \times \mathbb{R} \to \mathbb{R}$ satisfying the conditions:

(a) $j(\cdot, t) : \Omega \to \mathbb{R}$ is measurable, $\forall t \in \mathbb{R}$;

(b) $j(x, \cdot) : \mathbb{R} \to \mathbb{R}$ is locally Lipschitz, a.e. $x \in \Omega$;

(c) $j(\cdot, 0) \in L^1(\Omega)$ and $\int_\Omega j(x,0)\,dx + \alpha(0) \leq 0$.

We use the notation $j^0(x,\cdot;\cdot)$ for the generalized directional derivative in the sense of Clarke ([6], p. 25) with respect to the second variable, namely

$$j^0(x,y;z) = \limsup_{\substack{w \to y \\ t \downarrow 0}} \frac{1}{t}[j(x,w+tz) - j(x,w)], \tag{9.2}$$

for a.e. $x \in \Omega$ and all $y, z \in \mathbb{R}$. The generalized gradient in the sense of Clarke ([6], p. 27) $\partial j(x,\cdot)$ is given by

$$\partial j(x,y) = \{w \in \mathbb{R} : \ wz \leq j^0(x,y;z), \ \forall z \in \mathbb{R}\}, \tag{9.3}$$

for a.e. $x \in \Omega$ and all $y, z \in \mathbb{R}$. In addition, making use of (9.2) and (9.3), we assume:

(H1) $|z| \leq c(1 + |t|^{p-1})$, $\forall z \in \partial j(x,t)$, *a.e.* $x \in \Omega$, $\forall t \in \mathbb{R}$, *for some constant* $c > 0$;

(H2) $\mu j(x,t) \geq j^0(x,t;t)$, *a.e.* $x \in \Omega$, $\forall t \in \mathbb{R}$, *for some constant* $\mu > 2$;

(H3) $\liminf_{t \to 0} \frac{j(x,t)}{t^2} \geq -\nu$ *uniformly with respect to* $x \in \Omega$, *where the constant* ν *satisfies* $0 < \nu < \frac{\alpha_0}{2c_2^2}$;

(H4) *there exists* $u_0 \in X$ *such that* $\int_\Omega j(x, u_0(x))\,dx < 0$;

(H5) $(1+\frac{1}{\mu})\alpha(v) - \frac{1}{\mu}\alpha(2v) \geq -a_1\|v\|^\sigma - a_2$, $\forall v \in \mathrm{dom}(\alpha)$, *where either* $0 \leq \sigma < 2$ *and* $a_1 \geq 0$, $a_2 \geq 0$, *or* $\sigma = 2$ *and* $0 \leq a_1 < (\frac{1}{2} - \frac{1}{\mu})\alpha_0$, $a_2 \geq 0$;

$(H6)$ *there exists* $R > 0$ *such that* $\alpha(v) \geq \int_\Omega |j(x,0)|\, dx$, $\forall v \in \mathrm{dom}(\alpha)$, $\|v\| < R$;

$(H7)$ $\liminf_{t \to +\infty} \dfrac{\alpha(tu_0)}{t^\mu} = 0$.

Our goal is to study the existence of solutions to the following variational-hemivariational inequality problem:

(P) *Find* $u \in \mathrm{dom}(\alpha)$ *such that*

$$a(u, v-u) + \int_\Omega j^0(x, u(x); v(x) - u(x))dx + \alpha(v) - \alpha(u) \geq 0, \ \forall v \in \mathrm{dom}(\alpha).$$

Our main existence result for problem (P) is the following.

Theorem 9.1 (Motreanu [18]) Under assumptions $(H1)$-$(H7)$, problem (P) has at least a nontrivial solution $u \in X$, i.e. a solution u of (P) with $u \in X \setminus \{0\}$.

Proof. The idea of the proof is to apply the version of Mountain Pass Theorem formulated in Corollary 3.1 for a suitable functional $f : X \to I\!R \cup \{+\infty\}$ associated to problem (P). To this end we introduce the nonsmooth functional $\Phi : X \to I\!R$ by

$$\Phi(v) = \frac{1}{2} a(v, v) + \int_\Omega j(x, v(x))\, dx, \quad \forall v \in X. \tag{9.4}$$

Hypothesis $(H1)$ ensures that Φ defined by (9.4) is locally Lipschitz. Consequently the functional $f : X \to I\!R \cup \{+\infty\}$ given by $f = \Phi + \alpha$ with Φ in (9.4) and α entering problem (P) has the form required in (H_f) in Section 2 of Chapter 3. We will show that f verifies the assumptions of Corollary 3.1. As soon as this is done, Corollary 3.1 applies yielding the existence of a nontrivial critical point for f as required in Section 2 of Chapter 3 (see also Chapter 2). Using a property in Clarke [6, p. 83-85] for the generalized gradient of integral functions, we infer that any critical point of f is a solution to problem (P). Thus we find a nontrivial solution to (P), which is the conclusion of Theorem 9.1.

First, we check that the function f satisfies the Palais-Smale condition $(\mathrm{PS})_{f,c}$ at any level $c \in I\!R$. Let $\{u_n\} \subset \mathrm{dom}(\alpha)$ be a sequence such that

$$|f(u_n)| \leq M, \ \forall n \geq 1, \tag{9.5}$$

$$\Phi^0(u_n; v - u_n) + \alpha(v) - \alpha(u_n) \geq -\varepsilon_n \|v - u_n\|, \tag{9.6}$$

for all $v \in \text{dom}(\alpha)$ and all $n \geq 1$, for some constant $M > 0$ and a sequence $\varepsilon_n \downarrow 0$. In particular, (9.6) reads

$$a(u_n, v - u_n) + \int_\Omega j^0(x, u_n(x); v(x) - u_n(x))\, dx + \alpha(v) - \alpha(u_n)$$

$$\geq -\varepsilon_n \|v - u_n\|, \ \forall v \in \text{dom}(\alpha), \ \forall n \geq 1. \tag{9.7}$$

Using the property of the domain of α to be a cone, one can set $v = 2u_n \in \text{dom}(\alpha)$ in (9.7) to obtain for n sufficiently large, say $n \geq n_0$, that

$$\|u_n\| \geq -a(u_n, u_n) - \int_\Omega j^0(x, u_n(x); u_n(x))\, dx - \alpha(2u_n) + \alpha(u_n). \tag{9.8}$$

Summing up the inequality $f(u_n) \leq M$ in (9.5) with (9.8) multiplied by $\frac{1}{\mu}$ and using (9.1), $(H2)$, $(H5)$ and the fact that $\mu > 2$, we derive

$$M + \frac{1}{\mu}\|u_n\| \geq (\frac{1}{2} - \frac{1}{\mu})a(u_n, u_n)$$

$$+ \int_\Omega (j(x, u_n(x)) - \frac{1}{\mu} j^0(x, u_n(x); u_n(x)))\, dx + (1 + \frac{1}{\mu})\alpha(u_n) - \frac{1}{\mu}\alpha(2u_n)$$

$$\geq (\frac{1}{2} - \frac{1}{\mu})\alpha_0 \|u_n\|^2 - a_1 \|u_n\|^\sigma - a_2, \ \forall n \geq n_0. \tag{9.9}$$

By the properties of μ and σ, from (9.9) it follows that the sequence $\{u_n\}$ is bounded in X. Using the reflexivity of X and the compactness of the embedding of X in $L^p(\Omega)$ we deduce that along a subsequence, denoted again by $\{u_n\}$, one has

$$u_n \to u \ \text{ weakly in } X \text{ as } n \to \infty, \tag{9.10}$$

$$u_n \to u \ \text{ strongly in } L^p(\Omega) \text{ as } n \to \infty. \tag{9.11}$$

As the domain of α is convex and closed, the convergence in (9.10) yields $u \in \text{dom}(\alpha)$.

Define the function $J : L^p(\Omega) \to I\!R$ by $J(v) = \int_\Omega j(x, v)\, dx$, $\forall v \in L^p(\Omega)$. By assumption $(H1)$ we infer that J is well-defined and locally Lipschitz on $L^p(\Omega)$. Setting $v = u$ in (9.6) it results that

$$a(u_n, u_n) \leq a(u_n, u) + J^0(u_n; u - u_n)$$

$$+\alpha(u) - \alpha(u_n) + \varepsilon_n \|u - u_n\|, \ \forall n \geq 1. \tag{9.12}$$

Letting the upper limit as $n \to \infty$ in (9.12) and making use of (9.10) we derive

$$\limsup_{n\to\infty} a(u_n, u_n) \leq a(u, u) + \limsup_{n\to\infty} J^0(u_n; u - u_n)$$

$$+\alpha(u) - \liminf_{n\to\infty} \alpha(u_n). \tag{9.13}$$

Taking into account that J^0 is upper semicontinuous on $L^p(\Omega)$, α is convex, l.s.c. and the convergences in (9.10), (9.11), inequality (9.13) implies $\limsup_{n\to\infty} a(u_n, u_n) \leq a(u, u)$. Since the Banach space X endowed with the inner product $a : X \times X \to I\!R$ becomes a Hilbert space, the previous inequality in conjunction with the weak convergence in (9.10) ensures that $u_n \to u$ strongly in X. Thus the function f verifies the Palais-Smale condition, which is a part of assumption (f_5) in Corollary 3.1.

We pass now to the verification of condition (f_4) in Corollary 3.1 with $a = 0$ for the function $f = \Phi + \alpha$ with Φ entering (9.4) and α given in problem (P). First, the second part of condition (c) ensures that $f(0) \leq 0$.

Due to the continuity of the embedding of X in $L^p(\Omega)$ and using the inequality $p > 2$, there exist constants $c_2 > 0$ and $c_p > 0$ such that

$$\|v\|_{L^2(\Omega)} \leq c_2\|v\| \quad \text{and} \quad \|v\|_{L^p(\Omega)} \leq c_p\|v\|, \quad \forall v \in X. \tag{9.14}$$

Let us fix a number ε with $0 < \varepsilon < \frac{\alpha_0}{2c_2^2} - \nu$. Assumption $(H3)$ allows us to find $\delta > 0$ such that

$$j(x,t) \geq -(\nu + \varepsilon)t^2, \quad \text{a.e. } x \in \Omega, \ \forall |t| < \delta. \tag{9.15}$$

Using Lebourg's mean value theorem and hypothesis $(H1)$, we can write

$$j(x,t) = j(x,t) - j(x,0) + j(x,0) \geq -c(1 + |t|^{p-1})|t| + j(x,0)$$

$$\geq -c(\frac{1}{\delta^{p-1}} + 1)|t|^p - |j(x,0)|, \ \forall |t| \geq \delta. \tag{9.16}$$

Combining (9.15) and (9.16) we obtain that

$$j(x,t) \geq -(\nu + \varepsilon)t^2 - c(\frac{1}{\delta^{p-1}} + 1)|t|^p - |j(x,0)|, \ \forall t \in I\!R. \tag{9.17}$$

Integrating in (9.17) over Ω and using (9.14) it follows that

$$\int_\Omega j(x, v(x))\, dx \geq -(\nu+\varepsilon)\|v\|^2_{L^2(\Omega)} - c(\frac{1}{\delta^{p-1}}+1)\|v\|^p_{L^p(\Omega)} - \int_\Omega |j(x,0)|\, dx$$

$$\geq -c_2^2(\nu + \varepsilon)\|v\|^2 - c_p^p c(\frac{1}{\delta^{p-1}} + 1)\|v\|^p - \int_\Omega |j(x,0)|\, dx, \ \forall v \in X. \tag{9.18}$$

In view of (9.1) and (9.18) we have that

$$f(v) = \Phi(v) + \alpha(v) = \frac{1}{2}a(v,v) + \int_\Omega j(x,v)\, dx + \alpha(v)$$

$$\geq \|v\|^2(\frac{1}{2}\alpha_0 - c_2^2\nu - c_2^2\varepsilon - c_p^p c(\frac{1}{\delta^{p-1}} + 1)\|v\|^{p-2})$$

$$- \int_\Omega |j(x,0)|\, dx + \alpha(v). \tag{9.19}$$

Since $p > 2$, by the choice of ε, there exists $0 < r < R$ such that

$$\frac{1}{2}\alpha_0 - c_2^2\nu - c_2^2\varepsilon - c_p^p c(\frac{1}{\delta^{p-1}} + 1) r^{p-2} > 0. \tag{9.20}$$

Inequality (9.19) in conjunction with assumption $(H6)$ yields

$$f(v) \geq \overline{a} := r^2(\frac{1}{2}\alpha_0 - c_2^2\nu - c_2^2\varepsilon - c_p^p c(\frac{1}{\delta^{p-1}} + 1) r^{p-2}), \tag{9.21}$$

for all $v \in X$, $\|v\| = r$. The choice of r in (9.20) ensures that $\overline{a} > 0$. This remark together with (9.21) proves that the function f satisfies the second part of the double inequality required in (f_4) in Corollary 3.1, with strict inequality sign.

We show that f verifies assumption (f_4) in Corollary 3.1 with some $x_1 \in X$ for which one has $\|x_1\| > r$, where r has been already determined. Using calculus with generalized gradient (see Clarke [6]), we have

$$\partial_t(t^{-\mu} j(x,ty))(t)$$

$$= t^{-\mu-1}(-\mu j(x,ty) + \partial j(x,ty)ty), \ \forall t > 0, \ y \in \mathbb{R}. \tag{9.22}$$

By means of Lebourg's mean value theorem, relation (9.22) and hypothesis $(H2)$, we deduce that for every $t > 1$, there is $\tau \in (1,t)$ such that

$$t^{-\mu} j(x,ty) - j(x,y)$$

$$\leq -\tau^{-\mu-1}(\mu j(x,\tau y) - j^0(x,\tau y;\tau y))(t-1) \leq 0. \tag{9.23}$$

For u_0 introduced in $(H4)$ and $t > 1$ we see from (9.23) that

$$f(tu_0) = \frac{t^2}{2} a(u_0,u_0) + \int_\Omega j(x, tu_0(x))\, dx + \alpha(tu_0)$$

$$\leq t^\mu(\frac{t^{2-\mu}}{2} a(u_0,u_0) + \int_\Omega j(x,u_0(x))\, dx + \frac{1}{t^\mu}\alpha(tu_0)). \tag{9.24}$$

By $(H7)$ along a sequence $\{t_n\}$ with $t_n \to +\infty$ one has $\lim_{n\to\infty} \frac{1}{t_n^\mu}\alpha(t_n u_0) = 0$. Then, since $\mu > 2$, the use of $(H4)$ in (9.24) implies that $f(t_n u_0) \leq 0$, for any n sufficiently large. Assumption (f_4) in Corollary 3.1 is verified with $x_1 = t_n u_0$, where n is taken having in addition the property $\|t_n u_0\| > r$. Since in our situation we are in the case $c > a$ with the notation of Theorem 2.5 and Corollary 3.1, we may apply Corollary 3.1. The proof of Theorem 9.1 is complete. ∎

The example below provides a concrete situation where the data entering problem (P) fulfill the hypotheses of Theorem 9.1.

Example 9.1. Let Ω be a bounded domain of $\mathbb{R}^N$ with smooth boundary, $X = H_0^1(\Omega)$, which is known to be compactly embedded in $L^p(\Omega)$ for $p < 2^* := \frac{2N}{N-2}$. Fix a number $2 < p < 2^*$. Let the bilinear form $a : H_0^1(\Omega) \times H_0^1(\Omega) \to \mathbb{R}$ given by $a(u,v) = \int_\Omega \nabla u(x) \cdot \nabla v(x)\, dx$, $\forall u, v \in H_0^1(\Omega)$. Consider the function $j : \mathbb{R} \to \mathbb{R}$ defined by $j(t) = \max\{-\frac{|t|^q}{q}, -\frac{|t|^s}{s}\}$, $\forall t \in \mathbb{R}$, for some constants $2 < q < p$ and $2 < s < p$. If K is a nonempty, closed, convex cone in $H_0^1(\Omega)$ let us choose the function $\alpha : H_0^1(\Omega) \to \mathbb{R} \cup \{+\infty\}$ as $\alpha = I_K$, where I_K is the indicator function of K, or, as another choice, $\alpha(u) = \|u\|$ if $u \in K$ and $\alpha(u) = +\infty$ if $u \notin K$. One can verify that all the assumptions of Theorem 9.1 are satisfied, leading to the existence of at least a nontrivial solution to problem (P) written for the aforementioned data.

We now present a version of Theorem 9.1 for treating the following hemivariational inequality problem:

(P') *Find* $u \in X$ *such that*

$$a(u,v) + \int_\Omega j^0(x, u(x); v(x))dx \geq \langle g, v\rangle_{X^*,X},\ \forall v \in X.$$

The data X and a have the same meaning as in Theorem 9.1, while

$$g \in X^*. \tag{9.25}$$

The function $j : \Omega \times \mathbb{R}^m \to \mathbb{R}$ is supposed to satisfy the conditions (a), (b) as above and

(c') $j(\cdot, 0) \in L^\infty(\Omega)$ and $\int_\Omega j(x,0)dx \leq 0$.

In addition, the function $j : \Omega \times \mathbb{R}^m \to \mathbb{R}$ is assumed to fulfill the following hypotheses:

$(A1)$ *there exists a constant* $c > 0$ *such that*

$$|w| \leq c(1 + |y|^{p-1}),\ \forall w \in \partial j(x,y),\quad \text{a.e } x \in \Omega,\ \forall y \in \mathbb{R}^m;$$

$(A2)$ *there exist constants* $\mu > 2$, $a_1 \geq 0$, $a_2 \geq 0$ *and* $0 \leq \sigma < 2$ *such that*

$$\mu j(x,y) - j^0(x,y;y) \geq -a_1|y|^\sigma - a_2,\quad \text{a.e. } x \in \Omega,\ \forall y \in \mathbb{R}^m;$$

$(A3)$ $\liminf_{y\to 0} \frac{j(x,y)}{|y|^2} \geq 0$ *uniformly with respect to* $x \in \Omega$;

$(A4)$ *there exists* $v_0 \in X \setminus \{0\}$ *such that*

$$\liminf_{s\to+\infty} s^{-\sigma} \int_\Omega j(x, sv_0(x))dx < \frac{a_1}{\sigma - \mu} \int_\Omega |v_0(x)|^\sigma dx.$$

Notice first that $(A1)$ ensures that the integral in (P') exists. Indeed, by $(A1)$ and a basic property of generalized gradients (see [6], p. 27) we can write

$$|j^0(x, u(x); v(x))| = |\max\{w \cdot v(x) : \ w \in \partial j(x, u(x))\}|$$
$$\leq c(1 + |u(x)|^{p-1})|v(x)| \ \text{ a.e. } x \in \Omega, \ \forall u, v \in X.$$

According to the embedding of X in $L^p(\Omega)$ one has $|u|^{p-1} \in L^{\frac{p}{p-1}}(\Omega)$ and $|v| \in L^p(\Omega)$, thus the integral in (P') is finite. We remark also that the integrals in $(A4)$ make sense. This is easily seen from (c'), Lebourg's mean value theorem for locally Lipschitz functionals (see [6], p. 41) and the growth condition $(A1)$.

The existence of solutions to problem (P') is established in

Theorem 9.2 (Haslinger and Motreanu [12]) Assume that conditions $(A1)$-$(A4)$ are satisfied. Then problem (P') possesses at least one solution $u \in X \setminus \{0\}$ whenever $\|g\|_{X^*}$ is sufficiently small, say $\|g\|_{X^*} \leq B$ for a constant $B > 0$ which can be estimated a priori. Moreover, defining $f : X \to I\!R$ by

$$f(v) = \frac{1}{2}a(v,v) + \int_\Omega j(x, v(x))dx - \langle g, v\rangle_{X^*,X}, \ \forall v \in X,$$

the value $f(u)$ of f at the solution $u \in X$ admits the following minimax characterization: there exists $t_0 > 0$ such that

$$f(u) = \inf\{\max_{0\leq t\leq 1} f(g(t)) : \ g \in C([0,T], X), \ g(0) = 0, \ g(1) = t_0 v_0\},$$

with $v_0 \in X$ given in $(A4)$. In addition, there exists a constant $b > 0$ depending only on $B > 0$ such that $u \equiv u(g)$ satisfies the following uniform estimate

$$b \leq \frac{1}{2}a(u,u) + \int_\Omega j(x, u(x))dx - \langle g, u\rangle_{X^*,X}$$

$$\leq \max_{0\leq t\leq 1}\left[\frac{1}{2}t^2 t_0^2 a(v_0, v_0) + \int_\Omega j(x, tt_0 v_0(x))dx - tt_0\langle g, v_0\rangle_{X^*,X}\right],$$

for all $g \in X^*$ with $\|g\|_{X^*} \le B$.

Proof. The method of proof follows the pattern of the proof of Theorem 9.1. For the sake of clarity we give the complete proof.

In view of the application of a variational approach we consider the functional $f : X \to \mathbb{R}$ given by (9.26). Let us introduce the functional $J : L^p(\Omega; \mathbb{R}^m) \to \mathbb{R}$ by

$$J(v) = \int_\Omega j(x, v(x))dx, \ \forall v \in L^p(\Omega; \mathbb{R}^m).$$

Assumption $(A1)$ guarantees that J is Lipschitz continuous on bounded subsets of $L^p(\Omega; \mathbb{R}^m)$ and its generalized gradient $\partial J(v) \subset L^{p'}(\Omega; \mathbb{R}^m)$, where $(1/p) + (1/p') = 1$, has the property

$$\partial J(v) \subset \int_\Omega \partial j(x, v(x))dx, \ \forall v \in L^p(\Omega; \mathbb{R}^m) \tag{9.26}$$

(see [6], p. 83). The inclusion in (9.26) is interpreted as follows: every element $z \in \partial J(v) \subset L^{p'}(\Omega; \mathbb{R}^m)$, verifying

$$\langle z, v \rangle = \int_\Omega z(x) \cdot v(x)dx, \ \forall v \in L^p(\Omega; \mathbb{R}^m), \tag{9.27}$$

satisfies

$$z(x) \in \partial j(x, v(x)) \text{ for a.e. } x \in \Omega. \tag{9.28}$$

It is clear that the locally Lipschitz functional $f : X \to \mathbb{R}$ is expressed by

$$f(v) = \frac{1}{2}a(v, v) + (J|_X)(v) - \langle g, v \rangle_{X^*, X}, \ \forall v \in X, \tag{9.29}$$

and its generalized gradient $\partial f(v) \subset X^*$ satisfies

$$\partial f(v) = Av + i^* \partial J(v) - g, \ \forall v \in X, \tag{9.30}$$

where $A : X \to X^*$ is the continuous linear operator corresponding to the bilinear form $a : X \times X \to X$:

$$\langle Av, w \rangle_{X^*, X} = a(v, w), \ \forall v, w \in X,$$

and $i : X \to L^p(\Omega; \mathbb{R}^m)$ is the embedding map.

Our goal is to show that the functional $f : X \to \mathbb{R}$ has a critical point $u \in X$ in the sense of Chang [5] (see Chapter 2), i.e.

$$0 \in \partial f(u). \tag{9.31}$$

To this end we shall apply Chang's version of Mountain Pass Theorem (Corollary 3.1 for $\alpha = 0$ and $c > a$ with the notation therein).

We first verify that f satisfies the Palais-Smale condition in the sense of Chang [5] (see Chapter 2) . For checking this assertion, let $\{v_n\} \subset X$ be a sequence such that

$$|f(v_n)| \leq M, \ \forall n \geq 1, \tag{9.32}$$

with a constant $M > 0$, and let $\{w_n\} \subset X^*$ be a sequence satisfying

$$w_n \in \partial f(v_n) \ \forall n \geq 1, \tag{9.33}$$

and

$$w_n \to 0 \text{ in } X^* \text{ as } n \to \infty. \tag{9.34}$$

By (9.30) and (9.33) one finds

$$z_n \in \partial J(v_n) \subset L^{\frac{p}{p-1}}(\Omega; I\!R^m) \ \forall n \geq 1, \tag{9.35}$$

such that

$$w_n = Av_n + i^* z_n - g, \ \forall n \geq 1. \tag{9.36}$$

For n sufficiently large, by (9.32), (9.34), (9.29) and (9.36), it holds

$$\begin{aligned} M + \|v_n\|_X &\geq f(v_n) - \frac{1}{\mu}\langle w_n, v_n\rangle_{X^*,X} \\ &= \left(\frac{1}{2} - \frac{1}{\mu}\right) a(v_n, v_n) - \left(1 - \frac{1}{\mu}\right) \langle g, v_n\rangle_{X^*,X} \\ &+ \frac{1}{\mu}\int_\Omega [\mu j(x, v_n(x)) - z_n(x) \cdot v_n(x)]dx, \end{aligned} \tag{9.37}$$

where $\mu > 2$ is given in $(A2)$. Using (1.2), (9.28), $(A2)$ and the continuity of embedding of X in $L^p(\Omega)$, we obtain from (9.37) that there exist constants $b_1 \geq 0$ and $b_2 \geq 0$ such that

$$\begin{aligned} &M + \|v_n\|_X \\ &\geq \alpha\left(\frac{1}{2} - \frac{1}{\mu}\right) \|v_n\|_X^2 - \left(1 - \frac{1}{\mu}\right) \|g\|_{X^*}\|v_n\|_X - b_1\|v_n\|_X^\sigma - b_2. \end{aligned} \tag{9.38}$$

Since $\mu > 2$ and $\sigma < 2$, estimate (9.38) enables us to deduce that the sequence $\{v_n\}$ is bounded in X. The reflexivity of X and the continuity of embedding of X in $L^p(\Omega)$ ensure the existence of a subsequence and $v \in X$ with the property that

$$v_n \rightharpoonup v \text{ weakly in } X \text{ as } n \to \infty$$

and

$$v_n \to v \text{ (strongly) in } L^p(\Omega; \mathbb{R}^m) \text{ as } n \to \infty. \tag{9.39}$$

Since $J : L^p(\Omega; \mathbb{R}^m) \to \mathbb{R}$ is locally Lipschitz, we see from (9.35) and (9.39) that

$$\{z_n\} \text{ is bounded in } L^{\frac{p}{p-1}}(\Omega; \mathbb{R}^m). \tag{9.40}$$

Further the continuity of embedding of X in $L^p(\Omega)$ and (9.40) imply that for (possibly) another subsequence

$$\{i^* z_n\} \text{ converges strongly in } X^*. \tag{9.41}$$

From (9.34), (9.36) and (9.41) we see that for a subsequence of $\{v_n\}$ denoted again by $\{v_n\}$, $\{Av_n\}$ converges strongly in X^*. The linear operator $A : X \to X^*$ is a topological isomorphism implying that a strongly convergent subsequence of $\{v_n\}$ exists. The Palais-Smale condition for the locally Lipschitz functional f is verified.

As the next step of the proof show that

$$\lim_{t \to +\infty} f(tv_0) = -\infty, \tag{9.42}$$

where $v_0 \in X$ is the element from $(A4)$. In order to prove (9.42) we remark that, given $x \in \Omega$ and $y \in \mathbb{R}^m$, the following differentiation formula holds

$$\frac{d}{d\tau}(\tau^{-\mu} j(x, \tau y))$$

$$= \mu \tau^{-\mu-1}[-j(x, \tau y) + \frac{1}{\mu} j'_y(x, \tau y)(\tau y)], \quad \text{a.e. } \tau \in \mathbb{R}, \tag{9.43}$$

where j'_y denotes the differential with respect to y (it exists a.e. because $j(x, \cdot)$ is locally Lipschitz). Using the equality

$$t^{-\mu} j(x, ty) - j(x, y) = \int_1^t \frac{d}{d\tau}(\tau^{-\mu} j(x, \tau y)) d\tau,$$

together with (9.43) and taking into account that the differential always belongs to the generalized gradient (see [6], p. 32) we see that

$$t^{-\mu} j(x, ty) - j(x, y) \le -\int_1^t \tau^{-\mu-1}[\mu j(x, \tau y) - j^0(x, \tau y; \tau y)] d\tau, \tag{9.44}$$

for all $t > 1$, a.e. $x \in \Omega$, $\forall y \in \mathbb{R}^m$. Then (9.44) and $(A2)$ yield

$$t^{-\mu} j(x, ty) - j(x, y) \le \int_1^t \tau^{-\mu-1}(a_1 \tau^\sigma |y|^\sigma + a_2) d\tau$$

$$= a_1|y|^\sigma \frac{1}{\sigma-\mu}(t^{\sigma-\mu}-1) - \frac{a_2}{\mu}(t^{-\mu}-1)$$

$$\leq \frac{a_1}{\mu-\sigma}|y|^\sigma + \frac{a_2}{\mu}, \ \forall t > 1, \text{ a.e. } x \in \Omega, \ \forall y \in \mathbb{R}^m. \tag{9.45}$$

Setting $y = sv_0(x)$ in (9.45), $x \in \Omega$ and $s > 0$, it turns out that

$$j(x, tsv_0(x)) \leq t^\mu \left[j(x, sv_0(x)) + \frac{a_1}{\mu-\sigma}s^\sigma|v_0(x)|^\sigma + \frac{a_2}{\mu}\right], \tag{9.46}$$

for all $t > 1$, $s > 0$, a.e. $x \in \Omega$. On the basis of (9.29) and (9.46) we get

$$f(tsv_0) \leq \frac{1}{2}t^2s^2a(v_0,v_0) - ts\langle g, v_0\rangle_{X^*,X} + t^\mu s^\sigma \left[s^{-\sigma}\int_\Omega j(x, sv_0(x))dx\right.$$

$$\left.+\frac{a_1}{\mu-\sigma}\int_\Omega |v_0(x)|^\sigma dx + \frac{a_2}{\mu}|\Omega|s^{-\sigma}\right], \ \forall t > 1, \ s > 0, \text{ a.e. } x \in \Omega. \tag{9.47}$$

By assumption $(A4)$ there is a number $s > 0$ such that

$$C := s^{-\sigma}\int_\Omega j(x, sv_0(x))dx$$

$$+\frac{a_1}{\mu-\sigma}\int_\Omega |v_0(x)|^\sigma dx + \frac{a_2}{\mu}|\Omega|s^{-\sigma} < 0. \tag{9.48}$$

Fixing $s > 0$ in (9.48), passing to the limit in (9.47) as $t \to +\infty$, we arrive at

$$\limsup_{t\to+\infty} f(tv_0) = \limsup_{t\to+\infty} f(tsv_0)$$

$$\leq \limsup_{t\to+\infty} t^\mu \left[\frac{1}{2}\frac{1}{t^{\mu-2}}s^2a(v_0,v_0) - \frac{s}{t^{\mu-1}}\langle g, v_0\rangle_{X^*,X} + Cs^\sigma\right] = -\infty.$$

This justifies (9.42).

We now check that there exist constants $b > 0$ and $\rho > 0$ for which

$$f(v) \geq b, \ \forall v \in X, \ \|v\|_X = \rho, \tag{9.49}$$

and, furthermore, this result is uniform with respect to the elements g whose norm is small enough. To this end we make use of $(A3)$. Given $\varepsilon > 0$, by $(A3)$ there is $\delta \equiv \delta(\varepsilon) > 0$ such that

$$j(x,y) \geq -\varepsilon|y|^2, \text{ a.e. } x \in \Omega, \ \forall y \in \mathbb{R}^m, \ |y| \leq \delta. \tag{9.50}$$

By $(A1)$, the function $j : \Omega \times \mathbb{R}^m \to \mathbb{R}$ can be estimated by

$$|j(x,y)| \leq c_1|y|^p + c_2, \text{ a.e. } x \in \Omega, \ \forall y \in \mathbb{R}^m,$$

with constants $c_1 > 0$ and $c_2 > 0$ implying that

$$|j(x,y)| \leq \left(c_1 + \frac{c_2}{\delta^p}\right) |y|^p, \text{ a.e. } x \in \Omega, \ \forall y \in \mathbb{R}^m, \ |y| \geq \delta. \tag{9.51}$$

Relations (9.50) and (9.51) lead to

$$j(x,y) \geq -\varepsilon|y|^2 - \left(c_1 + \frac{c_2}{\delta^p}\right) |y|^p, \text{ a.e. } x \in \Omega, \ \forall y \in \mathbb{R}^m. \tag{9.52}$$

From the continuity of embedding of X in $L^p(\Omega)$ and (9.52) we see that constants $c_0 > 0$ and $\bar{c} > 0$ exist such that

$$f(v) \geq \frac{1}{2}\alpha\|v\|_X^2 - \|g\|_{X^*}\|v\|_X - \varepsilon c_0\|v\|_X^2$$

$$-\bar{c}\left(c_1 + \frac{c_2}{\delta^p}\right) \|v\|_X^p \ \forall v \in X.$$

This and the Young's inequality yield

$$f(v) \geq \left[\frac{1}{2}\alpha - c_0\varepsilon - \left(\frac{1}{p} + \bar{c}\left(c_1 + \frac{c_2}{\delta^p}\right)\right) \|v\|_X^{p-2}\right] \|v\|_X^2$$

$$-\frac{p-1}{p}\|g\|_{X^*}^{\frac{p}{p-1}}, \ \forall v \in X. \tag{9.53}$$

Choosing $\varepsilon > 0$ sufficiently small and using that $p > 2$, estimate (9.53) guarantees that a number $\rho > 0$ can be found such that

$$f(v) \geq E\rho^2 - \frac{p-1}{p}\|g\|_{X^*}^{\frac{p}{p-1}}, \quad \forall v \in X, \ \|v\|_X = \rho, \tag{9.54}$$

where $E > 0$ is a constant which is independent of $v \in X$, $g \in X^*$ and $\rho > 0$.

Bound (9.54) establishes assertion (9.49). It allows to get constants $B > 0$ and $b > 0$ such that the following uniform estimate holds

$$f(v) \geq b \text{ whenever } \|v\|_X = \rho \text{ and } \|g\|_{X^*} \leq B. \tag{9.55}$$

On the other hand, by assumption (c'), we know that

$$f(0) \leq 0. \tag{9.56}$$

From (9.42) one can find a number $t_0 > 0$ satisfying

$$t_0\|v_0\|_X > \rho \text{ and } f(t_0 v_0) < 0, \tag{9.57}$$

where $\rho > 0$ is as in (9.55).

The Palais-Smale condition, (9.55), (9.56) and (9.57) enable us to apply Corollary 3.1 with $\alpha = 0$ and $c > a$ with the notations therein. It says that the number

$$c := \inf\{\max_{0\le t\le 1} f(\gamma(t)) : \gamma \in C([0,T],X),\ \gamma(0) = 0,\ \gamma(1) = t_0 v_0\} \quad (9.58)$$

is a critical value of f, i.e. there exists $u \in X$ such that

$$0 \in \partial f(u) \text{ and } f(u) = c. \quad (9.59)$$

In addition

$$f(u) = c \ge b \quad \text{wherever } \|g\|_{X^*} \le B. \quad (9.60)$$

The final estimate in the statement of Theorem 9.2 follows from (9.58) and (9.60).

We now show that the inclusion in (9.59) (or (9.31)) implies that u solves problem (P'). Indeed, in view of (9.30), relation (9.59) becomes

$$Au + i^* z - g = 0 \text{ for some } z \in \partial J(u). \quad (9.61)$$

Taking into account (9.26), (9.27), (9.28), equality (9.61) yields

$$0 = a(u,v) + \int_\Omega z(x)\cdot v(x)dx - \langle g, v\rangle_{X^*,X}$$

$$\le a(u,v) + \int_\Omega j^0(x,u(x);v(x))dx - \langle g, v\rangle_{X^*,X},\ \forall v \in X.$$

This completes the proof of Theorem 9.2. ■

The following example provides a nonsmooth function j satisfying $(A1)$-$(A4)$. To simplify the exposition we consider only the dependence with respect to y, i.e. $j(x,y) \equiv j(y)$.

Example 9.2. Let the function $j : \mathbb{R} \to \mathbb{R}$ which has been considered in Example 9.1, i.e.

$$j(y) = \max\left\{-\frac{1}{q}|y|^q, -\frac{1}{r}|y|^r\right\},\ \forall y \in \mathbb{R},$$

with numbers q and r satisfying $q, r \in (2,p]$, for $2 < p < 2^*$, where 2^* denotes the Sobolev critical exponent for the embedding $X = H_0^1(\Omega) \subset L^p(\Omega)$. It is clear that the function j is locally Lipschitz and using the differentiation formula for the generalized gradient of the maximum of finitely many functions, one sees that

$$j^0(y;z) = \max\{-|y|^{q-2}yz, -|y|^{r-2}yz\},\ \forall y, z \in \mathbb{R}.$$

Thus the generalized gradient of j verifies the growth condition $(A1)$. A direct computation shows that $(A2)$ is satisfied with an arbitrary $\sigma \geq 0$ by choosing any number μ such that $2 < \mu \leq \min\{q, r\}$. To check $(A3)$ and $(A4)$ let us make the choice $q \geq r$. Then $j(y) = -\frac{1}{q}|y|^q$ if $|y| < 1$, hence $\liminf_{y \to 0} y^{-2} j(y) = 0$ and $(A3)$ holds, too. Since $q \geq r$, we have for all $v_0 \in X \setminus \{0\}$ and $\sigma < r$

$$\liminf_{s \to +\infty} s^{-\sigma} \int_\Omega j(sv_0(x))dx$$

$$\leq \liminf_{s \to +\infty} s^{-\sigma + r} \int_\Omega \max\left\{-\frac{1}{q}|v_0(x)|^q, -\frac{1}{r}|v_0(x)|^r\right\} dx = -\infty.$$

Therefore assumption $(A4)$ is verified and Theorem 9.2 can be applied to problem (P').

The next example sets forth a function j determining the nonlinear part of problem (P') which is vector-valued, contains both superlinear and sublinear terms and verifies assumptions $(A1)$-$(A4)$.

Example 9.3. Consider the function $j : I\!R^2 \to I\!R$ defined by

$$j(y) = -\frac{1}{p}|y_1|^p + \int_0^{y_2} h(t)dt, \ \forall y = (y_1, y_2) \in I\!R^2,$$

with $2 < p < 2^*$, and a function $h : I\!R \to I\!R$ satisfying $h \in L^\infty_{loc}(I\!R)$, $th(t) \geq 0$ for $t \in I\!R$ near 0 and $|h(t)| \leq c(1 + |t|^\gamma)$, $\forall t \in I\!R$, where $c > 0$ and $0 \leq \gamma < 1$. It is readily seen that the function j is locally Lipschitz and its generalized directional derivative is given by

$$j^0(y; z) = -|y_1|^{p-2} y_1 z_1 + \max\{h^-(y_2)z_2, h^+(y_2)z_2\}, \ \forall y, z \in I\!R^2$$

where $h^-(t)$ and $h^+(t)$ denote the essential supremum and essential infimum of h at t, respectively (see Clarke [6], p. 34, or Chang [5]). Clearly, it satisfies the growth condition $(A1)$. Assumption $(A2)$ is verified for any $2 < \mu \leq p$ and $\sigma = \gamma + 1$. The assumption $th(t) \geq 0$ for $t \in I\!R$ near 0 implies that $(A3)$ is valid. Taking $v_0 = (v_1, 0) \in X \setminus \{0\}$, we note that $\lim_{s \to +\infty} s^{-(\gamma+1)} \int_\Omega j(sv_0(x))dx = -\infty$ since $\gamma + 1 < p$. Condition $(A4)$ is verified and Theorem 9.2 can be applied to problem (P').

Remark 9.1 If $g = 0$ problem (P') is a particular case o problem (P) and thus Theorem 9.2 is contained in Theorem 9.1 (with $\alpha = 0$). If the term g entering problem (P') is not 0, then Theorem 9.2 cannot be deduced from Theorem 9.1 (with $\alpha = 0$).

2. Location of Eigensolutions to Variational-Hemivariational Inequalities

In the following we need the nonsmooth minimax result formulated below.

Lemma 9.1 Let X be a Banach space, let $\Phi : X \times I\!R \to I\!R$ be a locally Lipschitz functional and let $\Psi : X \times I\!R \to I\!R \cup \{+\infty\}$ be a proper, convex and lower semicontinuous functional. Assume that two positive numbers $\rho < r$ are given such that the functional $f = \Phi + \Psi : X \times I\!R \to I\!R \cup \{+\infty\}$ satisfies relations

$$f(0,0) = f(0,r) = 0$$

and

$$\inf_{v \in X} f(v,\rho) > 0.$$

Consider the minimax value

$$c := \inf\{ \sup_{t \in [0,1]} f(g(t)) : g \in C([0,1], X \times I\!R),\ g(0) = (0,0),\ g(1) = (0,r)\}$$

and assume the Palais-Smale condition $(\mathrm{PS})_{f,c}$ (with $X \times I\!R$ in place of X) in Definition 3.3. Then the minimax value c is a critical value of $f : X \times I\!R \to I\!R \cup \{+\infty\}$, i.e. there exists a critical point of f (see Chapter 2 and Section 2 in Chapter 3). Moreover, one has the estimate

$$\inf_{v \in X} f(v,\rho) \leq c \leq \sup_{t \in [0,1]} f(0,tr).$$

Proof. One applies the general nonsmooth minimax principle in Corollary 3.1 to the functional $f = \Phi + \Psi : X \times I\!R \to I\!R \cup \{+\infty\}$ (so $\alpha = \Psi$ in Corollary 3.1) and the sets $S = X \times \{\rho\}$ and $Q = \{(0,tr) : t \in [0,1]\}$. Notice that f is bounded from above on Q since Q is compact, Φ is continuous and Ψ is convex (thus $\Psi(0,rt) \leq (1-t)\Psi(0,0) + t\Psi(0,r) \leq \max\{\Psi(0,0), \Psi(0,r)\} < +\infty$, $\forall\, t \in [0,1]$). The linking property is true because we imposed $0 < \rho < r$. The stated conclusion follows from Corollary 3.1. ■

We now describe the framework where the location of solutions will be studied. Let V be a real Hilbert space endowed with the inner product $(\cdot,\cdot)_V$ and the associated norm $|\cdot|_V$. Let $A : V \to V$ be a symmetric, linear and continuous operator for which there is a constant $m > 0$ such that

$$(Av, v)_V \geq m|v|_V^2 \quad \forall v \in V. \tag{9.62}$$

For the rest of this Section we fix a number $\bar{\lambda} > 0$.

Let $J : V \to I\!R$ be a locally Lipschitz functional and let $\psi : V \to I\!R \cup \{+\infty\}$ be a convex and lower semicontinuous function.

We state the following conditions :

(I_1) $-J(0) + \psi(0) = 0$ and $0 \in V$ is not a critical point of the functional $-J + \psi : V \to I\!R \cup \{+\infty\}$ in the sense of Section 2 of Chapter 3, i.e. there is some $v_0 \in V$ such that

$$(-J)^0(0; v_0) + \psi(v_0) - \psi(0) < 0.$$

(I_2) There are constants $a_1 > 0$, $a_2 > 0$ and $q \geq 2$ such that

$$-J(v) + \psi(v) \geq -a_1 - a_2 \, |v|_V^q \quad \forall v \in V.$$

(I_3) If $u_n \rightharpoonup u$ weakly in V, then there exists a subsequence of $\{u_n\}$ denoted again by $\{u_n\}$ such that

$$\limsup_{n \to \infty} (-J)^0(u_n; u - u_n) \leq 0.$$

(I_4) If $\{u_n\} \subset V$ is a sequence such that

$$-J(u_n) + \psi(u_n) + \frac{\bar{\lambda}}{2} \, (Au_n, u_n)_V$$

is bounded and

$$(-J)^0(u_n; v - u_n) + \psi(v) - \psi(u_n) + \bar{\lambda}(Au_n, v - u_n)_V$$

$$\geq -\varepsilon_n \, |v - u_n|_V \, , \quad \forall v \in V,$$

for a sequence $\{\varepsilon_n\} \subset I\!R^+$ with $\varepsilon_n \to 0$, then $\{u_n\} \subset V$ contains a bounded subsequence.

In this Section we deal with the following general nonsmooth eigenvalue problem.

(EP) *Find* $\lambda \in I\!R$ *and* $u \in V \setminus \{0\}$ *such that*

$$\begin{cases} \lambda(Au, v - u)_V + (-J)^0(u; v - u) + \psi(v) - \psi(u) \geq 0, & \forall v \in V, \\ \lambda > \bar{\lambda}. \end{cases}$$

Problem (EP) is seen to be an eigenvalue problem (with the eigensolution $(u, \lambda) \in V \times I\!R$) expressed as a variational-hemivariational inequality in the sense of Panagiotopoulos [22], [23]. Our main result for problem (EP) will be formulated by means of a function $\beta \in C^1(\mathbb{R})$ and numbers $0 < \rho < r$ such that these data will permit the location of eigensolutions (u, λ) of problem (EP).

For a given function $\beta \in C^1(\mathbb{R})$ and positive numbers $\rho < r$ we impose the following conditions :

(β_1) $\beta(0) = \beta(r) = 0$;

(β_2) $m^{\frac{q}{2}} \rho^{q+1} \geq a_2 q$ and $\frac{q+1}{q} \beta(\rho) > a_1$;

(β_3) $\lim_{|t| \to +\infty} \beta(t) = +\infty$;

(β_4) $\beta'(t) < 0 \Longleftrightarrow t < 0$ or $\rho < t < r$;

(β_5) $\beta'(t) = 0 \Longrightarrow t \in \{0, \rho, r\}$.

An example of such function β is given by :

$$\beta(t) = \begin{cases} t^2 & \text{if} \quad t \leq 0, \\ \frac{-2}{\rho^2} t^3 + \frac{3}{\rho} t^2 & \text{if} \quad t \in [0, \rho], \\ \frac{-2\rho}{(\rho - r)^3} t^3 + \frac{3\rho(r+\rho)}{(\rho - r)^3} t^2 - \frac{6r\rho^2}{(\rho - r)^3} t + \frac{r^2 \rho(-r+3\rho)}{(\rho - r)^3} & \text{if} \quad t \in [\rho, r], \\ (t - r)^2 & \text{if} \quad t \geq r. \end{cases}$$

The graph of such a function β is given below.

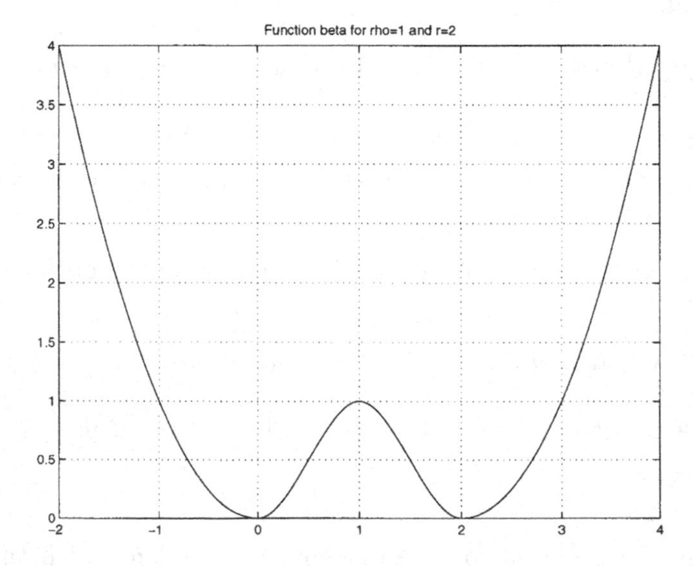

In the theorem below it is formulated the main location result for the eigensolutions of problem (EP).

Theorem 9.3 (Adly and Motreanu [1]) Assume that conditions (I_1)-(I_4) are satisfied and fix a function $\beta \in C^1(I\!R)$ together with positive numbers $\rho < r$ for which (β_1)-(β_5) are verified. Then the following alternative holds :

either

(i) there exists $u \in V \setminus \{0\}$ such that

$$\bar{\lambda}(Au, v-u)_V + (-J)^0(u; v-u) + \psi(v) - \psi(u) \geq 0, \quad \forall v \in V, \tag{9.63}$$

i.e. $\bar{\lambda}$ is an eigenvalue for the inequality in the eigenvalue problem (EP), and

$$\frac{q+1}{q}\beta(\rho) - a_1 \leq -J(u) + \psi(u) + \frac{\bar{\lambda}}{2}(Au, u)_V \leq \frac{q+1}{q}\beta(\rho); \tag{9.64}$$

or

(ii) there exist an eigensolution (u, λ) of problem (EP) and a number $s \in]\rho, r[$ such that

$$(Au, u)_V^{\frac{1}{2}} = s^{-1}(-\beta'(s))^{\frac{1}{q}}, \tag{9.65}$$

$$\lambda = \bar{\lambda} + s^3(-\beta'(s))^{\frac{q-2}{q}}, \tag{9.66}$$

$$\begin{aligned} &\tfrac{q+1}{q}(\beta(\rho) - \beta(s)) - a_1 - \tfrac{1}{q}s^{q+1}(Au, u)_V^{\frac{q}{2}} \leq -J(u) + \psi(u) \\ &+\tfrac{\bar{\lambda}}{2}(Au, u)_V \leq \tfrac{q+1}{q}(\beta(\rho) - \beta(s)) - \tfrac{1}{q}s^{q+1}(Au, u)_V^{\frac{q}{2}}. \end{aligned} \tag{9.67}$$

Proof. The basic idea of the proof is to apply Lemma 9.1 (with $X = V$) to an appropriate functional on the product space $V \times I\!R$. To this end let us introduce the functional $f : V \times I\!R \to I\!R \cup \{+\infty\}$ by

$$f(v,t) = \frac{1}{q}|t|^{q+1}(Av, v)_V^{\frac{q}{2}} + \frac{q+1}{q}\beta(t) - J(v) + \psi(v) + \frac{\bar{\lambda}}{2}(Av, v)_V \tag{9.68}$$

for all $(v, t) \in V \times I\!R$.

The data entering (9.68) are those described in the statement of Theorem 9.3. It is clear that

$$f = \Phi + \Psi, \tag{9.69}$$

where $\Phi : V \times I\!R \to I\!R$ is the locally Lipschitz functional given by

$$\Phi(v,t) = \frac{1}{q}|t|^{q+1}(Av, v)_V^{\frac{q}{2}} + \frac{q+1}{q}\beta(t) - J(v) + \frac{\bar{\lambda}}{2}(Av, v)_V \tag{9.70}$$

for all $(v,t) \in V \times I\!R$, and $\Psi : V \times I\!R \to I\!R \cup \{+\infty\}$ is the convex, proper and lower semicontinuous functional defined by

$$\Psi(v,t) = \psi(v), \quad \forall (v,t) \in V \times I\!R.$$

In order to motivate the expression of the functional f in (9.68) we stress that the critical points of f (in the sense of Section 2 of Chapter 3) will determine the eigensolutions to problem (EP) with the required properties. According to assumptions (I_1) and (β_1) we see from (9.68) that

$$f(0,0) = f(0,r) = 0. \tag{9.71}$$

By assumptions (I_2) and (β_2) in conjunction with (9.68) and (9.62) we obtain the estimate

$$f(v,\rho) \geq \left(\frac{1}{q}\rho^{q+1} m^{\frac{q}{2}} - a_2\right) |v|_V^q + \frac{q+1}{q}\beta(\rho) - a_1 \tag{9.72}$$

$$\geq \frac{q+1}{q}\beta(\rho) - a_1, \quad \forall v \in V.$$

We claim that the functional $f : V \times I\!R \to I\!R \cup \{+\infty\}$ satisfies the Palais-Smale condition $(\mathrm{PS})_{f,c}$ for any $c \in I\!R$ in the sense of Definition 3.3. Towards this let $\{(u_n, t_n)\} \subset V \times I\!R$ be a sequence such that there is a constant $M > 0$ with

$$|f(u_n,t_n)| \leq M, \quad \forall n \geq 1 \tag{9.73}$$

and there is a sequence $\{\varepsilon_n\} \subset I\!R^+$, $\varepsilon_n \to 0^+$, for which one has

$$\Phi^0(u_n,t_n; v-u_n, t-t_n) + \psi(v) - \psi(u_n)$$

$$\geq -\varepsilon_n(|v-u_n|_V + |t-t_n|), \quad \forall (v,t) \in V \times I\!R. \tag{9.74}$$

In view of relations (9.73) and (9.68) we derive that

$$M \geq |\frac{1}{q}|t_n|^{q+1}(Au_n,u_n)_V^{\frac{q}{2}} + \frac{q+1}{q}\beta(t_n)$$

$$-J(u_n) + \psi(u_n) + \frac{\bar{\lambda}}{2}(Au_n,u_n)_V|. \tag{9.75}$$

Combining (9.75) and assumption (I_2) we get the estimate

$$M \geq \left(\frac{1}{q}|t_n|^{q+1} m^{\frac{q}{2}} - a_2\right) |u_n|_V^q + \frac{q+1}{q}\beta(t_n) - a_1, \quad \forall n \geq 1. \tag{9.76}$$

In view of the coerciveness condition (β_3) it is clear that (9.76) implies the boundedness of $\{t_n\}$ in $I\!R$.

Two cases can appear:

Case 1. The sequence $\{t_n\}$ fulfills $t_n \to 0$ in $\mathbb{R}$ as $n \to \infty$.
By (9.74) and (9.70) we know that

$$|t_n|^{q+1}\,(Au_n,u_n)_V^{\frac{q}{2}-1}\,(Au_n,v-u_n)_V + (-J)^0(u_n;v-u_n)$$

$$+\bar{\lambda}(Au_n,v-u_n)_V + \frac{q+1}{q}(|t_n|^{q-1}t_n(Au_n,u_n)_V^{\frac{q}{2}} + \beta'(t_n))(t-t_n)$$

$$+\psi(v)-\psi(u_n) \geq -\varepsilon_n(|v-u_n|_V + |t-t_n|), \quad \forall\,(v,t)\in V\times\mathbb{R}. \tag{9.77}$$

Setting $v = u_n$ in (9.77) it turns out that

$$\frac{q+1}{q}(|t_n|^{q-1}t_n(Au_n,u_n)_V^{\frac{q}{2}} + \beta'(t_n))\,t \geq -\varepsilon_n|t|, \quad \forall t\in\mathbb{R}$$

and, therefore,

$$|\,|t_n|^{q-1}t_n(Au_n,u_n)_V^{\frac{q}{2}} + \beta'(t_n)| \leq \varepsilon_n. \tag{9.78}$$

Since $\varepsilon_n \to 0$ as $n\to\infty$, we infer from (9.78) that

$$|t_n|^{q-1}t_n(Au_n,u_n)_V^{\frac{q}{2}} + \beta'(t_n) \to 0 \quad \text{as} \quad n\to\infty. \tag{9.79}$$

Setting now $t = t_n$ in (9.77) we arrive at

$$(|t_n|^{q+1}\,(Au_n,u_n)_V^{\frac{q}{2}-1} + \bar{\lambda})(Au_n,v-u_n)_V + (-J)^0(u_n;v-u_n)$$

$$+\psi(v)-\psi(u_n) \geq -\varepsilon_n\,|v-u_n|_V\,, \quad \forall\, v\in V.$$

This implies that

$$(-J)^0(u_n;v-u_n)+\psi(v)-\psi(u_n)+\bar{\lambda}(Au_n,v-u_n)_V$$

$$\geq -(\varepsilon_n + |t_n|^{q+1}(Au_n,u_n)_V^{\frac{q}{2}-1}|Au_n|_V)|v-u_n|_V. \tag{9.80}$$

Using that $t_n \to 0$ and (9.79) we derive

$$t_n(Au_n,u_n)_V^{\frac{1}{2}} \to 0 \quad \text{as} \quad n\to\infty. \tag{9.81}$$

Due to (9.62), property (9.81) is equivalent to

$$t_n\,u_n \to 0 \text{ in } V \text{ as } n\to\infty. \tag{9.82}$$

Then (9.75) and (9.81) yield

$$-J(u_n)+\psi(u_n)+\frac{\bar{\lambda}}{2}\,(Au_n,u_n)_V \;\text{ is bounded.} \tag{9.83}$$

On the other hand, relations (9.81), (9.82) and $t_n \to 0$ ensure

$$|t_n|^{q+1}(Au_n, u_n)_V^{\frac{q}{2}-1}\ |Au_n|_V \to 0 \quad \text{as} \quad n \to \infty.$$

Therefore, on the basis of (9.80) and because $\varepsilon_n \to 0$, we are allowed to make use of assumption (I_4) in conjunction with (9.83) to deduce that

$$\{u_n\} \text{ contains a bounded subsequence.} \tag{9.84}$$

By (9.84) we may suppose that along a subsequence and for certain $u \in V$ one has

$$u_n \rightharpoonup u \ \text{ weakly in } V \ \text{ as } \ n \to \infty. \tag{9.85}$$

Let us put $v = u$ and $t = 0$ in (9.77). It results the inequality

$$\begin{aligned} |t_n|^{q+1}\ (Au_n, u_n)_V^{\frac{q}{2}-1}(Au_n, u - u_n)_V + (-J)^0(u_n; u - u_n) \\ +\bar{\lambda}(Au_n, u-u_n)_V - \frac{q+1}{q}(|t_n|^{q-1}t_n(Au_n, u_n)_V^{\frac{q}{2}} + +\beta'(t_n))t_n + \psi(u) - \psi(u_n) \\ \geq -\varepsilon_n(|u - u_n|_V + |t_n|), \quad \forall n \geq 1. \end{aligned} \tag{9.86}$$

We pass to the limit in (9.86) as $n \to \infty$ by using $t_n \to 0$, $\varepsilon_n \to 0$, property (9.85), the lower semicontinuity of ψ and assumption (I_3). We get

$$\begin{aligned} 0 &\leq \limsup_{n\to+\infty}\Big((-J)^0(u_n; u - u_n) + \bar{\lambda}(Au_n, u - u_n)_V\Big) \\ &\leq -\bar{\lambda}\liminf_{n\to+\infty}\ (A(u_n - u), u_n - u)_V. \end{aligned}$$

Along a subsequence and using (9.62) we have that

$$\bar{\lambda}m \limsup_{n\to\infty} |u_n - u|_V^2 \leq \bar{\lambda}m \lim_{n\to\infty}(A(u_n - u), u_n - u)_V \leq 0.$$

In view of (9.85) this amounts to saying that $\{u_n\}$ possesses a strongly convergent subsequence in V. Hence, the Palais-Smale condition is verified for the functional $f : V \times I\!R \to I\!R \cup \{+\infty\}$ of (9.68) in Case 1.

Case 2. Suppose now that the sequence $\{t_n\} \subset I\!R$ is bounded away from 0 along a subsequence, i.e. there is a constant $b > 0$ with

$$|t_n| \geq b, \quad \forall\, n \geq 1, \tag{9.87}$$

where in fact a relabelled subsequence has been considered.

Firstly, we note that property (9.79) does not depend on the fact whether $t_n \to 0$ or not. Thus (9.79) can be used also in Case 2. It ensures, in conjunction with (9.87), that $(Au_n, u_n)_V$ is bounded in V, so $\{u_n\}$ is bounded in V. Consequently, we may admit that (9.85) holds for some $u \in V$ and $t_n \to t$ as $n \to \infty$.

Setting $v = u$ in (9.77) (which is also independent of the cases 1 and 2 for $\{t_n\}$) and then letting $n \to \infty$ in (9.77), by (I_3) and (9.85) one sees that

$$\begin{aligned} 0 &\leq \limsup_{n\to\infty} \Big[\Big(|t_n|^{q+1}(Au_n, u_n)_V^{\frac{q}{2}-1} + \bar{\lambda}\Big)(Au_n, u - u_n)_V\Big] \\ &= -\liminf_{n\to\infty} \Big[\Big(|t_n|^{q+1}\,(Au_n, u_n)_V^{\frac{q}{2}-1} + \bar{\lambda}\Big)(A(u_n - u), u_n - u)_V\Big]. \end{aligned}$$

It follows that along a renamed subsequence one has

$$\begin{aligned} 0 &\geq \lim_{n\to\infty} \Big[\Big(|t_n|^{q+1}(Au_n, u_n)_V^{\frac{q}{2}-1} + \bar{\lambda}\Big)(A(u_n - u), u_n - u)_V\Big] \\ &\geq \bar{\lambda}\, m\, \limsup_{n\to\infty} |u_n - u|_V^2, \end{aligned}$$

where (9.62) has been used, too. This, in conjunction with (9.85), guarantees that $\{u_n\}$ possesses a strongly convergent subsequence in V.

Therefore the functional $f : V \times I\!R \to I\!R \cup \{+\infty\}$ introduced in (9.68) satisfies the Palais-Smale condition in the sense of Definition 3.3. Since the functional f in (9.68) verifies the properties (9.69), (9.71), (9.72) and the Palais-Smale condition as stated in Definition 3.3, we are in a position to apply Lemma 9.1 to the functional $f : V \times I\!R \to I\!R \cup \{+\infty\}$ in (9.68). We find that there exists $(u, s) \in V \times I\!R$ such that

$$\begin{aligned} &|s|^{q+1}(Au, u)_V^{\frac{q}{2}-1}(Au, v - u)_V + (-J)^0(u; v - u) + \bar{\lambda}(Au, v - u)_V \\ &\quad + \frac{q+1}{q}\Big(|s|^{q-1}s(Au, u)_V^{\frac{q}{2}} + \beta'(s)\Big)(t - s) \\ &\quad + \psi(v) - \psi(u) \geq 0, \quad \forall\, (v, t) \in V \times I\!R \end{aligned} \tag{9.88}$$

and

$$\frac{q+1}{q}\beta(\rho) - a_1 \leq f(u, s) \leq \max_{0\leq t\leq 1} f(0, tr) = \frac{q+1}{q}\beta(\rho) \tag{9.89}$$

(see also the second relation in (β_2)).

Setting $t = s$ and $v = u$ in (9.88), it can be readily seen that (9.88) gives rise to the next relations

$$\begin{aligned} &|s|^{q+1}\,(Au, u)_V^{\frac{q}{2}-1}(Au, v - u)_V + (-J)^0(u; v - u) \\ &\quad + \bar{\lambda}(Au, v - u)_V + \psi(v) - \psi(u) \geq 0, \quad \forall\, v \in V \end{aligned} \tag{9.90}$$

and

$$|s|^{q-1}s\,(Au,u)_V^{\frac{q}{2}} + \beta'(s) = 0. \tag{9.91}$$

Equality (9.91) yields

$$s\beta'(s) \le 0. \tag{9.92}$$

If $s = 0$, then formulas (9.90) and (9.89) imply (9.63) and (9.64), respectively. This means that part (i) in the alternative is valid.

We proceed now by noting that conditions (β_4) and (9.92) show that the case $s < 0$ cannot occur. Consequently, only the case $s > 0$ remains to be considered. If $s > 0$, relation (9.92) yields that $\beta'(s) \le 0$. Then conditions (β_4) and (β_5) insure that $\rho < s < r$ or $s = \rho$ or $s = r$.

When $s = \rho$ or $s = r$, from (9.91) and (9.62) it results that $u = 0$. Then (9.90) contradicts the final part of assumption (I_1). Thus we established that

$$\rho < s < r. \tag{9.93}$$

Let us then denote

$$\lambda = \bar{\lambda} + s^{q+1}(Au,u)_V^{\frac{q}{2}-1}. \tag{9.94}$$

By means of (9.94) one can express (9.90) in the following form

$$\lambda\,(Au, v-u)_V + (-J)^0(u; v-u) + \psi(v) - \psi(u) \ge 0, \quad \forall\, v \in V. \tag{9.95}$$

Comparing (9.95) with assumption (I_1) we derive that $u \neq 0$. According to (9.95) it is thus shown that the pair $(u, \lambda) \in V \times \mathbb{R}$, with λ given by (9.94), is an eigensolution of problem (EP).

Relations (9.91) and (9.93) lead to

$$(Au,u)_V^{\frac{q}{2}} = -s^{-q}\beta'(s).$$

This represents just formula (9.65) in the statement of Theorem 9.3. Combining (9.94) and (9.65) we obtain relation (9.66) in Theorem 9.3. Finally, the energy estimate in (9.67) is a consequence of (9.89). Part (ii) in the alternative is thus established. This completes the proof of Theorem 9.3. ■

Remark 9.2 (i) Assumption (I_2) is in fact a condition only for J because always ψ is bounded from below by an affine function.

(ii) Assumption (I_3) is weaker than the corresponding assumption in [15], [17]. Indeed, if $u_n \rightharpoonup u$ weakly in V, then $\{u_n\}$ is bounded in V and $(-J)^0(u_n; u - u_n) = (w_n, u_n - u)_V$ for some $w_n \in \partial J(u_n)$ (see [6], p. 27). Then (I_3) in [15], [17] implies that along a subsequence one has

$$\limsup_{n\to\infty}(-J)^0(u_n; u-u_n) = \limsup_{n\to\infty}(w_n, u_n - u)_V$$

$$= \lim_{n\to\infty} (w_n, u_n - u)_V = 0,$$

so our assumption (I_3) holds true.

Remark 9.3 Theorem 9.3 can be regarded as a result providing the location of eigensolutions (u, λ) to problem (EP) by means of the graph of the function β'. Theorem 9.3 can be also seen as a parametric representation of the eigensolution (u, λ) with the parameter s (see formulas (9.65), (9.66)).

We now illustrate the applicability of Theorem 9.3 in the study of qualitative properties of eigensolutions (u, λ) to problem (EP) by presenting a density result (Corollary 9.1) and an asymptotic result (Corollary 9.2).

Corollary 9.1 Under assumptions (I_1)-(I_4) of Theorem 9.3 we suppose that the given number $\bar{\lambda} > 0$ is not an eigenvalue λ of problem

$$\begin{cases} \lambda(Au, v-u)_V + (-J)^0(u; v-u) + \psi(v) - \psi(u) \geq 0, \quad \forall v \in V, \\ u \in V \setminus \{0\}. \end{cases}$$

Then, for all numbers $0 < \rho < r$ with $m^{\frac{q}{2}} \rho^{q+1} \geq a_2 q$, there exists an eigensolution (u, λ) of problem (P') such that

$$\rho^{q+1} < \frac{\lambda - \bar{\lambda}}{(Au, u)_V^{\frac{q}{2}-1}} < r^{q+1}. \tag{9.96}$$

In particular, with the notations in (9.96), one has

$$\lim_{r\to\rho} \frac{\lambda - \bar{\lambda}}{(Au, u)_V^{\frac{q}{2}-1}} = \rho^{q+1}.$$

Proof. For each pair of positive numbers $\rho < r$ as required in the statement of Corollary 9.1 let us choose some function $\beta \in C^1(I\!R)$ verifying the conditions (β_1)-(β_5). We apply Theorem 9.3 with this function β and the positive numbers ρ and r. Since, by hypothesis, part (i) in the alternative cannot happen, we conclude that an eigensolution $(u, \lambda) \in V \times I\!R$ of problem (EP) can be found such that the properties in part (ii) of the alternative are fulfilled. Moreover, let us consider the relation (9.94) which was established in the proof of Theorem 9.3. The fact that $s \in]\rho, r[$ (see (9.93)) and relation (9.94) enable us to achieve (9.96). The proof of Corollary 9.1 is thus complete. ∎

Corollary 9.2 Under the same assumptions as in Corollary 9.1, there exists a sequence $\{(u_n, \lambda_n)\}$ of eigensolutions of problem (EP) such that

$$u_n \to 0 \text{ in } V \quad \text{as} \quad n \to \infty, \tag{9.97}$$

$$\lambda_n \to +\infty \quad \text{as} \quad n \to \infty, \tag{9.98}$$

$$\lambda_n |u_n|_V^2 \to 0 \quad \text{as} \quad n \to \infty. \tag{9.99}$$

Proof. For every number $\varepsilon > 0$ we can find a function $\beta_\varepsilon \in C^1(I\!R)$ and positive numbers $\rho_\varepsilon < r_\varepsilon$ with

$$\rho_\varepsilon \to +\infty \quad \text{as} \quad \varepsilon \to 0 \tag{9.100}$$

such that the conditions (β_1)-(β_5) be satisfied for $\beta = \beta_\varepsilon$, $\rho = \rho_\varepsilon$, $r = r_\varepsilon$ and, in addition, to fulfill the requirement

$$|\beta_\varepsilon'(t)| \leq \varepsilon t^{-1}, \quad \forall\, t \geq \rho_\varepsilon. \tag{9.101}$$

We point out that the choice in (9.101) is compatible with the coerciveness condition (β_3). The construction in (9.101) can be locally accomplished because

$$|\beta_\varepsilon'(\rho_\varepsilon)| = 0 < \frac{\varepsilon}{\rho_\varepsilon}$$

and then we proceed by integration in (9.101) for $t > 0$ sufficiently large.

Theorem 9.3 applied for each choice $\beta = \beta_\varepsilon$, $\rho = \rho_\varepsilon$, $r = r_\varepsilon$, with an arbitrarily small $\varepsilon > 0$, supplies an eigensolution $(u_\varepsilon, \lambda_\varepsilon) \in V \times I\!R$ of problem (EP) satisfying

$$(Au_\varepsilon, u_\varepsilon)_V^{\frac{1}{2}} = s_\varepsilon^{-1}(-\beta_\varepsilon'(s_\varepsilon))^{\frac{1}{q}} \tag{9.102}$$

and

$$\lambda_\varepsilon = \bar{\lambda} + s_\varepsilon^3(-\beta_\varepsilon'(s_\varepsilon))^{\frac{q-2}{q}}. \tag{9.103}$$

By (9.62), (9.102) and (9.101) we infer that

$$\begin{aligned} m^{\frac{1}{2}}|u_\varepsilon|_V &\leq (Au_\varepsilon, u_\varepsilon)_V^{\frac{1}{2}} = s_\varepsilon^{-1}(-\beta_\varepsilon'(s_\varepsilon))^{\frac{1}{q}} \\ &\leq \varepsilon^{\frac{1}{q}} s_\varepsilon^{-\frac{q+1}{q}}. \end{aligned}$$

Since $s_\varepsilon > \rho_\varepsilon$ we get from (9.100) that property (9.97) holds.

Arguing by contradiction let us suppose that (9.98) is not valid, so we admit that $\lambda_\varepsilon > \bar{\lambda}$ is bounded as $\varepsilon \to 0$. We know that $(u_\varepsilon, \lambda_\varepsilon) \in V \times I\!R$ solves problem (EP), which means that one has

$$\lambda_\varepsilon(Au_\varepsilon, v - u_\varepsilon)_V + (-J)^0(u_\varepsilon; v - u_\varepsilon) + \psi(v) - \psi(u_\varepsilon) \geq 0, \quad \forall v \in V.$$

Then we pass to the limit as $\varepsilon \to 0$ in the inequality above by using (9.97), the boundedness of λ_ε and the upper semicontinuity of $-\psi$ and $(-J)^0$ (see [6], p. 26). We arrive at

$$(-J)^0(0; v) + \psi(v) - \psi(0) \geq 0, \quad \forall v \in V.$$

This contradicts the final assumption in (I_1). The achieved contradiction justifies that, along a subsequence, property (9.98) holds true.

On the basis of (9.103), (9.102) and (9.101) it results that

$$0 < (\lambda_\varepsilon - \bar{\lambda})\,(Au_\varepsilon, u_\varepsilon)_V = s_\varepsilon(-\beta'_\varepsilon(s_\varepsilon)) \leq \varepsilon.$$

Since (9.97) was already shown, we may conclude that

$$\lambda_\varepsilon(Au_\varepsilon, u_\varepsilon)_V \to 0 \quad \text{as} \quad \varepsilon \to 0.$$

According to (9.62) this is equivalent to (9.99), which is thus established. The proof of Corollary 9.2 is thereby complete. ∎

Remark 9.4 (i) In the case where $\psi = 0$ and $J \in C^1(V; \mathbb{R})$ the above results have been obtained in [15]. In the case where $\psi = 0$ and $J : V \to \mathbb{R}$ is locally Lipschitz these results have been established in [17]. Theorem 9.3 and Corollaries 9.2, 9.3 are from [1]. In fact, just for these particular cases the stated results slightly improve the previous mentioned results.

(ii) The rate of convergence in (9.99) cannot be improved to $\lambda_n |u_n|_V \to 0$ as $n \to \infty$ together with (9.97) and (9.98) (see [1]).

(iii) An important situation for asymptotically linear elliptic boundary value problems which is related to the property (9.97) in Corollary 9.2 has been pointed out in [4] and [13] (see also [1]).

3. Location of Solutions to Nonlinear Dirichlet Problems

This Section is devoted to the application of the abstract results in Section 2 to nonlinear eigenvalue problems in terms of variational-hemivariational inequalities. To this end, let V be a real Hilbert space with a compact and dense embedding

$$V \subset L^q(\Omega) \tag{9.104}$$

for a bounded domain Ω in $\mathbb{R}^N$ and for some $q \geq 2$. Let $A : V \to V$ be a symmetric, linear, continuous operator satisfying the coerciveness assumption (9.62), let $g \in V$ and let the convex and closed subset K of V given by

$$K := \{v \in V \;:\; v \geq 0 \text{ a.e. in } \Omega\}. \tag{9.105}$$

Let $j : \Omega \times \mathbb{R} \to \mathbb{R}$ be a Carathéodory function $j(x,t)$ which is locally Lipschitz with respect to the second variable $t \in \mathbb{R}$.

For a prescribed number $\bar{\lambda} > 0$ we formulate the following nonlinear and nonsmooth eigenvalue problem with constraints:

$(EP)'$ *Find* $u \in K \setminus \{0\}$ *and* $\lambda \in \mathbb{R}$ *such that*

$$\begin{cases} \lambda(Au, v-u)_V + \int_\Omega (-j)_t^0(x, u(x); v(x) - u(x))dx \\ \qquad\qquad \geq (g, v-u)_V\,, \quad \forall v \in K, \\ \\ \lambda > \bar{\lambda}. \end{cases}$$

In the statement of problem $(EP)'$ the notation $(-j)_t^0$ stands for the generalized directional derivative of function $-j(x,t)$ with respect to t-variable. Problem $(EP)'$ is a variational-hemivariational inequality in the sense of Panagiotopoulos [19], [23], where we added the constraint $\lambda > \bar{\lambda}$ for the eigenvalue unknown λ. The inequality in problem $(EP)'$ contains in particular the case of variational inequalities (see, e.g., [3]). General differential inclusion problems of the type above can be found in [8].

For the function $j : \Omega \times \mathbb{R} \to \mathbb{R}$ we further impose the hypotheses below:

(J_1) $j(x,0) = 0$, $x \in \Omega$, and $-g \notin \int_\Omega \partial_t j(x,0)dx$, where the integral term in the right-hand side of the relation above is considered in the sense of Clarke [6], p. 83 (which is guaranteed by (9.104) and the growth condition in assumption (J_2) below), that is, there are $v \in V$ and a selection $\xi : \Omega \to L^{\frac{q}{q-1}}(\Omega)$ of $x \mapsto \partial_t j(x,0)$ such that $-(g,v)_V \neq \int_\Omega \xi(x)v(x)dx$;

(J_2) there is a constant $C > 0$ such that $|z| \leq C(1 + |t|^{q-1})$ for a.e. $x \in \Omega$, $\forall\, t \in \mathbb{R}$, $\forall\, z \in \partial_t j(x,t)$;

(J_3) there exist constants $c_1 \geq 0$, $c_2 \geq 0$, $1 \leq \sigma < 2$ and $\mu > 2$ such that

$$\frac{1}{\mu}(-j)^0(x,t;t) + j(x,t) \leq c_1 + c_2 t^\sigma$$

for a.e. $x \in \Omega$, $\forall\, t \geq 0$.

In order to apply Theorem 9.3 as well as Corollaries 9.1 and 9.2 we introduce the locally Lipschitz functional $\tilde{J} : L^q(\Omega) \to \mathbb{R}$ by

$$\tilde{J}(v) = \int_\Omega j(x, v(x))dx, \quad \forall v \in L^q(\Omega) \tag{9.106}$$

and then the locally Lipschitz functional $J : V \to \mathbb{R}$ equal to

$$J = \tilde{J}|_V - (g, \cdot)_V. \tag{9.107}$$

Corresponding to the set K in (9.105) we take $\psi = I_K$ which denotes the indicator function of K, i.e.,

$$I_K(v) = \begin{cases} 0 & \forall v \in K, \\ +\infty & \text{otherwise.} \end{cases}$$

Clearly, $\psi : V \to I\!R \cup \{+\infty\}$ is proper, convex and lower semicontinuous.

Let us check that assumptions (I_1)-(I_4) are satisfied. Assumption (J_2) ensures that Theorem 2.7.5 in [6] can be applied to derive that

$$\partial J(v) \subset \int_\Omega \partial_t j(x, v(x))dx, \quad \forall v \in L^q(\Omega), \tag{9.108}$$

where the inclusion is understood in $L^{\frac{q}{q-1}}(\Omega)$ in a sense discussed in the statement of (J_1). Then we see from (J_1) that condition (I_1) is verified.

A straightforward computation based on the growth condition (J_2), the first part in (J_1), Lebourg's mean value theorem (see [6] p. 41) and (9.104) show that

$$\begin{aligned} |J(v)| &\leq \textstyle\int_\Omega |j(x, v(x)|dx + |g|_V |v|_V \\ &\leq C \textstyle\int_\Omega (1 + |v(x)|^{q-1})|v(x)|dx + |g|_V |v|_V \\ &\leq b_1 + b_2 |v|_V^q, \quad \forall v \in V, \end{aligned}$$

for constants $b_1 \geq 0$, $b_2 \geq 0$. It follows that condition (I_2) is verified.

To justify assumption (I_3) let $\{u_n\} \subset V$ and $u \in V$ be such that $u_n \rightharpoonup u$ weakly in V as $n \to \infty$. The compactness of the embedding (9.104) allows to admit that $u_n \to u$ strongly in $L^q(\Omega)$ along a subsequence denoted again by $\{u_n\}$. Using (9.106), (9.107) and the upper semicontinuity of $\tilde{J}^0$ one obtains that

$$\limsup_{n\to\infty} J^0(u_n; u - u_n) \leq \limsup_{n\to\infty} [\tilde{J}^0(u_n; u - u_n) - (g, u - u_n)_V] = 0.$$

Hence assumption (I_3) is satisfied.

It remains to verify assumption (I_4). To this end let $\{u_n\} \subset K$ be such that

$$|- J(u_n) + \frac{\bar{\lambda}}{2} (Au_n, u_n)_V| \leq M, \quad \forall n \geq 1$$

and

$$(-J)^0(u_n; v - u_n) + \bar{\lambda}\, (Au_n, v - u_n)_V \geq -\varepsilon_n |v - u_n|_V, \quad \forall v \in K, \tag{9.109}$$

for a sequence $\{\varepsilon_n\} \subset I\!R^+$ with $\varepsilon_n \to 0$ as $n \to \infty$. We may assume that $\varepsilon_n \leq 1$. Setting $v = 2u_n$ in (9.109) and in conjunction with (b) in Proposition 2.12 of [6], we derive that

$$\begin{aligned} &\int_\Omega (-j)_t^0(x, u_n(x); u_n(x))dx - (g, u_n)_V \\ &+\bar{\lambda}(Au_n, u_n)_V \geq -|u_n|_V, \quad \forall n \geq 1. \end{aligned}$$

Then (9.108), (9.109), (9.104), (9.62) and assumption (J_3) yield

$$M + \frac{1}{\mu}|u_n|_V \geq \left(\frac{1}{2} - \frac{1}{\mu}\right) \bar{\lambda}(Au_n, u_n)_V -$$

$$- \int_\Omega \left[j(x, u_n(x)) + \frac{1}{\mu}(-j)^0(x, u_n(x); u_n(x))\right] dx + (1 + \frac{1}{\mu})(g, u_n)_V$$

$$\geq \left(\frac{1}{2} - \frac{1}{\mu}\right) \bar{\lambda} m |u_n|_V^2 - \int_\Omega (c_1 + c_2(u_n(x))^\sigma) dx - (1 + \frac{1}{\mu})|g|_V \, |u_n|_V$$

$$\geq \left(\frac{1}{2} - \frac{1}{\mu}\right) \bar{\lambda} m |u_n|_V^2 - (1 + \frac{1}{\mu})|g|_V \, |u_n|_V - c_1|\Omega| - c|u_n|_V^\sigma, \quad \forall n \geq 1$$

with a constant $c > 0$. Since $\mu > 2$ and $1 \leq \sigma < 2$, we deduce that the sequence $\{u_n\}$ is bounded in V. Therefore assumption (I_4) holds, too. Thus we can apply to the functional J in (9.107) and $\psi = I_K$ our results in Theorem 9.3 and Corollaries 9.1 and 9.2. Using (9.107) it is easy to see that these results lead to corresponding properties for the eigensolutions of problem $(EP)'$. Indeed, in the framework of our functionals J (given in (9.107)) and $\psi = I_K$, every eigensolution (u, λ) of problem (EP) becomes an eigensolution (u, λ) of problem $(EP)'$. Consequently, the qualitative properties of the eigensolutions of problem (EP) supplied by our results in Theorem 9.3 and Corollaries 9.1 and 9.2 can be transferred to the set of eigensolutions of problem $(EP)'$.

We end this chapter with some comments on assumptions (J_1)-(J_3) as well as on the relationship between our theory and a number of results which can be find in the literature.

Remark 9.5 (i) The example above covers the situation of eigenvalue problems for the semilinear elliptic boundary value problems

$$\begin{cases} -\lambda \Delta u = h(x, u) + g(x), & \forall x \in \Omega, \\ u = 0 \text{ on } \partial\Omega \\ \lambda > \bar{\lambda}, \end{cases}$$

with appropriate assumptions for h, taking $V = H_0^1(\Omega)$ as studied in [15], and

$$\begin{cases} \lambda(Au, v)_V \leq \int_\Omega j^0(x, u(x); v(x)) dx + (g, v)_V, & \forall v \in V, \\ \lambda > \bar{\lambda} \end{cases}$$

as studied in [17]. The variational method developed in Sections 2 and 3 has been applied in [6], [7] and [20] to nonlinear Dirichlet boundary value problems involving the p-Laplacian.

(ii) An interesting situation in the setting of semilinear elliptic boundary value problem where (I_1) (and, specifically, (J_1)) holds is provided in [4] and [13].

(iii) Different other choices for the set of constraints K are possible in place of (9.105). For instance, one can consider an obstacle or a constraint involving the gradient.

(iv) Condition (J_3) extends, just in the case where $j(x,t)$ is smooth, the celebrated assumption (p_5) in [2] or (p_4) in [24] by dropping the sign condition for $j(x,t)$ and by permitting to cover the subquadratic case (in addition to the superquadratic case) due to the presence of the term $c_1 + c_2|t|^\sigma$ with $1 \leq \sigma < 2$ in the growth condition (J_2). For example, taking $j(x,t) = |t|^p$, we see that (J_3) is satisfied for any $0 \leq p < 2$ and $p > 2$. It is readily seen that for $p = 2$ condition (J_3) is not satisfied.

(v) Conditions (I_1)-(I_4) are mutually independent. If only one is missing, then generally the result does not hold. For example, let $A = id : \mathbb{R}^2 \to \mathbb{R}^2$, $J : \mathbb{R}^2 \to \mathbb{R}$ be given by

$$J(x,y) = -x^2 - 2y^2 - x, \quad \forall (x,y) \in \mathbb{R}^2,$$

and $\psi = I_K$, with

$$K = \mathbb{R}^2_+ = \{(x,y) \in \mathbb{R}^2 : x, y \geq 0\}.$$

Then problem (EP) (or $(EP)'$) reads

$$\begin{cases} (\nabla J(u) - \lambda u, v - u)_V \leq 0, & \forall v \in K, \\ \lambda > \bar{\lambda} \end{cases}$$

with some $\bar{\lambda} > 0$. A direct computation shows that the problem above does not admit a solution. We observe that all assumptions (I_2)-(I_4) are satisfied, while hypothesis (I_1) is violated because

$$(-\nabla J(0), v)_V = x \geq 0, \quad \forall v = (x,y) \in K.$$

(vi) Assertion (9.99) in Corollary 9.2 cannot be generally improved to

$$\lambda_n |u_n|_V \to 0, \quad \text{as} \quad n \to \infty$$

(keeping (9.97) and (9.98) as stated). For example, let $J : \mathbb{R}^2 \to \mathbb{R}$ be defined by

$$J(x,y) = -x^2 - 2y^2 + x, \quad \forall (x,y) \in \mathbb{R}^2,$$

and the data A and K (thus ψ) be as in (v) of the present remark. For our problem (EP) we find that the eigensolutions $(u = u_\lambda, \lambda)$ (with $\lambda > \bar{\lambda} > 0$) are $u_\lambda = \left(\frac{1}{\lambda+2}, 0\right) \in K$ and $\lambda > \bar{\lambda} > 0$. We see that for any $\lambda_n \to +\infty$ we have $u_{\lambda_n} \to 0$ and $\lambda_n |u_{\lambda_n}|_V \to 1$ as $n \to \infty$.

References

[1] S. Adly and D. Motreanu, Location of eigensolutions to variational-hemivariational inequalities, *J. Nonlinear Convex Anal.* **1** (2000), 255-270.

[2] A. Ambrosetti and P. H. Rabinowitz, Dual variational methods in critical point theory and applications, *J. Func. Anal.* **14** (1973), 349-381.

[3] C. Baiocchi and A. Capelo, *Variational and Quasivariational Inequalities. Applications to Free Boundary Problems*, John Wiley and Sons, New York, 1984.

[4] H. Brézis and L. Nirenberg, *Functional Analysis and Applications to Partial Differential Equations*, in preparation.

[5] K. C. Chang, Variational methods for non-differentiable functionals and their applications to partial differential equations, *J. Math. Anal. Appl.* **80** (1981), 102-129.

[6] F. H. Clarke, *Optimization and Nonsmooth Analysis*, John Wiley & Sons, New York (1983).

[7] S. Dăbuleanu and D. Motreanu, Existence results for a class of eigenvalue quasilinear problems with nonlinear boundary condition, *Adv. Nonlinear Var. Inequal.* **2** (1999), 41-54.

[8] G. Dincă, P. Jebelean and D. Motreanu, Existence and approximation for a general class of differential inclusions, *Houston J. Math.* **28** (2002), 193-215.

[9] D. Goeleven, D. Motreanu, Y. Dumont and M. Rochdi, *Variational and Hemivariational Inequalities, Theory, Methods and Applications*, Volume I: Unilateral Analysis and Unilateral Mechanics, Kluwer Academic Publishers, Dordrecht / Boston / London, to appear.

[10] D. Goeleven and D. Motreanu, *Variational and Hemivariational Inequalities, Theory, Methods and Applications*, Volume II: Unilateral Problems, Kluwer Academic Publishers, Dordrecht / Boston / London, to appear.

[11] J. Haslinger, M. Miettinen and P. D. Panagiotopoulos, *Finite Element Method for Hemivariational Inequalities. Theory, Methods and Applications*, Kluwer Academic Publishers, Nonconvex Optimization and Its Applications, Vol. 35, Dordrecht / Boston / London, 1999.

[12] J. Haslinger and D. Motreanu, Hemivariational inequalities with a general growth condition: existence and approximation, *Appl. Anal.*, to appear.

[13] P. Mironescu and V. Rădulescu, The study of a bifurcation problem associated to an asymptotically linear function, *Nonlinear Anal.* **26** (1996), 857-875.

[14] E. S. Mistakidis and G. E. Stavroulakis, *Nonconvex Optimization in Mechanics*, Nonconvex Optimization and Its Applications, Vol. 21, Kluwer Academic Publishers, Dordrecht / Boston / London, 1998.

[15] D. Motreanu, A saddle-point approach to nonlinear eigenvalues problems, *Math. Slovaca* **47** (1997), 463-477.

[16] D. Motreanu, Eigenvalue problems for variational-hemivariational inequalities in the sense of P. D. Panagiotopoulos, *Nonlinear Anal.* **47** (2001), 5101-5112.

[17] D. Motreanu, Location of solutions to eigenvalue problems for hemivariational inequalities, Chapter 12, p. 263-276 in: *Nonsmooth/Nonconvex Mechanics: Modeling, Analysis and Numerical Methods, A Volume dedicated to the memory of Professor P.D. Panagiotopoulos*, D. Gao, R. W. Ogden and G. E. Stavroulakis (eds.), Kluwer Academic Publishers, Dordrecht, Boston, London, 2001.

[18] D. Motreanu, Existence and multiplicity results for variational-hemivariational inequalities in the sense of P. D. Panagiotopoulos, in: Proceedings of the International Conference on Nonsmooth/Nonconvex Mechanics with Applications in Engineering, In memoriam of Professor P. D. Panagiotopoulos, 5-6 July 2002, Thessaloniki, Greece, pp. 23-30.

[19] D. Motreanu and P. D. Panagiotopoulos, *Minimax Theorems and Qualitative Properties of the Solutions of Hemivariational Inequalities and Applications*, Kluwer Academic Publishers, Nonconvex Optimization and Its Applications, Vol. 29, Dordrecht/Boston/London, 1999.

[20] D. Motreanu and V. Radulescu, Existence theorems for some classes of boundary value problems involving the p-Laplacian, *Panam. Math. J.* **7** (1997), 53-66.

[21] Z. Naniewicz and P. D. Panagiotopoulos, *Mathematical Theory of Hemivariational Inequalities and Applications*, Marcel Dekker, Inc., New York (1995).

[22] P. D. Panagiotopoulos, *Inequality Problems in Mechanics and Applications. Convex and Nonconvex Energy Functions*, Birkhäuser Verlag, Basel, 1985.

[23] P. D. Panagiotopoulos, *Hemivariational Inequalities. Applications in Mechanics and Engineering*, Springer-Verlag, Berlin, New York, 1993.

[24] P.H. Rabinowitz, *Minimax Methods in Critical Point Theory with Applications to Differential Equations, CBMS Reg. Conf. Ser. Math.* **65**, Amer. Math. Soc., Providence, R.I., 1996.

Chapter 10

NONSMOOTH EVOLUTION PROBLEMS

In this Chapter one discusses existence, uniqueness, Lipschitz continuous dependence on initial conditions and stability of solutions for different evolution initial value problems written in the form of variational inequalities or equalities. Section 1 concerns the study of the Cauchy problem for a first order dynamical variational inequality. Section 2 contains an existence result for the solutions of a Cauchy problem for a second order evolution variational equation. In Section 3 one presents stability, asymptotic stability and unstability results for first order evolution variational inequalities.

1. First Order Evolution Variational Inequalities

We start with the study of a first order evolution quasivariational inequality encountered in the unilateral mechanics. For the general theory of variational inequalities we refer to [1], [2], [8], [10], [11], [24], [25]. To this end we set up the functional framework where this problem is considered.

Let V be a real Hilbert space endowed with the inner product $(\cdot,\cdot)_V$ and the associated norm $|\cdot|_V$ and let a prescribed number $T > 0$. In the following we use the Sobolev space $W^{1,\infty}(0,T;V)$ with the norm

$$|u|_{W^{1,\infty}(0,T;V)} = |u|_{L^\infty(0,T;V)} + |\dot{u}|_{L^\infty(0,T;V)},$$

where a dot represents the weak derivative with respect to the time variable.

We are concerned with the following initial value problem:

(IP) Find $u : [0,T] \to V$ *such that*

$$\begin{cases} a(u(t), v - \dot{u}(t)) + j(u(t), v) - j(u(t), \dot{u}(t)) \geq (f(t), v - \dot{u}(t))_V \\ \qquad\qquad\qquad\qquad\qquad\qquad\qquad\qquad \forall v \in V, \ \textit{a.e. } t \in]0,T[, \\ u(0) = u_0. \end{cases}$$

The meaning of the data entering initial problem (IP) and the basic assumptions are made precise in the following:

(i_1) $a : V \times V \to I\!R$ is a coercive, continuous, bilinear, symmetric form, thus there exist constants $M > 0$ and $m > 0$ such that

$$|a(u,v)| \leq M|u|_V|v|_V, \quad \forall\, u, v \in V;$$

$$a(v,v) \geq m|v|_V^2, \quad \forall v \in V.$$

(i_2) $j : V \times V \to I\!R$ and for every $\eta \in V$, $j(\eta, \cdot) : V \to I\!R$ is a positively homogeneous, subadditive functional, i.e.

$$j(\eta, \lambda u) = \lambda j(\eta, u), \quad \forall u \in V, \ \lambda \in I\!R_+;$$

$$j(\eta, u + v) \leq j(\eta, u) + j(\eta, v), \quad \forall u, v \in V.$$

(i_3) $f \in W^{1,\infty}(0,T;V)$.

(i_4) $u_0 \in V$ with $a(u_0, v) + j(u_0, v) \geq (f(0), v)_V$, $\forall v \in V$.

Assumption (i_2) guarantees that, for all $\eta \in V$, $j(\eta, \cdot) : V \to I\!R$ is a convex function. Therefore, there exists the directional derivative j_2' given by

$$j_2'(\eta, u; v) = \lim_{\lambda \to 0+} \frac{1}{\lambda}\Big[j(\eta, u + \lambda v) - j(\eta, u)\Big], \quad \forall \eta,\ u,\ v \in V.$$

We formulate now the following additional assumptions on the functional $j : V \times V \to I\!R$.

(j_1) For every sequence $\{u_n\} \subset V$ with $\|u_n\|_V \to \infty$, every sequence $\{t_n\} \subset [0,1]$ and each $\overline{u} \in V$, one has

$$\liminf_{n \to \infty} \Big[\frac{1}{\|u_n\|_V^2} j_2'(t_n u_n, u_n - \overline{u}; -u_n)\Big] < m.$$

(j_2) For every sequence $\{u_n\} \subset V$ with $\|u_n\|_V \to \infty$, every bounded sequence $\{\eta_n\} \subset V$ and each $\overline{u} \in V$ one has

$$\liminf_{n \to \infty} \Big[\frac{1}{\|u_n\|_V^2} j_2'(\eta_n, u_n - \overline{u}; -u_n)\Big] < m.$$

(j_3) For all sequences $\{u_n\} \subset V$ and $\{\eta_n\} \subset V$ such that $u_n \rightharpoonup u$ weakly in V, $\eta_n \rightharpoonup \eta$ weakly in V, with some $u, \eta \in V$, and for every $v \in V$, the inequality below holds

$$\limsup_{n\to\infty} [j(\eta_n, v) - j(\eta_n, u_n)] \leq j(\eta, v) - j(\eta, u).$$

(j_4) There exists $c_0 \in]0, m[$ such that

$$j(u, v-u) - j(v, v-u) \leq c_0 \|u - v\|_V^2, \quad \forall u, v \in V.$$

(j_5) There exist two functions $a_1 : V \to \mathbb{R}$ and $a_2 : V \to \mathbb{R}$ which map bounded sets in V into bounded sets in $\mathbb{R}$ such that $a_1(0) < m - c_0$ and

$$|j(\eta, u)| \leq a_1(\eta)\|u\|_V^2 + a_2(\eta), \quad \forall \eta, u \in V.$$

(j_6) For every sequence $\{\eta_n\} \subset V$ with $\eta_n \rightharpoonup \eta \in V$ weakly in V and every bounded sequence $\{u_n\} \subset V$ one has

$$\lim_{n\to\infty} [j(\eta_n, u_n) - j(\eta, u_n)] = 0.$$

(j_7) For every $s \in]0, T]$ and every functions $u, v \in W^{1,\infty}(0, T; V)$ with $u(0) = v(0)$, $u(s) \neq v(s)$, the inequality below holds

$$\int_0^s [j(u(t), \dot{v}(t)) - j(u(t), \dot{u}(t)) + j(v(t), \dot{u}(t)) - j(v(t), \dot{v}(t))]\, dt$$
$$< \frac{m}{2}\|u(s) - v(s)\|_V^2.$$

(j_8) There exists $\alpha \in]0, \frac{m}{2}[$ such that for every $s \in]0, T]$ and every functions $u, v \in W^{1,\infty}(0, T; V)$ with $u(s) \neq v(s)$, one has

$$\int_0^s [j(u(t), \dot{v}(t)) - j(u(t), \dot{u}(t)) + j(v(t), \dot{u}(t)) - j(v(t), \dot{v}(t))]\, dt$$
$$< \alpha\|u(s) - v(s)\|_V^2.$$

We need the following preliminary result.

Lemma 10.1 (Motreanu and Sofonea [15]) Let (i_1), (i_2), (j_1)-(j_3) hold. Then, for all $f \in V$ there exists at least an element $u \in V$ such that

$$a(u, v-u) + j(u, v) - j(u, u) \geq (f, v-u)_V, \quad \forall v \in V.$$

The proof of Lemma 10.1 can be found in [15].

We state the main result in studying problem (IP).

Theorem 10.1 (D. Motreanu and M. Sofonea [14]) Assume (i_1)-(i_4).

(i) If conditions (j_1)-(j_6) hold then there exists at least a solution $u \in W^{1,\infty}(0,T;V)$ to problem (IP).

(ii) If conditions (j_1)-(j_7) hold then there exists a unique solution $u \in W^{1,\infty}(0,T;V)$ to problem (IP).

(iii) Under the assumptions (j_1)-(j_6) and (j_8) there exists a unique solution $u = u(f, u_0) \in W^{1,\infty}(0,T;V)$ to problem (IP) and the mapping $(f, u_0) \mapsto u$ is Lipschitz continuous from $W^{1,\infty}(0,T;V) \times V$ to $L^\infty(0,T;V)$.

Proof. We suppose that (j_1)-(j_6) are fulfilled. Let $n \in \mathbb{N}$. We consider the following implicit scheme: find $u_n^{i+1} \in V$ such that

$$a(u_n^{i+1}, v - \frac{n}{T}(u_n^{i+1} - u_n^i)) + j(u_n^{i+1}, v) - j(u_n^{i+1}, \frac{n}{T}(u_n^{i+1} - u_n^i))$$

$$\geq (f(\frac{T(i+1)}{n}), v - \frac{n}{T}(u_n^{i+1} - u_n^i))_V, \quad \forall v \in V, \tag{10.1}$$

where $u_n^0 = u_0$, $i = 0, 1, ..., n-1$.

In the first step we prove the solvability of the quasivariational inequality (10.1) and we provide estimates of the solution to this problem.

Step 1: *There exists at least a solution u_n^{i+1} to the quasivariational inequality* (10.1), *for $i = 0, 1, ..., n-1$. Moreover, the solution satisfies:*

$$|u_n^{i+1}|_V^2 \leq \frac{1}{m - c_0 - a_1(0)}(|f(\frac{T(i+1)}{n})|_V |u_n^{i+1}|_V + a_2(0)), \tag{10.2}$$

$$|u_n^{i+1} - u_n^i|_V \leq \frac{1}{m - c_0}|f(\frac{T(i+1)}{n}) - f(\frac{Ti}{n})|_V, \tag{10.3}$$

for all $i = 0, 1, ..., n-1$.

Let us check the assertions in Step 1. Let $i \in \{0, 1, ..., n-1\}$. Using (i_2) and setting $w = \frac{T}{n}v + u_n^i$ it follows that (10.1) is equivalent to the inequality

$$a(u_n^{i+1}, w - u_n^{i+1}) + j(u_n^{i+1}, w - u_n^i) - j(u_n^{i+1}, u_n^{i+1} - u_n^i)$$

$$\geq (f(\frac{T(i+1)}{n}), w - u_n^{i+1})_V, \quad \forall w \in V. \tag{10.4}$$

Lemma 10.1 implies the existence of the solution to (10.4), while the equivalence of problems (10.1) and (10.4) yields the existence part in Step 1.

Taking now $w = 0$ in (10.4) and using (i_2) we find

$$a(u_n^{i+1}, u_n^{i+1}) \leq (f(\frac{T(i+1)}{n}), u_n^{i+1})_V + j(u_n^{i+1}, -u_n^i) - j(u_n^{i+1}, u_n^{i+1} - u_n^i)$$

$$\leq |f(\frac{T(i+1)}{n})|_V |u_n^{i+1}|_V + j(u_n^{i+1}, -u_n^{i+1}),$$

and (i_1) yields

$$m|u_n^{i+1}|_V^2 \leq |f(\frac{T(i+1)}{n})|_V |u_n^{i+1}|_V + j(u_n^{i+1}, -u_n^{i+1}). \tag{10.5}$$

Taking $u = u_n^{i+1}$ and $v = 0$ in (j_4) and using (j_5) with $\eta = 0$ we obtain

$$j(u_n^{i+1}, -u_n^{i+1}) \leq c_0 |u_n^{i+1}|_V^2 + j(0, -u_n^{i+1})$$

$$\leq (c_0 + a_1(0))|u_n^{i+1}|_V^2 + a_2(0). \tag{10.6}$$

Since $a_1(0) < m - c_0$, estimate (10.2) results from (10.5) and (10.6).

Using again (i_2) it follows that $j(u, 0) = \lambda j(u, 0)$ for all $u \in V$ and $\lambda > 0$, which implies

$$j(u, 0) = 0, \quad \forall u \in V. \tag{10.7}$$

Setting $w = u_n^i$ in (10.4) and using (10.7) it follows that

$$a(u_n^{i+1}, u_n^{i+1} - u_n^i) \leq (f(\frac{T(i+1)}{n}), u_n^{i+1} - u_n^i)_V$$

$$-j(u_n^{i+1}, u_n^{i+1} - u_n^i), \quad \forall i = 0, 1, ..., n-1. \tag{10.8}$$

Using again (10.4) with $i - 1$ in place of i and $w = u_n^{i+1}$ we find

$$a(u_n^i, u_n^{i+1} - u_n^i) + j(u_n^i, u_n^{i+1} - u_n^{i-1}) - j(u_n^i, u_n^i - u_n^{i-1})$$

$$\geq (f(\frac{Ti}{n}), u_n^{i+1} - u_n^i)_V, \quad \forall i = 1, ..., n-1.$$

By (i_2) and (i_4), we obtain

$$-a(u_n^i, u_n^{i+1} - u_n^i) \leq (-f(\frac{Ti}{n}), u_n^{i+1} - u_n^i)_V + j(u_n^i, u_n^{i+1} - u_n^i) \tag{10.9}$$

for all $i = 0, 1, ..., n-1$. It follows now from (i_1), (10.8), (10.9) and (j_4) that

$$m|u_n^{i+1} - u_n^i|_V^2 \leq a(u_n^{i+1} - u_n^i, u_n^{i+1} - u_n^i)$$

$$\leq (f(\frac{T(i+1)}{n})-f(\frac{Ti}{n}), u_n^{i+1}-u_n^i)_V - j(u_n^{i+1}, u_n^{i+1}-u_n^i)+j(u_n^i, u_n^{i+1}-u_n^i)$$

$$\leq |f(\frac{T(i+1)}{n})-f(\frac{Ti}{n})|_V |u_n^{i+1}-u_n^i|_V + c_0|u_n^{i+1}-u_n^i|_V^2, \quad \forall i=0,1,...,n-1,$$

which implies (10.3). The claim in Step 1 is proved.

We now consider the functions $u_n : [0,T] \to V$ and $\tilde{u}_n : [0,T] \to V$ defined as follows:

$$\begin{aligned} &u_n(0) = u_0, \\ &u_n(t) = u_n^i + \tfrac{nt-Ti}{T}(u_n^{i+1} - u_n^i), \quad \forall t \in]\tfrac{Ti}{n}, \tfrac{T(i+1)}{n}], \end{aligned} \tag{10.10}$$

$$\tilde{u}_n(0) = u_0, \ \tilde{u}_n(t) = u_n^{i+1}, \quad \forall t \in]\frac{Ti}{n}, \frac{T(i+1)}{n}], \tag{10.11}$$

where $u_n^0 = u_0$, u_n^{i+1} solves (10.1) and $i = 0, 1, ..., n-1$.

In the next step we provide convergence results involving the sequences $\{u_n\}$ and $\{\tilde{u}_n\}$.

Step 2: *There exist an element $u \in W^{1,\infty}(0,T;V)$ and subsequences of the sequences $\{u_n\}$ and $\{\tilde{u}_n\}$, again denoted $\{u_n\}$ and $\{\tilde{u}_n\}$, respectively, such that:*

$$u_n \overset{*}{\rightharpoonup} u \quad \textit{weakly}* \ \textit{in} \ \ L^\infty(0,T;V), \tag{10.12}$$

$$\dot{u}_n \overset{*}{\rightharpoonup} \dot{u} \quad \textit{weakly}* \ \textit{in} \ \ L^\infty(0,T;V), \tag{10.13}$$

$$\tilde{u}_n(t) \rightharpoonup u(t) \quad \textit{weakly in} \ \ V, \quad \textit{a.e.} \ \ t \in]0,T[. \tag{10.14}$$

Let us show the assertions in Step 2. Let $n \in I\!N$. Using (10.10) it follows that $u_n : [0,T] \to V$ is an absolutely continuous function and its derivative is given by

$$\dot{u}_n(t) = \frac{n}{T}(u_n^{i+1} - u_n^i) \ \text{ a.e. } \ t \in]\frac{Ti}{n}, \frac{T(i+1)}{n}[, \ i = 0,1,...,n-1. \tag{10.15}$$

Therefore, from (10.10), (10.15), (10.2) and (10.3) we deduce

$$|u_n(t)|_V \leq |u_0|_V + \frac{1}{m-c_0}|f(\frac{T}{n}) - f(0)|_V \ \text{ a.e. } \ t \in]0, \frac{T}{n}[,$$

$$|u_n(t)|_V \leq \frac{1}{(m-c_0-a_1(0))^{1/2}}(|f(\frac{Ti}{n})|_V |u_n^i|_V + a_2(0))^{1/2}$$

$$+\frac{1}{m-c_0}|f(\frac{T(i+1)}{n}) - f(\frac{Ti}{n})|_V \ \text{ a.e. } \ t \in]\frac{Ti}{n}, \frac{T(i+1)}{n}[, \ i = 1, ..., n-1,$$

$$|\dot{u}_n(t)|_V \leq \frac{1}{m-c_0}\frac{n}{T}|f(\frac{T(i+1)}{n}) - f(\frac{Ti}{n})|_V \ \text{ a.e. } \ t \in]\frac{Ti}{n}, \frac{T(i+1)}{n}[,$$

for all $i = 0, 1, ..., n-1$.

Taking into account (i_3) and estimate (10.2), from the previous inequalities it follows that $u_n \in W^{1,\infty}(0,T;V)$ and

$$|u_n|_{W^{1,\infty}(0,T;V)} \leq C, \tag{10.16}$$

for a constant $C > 0$.

The existence of an element $u \in W^{1,\infty}(0,T;V)$ as well as the convergences (10.12) and (10.13) follow from standard compactness arguments.

We turn now to the proof of (10.14). To this end we remark that the convergence results (10.12) and (10.13) imply

$$u_n(t) \rightharpoonup u(t) \ \text{ weakly in } \ V, \ \text{ for all } \ t \in [0,T]. \tag{10.17}$$

Moreover, using again (10.10), (10.11) and (10.3) we find

$$|u_n(t) - \tilde{u}_n(t)|_V = (1 - \frac{nt - Ti}{T})|u_n^{i+1} - u_n^i)|_V$$

$$\leq \frac{1}{m-c_0}|f(\frac{T(i+1)}{n}) - f(\frac{Ti}{n})|_V, \ \forall t \in]\frac{Ti}{n}, \frac{T(i+1)}{n}], \ i = 0, 1, ..., n-1$$

and, by the regularity (i_3), we deduce

$$|u_n - \tilde{u}_n|_{L^\infty(0,T;V)} \leq \frac{1}{m-c_0} \cdot \frac{T}{n}|\dot{f}|_{L^\infty(0,T;V)}.$$

This inequality proves that

$$u_n - \tilde{u}_n \to 0 \ \text{ in } \ L^\infty(0,T;V) \tag{10.18}$$

and therefore

$$u_n(t) - \tilde{u}_n(t) \to 0 \ \text{ a.e. } \ t \in]0,T[. \tag{10.19}$$

The convergence (10.14) is now a consequence of (10.17) and (10.19). Step 2 is justified.

In the next two steps we prove additional convergence and semicontinuity results. To this end, for every $n \in \mathbb{N}$ consider the function $f_n : [0,T] \to V$ defined as follows:

$$\begin{aligned} &f_n(0) = f(0), \\ &f_n(t) = f(\tfrac{T(i+1)}{n}), \ \ \forall t \in]\tfrac{Ti}{n}, \tfrac{T(i+1)}{n}], \ i = 0, 1, ..., n-1. \end{aligned} \tag{10.20}$$

Everywhere in the sequel u will denote the element of $W^{1,\infty}(0,T;V)$ whose existence was proved in Step 2 and $\{u_n\}$, $\{\tilde{u}_n\}$, $\{f_n\}$ will represent appropriate subsequences of the sequences $\{u_n\}$, $\{\tilde{u}_n\}$ and $\{f_n\}$, respectively.

Step 3: *The following properties hold:*

$$\lim_{n\to\infty}\int_0^T a(\tilde{u}_n(t), g(t))dt$$

$$= \int_0^T a(u(t), g(t))dt, \quad \forall g \in L^2(0,T;V), \tag{10.21}$$

$$\liminf_{n\to\infty}\int_0^T a(\tilde{u}_n(t), \dot{u}_n(t))dt \geq \int_0^T a(u(t), \dot{u}(t))dt, \tag{10.22}$$

$$\lim_{n\to\infty}\int_0^T (f_n(t), g(t) - \dot{u}_n(t))_V dt = \int_0^T (f(t), g(t) - \dot{u}(t))_V dt, \tag{10.23}$$

for all $g \in L^2(0,T;V)$.

Let us establish the properties in Step 3. It follows from (10.12) and (10.18) that $\tilde{u}_n \rightharpoonup u$ weakly in $L^2(0,T;V)$ and therefore, keeping in mind (i_1), we deduce (10.21). Using (i_1), (10.18) and (10.16) we find

$$\lim_{n\to\infty}\int_0^T a(\tilde{u}_n(t) - u_n(t), \dot{u}_n(t))dt = 0 \tag{10.24}$$

and, from (10.17), $u_n(0) = u_0$ and standard semicontinuity arguments, we obtain

$$\liminf_{n\to\infty}\int_0^T a(u_n(t), \dot{u}_n(t))dt \geq \int_0^T a(u(t), \dot{u}(t))dt. \tag{10.25}$$

Inequality (10.22) is now a consequence of (10.24) and (10.25).

Finally, from (i_3) and (10.20) we obtain that the sequence $\{f_n\}$ converges uniformly to f on $[0,T]$, i.e.

$$\max_{t\in[0,T]} |f_n(t) - f(t)|_V \to 0, \quad \forall t \in [0,T]. \tag{10.26}$$

The convergence (10.23) is now a consequence of (10.13) and (10.26). The proof of Step 3 is complete.

Step 4: *The following properties hold:*

$$\limsup_{n\to\infty}\int_0^T j(\tilde{u}_n(t), g(t))dt$$

$$\leq \int_0^T j(u(t), g(t))dt, \quad \forall g \in L^2(0,T;V), \tag{10.27}$$

$$\limsup_{n\to\infty}\int_0^T [j(u(t), \dot{u}_n(t)) - j(\tilde{u}_n(t), \dot{u}_n(t))]dt \leq 0, \tag{10.28}$$

$$\liminf_{n\to\infty} \int_0^T j(u(t), \dot{u}_n(t))dt \geq \int_0^T j(u(t), \dot{u}(t))dt. \qquad (10.29)$$

To prove Step 4 let $g \in L^2(0,T;V)$. Using (10.11), (10.2) and (i_3) it follows that $\{\tilde{u}_n(t)\}$ is a bounded sequence in V, for all $t \in [0,T]$. Therefore, by assumption (j_5) we deduce that there exists a constant $C_1 > 0$ such that

$$|j(\tilde{u}_n(t), g(t))| \leq C_1(|g(t)|_V^2 + 1) \ \text{ a.e. } \ t \in]0,T[, \ \forall n \in I\!N.$$

This inequality allows us to apply Fatou's lemma to obtain

$$\limsup_{n\to\infty} \int_0^T j(\tilde{u}_n(t), g(t))dt \leq \int_0^T \limsup_{n\to\infty} j(\tilde{u}_n(t), g(t))dt. \qquad (10.30)$$

We make use of (10.14) and assumption (j_6) to find

$$\lim_{n\to\infty} j(\tilde{u}_n(t), g(t)) = j(u(t), g(t)) \ \text{ a.e. } \ t \in]0,T[. \qquad (10.31)$$

Inequality (10.27) is now a consequence of (10.30) and (10.31).

Using assumption (j_5) and (10.16) we deduce that there exists $C > 0$ such that

$$|j(u(t), \dot{u}_n(t)) - j(\tilde{u}_n(t), \dot{u}_n(t))| \leq C \ \text{ a.e. } \ t \in]0,T[, \ \forall n \in I\!N.$$

This inequality allows us to apply again Fatou's lemma to obtain

$$\limsup_{n\to\infty} \int_0^T [j(u(t), \dot{u}_n(t)) - j(\tilde{u}_n(t), \dot{u}_n(t))]dt$$

$$\leq \int_0^T \limsup_{n\to\infty} [j(u(t), \dot{u}_n(t)) - j(\tilde{u}_n(t), \dot{u}_n(t))]dt. \qquad (10.32)$$

Moreover, using (10.14), (10.16) and assumption (j_6), we derive

$$\lim_{n\to\infty} [j(u(t), \dot{u}_n(t)) - j(\tilde{u}_n(t), \dot{u}_n(t))] = 0 \ \text{ a.e. } \ t \in]0,T[. \qquad (10.33)$$

Inequality (10.28) follows now from (10.32) and (10.33).

Finally, inequality (10.29) is obtained from standard semicontinuity arguments, employing (i_2), (j_5) and (10.13). Step 4 is thus established.

Step 5: *Completion of the proof of Theorem* 10.1.

(i) Using (10.1), (10.11), (10.15) and (10.20) we obtain

$$a(\tilde{u}_n(t), v - \dot{u}_n(t))_V + j(\tilde{u}_n(t), v) - j(\tilde{u}_n(t), \dot{u}_n(t))$$

$$\geq (f_n(t), v - \dot{u}_n(t))_V, \quad \forall v \in V, \quad \text{a.e. } t \in]0, T[.$$

This inequality and assumption (j_5) yield

$$\int_0^T a(\tilde{u}_n(t), g(t) - \dot{u}_n(t))dt + \int_0^T j(\tilde{u}_n(t), g(t))dt - \int_0^T j(\tilde{u}_n(t), \dot{u}_n(t))dt$$

$$\geq \int_0^T (f_n(t), g(t) - \dot{u}_n(t)_V)dt, \quad \forall g \in L^2(0, T; V). \qquad (10.34)$$

Using now (10.21)-(10.23), (10.27)-(10.29) and (10.34) we find

$$\int_0^T a(u(t), g(t) - \dot{u}(t))dt + \int_0^T j(u(t), g(t))dt - \int_0^T j(u(t), \dot{u}(t))dt$$

$$\geq \int_0^T (f(t), g(t) - \dot{u}(t))_V dt, \quad \forall g \in L^2(0, T; V). \qquad (10.35)$$

By (10.35) and a classical application of Lebesgue point for L^1 functions, we obtain that $u \in W^{1,\infty}(0, T; V)$ satisfies the inequality in problem (IP), while from (10.10) and (10.17) we deduce the initial condition in problem (IP), which concludes the proof of part (i).

(ii) Consider two solutions $u_1, u_2 \in W^{1,\infty}(0, T; V)$ to the Cauchy problem (IP). The inequalities below hold for all $v \in V$ and a.e. $t \in]0, T[$:

$$a(u_1(t), v - \dot{u}_1(t)) + j(u_1(t), v) - j(u_1(t), \dot{u}_1(t)) \geq (f(t), v - \dot{u}_1(t))_V,$$

$$a(u_2(t), v - \dot{u}_2(t)) + j(u_2(t), v) - j(u_2(t), \dot{u}_2(t)) \geq (f(t), v - \dot{u}_2(t))_V.$$

We set $v = \dot{u}_2(t)$ in the first inequality, $v = \dot{u}_1(t)$ in the second inequality. Adding the corresponding inequalities and using (i_1) we obtain

$$\frac{1}{2}\frac{d}{dt} a(u_1(t) - u_2(t), u_1(t) - u_2(t)) \leq j(u_1(t), \dot{u}_2(t)) - j(u_1(t), \dot{u}_1(t))$$

$$+ j(u_2(t), \dot{u}_1(t)) - j(u_2(t), \dot{u}_2(t)) \quad \text{a.e. } t \in]0, T[. \qquad (10.36)$$

Moreover, from the initial condition in problem (IP) we have

$$u_1(0) = u_2(0) = u_0. \qquad (10.37)$$

Arguing by contradiction, let us suppose that $u_1 \neq u_2$. Then there exists $s \in]0, T]$ such that

$$u_1(s) \neq u_2(s). \qquad (10.38)$$

Integrating (10.36) over $[0, s]$, by using (i_1) and (10.37) yields

$$\int_0^s [j(u_1(t), \dot{u}_2(t)) - j(u_1(t), \dot{u}_1(t)) + j(u_2(t), \dot{u}_1(t)) - j(u_2(t), \dot{u}_2(t))]dt$$

$$\geq \frac{m}{2}|u_1(s) - u_2(s)|_V^2. \tag{10.39}$$

In view of (10.37), (10.38) and assumption (j_7), inequality (10.39) leads to a contradiction, which concludes the proof of part (ii).

(iii) The unique solvability of the Cauchy problem (IP) follows from (ii) since assumption (j_8) implies (j_7). Let now $f_i \in W^{1,\infty}(0,T;V)$ and $u_{0i} \in V$ be such that the inequality in (i_4) holds for u_{0i} in place of u_0, $i = 1,2$. We denote in the sequel by $u_i \in W^{1,\infty}(0,T;V)$ the solution of the Cauchy problem (IP) for the data f_i and u_{0i}. A computation similar to the one in (10.36) leads to the inequality

$$\frac{1}{2}\frac{d}{dt}a(u_1(t) - u_2(t), u_1(t) - u_2(t))$$

$$\leq j(u_1(t), \dot{u}_2(t)) - j(u_1(t), \dot{u}_1(t)) + j(u_2(t), \dot{u}_1(t)) - j(u_2(t), \dot{u}_2(t))$$
$$+(f_1(t) - f_2(t), \dot{u}_1(t) - \dot{u}_2(t))_V \quad \text{a.e. } t \in]0,T[.$$

We suppose in the sequel that $u_1 \neq u_2$ and let $s \in]0,T]$ be such that $u_1(s) \neq u_2(s)$. Integrating over $[0,s]$ the previous inequality, using the initial conditions $u_i(0) = u_{0i}$ and (i_1), yields

$$\frac{m}{2}|u_1(s) - u_2(s)|_V^2 \leq \frac{M}{2}|u_{01} - u_{02}|_V^2$$

$$+ \int_0^s [j(u_1(t), \dot{u}_2(t)) - j(u_1(t), \dot{u}_1(t)) + j(u_2(t), \dot{u}_1(t)) - j(u_2(t), \dot{u}_2(t))]dt$$

$$+ \int_0^s (f_1(t) - f_2(t), \dot{u}_1(t) - \dot{u}_2(t))_V dt.$$

In view of assumption (j_8) we obtain

$$(\frac{m}{2} - \alpha)|u_1(s) - u_2(s)|_V^2$$

$$\leq \frac{M}{2}|u_{01} - u_{02}|_V^2 + \int_0^s (f_1(t) - f_2(t), \dot{u}_1(t) - \dot{u}_2(t))_V dt. \tag{10.40}$$

Let $\delta \in (0, m - 2\alpha)$. Using the inequality

$$ab \leq \frac{a^2}{2\delta} + \frac{\delta b^2}{2}$$

we infer that

$$\int_0^s (f_1(t) - f_2(t), \dot{u}_1(t) - \dot{u}_2(t))_V dt$$

$$= (f_1(s) - f_2(s), u_1(s) - u_2(s))_V - (f_1(0) - f_2(0), u_{01} - u_{01})_V$$

$$- \int_0^s (\dot{f}_1(t) - \dot{f}_2(t), u_1(t) - u_2(t))_V dt$$

$$\leq \frac{1}{2\delta}|f_1(s) - f_2(s)|_V^2 + \frac{\delta}{2}|u_1(s) - u_2(s)|_V^2$$

$$+\frac{1}{2\delta}|f_1(0) - f_2(0)|_V^2 + \frac{\delta}{2}|u_{01} - u_{02}|_V^2$$

$$+\frac{1}{2\delta}\int_0^s |\dot{f}_1(t) - \dot{f}_2(t)|_V^2 dt + \frac{\delta}{2}\int_0^s |u_1(t) - u_2(t)|_V^2 dt$$

$$\leq \frac{T+2}{2\delta}|f_1 - f_2|^2_{W^{1,\infty}(0,T;V)}$$

$$+\frac{\delta}{2}|u_1(s) - u_2(s)|_V^2 + \frac{\delta}{2}|u_{01} - u_{02}|_V^2 + \frac{\delta}{2}\int_0^s |u_1(t) - u_2(t)|_V^2 dt.$$

By (10.40) and the previous inequality we get

$$|u_1(s) - u_2(s)|_V^2 \leq c_1\Big(|u_{01} - u_{02}|_V^2 + |f_1 - f_2|^2_{W^{1,\infty}(0,T;V)}\Big)$$

$$+c_2\int_0^s |u_1(t) - u_2(t)|_V^2 dt \tag{10.41}$$

where c_1, $c_2 > 0$ depend on M, m, α, δ and T. Clearly inequality (10.41) holds for all $s \in [0, T]$. Using now a Gronwall-type argument, from (10.41) we obtain

$$|u_1(t) - u_2(t)|_V^2 \leq C_0\Big(|u_{01} - u_{02}|_V^2 + |f_1 - f_2|^2_{W^{1,\infty}(0,T;V)}\Big), \quad \forall s \in [0, T]$$

where $C_0 > 0$. This completes the proof. ■

Theorem 10.1 yields the following version of Proposition II. 9 in [2].

Corollary 10.1 Let (i_1), (i_3) hold, let $\varphi : V \to I\!R_+$ be a continuous seminorm and let us suppose that $u_0 \in V$ satisfies the condition

$$a(u_0, v) + \varphi(v) \geq (f(0), v)_V, \quad \forall v \in V.$$

Then, there exists a unique function $u \in W^{1,\infty}(0, T; V)$ such that

$$\begin{cases} a(u(t), v - \dot{u}(t)) + \varphi(v) - \varphi(\dot{u}(t)) \\ \qquad\qquad \geq (f(t), v - \dot{u}(t))_V, \quad \forall v \in V, \quad \text{a.e. } t \in]0, T[, \\ u(0) = u_0. \end{cases}$$

Moreover, the mapping $(f, u_0) \mapsto u$ is Lipschitz continuous from $W^{1,\infty}(0, T; V) \times V$ to $L^\infty(0, T; V)$.

Proof. Since $\varphi : V \to I\!R_+$ is a continuous seminorm then the functional j defined by $j(u,v) = \varphi(v)$ for all $u, v \in V$ satisfies the assumptions (i_2), (j_1)-(j_8). Theorem 10.1 insures the desired conclusion. ∎

Applications in contact mechanics of Theorem 10.1 and Corollary 10.1 are given in [5]-[7], [10], [12], [13]. Related applications in Mechanics can be found in [21]-[23].

2. Second Order Evolution Variational Equations

This Section concerns the study of a class of abstract second order variational equations. Precisely, we treat the following Cauchy problem

(CP) *Find* $u : [0,T] \to V$ *such that*

$$\begin{cases} \langle \ddot{u}(t), v\rangle_{V',V} + a(u(t), v) + \langle G((u(t)), v\rangle_{V',V} = \langle f(t), v\rangle_{V',V} \\ \qquad\qquad\qquad\qquad\qquad\qquad \forall v \in V, \;\; a.e.\ t \in]0,T[, \\ u(0) = u_0, \quad \dot{u}(0) = u_1. \end{cases}$$

In the statement of Cauchy problem (CP), V is a real separable reflexive Banach space and $\langle \cdot, \cdot \rangle_{V',V}$ denotes the duality pairing between V and its dual V'. Here the final time $T > 0$ is fixed and the dots above represent the derivative with respect to the time variable t, that is

$$\dot{u} = \frac{du}{dt}, \quad \ddot{u} = \frac{d^2u}{dt^2}.$$

In addition to V, we consider a real Hilbert space H. Let $|\cdot|_V$, $(\cdot,\cdot)_H$ and $|\cdot|_H$ denote the norm on V, the inner product and the associated norm on H, respectively. We suppose that V is dense in H and the injection of V into H is continuous, i.e.

$$|v|_H \leq c|v|_V, \quad \forall v \in V, \tag{10.42}$$

for a constant $c > 0$. We identify H with its dual and with a subspace of the dual V' of V, i.e. $V \subset H \subset V'$ algebraically and topologically. If $p \in [1, +\infty]$ and $k \in I\!N$ we use the standard notation for the spaces $L^p(0,T;X)$ and $W^{k,p}(0,T;X)$, where X is a real Banach space.

In the study of (CP) we consider the following assumptions:

(a) $a : V \times V \to I\!R$ is a continuous, bilinear, symmetric form, thus there exists a constant $M_a > 0$ such that

$$|a(u,v)| \leq M_a |u|_V |v|_V, \quad \forall\, u, v \in V;$$

(b) there exist constants $m_a > 0$ and $\lambda_a > 0$ such that

$$a(v,v) + \lambda_a |v|_H^2 \geq m_a |v|_V^2, \quad \forall\, v \in V;$$

(c) $f \in W^{1,\infty}(0,T;V)$;

(d) $u_0 \in V, \ u_1 \in H$.

We assume the following conditions on the nonlinear operator $G : V \to V'$:

(G_1) G is a potential operator, i.e. there exists a continuous, Gâteaux differentiable function $g : V \to \mathbb{R}$ such that $G = \nabla g$;

(G_2) there exists constants $c_0 > 0$ and $c_1 > 0$ such that

$$g(v) \geq -c_0|v|_V^2 - c_1, \quad \forall v \in V;$$

(G_3)

$$c_0 < m_a;$$

(G_4) $G : V \to V'$ is weakly-weakly$*$ continuous, i.e. for every sequence $\{u_n\} \subset V$ with $u_n \rightharpoonup u \in V$ weakly one has $G(u_n) \overset{*}{\rightharpoonup} G(u)$ in V' weakly $*$.

In (G_1) the notation ∇g represents the Gâteaux derivative of the functional g, so

$$\langle \nabla g(u), v \rangle_{V',V} = \lim_{t \to 0} \frac{g(u+tv) - g(u)}{t}, \quad \forall u, v \in V.$$

The existence of solutions to Cauchy problem (CP) under the assumptions mentioned above is established in the following result.

Theorem 10.2 (D. Motreanu and M. Sofonea [16]) Assume (i_1)-(i_4) and (G_1)-(G_4). Then there exists at least a solution of problem (CP). Moreover, this solution satisfies

$$u \in L^\infty(0,T;V), \quad \dot{u} \in L^\infty(0,T;H), \quad \ddot{u} \in L^\infty(0,T;V'). \tag{10.43}$$

Proof. The proof will be carried out by using the Galerkin approximation method. The separability of V ensures the existence of a sequence $\{w_i\} \subset V$ such that $w_1, w_2, \ldots, w_n$ are linearly independent elements of V, for all $n \in \mathbb{N}$ and

$$V = \overline{\bigcup_{n=1}^{\infty} V_n}, \tag{10.44}$$

where V_n denotes the linear subspace of V spanned by the vectors $w_1, w_2, \ldots, w_n$.

Taking into account assumption (d), the density of the inclusion $V \subset H$ and (10.44), we can find sequences $u_{n0} \in V_n$ and $u_{n1} \in V_n$ such that

$$u_{n0} \to u_0 \text{ strongly in } V \text{ as } n \to \infty, \tag{10.45}$$

$$u_{n1} \to u_1 \text{ in } H \text{ as } n \to \infty. \tag{10.46}$$

Let $n \in \mathbb{N}$. It follows from the theory of systems of ordinary differential equations that there exist $0 < T_n \leq T$ and $u_n \in C^2([0,T_n]; V_n)$ such that

$$(\ddot{u}_n(t), w)_{V',V} + a(u_n(t), w) + \langle G(u_n(t)), w\rangle_{V',V} \tag{10.47}$$

$$= \langle f(t), w\rangle_{V',V}\,, \quad \forall w \in V_n,\ t \in [0, T_n],$$

$$u_n(0) = u_{n0}, \quad \dot{u}_n(0) = u_{n1}. \tag{10.48}$$

We proceed to obtain a priori estimates on the solution u_n that ultimately will show that $T_n = T$. To this end we put $w = \dot{u}_n(t)$ in (10.47), integrate the resulting equality on $[0, s]$ and use (10.48), (a) and (G_1) to obtain

$$\frac{1}{2}\,|\dot{u}_n(s)|_H^2 - \frac{1}{2}\,|u_{n1}|_H^2 + \frac{1}{2}\,a(u_n(s), u_n(s)) - \frac{1}{2}\,a(u_{n0}, u_{n0})$$

$$+g(u_n(s)) - g(u_{n0}) = \int_0^s \langle f(t), \dot{u}_n(t)\rangle_{V',V} dt \quad \forall s \in [0, T_n].$$

Integrating by parts in the right-hand side of the previous equality and using (b) we deduce

$$\frac{1}{2}\,|\dot{u}_n(s)|_H^2 + m_a|u_n(s)|_V^2 - \lambda_a|u_n(s)|_H^2$$

$$\leq \frac{1}{2}\,|u_{n1}|_H^2 + \frac{1}{2}\,M_a|u_{n0}|_V^2 - g(u_n(s)) + g(u_{n_0})$$

$$+\langle f(s), u_n(s)\rangle_{V'\times V} - \langle f(0), u_{n0}\rangle_{V',V}$$

$$-\int_0^s \langle \dot{f}(t), u_n(t)\rangle_{V',V} dt \quad \forall s \in [0, T_n].$$

Using now (10.42), (c), (10.45), (10.46), (G_2), (G_3), Cauchy-Schwarz and Young inequalities, from the previous inequalities we infer that

$$\frac{1}{2}\,|\dot{u}_n(s)|_H^2 + m_a|u_n(s)|_V^2 - c_0|u_n(s)|_V^2 \tag{10.49}$$

$$\leq \lambda_a|u_n(s)|_H^2 + \frac{1}{2}|u_{n1}|_H^2 + \frac{1}{2}\,M_a|u_{n0}|_V^2 + g(u_{n0}) + c_1$$

$$+|f|_{L^\infty(0,T;V')}|u_n(s)|_V-\langle f(0),u_{n0}\rangle_{V',V}+\frac{1}{2}|\dot{f}|^2_{L^2(0,T;V')}+\frac{1}{2}\int_0^s|u_n(t)|_V^2dt$$

$$\leq \lambda_a\left[|u_{n0}|_H+\int_0^s|\dot{u}_n(t)|_Hdt\right]^2+\frac{1}{2}|u_{n1}|_H^2+\frac{1}{2}M_a|u_{n0}|_V^2+g(u_{n0})+c_1$$

$$+\frac{1}{2(m_a-c_0)}|f|^2_{L^\infty(0,T;V')}+\frac{m_a-c_0}{2}|u_n(s)|_V^2$$

$$-\langle f(0),u_{n0}\rangle_{V',V}+\frac{1}{2}|\dot{f}|^2_{L^2(0,T;V')}+\frac{1}{2}\int_0^s|u_n(t)|_V^2dt$$

$$\leq\frac{m_a-c_0}{2}|u_n(s)|_V^2+C\left[1+\int_0^s\left(|u_n(t)|_V^2+|\dot{u}_n(t)|_H^2\right)dt\right],\quad\forall s\in[0,T_n],$$

with a constant $C>0$. We derive from (10.49) that there exist some constants $c_2,c_3>0$, which are independent on n, such that

$$|u_n(s)|_V^2+|\dot{u}_n(s)|_H^2\leq c_2+c_3\int_0^s\left(|u_n(t)|_V^2+|\dot{u}_n(t)|_H^2\right)dt,\quad\forall s\in[0,T_n].$$

Using now a Gronwall type argument, from the previous inequality we obtain that

$$|u_n(s)|_V\leq c_4,\quad\forall s\in[0,T_n],\tag{10.50}$$

$$|\dot{u}_n(s)|_H\leq c_5,\quad\forall s\in[0,T_n],\tag{10.51}$$

where the constants $c_4>0$ and $c_5>0$ do not depend on n. Since $u_n(s)\in V_n$, $\forall s\in[0,T_n]$, estimates (10.50) and (10.51) ensure that we have

$$T_n=T.\tag{10.52}$$

Since $n\in I\!N$ was arbitrarily chosen, we conclude from (10.50)-(10.52) that there exists a function $u:[0,T]\to V$ and a subsequence of $\{u_n\}$, again denoted $\{u_n\}$, such that

$$u\in L^\infty(0,T;V),\quad \dot{u}\in L^\infty(0,T;H),\tag{10.53}$$

$$u_n\overset{*}{\rightharpoonup}u\ \text{ weakly}*\ \text{ in }\ L^\infty(0,T;V)\ \text{ as }\ n\to\infty,\tag{10.54}$$

$$\dot{u}_n\overset{*}{\rightharpoonup}\dot{u}\ \text{ weakly}*\ \text{ in }\ L^\infty(0,T;H)\ \text{ as }\ n\to\infty.\tag{10.55}$$

Notice that assumption (G_4) guarantees that the operator $G:V\to V'$ maps bounded sets into bounded sets. This property, (10.50) and (10.52) imply that there exists $c_6>0$ such that

$$|G(u_n(s))|_{V'}\leq c_6,\quad\forall s\in[0,T],\ \forall n\in I\!N.$$

This inequality enables us to find an element $\xi \in L^\infty(0,T;V')$ and a subsequence of $\{u_n\}$, again denoted $\{u_n\}$, such that

$$G(u_n) \overset{*}{\rightharpoonup} \xi \text{ weakly} * \text{ in } L^\infty(0,T;V') \text{ as } n \to \infty. \tag{10.56}$$

Let $m \in I\!N$, $w \in V_m$ and $\varphi \in C_0^\infty(0,T)$. Since $V_m \subset V_n$ whenever $n \geq m$, from (10.47) we deduce that

$$\int_0^T (\ddot{u}_n(t), \varphi(t)w)_H dt + \int_0^T a(u_n(t), \varphi(t)w)dt$$

$$+ \int_0^T \langle G(u_n(t)), \varphi(t)w\rangle_{V'\times V} dt = \int_0^T \langle f(t), \varphi(t)w\rangle_{V'\times V} dt \quad \forall n \geq m$$

or, equivalently,

$$- \int_0^T (\dot{u}_n(t), \dot{\varphi}(t)w)_H dt + \int_0^T a(u_n(t), \varphi(t)w)dt$$

$$+ \int_0^T \langle G(u_n(t)), \varphi(t)w\rangle_{V',V} dt = \int_0^T \langle f(t), \varphi(t)w\rangle_{V',V} dt, \quad \forall n \geq m.$$

Letting $n \to \infty$ in the previous inequality and making use of (10.54)-(10.56) we find

$$- \int_0^T (\dot{u}(t), w)_H \dot{\varphi}(t) dt + \int_0^T a(u(t), w)\varphi(t)dt \tag{10.57}$$

$$+ \int_0^T \langle \xi(t), w\rangle_{V',V} \varphi(t) dt = \int_0^T \langle f(t), w\rangle_{V',V} \varphi(t)dt.$$

It follows from properties (a),(b) that there exists a linear continuous operator $A : V \to V'$ such that

$$\langle Au, v\rangle_{V',V} = a(u,v), \quad \forall u, v \in V. \tag{10.58}$$

Since $\bigcup_{n=1}^{\infty} V_n$ is dense in V, using (10.57) and (10.58) we obtain

$$- \int_0^T (\dot{u}(t), w)_H \dot{\varphi}(t) dt + \int_0^T \langle Au(t), w\rangle_{V',V} \varphi(t)dt$$

$$+ \int_0^T \langle \xi(t), w\rangle_{V',V} \varphi(t) dt = \int_0^T \langle f(t), w\rangle_{V',V} \varphi(t)dt, \quad \forall w \in V.$$

Keeping in mind that φ is an arbitrary element of the space $C_0^\infty(0,T)$, the previous inequality implies

$$\ddot{u} + Au + \xi = f \text{ in } \mathcal{D}'(0,T;V'). \tag{10.59}$$

Using (10.53) and (c), it follows that $Au \in L^\infty(0,T;V')$ and $f \in L^\infty(0,T;V')$. Therefore, since $\xi \in L^\infty(0,T;V')$, equality (10.59) implies that

$$\ddot{u} \in L^\infty(0,T;V'). \tag{10.60}$$

This regularity shows that (10.59) is in fact an equality in $L^\infty(0,T;V')$, hence

$$\ddot{u}(t) + Au(t) + \xi(t) = f(t) \text{ a.e. } t \in (0,T). \tag{10.61}$$

On the other hand, combining (10.55), (10.55) and the continuous embedding (10.42), we obtain that

$$u_n(t) \rightharpoonup u(t) \text{ weakly in } H \text{ as } n \to \infty, \text{ for all } t \in [0,T]. \tag{10.62}$$

Now, relations (10.48), (10.45), (10.42) and (10.62) imply that

$$u(0) = u_0. \tag{10.63}$$

To verify the initial condition $\dot{u}(0) = u_1$ we consider an arbitrary element $v \in V$ and we remark that (10.55) implies

$$(\dot{u}_n, v)_H \overset{*}{\rightharpoonup} (\dot{u}, v)_H \text{ weakly} * \text{ in } L^\infty(0,T) \text{ as } n \to \infty. \tag{10.64}$$

Thus, we deduce that

$$(\dot{u}_n, v)_H \to (\dot{u}, v)_H \text{ in } \mathcal{D}'(0,T) \text{ as } n \to \infty$$

and therefore it follows that

$$\langle \ddot{u}_n, v\rangle_{V',V} \to \langle \ddot{u}, v\rangle_{V',V} \text{ in } \mathcal{D}'(0,T) \text{ as } n \to \infty.$$

This last convergence shows that

$$\int_0^T \langle \ddot{u}_n(t), v\rangle_{V',V}\varphi(t)dt \to \int_0^T \langle \ddot{u}(t), v\rangle_{V',V}\varphi(t)dt \text{ as } n \to \infty, \tag{10.65}$$

for all $\varphi \in C_0^\infty(0,T)$. Since $C_0^\infty(0,T)$ is dense in $L^1(0,T)$, using (10.60) and (10.65) we conclude that

$$\langle \ddot{u}_n, v\rangle_{V',V} \overset{*}{\rightharpoonup} \langle \ddot{u}, v\rangle_{V',V} \text{ weakly} * \text{ in } L^\infty(0,T) \text{ as } n \to \infty. \tag{10.66}$$

It follows now from (10.64) and (10.66) that

$$(\dot{u}_n(t), v)_H \to (\dot{u}(t), v)_H \text{ as } n \to \infty, \text{ for all } t \in [0,T],$$

and, since v is an arbitrary element in V and V is dense in H, we find

$$\dot{u}_n(t) \rightharpoonup \dot{u}(t) \text{ weakly in } H \text{ as } n \to \infty, \text{ for all } t \in [0,T]. \tag{10.67}$$

Using (10.12), (10.46) and (10.67) we obtain

$$\dot{u}(0) = u_1. \tag{10.68}$$

Fix now $t \in [0, T]$. By (10.50), (10.42) and (10.62) it follows that along a subsequence of $\{u_n\}$, again denoted $\{u_n\}$, we have

$$u_n(t) \rightharpoonup u(t) \quad \text{weakly in } V \text{ as } n \to \infty$$

and, from (G_4), we find

$$G(u_n(t)) \overset{*}{\rightharpoonup} G(u(t)) \quad \text{weakly} * \text{ in } V' \text{ as } n \to \infty. \tag{10.69}$$

A standard argument based on the Lebesgue point of an L^1 function and properties (10.56), (10.69) lead to

$$\xi(t) = G(u(t)) \quad \text{a.e. } t \in (0, T). \tag{10.70}$$

It turns out now from (10.61), (10.70) and (10.58) that u satisfies the variational equality in (CP). Using now (10.63) and (10.68) we obtain that u fulfills the initial condition in (CP) and, using (10.53), (10.60), we deduce (10.43), which completes the proof. ∎

3. Stability Properties for Evolution Variational Inequalities

In this Section we study the stability of stationary solutions of evolution variational inequalities in Hilbert spaces. The exposition follows the development in D. Goeleven, D. Motreanu and V. V. Motreanu [9].

We first recall the result of Kato for a general nonlinear Cauchy problem (see [2]) that we need in the sequel.

Lemma 10.2 Let H be a real Hilbert space and let $A : D(A) \subset H \to 2^H$ be a maximal monotone operator. Let $T > 0$ be given. Then for any $\sigma \in I\!R$, $u_0 \in D(A)$ and $f : [0, T] \to H$ satisfying

$$f \in C^0([0, T]; H), \; \frac{df}{dt} \in L^1(0, T; H),$$

there exists a unique $u \in C^0([0, T]; H)$ satisfying

$$\frac{du}{dt} \in L^\infty(0, T; H);$$

$$u \text{ is right-differentiable on } [0, T);$$

$$u(0) = u_0;$$

$$u(t) \in D(A), \ 0 \leq t \leq T;$$

$$\sigma u(t) + f(t) \in \frac{du}{dt}(t) + Au(t), \ \text{a.e. } 0 \leq t \leq T.$$

Lemma 10.2 admits the following variant.

Lemma 10.3 Let H be a real Hilbert space and let $A : D(A) \subset H \to 2^H$ be a maximal monotone operator. Let $t_0 \in I\!R$, $\sigma \in I\!R$, $u_0 \in D(A)$ be given and suppose that $f : [t_0, +\infty[\to H$ satisfies

$$f \in C^0([t_0, +\infty[; H), \ \frac{df}{dt} \in L^1_{\text{loc}}(t_0, +\infty; H).$$

Then there exists a unique $u \in C^0([t_0, +\infty[; H)$ satisfying

$$\frac{du}{dt} \in L^\infty_{\text{loc}}(t_0, +\infty; H);$$

$$u \text{ is right-differentiable on } [t_0, +\infty[;$$

$$u(t) \in D(A), \ t \geq t_0;$$

$$u(t_0) = u_0;$$

$$\sigma u(t) + f(t) \in \frac{du}{dt}(t) + Au(t), \ \text{a.e. } t \geq t_0.$$

We turn now to evolution variational inequalities associated with the Cauchy problem of Lemma 10.3. Throughout this Section, for a convex function $\varphi : H \to I\!R \cup \{+\infty\}$ the notations $D(\varphi)$ and $D(\partial\varphi)$ stand for the domain of φ and the domain of the subdifferential $\partial\varphi$ of φ, respectively.

Let $T : H \to H$, $\Phi : H \to I\!R$ and $\varphi : H \to I\!R \cup \{+\infty\}$ be given. The assumptions (h) described below will be employed:

(h_1) T is monotone and hemicontinuous

(h_2) $\Phi \in C^1(H; I\!R)$ and is convex

(h_3) φ is proper (i.e. $D(\varphi) \neq \emptyset$), convex and lower semicontinuous.

Remark 10.1. The operator $A : H \to 2^H$ defined by

$$A = T + \Phi' + \partial\varphi$$

is maximal monotone. Indeed, the operator A is monotone as sum of monotone operators. Moreover, $T + \Phi'$ is monotone and hemicontinuous, thus maximal monotone. On the other hand, it is known that $\overline{D(\partial\varphi)} = \overline{D(\varphi)}$ (see, e.g., Brézis [3], p. 39), hence $D(\partial\varphi) \neq \emptyset$. Since

$$D(\partial\varphi) \cap \mathrm{int}D(T + \Phi') = D(\partial\varphi) \cap H = D(\partial\varphi) \neq \emptyset,$$

we may apply Rockafellar's theorem to conclude that $A = T + \Phi' + \partial\varphi$ is maximal monotone and, in addition,

$$D(A) = D(\partial\varphi).$$

In the sequel the scalar product on H is denoted by $(\cdot,\cdot)$ (with the associated norm $\|\cdot\|$). Using Lemma 10.3 together with Remark 10.1, we get the following existence and uniqueness result for evolution variational inequalities.

Proposition 10.1. Let H be a real Hilbert space and let $T : H \to H$, $\Phi : H \to \mathbb{R}$, $\varphi : H \to \mathbb{R} \cup \{+\infty\}$ satisfy conditions (h). Given $t_0 \in \mathbb{R}$, $\sigma \in \mathbb{R}$, $u_0 \in D(\partial\varphi)$, suppose that $f : [t_0, +\infty[\to H$ satisfies

$$f \in C^0([t_0, +\infty[; H), \ \frac{df}{dt} \in L^1_{\mathrm{loc}}(t_0, +\infty; H).$$

Then there exists a unique $u \in C^0([t_0, +\infty[; H)$ such that

$$\frac{du}{dt} \in L^\infty_{\mathrm{loc}}(t_0, +\infty; H);$$

$$u \text{ is right-differentiable on } [t_0, +\infty[;$$

$$u(t) \in D(\partial\varphi), \ t \geq t_0;$$

$$u(t_0) = u_0;$$

$$\left(\frac{du}{dt}(t) + Tu(t) + \Phi'(u(t)) - f(t), v - u(t)\right)$$
$$+\varphi(v) - \varphi(u(t)) \geq (\sigma u(t), v - u(t)), \ \forall v \in H, \text{ a.e. } t \geq t_0.$$

We may now obtain from Proposition 10.1 the following useful existence and uniqueness result for evolution variational inequalities.

Corollary 10.2 Let H be a real Hilbert space and let $\Phi : H \to \mathbb{R}$ and $\varphi : H \to \mathbb{R} \cup \{+\infty\}$ satisfy conditions (h_2) and (h_3), respectively. Let $A : H \to H$ be a hemicontinuous operator such that for some $w_1 \geq 0$, $A + w_1 I$ is monotone. Let $B : H \to H$ be an operator such that

$$\|Bu - Bv\| \leq w_2\|u - v\|, \forall u, v \in H$$

for some $w_2 > 0$. Let $t_0 \in I\!R$ and $u_0 \in D(\partial\varphi)$ be given and suppose that $f : [t_0, +\infty[\to H$ satisfies

$$f \in C^0([t_0, +\infty[; H), \ \frac{df}{dt} \in L^1_{\text{loc}}(t_0, +\infty; H).$$

Then there exists a unique $u \in C^0([t_0, +\infty[; H)$ such that

$$\frac{du}{dt} \in L^\infty_{\text{loc}}(t_0, +\infty; H); \tag{10.71}$$

$$u \text{ is right-differentiable on } [t_0, +\infty[; \tag{10.72}$$

$$u(t) \in D(\partial\varphi), \ t \geq t_0; \tag{10.73}$$

$$u(t_0) = u_0; \tag{10.74}$$

$$\begin{aligned}&(\frac{du}{dt}(t) + Au(t) + Bu(t) + \Phi'(u(t)) - f(t), v - u(t))\\ &+\varphi(v) - \varphi(u(t)) \geq 0, \ \forall v \in H, \ \text{a.e. } t \geq t_0.\end{aligned} \tag{10.75}$$

Proof. Let us first remark that inequality (10.75) is equivalent to the following one

$$\begin{aligned}&(\frac{du}{dt}(t) + Au(t) + w_1u(t) + Bu(t) + w_2u(t) + \Phi'(u(t)) - f(t), v - u(t))\\ &+\varphi(v) - \varphi(u(t)) \geq ((w_1 + w_2)u(t), v - u(t)), \ \forall v \in H, \ \text{a.e. } t \geq t_0.\end{aligned}$$

We set

$$T_1 = A + w_1 I$$

and

$$T_2 = B + w_2 I.$$

It is clear that T_1 is monotone and hemicontinuous. The operator T_2 is Lipschitz continuous and thus hemicontinuous. In addition, it is seen that

$$\begin{aligned}(T_2x - T_2y, x - y) &= (Bx - By, x - y) + w_2\|x - y\|^2\\ &\geq -\|Bx - By\| \, \|x - y\| + w_2\|x - y\|^2\\ &\geq -w_2\|x - y\|^2 + w_2\|x - y\|^2 = 0, \ \forall x, y \in H\end{aligned}$$

thus T_2 is monotone.
We may now apply Proposition 10.1 with $T = T_1 + T_2$ which is monotone and hemicontinuous and $\sigma = w_1 + w_2$ to conclude that a unique map $u \in C^0([t_0, +\infty[; H)$ can be found to fulfill (10.71)-(10.75). ∎

Let us now specify the general mathematical framework for our stability theory. We formulate assumptions (H):

(H$_1$) H is a real Hilbert space

(H$_2$) $\varphi : H \to I\!R \cup \{+\infty\}$ is a convex, lower semicontinuous function such that

$$0 \in D(\partial\varphi)$$

(H$_3$) $A : H \to H$ is a hemicontinuous operator such that for $w_1 \geq 0$,

$$A + w_1 I \text{ is monotone}$$

(H$_4$) $B : H \to H$ is an operator such that for some $w_2 > 0$,

$$\|Bu - Bv\| \leq w_2 \|u - v\|, \ \forall u, v \in H$$

(H$_5$) $\Phi \in C^1(H; I\!R)$, convex

(H$_6$) $f \in C^0([t_0, +\infty[; H)$, $\frac{df}{dt} \in L^1_{\text{loc}}(t_0, +\infty; H)$

(H$_7$) $(A(0) + B(0) + \Phi'(0) - f(t), v) + \varphi(v) - \varphi(0) \geq 0$, $\forall v \in H$, $\forall t \geq t_0$.

Condition (H$_7$) is equivalent to

$$f(t) - (A(0) + B(0) + \Phi'(0)) \in \partial\varphi(0), \ \forall t \geq t_0.$$

For example, let $g \in C^0([t_0, +\infty[; I\!R)$ with $\frac{dg}{dt} \in L^1_{\text{loc}}(t_0, +\infty; I\!R)$ and $0 \leq g(t) \leq 1$, $\forall t \geq t_0$. Then, for all $F_0, F_1 \in \partial\varphi(0)$ one has that $f(t) := (A(0) + B(0) + \Phi'(0)) + (1 - g(t))F_0 + g(t)F_1$, $\forall t \geq t_0$, verifies (H$_6$) and (H$_7$).

Corollary 10.2 implies that for each $u_0 \in D(\partial\varphi)$ problem $P(t_0, u_0)$:

$$(\frac{du}{dt}(t) + Au(t) + Bu(t) + \Phi'(u(t)) - f(t), v - u(t))$$

$$+\varphi(v) - \varphi(u(t)) \geq 0, \ \forall v \in H, \text{ a.e. } t \geq t_0 \quad (10.76)$$

$$u(t) \in D(\partial\varphi), \ t \geq t_0 \quad (10.77)$$

$$u(t_0) = u_0 \quad (10.78)$$

has a unique solution $t \to u(t; t_0, u_0)$ $(t \geq t_0)$ with $u \in C^0([t_0, +\infty[; H)$, $\frac{du}{dt} \in L^\infty_{\text{loc}}(t_0, +\infty; H)$, and u right-differentiable on $[t_0, +\infty[$.

Moreover, conditions $0 \in D(\partial\varphi)$ and (H$_7$) ensure that

$$u(t; t_0, 0) = 0, \ t \geq t_0,$$

i.e. the trivial solution 0 is the unique solution of problem $P(t_0, 0)$. This solution is called stationary solution because for the unilateral system modeled by (10.76) the trajectory remains in the same position 0 for all times $t \geq t_0$.

Note that if $\varphi = \Psi_K$, where K is a closed convex subset of H such that $0 \in K$ and Ψ_K denotes the indicator function of K, then problem $P(t_0, u_0)$ $(t_0 \in I\!R, u_0 \in K)$ reads

$$(\frac{du}{dt}(t) + Au(t) + Bu(t) + \Phi'(u(t)) - f(t), v - u(t))$$

$$\geq 0, \ \forall v \in K, \ \text{a.e. } t \geq t_0, \tag{10.79}$$

$$u(t) \in K, \ t \geq t_0 \tag{10.80}$$

$$u(t_0) = u_0. \tag{10.81}$$

This last model appears frequently in applications. Note that in this case $(\mathtt{H}_2)$ is satisfied and $(\mathtt{H}_7)$ takes the form

$$(f(t) - (A(0) + B(0) + \Phi'(0)), v) \leq 0, \ \forall v \in K, \ \forall t \geq t_0, \tag{10.82}$$

which can be expressed equivalently

$$f(t) - (A(0) + B(0) + \Phi'(0)) \in N_K(0), \ \forall t \geq t_0,$$

with the normal cone

$$N_K(x) = \{w \in H : (w, y - x) \leq 0, \ \forall y \in K\}, \ x \in K.$$

We may now define the stability of the trivial solution in our setting of evolution variational inequalities (for the corresponding definitions in the case of ordinary differential equations see, e.g., [27]). The stationary solution 0 is called stable if small perturbations of the initial condition $u(t_0) = 0$ lead to solutions which remain in the neighborhood of 0 for all $t \geq t_0$, precisely:

Definition 10.1 The solution 0 is said to be stable (in the sense of Lyapunov) if for every $\varepsilon > 0$ there exists $\eta = \eta(\varepsilon) > 0$ such that for any $u_0 \in D(\partial\varphi)$ with $\|u_0\| \leq \eta$ the solution $u(\cdot; t_0, u_0)$ of problem $P(t_0, u_0)$ satisfies

$$\|u(t; t_0, u_0)\| \leq \varepsilon, \ \forall t \geq t_0.$$

If in addition the trajectories of the perturbed solutions are attracted by 0 then we say that the stationary solution is asymptotically stable, precisely:

Definition 10.2 We say that the solution 0 is asymptotically stable if it is stable and there exists $\delta > 0$ such that for any $u_0 \in D(\partial\varphi)$ with $\|u_0\| \leq \delta$ the solution $u(\cdot; t_0, u_0)$ of problem $P(t_0, u_0)$ fulfills

$$\lim_{t \to +\infty} \|u(t; t_0, u_0)\| = 0.$$

The notion of unstability is given below.

Definition 10.3 We say that the solution 0 is unstable if it is not stable (see Definition 10.1), i.e. there exists $\varepsilon > 0$ such that for any $\eta > 0$, one may find $u_0 \in D(\partial\varphi)$ with $\|u_0\| \leq \eta$ and $\bar{t} \geq t_0$ such that the solution $u(\cdot; t_0, u_0)$ of problem $P(t_0, u_0)$ verifies

$$\|u(\bar{t}; t_0, u_0)\| > \varepsilon.$$

We need two technical results.

Lemma 10.4 Given $t_0 \in I\!R$ and $\delta > 0$, let $a \in L^1(t_0, t_0 + \delta)$ and $V \in W^{1,1}(t_0, t_0 + \delta)$ satisfy

$$V'(t)\left(\begin{smallmatrix} \leq \\ \geq \end{smallmatrix} \right) a(t)V(t), \text{ a.e. } t \in [t_0, t_0 + \delta].$$

Then one has

$$V(t)\left(\begin{smallmatrix} \leq \\ \geq \end{smallmatrix} \right) V(t_0)e^{\int_{t_0}^t a(\tau)d\tau}, \ \forall t \in [t_0, t_0 + \delta].$$

Proof. Let us define the function $z : [t_0, t_0 + \delta] \to I\!R$ by

$$z(t) = V(t)e^{-\int_{t_0}^t a(\tau)d\tau}, \ \forall t \in [t_0, t_0 + \delta].$$

Then z is absolutely continuous on $[t_0, t_0 + \delta]$ and using the hypothesis it follows that

$$z'(t) = (V'(t) - a(t)V(t))e^{-\int_{t_0}^t a(\tau)d\tau}\left(\begin{smallmatrix} \leq \\ \geq \end{smallmatrix} \right)0, \text{ a.e. } t \in [t_0, t_0 + \delta].$$

This yields

$$z(t) = z(t_0) + \int_{t_0}^t z'(s)ds\left(\begin{smallmatrix} \leq \\ \geq \end{smallmatrix} \right)z(t_0) = V(t_0), \ \forall t \in [t_0, t_0 + \delta],$$

which completes the proof. ■

Lemma 10.5 Assume that conditions (H$_1$)-(H$_6$) hold and suppose that there exist $R > 0$, $a > 0$ and $V \in C^1(H; I\!R)$ such that

(1) $V(x) \geq a,\ x \in D(\partial\varphi),\ \|x\| = R$;

(2) $(Ax + Bx + \Phi'(x) - f(t), V'(x)) + (w, V'(x)) \geq 0$, for all $x \in D(\partial\varphi)$, $\|x\| \leq R$, $w \in \partial\varphi(x)$, $t \geq t_0$.

Then, for any $u_0 \in D(\partial\varphi)$ with $\|u_0\| < R$ and $V(u_0) < a$, the solution $u(\cdot; t_0, u_0)$ of problem $P(t_0, u_0)$ satisfies

$$\|u(t; t_0, u_0)\| < R,\ t \geq t_0. \tag{10.83}$$

Proof. Fix $u_0 \in D(\partial\varphi)$ with $\|u_0\| < R$ and $V(u_0) < a$. It suffices to show (10.83) for $t > t_0$. Suppose that (10.83) is not true, i.e. there exists $t^* > t_0$ such that $\|u(t^*; t_0, u_0)\| \geq R$. The application $u(\cdot; t_0, u_0)$ is continuous and, by (10.78), $\|u(t_0; t_0, u_0)\| = \|u_0\| < R$. We may thus find some $\bar{t} > t_0$ such that

$$\|u(t; t_0, u_0)\| < R,\ t \in [t_0, \bar{t})$$

and

$$\|u(\bar{t}; t_0, u_0)\| = R.$$

We set

$$V^*(t) = V(u(t; t_0, u_0)).$$

The function V^* is absolutely continuous on $[t_0, \bar{t}]$, thus a.e. strongly differentiable.

Setting $u(t) \equiv u(t; t_0, u_0)$, we obtain

$$\frac{dV^*}{dt}(t) = (V'(u(t)), \frac{du}{dt}(t)),\ \text{a.e. } t \in [t_0, \bar{t}].$$

It results from (10.76) that there exists an application $w : [t_0, +\infty[\to H$ such that

$$\frac{du}{dt}(t) + Au(t) + Bu(t) + \Phi'(u(t)) - f(t) + w(t) = 0,\ \text{a.e. } t \geq t_0$$

and

$$w(t) \in \partial\varphi(u(t)),\ \text{a.e. } t \geq t_0.$$

We obtain

$$\begin{aligned}(\frac{du}{dt}(t), V'(u(t))) &= (Au(t) + Bu(t) + \Phi'(u(t)) - f(t), -V'(u(t))) \\ &\quad + (w(t), -V'(u(t))),\ \text{a.e. } t \geq t_0.\end{aligned}$$

Thanks to assumption (2) we have

$$(\frac{du}{dt}(t), V'(u(t))) \leq 0, \text{ a.e. } t \in [t_0, \bar{t}],$$

i.e.

$$\frac{dV^*}{dt}(t) \leq 0, \text{ a.e. } t \in [t_0, \bar{t}].$$

Using now Lemma 10.4, we see that

$$V^*(t) \leq V^*(t_0), \ t \in [t_0, \bar{t}].$$

The choice of u_0 implies

$$V^*(\bar{t}) \leq V^*(t_0) = V(u_0) < a.$$

However, assumption (1) in conjunction with (10.77) yields

$$V^*(\bar{t}) = V(u(\bar{t}; t_0, u_0)) \geq a$$

which is a contradiction. ■

The following remark is useful in checking assumption (2) in Lemma 10.5 as well as other related inequalities.

Remark 10.2 If $x - V'(x) \in D(\varphi)$ and

$$(Ax + Bx + \Phi'(x) - f(t), V'(x)) + \varphi(x) - \varphi(x - V'(x)) \geq 0$$

hold for all $x \in D(\partial\varphi)$, $\|x\| \leq R$, $t \geq t_0$, then condition (2) in Lemma 10.5 is satisfied. Indeed, for an x as above one has

$$w \in \partial\varphi(x)$$

if and only if

$$\varphi(v) - \varphi(x) \geq (w, v - x), \ \forall v \in H.$$

Setting $v = x - V'(x)$ one obtains that

$$\varphi(x - V'(x)) - \varphi(x) \geq -(w, V'(x))$$

or, equivalently,

$$(w, V'(x)) \geq \varphi(x) - \varphi(x - V'(x)).$$

Combining this with the assumption in Remark 10.2 yields the desired conclusion.

Suggested by the approach in [20], we are now in a position to prove general abstract theorems of stability and asymptotic stability in terms of Lyapunov type function $V \in C^1(H; \mathbb{R})$ as required in Lemma 10.5.

Theorem 10.3 **(Stability)** Assume that conditions (H) hold and there exist $\sigma > 0$ and $V \in C^1(H; \mathbb{R})$ such that

(1)
$$V(x) \geq a(\|x\|),\ x \in D(\partial\varphi),\ \|x\| \leq \sigma,$$
with $a : [0, \sigma] \to \mathbb{R}$ satisfying $a(t) > 0$, $\forall t \in (0, \sigma)$;

(2) $V(0) = 0$;

(3) $(Ax + Bx + \Phi'(x) - f(t), V'(x)) + (w, V'(x)) \geq 0,\ x \in D(\partial\varphi)$, $\|x\| \leq \sigma$, $w \in \partial\varphi(x)$, $t \geq t_0$.

Then the (trivial) solution of problem $P(t_0, 0)$ is stable.

Proof. Without loss of generality, let $0 < \varepsilon < \sigma$. We have from assumption (1) that

$$V(x) \geq a(\varepsilon) > 0,\ x \in D(\partial\varphi),\ \|x\| = \varepsilon.$$

The function V is continuous and, by assumption (2), $V(0) = 0$. It results that there exists $\delta(\varepsilon) > 0$ such that

$$\|x_0\| \leq \delta(\varepsilon) \Rightarrow |V(x_0)| < a(\varepsilon).$$

We choose
$$0 < \eta(\varepsilon) < \min\{\varepsilon, \delta(\varepsilon)\}.$$

Let us now apply Lemma 10.5 with $R = \varepsilon$ and $a = a(\varepsilon)$. It is clear that assumptions (1), (2) in Lemma 10.5 are satisfied. On the other hand we notice that if $u_0 \in D(\partial\varphi)$ satisfies $\|u_0\| \leq \eta(\varepsilon)$, then $V(u_0) < a(\varepsilon)$ and $\|u_0\| < \varepsilon$. Thus the conclusion of Lemma 10.5 leads to

$$\|u(t; t_0, u_0)\| < \varepsilon,\ t \geq t_0,$$

which ensures that the trivial solution of $P(t_0, 0)$ is stable. ■

Let us recall that the tangent cone $T_K(x)$ of a subset K of H at $x \in K$ is defined by

$$T_K(x) = N_K(x)^- = \{w \in H : (w, v) \leq 0,\ \forall v \in N_K(x)\}.$$

In terms of tangent cone we derive the following.

Corollary 10.3 Assume condition (H_1) and let K be a closed convex subset of H with $0 \in K$. Assume that conditions (H_3)-(H_6) and (10.82) hold. Suppose that there exists $\sigma > 0$ and $V \in C^1(H; I\!R)$ such that

(1)

$$V(x) \geq a(\|x\|),\ x \in K,\ \|x\| \leq \sigma,$$

with $a : [0, \sigma] \to I\!R$ satisfying $a(t) > 0$, $\forall t \in (0, \sigma)$;

(2) $V(0) = 0$;

(3) $-V'(x) \in T_K(x),\ x \in \partial K,\ \|x\| \leq \sigma$;

(4) $(Ax + Bx + \Phi'(x) - f(t), V'(x)) \geq 0,\ x \in K,\ \|x\| \leq \sigma,\ t \geq t_0$.

Then the (trivial) solution of Cauchy problem (10.79)-(10.81) with $u_0 = 0$ is stable. In particular, the conclusion holds true if (3) is replaced by

(3)′ $x - V'(x) \in K,\ x \in \partial K,\ \|x\| \leq \sigma$.

Proof. Let us check that condition (3)′ implies (3). Assume that (3)′ is satisfied. Then for an x as in (3) and every $v \in N_K(x)$ we have

$$(v, y - x) \leq 0,\ \forall y \in K.$$

Therefore, taking $y = x - V'(x)$ (cf. (3)′), one obtains

$$(v, -V'(x)) \leq 0,\ \forall v \in N_K(x),$$

which means that $-V'(x) \in T_K(x)$, thus (3) is verified.

We apply Theorem 10.3 with $\varphi = \Psi_K$, i.e. φ is equal to the indicator function of K. Then (H_2) is clearly verified and (10.82) is equivalent to (H_7). Since (H_1), (H_3)-(H_6) as well as (1), (2) have been admitted, it remains to verify condition (3) in Theorem 10.3. Let $x \in K$ with $\|x\| \leq \sigma$. If $x \in \text{int}(K)$ then $\partial\Psi_K(x) = \{0\}$. If $x \in \partial K$ then $\partial\Psi_K(x) = N_K(x)$. Thus if $x \in \partial K$ and $w \in \partial\Psi_K(x)$ then, by (3), we obtain

$$(-V'(x), w) \leq 0.$$

It results that

$$(w, V'(x)) \geq 0,\ \forall x \in K,\ \|x\| \leq \sigma,\ w \in \partial\Psi_K(x).$$

Therefore (3) in Theorem 10.3 is satisfied. The conclusion follows from Theorem 10.3. ∎

A concrete situation showing the applicability of Corollary 10.3 is provided below.

Example 10.1. Let us consider problem (10.79)-(10.81) with $u_0 = 0$ where $H = I\!R^2$, $K = I\!R_+ \times I\!R_+$, $A \equiv 0$, $\Phi \equiv 0, f \equiv 0$ and

$$B(x_1, x_2) = (x_2, -\sin(x_1)),\ (x_1, x_2) \in I\!R^2.$$

It is clear that assumptions $(\mathtt{H_1})$, $(\mathtt{H_3})$-$(\mathtt{H_6})$ and (10.82) in Corollary 10.3 are satisfied. We choose

$$V(x_1, x_2) = \frac{x_2^2}{2} - \cos(x_1) + 1,\ \forall (x_1, x_2) \in I\!R^2.$$

We have for $|x_1|$ small enough that

$$1 - \cos(x_1) \geq \frac{x_1^2}{4}.$$

Thus, there exists $\sigma > 0$ such that

$$V(x_1, x_2) \geq \frac{x_1^2}{4} + \frac{x_2^2}{4},\ \|x\| \leq \sigma,$$

which establishes (1) in Corollary 10.3 taking $a(t) = \frac{t^2}{4}$ for $0 \leq t \leq \sigma$. Moreover, (2) in Corollary 10.3 is obviously satisfied. Since

$$V'(x_1, x_2) = (\sin(x_1), x_2)$$

it follows that

$$(B(x_1, x_2), V'(x_1, x_2)) = 0,\ (x_1, x_2) \in I\!R^2,$$

i.e. (4) in Corollary 10.3 holds. Using

$$(x_1, x_2) - V'(x_1, x_2) = (x_1 - \sin x_1, 0),$$

it turns out for σ above that

$$x = (x_1, x_2) \in \partial K,\ \|x\| \leq \sigma \Rightarrow (x_1, x_2) - V'(x_1, x_2) \in K,$$

i.e. $(3)'$ in Corollary 10.3 holds true. All assumptions of Corollary 10.3 are satisfied and we may conclude to the stability of the trivial solution of problem (10.79)-(10.81) with $u_0 = 0$.

We state now a basic abstract result concerning the asymptotic stability.

Theorem 10.4 **(Asymptotic Stability)** Assume that conditions (H) hold and there exist $\sigma > 0$, $\lambda > 0$ and $V \in C^1(H; I\!R)$ such that

(1)

$$V(x) \geq a(\|x\|), \ x \in D(\partial\varphi), \ \|x\| \leq \sigma,$$

with $a : [0, \sigma] \to \mathbb{R}$ satisfying $a(t) \geq ct^{\tau}$, $\forall t \in [0, \sigma]$, for some constants $c > 0$, $\tau > 0$;

(2) $V(0) = 0$;

(3) $(Ax + Bx + \Phi'(x) - f(t), V'(x)) + (w, V'(x)) \geq \lambda V(x)$, for all $x \in D(\partial\varphi)$, $\|x\| \leq \sigma$, $w \in \partial\varphi(x)$, $t \geq t_0$.

Then the (trivial) solution of $P(t_0, 0)$ is asymptotically stable.

Proof. The stability of the trivial solution follows from Theorem 10.3. In particular, there exists $\delta > 0$ such that every $u_0 \in D(\partial\varphi)$ with $\|u_0\| \leq \delta$ yields

$$\|u(t; t_0, u_0)\| \leq \sigma, \ t \geq t_0.$$

Here, using the same approach as in Lemma 10.5, we see that

$$\frac{dV^*}{dt}(t) = (V'(u(t; t_0, u_0)), \frac{d}{dt}u(t; t_0, u_0)) \leq -\lambda V^*(t), \ \text{a.e. } t \geq t_0.$$

Then, by Lemma 10.4 applied for every interval $[t_0, t_0 + h]$, $h > 0$, we obtain

$$V^*(t) \leq V^*(t_0)e^{-\lambda(t-t_0)}, \ t \geq t_0.$$

Thus, from hypothesis (1) and for all $u_0 \in D(\partial\varphi)$ with $\|u_0\| \leq \delta$, we get

$$a(\|u(t; t_0, u_0)\|) \leq V(u_0)e^{-\lambda(t-t_0)}, \ t \geq t_0.$$

Passing to the limit as $t \to +\infty$ we derive that

$$c \limsup_{t \to +\infty} \|u(t; t_0, u_0)\|^{\tau} \leq \limsup_{t \to +\infty} a(\|u(t; t_0, u_0)\|) \leq 0,$$

hence

$$\lim_{t \to +\infty} \|u(t; t_0, u_0)\| = 0.$$

■

The counterpart of Corollary 10.3 for asymptotic stability is formulated as follows.

Corollary 10.4 Assume condition (H_1) and let K be a closed convex set in H with $0 \in K$ provided that conditions (H_3)-(H_6) and (10.82) hold. Suppose that there exists $\lambda > 0$, $\sigma > 0$ and $V \in C^1(H; \mathbb{R})$ such that

(1)
$$V(x) \geq a(\|x\|),\ x \in K,\ \|x\| \leq \sigma,$$
with $a : [0,\sigma] \to I\!R$ satisfying $a(t) \geq ct^{\tau}$, $\forall t \in [0,\sigma]$, for some constants $c > 0$, $\tau > 0$;

(2) $V(0) = 0$;

(3) $-V'(x) \in T_K(x),\ x \in \partial K,\ \|x\| \leq \sigma$;

(4) $(Ax + Bx + \Phi'(x) - f(t), V'(x)) \geq \lambda V(x),\ x \in K,\ \|x\| \leq \sigma,\ t \geq t_0$.

Then the (trivial) solution of problem (10.79)-(10.81) with $u_0 = 0$ is asymptotically stable. In particular, the conclusion holds true if (3) is replaced by

(3)′ $x - V'(x) \in K,\ x \in \partial K,\ \|x\| \leq \sigma$.

Proof. One proceeds as in the proof of Corollary 10.3, applying Theorem 10.4 in place of Theorem 10.3. Specifically, one shows that condition (3) in Theorem 10.4 is fulfilled. Then one uses Theorem 10.4 with $\varphi = \Psi_K$. ■

We give now an unstability result.

Theorem 10.5 **(Unstability)** Assume that conditions (H) hold. Suppose in addition that 0 is a cluster point of $D(\partial\varphi)$. If there exist $V \in C^1(H; I\!R)$ and $\alpha > 0$ such that the following properties hold

(1)
$$V(x) \leq b(\|x\|),\ x \in D(\partial\varphi),$$
with $b : [0,+\infty[\to I\!R$ satisfying $b(t) \leq kt^{s}$, $\forall t \geq 0$, for some constants $k > 0$, $s > 0$;

(2) $V(x) > 0,\ x \in D(\partial\varphi),\ x \neq 0$ near 0;

(3) $(Ax + Bx + \Phi'(x) - f(t), V'(x)) + (w, V'(x)) \leq -\alpha V(x),\ x \in D(\partial\varphi),$ $w \in \partial\varphi(x),\ t \geq t_0$,

then the (trivial) solution of problem $P(t_0, 0)$ is unstable.

Proof. Let $\varepsilon > 0$. For a small $\eta > 0$ let us choose $u_0 \in D(\partial\varphi)$ such that $0 < \|u_0\| \leq \eta$. This choice is possible since 0 is a cluster point of $D(\partial\varphi)$. Let u be the unique solution of problem $P(t_0, u_0)$. We have from (10.76) that some mapping $w : [t_0, +\infty[\to H$ exists such that

$$(\frac{du}{dt}(t) + Au(t) + Bu(t) + \Phi'(u(t)) - f(t), v)$$

$$+(w(t), v) = 0, \ \forall v \in H, \text{ a.e. } t \geq t_0$$

and

$$w(t) \in \partial\varphi(u(t)), \text{ a.e. } t \geq t_0.$$

Using assumption (3) we obtain

$$0 \leq (\frac{du}{dt}(t), V'(u(t))) - \alpha V(u(t)), \text{ a.e. } t \geq t_0,$$

so that

$$\frac{d}{dt}V^*(t) \geq \alpha V^*(t), \text{ a.e. } t \geq t_0,$$

with $V^*(t) = V(u(t))$, $t \geq t_0$. From Lemma 10.4 applied for every bounded subinterval of $[t_0, +\infty[$ it is clear that

$$V^*(t) \geq V^*(t_0)e^{\alpha(t-t_0)} = V(u_0)e^{\alpha(t-t_0)}, \ t \geq t_0.$$

From assumption (2) we get $V(u_0) > 0$ while assumption (1) yields

$$k\|u(t)\|^s \geq b(\|u(t)\|) \geq V(u_0)e^{\alpha(t-t_0)}, \ t \geq t_0.$$

Choosing $\bar{t} > \max\{t_0, \frac{1}{\alpha}\ln\Big(\frac{k\varepsilon^s e^{\alpha t_0}}{V(u_0)}\Big)\}$, we see that

$$\|u(\bar{t})\| > \varepsilon.$$

The unstability of the trivial solution follows. ∎

Let us denote by K_∞ the recession cone of a nonempty, convex, closed set K in H, i.e.

$$K_\infty = \bigcap_{\lambda > 0} \lambda(K - x_0),$$

where x_0 is an arbitrary fixed element of K. The set K_∞ is a closed convex cone. Let us recall that

$$K + K_\infty \subset K.$$

A verifiable criterium of unstability is then available.

Corollary 10.5 Assume condition (H_1) and let K be a closed convex set in H with $0 \in K$ and $K \setminus \{0\} \neq \emptyset$. Suppose that ($H_3$)-($H_6$) and (10.82) hold. If there exist $V \in C^1(H; I\!R)$ and $\alpha > 0$ such that the properties below are valid

(1)

$$V(x) \leq b(\|x\|), \ x \in K,$$

with $b : [0, +\infty[\to \mathbb{R}$ satisfying $b(t) \leq kt^s$, $\forall t \geq 0$, for some constants $k > 0$, $s > 0$;

(2) $V(x) > 0$, $x \in K$, $x \neq 0$ near 0;

(3) $V'(x) \in K_\infty$, $x \in \partial K$;

(4) $(Ax + Bx + \Phi'(x) - f(t), V'(x)) \leq -\alpha V(x)$, $x \in K$, $t \geq t_0$,

then the (trivial) solution of (10.79)-(10.81) with $u_0 = 0$ is unstable.

Proof. We apply Theorem 10.5 with $\varphi = \Psi_K$. Notice that the imposed assumptions guarantee that 0 is a cluster point of K. Since assumption (H_2) is fulfilled with $\varphi = \Psi_K$ as well as (H_7), comparing the hypotheses of Corollary 10.5 and Theorem 10.5 it is seen that we have to check only condition (3) of Theorem 10.5. If $x \in \text{int}(K)$ then $\partial\Psi_K(x) = \{0\}$. If $x \in \partial K$ then

$$(w, v - x) \leq 0, \ \forall v \in K, \ w \in \partial\Psi_K(x).$$

By (3), for $x \in \partial K$ we have $V'(x) \in K_\infty$ and $x + V'(x) \in K$. Setting $v = x + V'(x)$, we obtain $(w, V'(x)) \leq 0$, $\forall w \in \partial\Psi_K(x)$. The relations above lead to

$$(w, V'(x)) \leq 0, \ \forall x \in K, \ w \in \partial\Psi_K(x).$$

This together with (4) implies that assumption (3) of Theorem 10.5 is satisfied. The conclusion follows from Theorem 10.5. ■

We pass now to applying our stability results to nonlinear variational inequalities in a Hilbert spaces using monotonicity type assumptions. A series of sufficient conditions are presented.

Corollary 10.6 Assume conditions (H) hold. If

(c) there exists $\sigma > 0$ such that $(A(x)-A(0), x)+(B(x)-B(0), x)+(\Phi'(x)-\Phi'(0), x) \geq 0$, $x \in D(\partial\varphi)$, $\|x\| \leq \sigma$

then the (trivial) solution of problem $P(t_0, 0)$ is stable.

Proof. In order to apply Theorem 10.3, we set

$$V(x) = \frac{1}{2}\|x\|^2, \ \forall x \in H.$$

Then $V'(x) = x$ and, using (H_7) and (c), we get

$$(A(x) + B(x) + \Phi'(x) - f(t), V'(x)) + \varphi(x) - \varphi(x - V'(x))$$

$$= (A(x) + B(x) + \Phi'(x) - f(t), x) + \varphi(x) - \varphi(0)$$
$$\geq (A(x) + B(x) + \Phi'(x), x) - (A(0) + B(0) + \Phi'(0), x)$$
$$\geq 0,\ x \in D(\partial\varphi),\ \|x\| \leq \sigma,\ t \geq t_0.$$

We have that $x - V'(x) = 0 \in D(\partial\varphi)$ since $(\mathtt{H}_2)$ holds. By Remark 10.2, we see that all the assumptions of Theorem 10.3 with $a(t) = \frac{1}{2}t^2$ are satisfied and the conclusion follows. ■

Remark 10.3 As an example of verifiable situation where assumption (c) in Corollary 10.6 is satisfied we indicate

$(\mathtt{c}_1)$ $\quad (A(x) - A(0), x) \geq 0,\ x \in D(\partial\varphi)$

$(\mathtt{c}_2)$ $\quad (B(x) - B(0), x) \geq 0,\ x \in D(\partial\varphi),$

in particular if A and B are monotone operators. Indeed, Φ being convex, conditions $(\mathtt{c}_1)$, $(\mathtt{c}_2)$ above imply (c) in Corollary 10.6.

Corollary 10.7 Assume conditions (H) hold. If

(c)′ there exist $\sigma > 0$ and $\lambda > 0$ such that
$(A(x) - A(0), x) + (B(x) - B(0), x) + (\Phi'(x) - \Phi'(0), x) \geq \lambda\|x\|^2,$
$x \in D(\partial\varphi),\ \|x\| \leq \sigma,$

then the (trivial) solution of problem $P(t_0, 0)$ is asymptotically stable.

Proof. Let us apply Theorem 10.4 through Remark 10.2. We set

$$V(x) = \frac{1}{2}\|x\|^2,\ \forall x \in H.$$

Then $V'(x) = x$ and, using $(\mathtt{H}_7)$ and (c)′, we derive

$$(A(x) + B(x) + \Phi'(x) - f(t), V'(x)) + \varphi(x) - \varphi(x - V'(x))$$
$$\geq (A(x) + B(x) + \Phi'(x), x) - (A(0) + B(0) + \Phi'(0), x)$$
$$\geq \lambda\|x\|^2 = 2\lambda V(x),\ x \in D(\partial\varphi),\ \|x\| \leq \sigma,\ t \geq t_0.$$

We have that $0 \in D(\varphi)$ as remarked in the proof of Corollary 10.6. By means of Remark 10.2, Theorem 10.4 can be applied which completes the proof. ■

Remark 10.4 An example of situation complying with assumption (c)′ in Corollary 10.7 is when we suppose

$(\mathtt{c}_1)'$ there exist $r > 0$ and $\alpha > 0$ such that

$$(A(x) - A(0), x) \geq \alpha\|x\|^2,\ x \in D(\partial\varphi),\ \|x\| \leq r;$$

(c$_2$)$'$ there exists $\theta > 0$ such that

$$(B(x) - B(0), x) \geq -\Psi(\|x\|)\|x\|^2, \ x \in D(\partial\varphi), \ \|x\| \leq \theta,$$

where $\Psi : \mathbb{R}_+ \to \mathbb{R}_+$ satisfies

$$\lim_{t \to 0^+} \Psi(t) = 0.$$

Indeed, there exists $\delta > 0$ such that

$$\Psi(\|x\|) \leq \frac{\alpha}{2}, \ x \in H, \ \|x\| \leq \delta.$$

Using (c$_1$)$'$, (c$_2$)$'$ and (H$_5$) we have

$$(A(x) - A(0), x) + (B(x) - B(0), x) + (\Phi'(x) - \Phi'(0), x)$$

$$\geq (\alpha - \Psi(\|x\|))\|x\|^2 \geq \frac{\alpha}{2}\|x\|^2, \ x \in D(\partial\varphi), \ \|x\| \leq \sigma, \ t \geq t_0,$$

where

$$\sigma = \min\{\delta, \theta, r\},$$

i.e. (c)$'$ in Corollary 10.7 holds true.

Corollary 10.8 Assume condition (H$_1$) and, for a closed convex cone K in H provided $K \setminus \{0\} \neq \emptyset$, assume that conditions (H$_3$), (H$_4$), (H$_6$) with $f(t) - (A(0) + B(0)) \in K^\perp$, $\forall t \geq t_0$, hold. If

(c)$''$ there exists $\alpha > 0$ such that
$(A(x) - A(0), x) + (B(x) - B(0), x) \leq -\alpha\|x\|^2, \ x \in K$

then the (trivial) solution of (10.79)-(10.81) with $u_0 = 0$ and $\Phi = 0$ is unstable.

Proof. We claim that the hypotheses of Corollary 10.5 with $\Phi = 0$ are verified. To this end, we set

$$V(x) = \frac{1}{2}\|x\|^2, \ \forall x \in H.$$

Assumption (1) in Corollary 10.5 is fulfilled with $b(t) = \frac{1}{2}t^2$, $\forall t \geq 0$, while assumption (2) is obvious. Since K is a cone one has $K_\infty = K$. It results that $V'(\partial K) = \partial K \subset K_\infty$, so (3) in Corollary 10.5 holds. Moreover, we may write

$$(A(x) + B(x) - f(t), x) = (A(x) + B(x), x) - (A(0) + B(0), x)$$

$$\leq -\alpha\|x\|^2, \ x \in K, \ t \geq t_0,$$

which yields (4) in Corollary 10.5. The conclusion follows from Corollary 10.5. ■

Remark 10.5 An example of situation described in assumption (c)″ of Corollary 10.8 is given by

$(c_1)''$ there exists $\bar{\alpha} > 0$ such that

$$(A(x) - A(0), x) \leq -\bar{\alpha}\|x\|^2, \; x \in K;$$

$(c_2)''$ there exists $\bar{\gamma} < \bar{\alpha}$ such that

$$(B(x) - B(0), x) \leq \bar{\gamma}\|x\|^2, \; x \in K.$$

Indeed, by $(c_1)''$ and $(c_2)''$, one has

$$(A(x) - A(0), x) + (B(x) - B(0), x) \leq (\bar{\gamma} - \bar{\alpha})\|x\|^2, \; x \in K.$$

The inequality $\bar{\gamma} - \bar{\alpha} < 0$ shows that condition (c)″ in Corollary 10.8 is fulfilled.

Remark 10.6. For related results of stability on a finite dimensional space as encountered in mechanics we refer to [4]. Many important problems in Mechanics like classical unilateral contact and Coulomb friction problems (see [12], [26]) involve inequalities in the context of stability. The subject of stability in unilateral mechanics has been dealt with numerous numerical studies (see [12] and the references cited therein).

We close the Section by showing how Quittner's conditions in [17]-[19] can be deduced from our results.

Corollary 10.9 Assume condition (H_1) and let K be a closed convex set in H with $0 \in K$. Assume that conditions (H_3)-(H_5) and (10.82) with a given $f \in H$ hold. Suppose in addition that

$$(Bx + \Phi'(x), x) \geq -\gamma(\|x\|)\|x\|^2, \; x \in K, \; \|x\| \leq \theta, \tag{10.84}$$

where $\theta > 0$ and $\gamma : I\!R_+ \to I\!R_+$ satisfies $\gamma(t) \to 0$ as $t \to 0^+$. In the case where $K \neq \{0\}$, we further assume that

$$\lambda_I := \liminf_{x \in K, \|x\| \to 0} \frac{(Ax - f, x)}{\|x\|^2} > 0. \tag{10.85}$$

Then the (trivial) solution of (10.79)-(10.81) with $u_0 = 0$ and $f(t) \equiv f \in H$ is asymptotically stable.

Proof. For applying Corollary 10.4, we set

$$V(x) = \frac{1}{2}\|x\|^2, \ \forall x \in H.$$

Then (1) and (2) in Corollary 10.4 are valid. We choose $0 < \sigma \leq \theta$ sufficiently small to have by (10.84) that

$$(Bx + \Phi'(x), x) \geq -\frac{\lambda_I}{4}\|x\|^2, \ x \in K, \ \|x\| \leq \sigma$$

and, making use of (10.85),

$$(Ax - f, x) \geq \frac{\lambda_I}{2}\|x\|^2, \ x \in K, \ \|x\| \leq \sigma.$$

Then for $x \in K$, $\|x\| \leq \sigma$, we infer that

$$(Ax + Bx + \Phi'(x) - f, V'(x)) \geq \frac{\lambda_I}{2}\|x\|^2 - \frac{\lambda_I}{4}\|x\|^2 = \frac{\lambda_I}{4}\|x\|^2 = \frac{\lambda_I}{2}V(x).$$

Thus (4) in Corollary 10.4 holds. Since condition $(3)'$ is evident, the conclusion follows from Corollary 10.4. ■

Remark 10.7. In particular, Corollary 10.9 holds if, in place of assumption (10.84), one assumes that

$$(Bx, x) \geq -\gamma(\|x\|)\|x\|^2, \ x \in K, \ \|x\| \leq \theta \tag{10.86}$$

and

$$-\Phi'(0) \in N_K(0). \tag{10.87}$$

This is true because taking into account (H_5), conditions (10.86) and (10.87) imply (10.84) since

$$\begin{aligned}(Bx + \Phi'(x), x) &= (Bx, x) + (\Phi'(x) - \Phi'(0), x) + (\Phi'(0), x) \\ &\geq -\gamma(\|x\|)\|x\|^2, \ x \in K, \ \|x\| \leq \theta.\end{aligned}$$

In the result below condition (10.84) is strengthened, while assumption (10.89) is more general then (10.85) (which is obtained from (10.89) taking L equal to the identity).

Corollary 10.10. Assume condition (H_1) and let K be a closed convex set in H with $0 \in K$. Assume that conditions (H_3)-(H_5) and (10.82) with a given $f \in H$ hold. Suppose in addition that

$$\|Bx + \Phi'(x)\| \leq \gamma(\|x\|)\|x\|, \ x \in K, \ \|x\| \leq \theta, \tag{10.88}$$

where $\theta > 0$ and $\gamma : I\!R_+ \to I\!R_+$ satisfies $\gamma(t) \to 0$ as $t \to 0^+$. Let $L : H \to H$ be a bounded linear operator such that $\mathrm{Ker}(L) = \{0\}$, $R(L)$ closed and $(I - L^*L)(\partial K) \subset K$. In the case where $K \neq \{0\}$, we further assume that

$$\lambda_{\mathrm{II}} := \liminf_{x \in K, \|x\| \to 0} \frac{(Ax - f, L^*Lx)}{\|Lx\|^2} > 0. \tag{10.89}$$

Then the (trivial) solution of (10.79)-(10.81) with $u_0 = 0$ $f(t) \equiv f \in H$ is asymptotically stable.

Proof. We apply Corollary 10.4 for

$$V(x) = \frac{1}{2}\|Lx\|^2, \; \forall x \in H.$$

The conditions $\mathrm{Ker}(L) = \{0\}$ and $R(L)$ closed yield the existence of a constant $c > 0$ such that

$$\|Lx\| \geq c\|x\|, \; x \in H.$$

Thus

$$V(x) \geq \frac{c^2}{2}\|x\|^2, \; x \in H.$$

Consequently, assumptions (1) and (2) in Corollary 10.4 are fulfilled. Moreover we derive that

$$V(x) = \frac{1}{2}(Lx, Lx) = \frac{1}{2}(L^*Lx, x)$$

and

$$V'(x) = L^*Lx.$$

Assumption $(I - L^*L)(\partial K) \subset K$ implies condition (3)' in Corollary 10.4. We choose $\sigma \in]0, \theta]$ small enough to have, according to (10.88),

$$\|Bx + \Phi'(x)\| \leq \frac{c\lambda_{\mathrm{II}}}{4\|L\|}\|x\|, \; x \in K, \; \|x\| \leq \sigma$$

and, making use of (10.89),

$$(Ax - f, L^*Lx) \geq \frac{\lambda_{\mathrm{II}}}{2}\|Lx\|^2, \; x \in K, \; \|x\| \leq \sigma.$$

Then, for $x \in K$ with $\|x\| \leq \sigma$ we can write

$$\begin{aligned}(Ax + Bx + \Phi'(x) - f, V'(x)) &= (Ax - f, L^*Lx) + (Bx + \Phi'(x), L^*Lx) \\ &\geq \frac{\lambda_{\mathrm{II}}}{2}\|Lx\|^2 + (L(Bx + \Phi'(x)), Lx)\end{aligned}$$

$$\geq \frac{\lambda_{\mathrm{II}}}{2}\|Lx\|^2 - \|L\|\,\|Bx + \Phi'(x)\|\,\|Lx\| \geq \frac{\lambda_{\mathrm{II}}}{2}\|Lx\|^2 - \|L\|\frac{c\lambda_{\mathrm{II}}}{4\|L\|}\|x\|\,\|Lx\|$$

$$\geq \frac{\lambda_{\mathrm{II}}}{2}\|Lx\|^2 - \frac{\lambda_{\mathrm{II}}}{4}\|Lx\|^2 = \frac{\lambda_{\mathrm{II}}}{4}\|Lx\|^2 = \frac{\lambda_{\mathrm{II}}}{2}V(x).$$

This shows that assumption (4) of Corollary 10.4 is satisfied. Corollary 10.4 yields the result. ■

References

[1] C. Baiocchi and A. Capelo, *Variational and Quasivariational Inequalities. Applications to Free Boundary Problems*, John Wiley and Sons, New York, 1984.

[2] H. Brézis, Problèmes unilatéraux, *J. Math. Pures Appl.* **51** (1972), 1-168.

[3] H. Brézis, *Opérateurs Maximaux Monotones et Semi-groupes de Contractions dans les Espaces de Hilbert*, North Holland, Amsterdam, 1973.

[4] B. Brogliato, *Absolute stability and the Lagrange-Dirichlet theorem with monotone multivalued mappings*, Internal Report, INRIA Rhônes-Alpes, 2002.

[5] O. Chau, D. Motreanu and M. Sofonea, Quasistatic frictional problems for elastic and viscoelastic materials, *Appl. Math.* **47** (2002), 341-360.

[6] C. Ciulcu, D. Motreanu and M. Sofonea, Analysis of an elastic contact problem with slip dependent coefficient of friction, *Math. Inequal. Appl.* **4** (2001), 465-479.

[7] C. Corneschi, T.-V. Hoarau-Mantel and M. Sofonea, A quasistatic contact problem with slip dependent coefficient of friction for elastic materials, *J. Appl. Anal.* **8** (2002), 59-80.

[8] G. Duvaut and J. L. Lions, *Inequalities in Mechanics and Physics*, Springer-Verlag, Berlin, 1976.

[9] D. Goeleven, D. Motreanu and V. V. Motreanu, On the stability of stationary solutions of first order evolution variational inequalities, *Adv. Nonlinear Var. Inequal.* **6** (2003), to appear.

[10] W. Han and M. Sofonea, *Quasistatic Contact Problems in Viscoelasticity and Viscoplasticity*, Studies in Advanced Mathematics, American Mathematical Society-International Press, to appear.

[11] D. Kinderlehrer and G. Stampacchia, *An Introduction to Variational Inequalities and Their Applications*, Academic Press, New York/London/Toronto/Sydney/San Francisco, 1980.

[12] J.A.C. Martins, S. Barbarin, M. Raous and A.P. da Costa, Dynamic stability of finite dimensional linearly elastic systems with unilateral contact and Coulomb friction, *Comput. Methods Appl. Mech. Eng.* **177** (1999), 289-328.

[13] A. Matei, V.V. Motreanu and M. Sofonea, A quasistatic antiplane contact problem with slip dependent friction, *Adv. Nonlinear Var. Inequal.* **4** (2001), 1-21.

[14] D. Motreanu and M. Sofonea, Evolutionary variational inequalities arising in quasistatic frictional contact problems for elastic materials, *Abstr. Appl. Anal.* **4** (1999) 255-279.

[15] D. Motreanu and M. Sofonea, Quasivariational inequalities and applications in frictional contact problems with normal compliance, *Adv. Math. Sci. Appl.* **10** (2000), 103-118.

[16] D. Motreanu and M. Sofonea, Second order variational equations and applications in dynamic contact problems for elastic materials, preprint.

[17] P. Quittner, On the principle of linearized stability for variational inequalities, *Math. Ann.* **283** (1989), 257-270.

[18] P. Quittner, On the stability of stationary solutions of parabolic variational inequalities, *Czech. Math. J.* **40** (1990), 472-474.

[19] P. Quittner, An instability criterion for variational inequalities, *Nonlinear Anal.* **15** (1990), 1167-1180.

[20] N. Rouche and J. Mawhin, *Equations Différentielles Ordinaires*, Tome 2, Masson and Cie, Paris, 1973.

[21] K. Tsilika, Study of an adhesively supported von Krmn plate. Existence and bifurcation of the solutions, in: Nonsmooth/nonconvex mechanics (Blacksburg, VA, 1999), 411–425, Nonconvex Optim. Appl.**50**, Kluwer Acad. Publ., Dordrecht, 2001.

[22] K. Tsilika, Buckling of a von Krmn plate adhesively connected to a rigid support allowing for delamination: existence and multiplicity results, *J. Global Optim.* **17** (2000), 387-402.

[23] K. Tsilika, On the buckling of an adhesively supported beam. A resonant eigenvalue problem for a hemivariational inequality, *Numer. Funct. Anal. Optim.* **23** (2002), 217-225.

[24] R. U. Verma, Nonlinear variational inequalities on convex subsets of Banach spaces, Appl. Math. Lett., 10 (1997), 25-27.

[25] R. U. Verma, On monotone nonlinear variational inequalities problems, Comment. Math. Univ. Carolinae 39 (1998), 91-98.

[26] D. Vola, M. Raous and J.A.C. Martins, Friction and instability of steady sliding: squeal of a rubber/glass contact, *Int. J. Numer. Methods Eng.* **46** (1999), 1699-1720.

[27] E. Zeidler, *Nonlinear Functional Analysis and its Applications*, I: Fixed-Point Theorems, Springer-Verlag, New York, 1986.

Chapter 11

INEQUALITY PROBLEMS IN BV AND GEOMETRIC APPLICATIONS

The theory of variational inequalities appeared in the middle 60's in connection with the notion of subdifferential in the sense of Convex Analysis (see e.g. [4], [10], [16] for the main aspects of this theory). All the inequality problems treated to the beginning 80's were related to convex energy functionals and therefore strictly connected to monotonicity: for instance, only monotone (possibly multivalued) boundary conditions and stress-strain laws could be studied. Nonconvex inequality problems first appeared in [18] in the setting of Global Analysis and were related to the subdifferential introduced in [7] (see A. Marino [17] for a survey of the developments in this direction).

In this Chapter we consider hemivariational inequalities containing both an area-type and a non-locally Lipschitz term. For this reason the Clarke theory does not apply and the main purpose is to show how multiplicity results can be obtained by means of nonsmooth techniques in the sense of the critical point theory introduced by Degiovanni (see Section 3 in Chapter 1). The energy functional associated to problems of this type does not satisfy the Palais-Smale condition in BV, which is its natural domain. After extending this functional in a larger space, the nonsmoothness of the energy increases. However, it is striking to remark that the new space is better behaved for the compactness properties. The Degiovanni critical point theory will show its force in this new framework for proving various existence and multiplicity theorems.

After giving some general abstract results involving critical point theory in the sense of Degiovanni we establish in Section 2 several properties of area type energy functionals. Finally, in Sections 3 and 4 we apply this general setting to obtain multiplicity results of Clark and Ambrosetti-

Rabinowitz type. We believe that our approach could be equally applied to other situations with different geometries.

1. The General Framework

In the setting of Continuum Mechanics, P. D. Panagiotopoulos started the study of nonconvex and nonsmooth potentials by using the Clarke subdifferential for locally Lipschitz functionals. Due to the lack of convexity, new types of inequality problems, called hemivariational inequalities, have been generated. Roughly speaking, mechanical problems involving nonmonotone stress-strain laws or boundary conditions derived by nonconvex superpotentials lead to hemivariational inequalities. We refer the reader to [23], [24] for the main aspects of this theory.

A typical feature of nonconvex problems is that, while in the convex case the stationary variational inequalities give rise to minimization problems for the potential or for the energy, in the nonconvex case the problem of the stationarity of the potential emerges and therefore it becomes reasonable to expect results also in the line of critical point theory.

For hemivariational inequalities, several contributions have been recently obtained by techniques of nonsmooth critical point theory (see [3], [11]-[15], [20]-[22], [25] and references therein). The associated functional f is typically of the form $f = f_0 + f_1$, where f_0 is the principal part satisfying some standard coerciveness condition and f_1 is locally Lipschitz. In such a setting, the main abstract tool is constituted by the nonsmooth critical point theory developed in [5] for locally Lipschitz functionals.

Our main purpose is to obtain existence and multiplicity results for hemivariational inequalities associated with functionals which come from the relaxation of

$$f(u) = \int_\Omega \sqrt{1 + |Du|^2}\, dx + \int_\Omega G(x, u)\, dx\,, \quad u \in W_0^{1,1}(\Omega; \mathbb{R}^N)\,,$$

where Ω is an open set in $\mathbb{R}^n$, $n \geq 2$. The first feature is that the functional f does not satisfy the Palais-Smale condition in $BV(\Omega; \mathbb{R}^N)$, the natural domain of f. Therefore we extend f to $L^{\frac{n}{n-1}}(\Omega; \mathbb{R}^N)$ with value $+\infty$ outside $BV(\Omega; \mathbb{R}^N)$. This larger space is better behaved for the compactness properties, but the nonsmoothness of the functional increases. The second feature is that the assumptions we impose on G imply the second term of f to be continuous on $L^{\frac{n}{n-1}}(\Omega; \mathbb{R}^N)$, but not locally Lipschitz. More precisely, the function $\{s \mapsto G(x, s)\}$ is supposed to be locally Lipschitz for a.e. $x \in \Omega$, but the growth conditions we impose do not ensure the corresponding property for the integral on

$L^{\frac{n}{n-1}}(\Omega; \mathbb{R}^N)$. Because of these facts, we will take advantage of the nonsmooth techniques developed by Degiovanni *et al.* and turn out to be suitable also for our setting.

We recall two critical point theorems we will apply later. The first one is an adaptation of a result of D. C. Clark (see [6] and [26], Theorem 9.1) to our setting.

Theorem 11.1 Let X be a Banach space and $f : X \to \mathbb{R} \cup \{+\infty\}$ an even lower semicontinuous function. Assume that

(*a*) f is bounded from below;

(*b*) for every $c < f(0)$, the function f satisfies $(PS)_c$ and $(epi)_c$ (see Definitions 1.11 and 1.12);

(*c*) there exist $k \geq 1$ and an odd continuous map $\psi : S^{k-1} \to X$ such that

$$\sup\left\{f(\psi(x)) : x \in S^{k-1}\right\} < f(0),$$

where S^{k-1} denotes the unit sphere in $\mathbb{R}^k$.

Then f admits at least k pairs $(u_1, -u_1), \ldots, (u_k, -u_k)$ of critical points with $f(u_j) < f(0)$.

Proof. See [9], Theorem 2.5. ■

The next result is an adaptation of the classical theorem of Ambrosetti and Rabinowitz [1], [26], [28].

Theorem 11.2 Let X be a Banach space and $f : X \to \mathbb{R} \cup \{+\infty\}$ an even lower semicontinuous function. Assume that there exists a strictly increasing sequence $\{V_h\}$ of finite-dimensional subspaces of X with the following properties:

(*a*) there exist a closed subspace Z of X, $\varrho > 0$ and $\alpha > f(0)$ such that $X = V_0 \oplus Z$ and

$$\forall u \in Z : \|u\| = \varrho \Longrightarrow f(u) \geq \alpha;$$

(*b*) there exists a sequence $\{R_h\}$ in $]\varrho, +\infty[$ such that

$$\forall u \in V_h : \|u\| \geq R_h \Longrightarrow f(u) \leq f(0);$$

(*c*) for every $c \geq \alpha$, the function f satisfies $(PS)_c$ and $(epi)_c$;

(*d*) we have $\left|d_{\mathbb{Z}_2}\mathcal{G}_f\right|(0, \lambda) \neq 0$ whenever $\lambda \geq \alpha$ (see Definition 1.9).

Then there exists a sequence $\{u_h\}$ of critical points of f with $f(u_h) \to +\infty$.

Proof. Because of assumption (c), the function $\mathcal{G}_f$ satisfies $(PS)_c$ for any $c \geq \alpha$. Then the assertion follows from [19], Theorem (2.7). ■

Let $n \geq 1$, $N \geq 1$, Ω be an open subset of $I\!R^n$ and $1 < p < \infty$. In the following, we will denote by $\|\cdot\|_q$ the usual norm in L^q $(1 \leq q \leq \infty)$. We now define the functional setting we are interested in.

Let $\mathcal{E} : L^p(\Omega; I\!R^N) \to I\!R \cup \{+\infty\}$ be a functional such that:

$(\mathcal{E}_1)$ $\mathcal{E}$ is convex, lower semicontinuous and $0 \in \mathcal{D}(\mathcal{E})$, where

$$\mathcal{D}(\mathcal{E}) = \left\{ u \in L^p(\Omega; I\!R^N) : \mathcal{E}(u) < +\infty \right\} ;$$

$(\mathcal{E}_2.1)$ there exists $\vartheta \in C_c(I\!R^N)$ with $0 \leq \vartheta \leq 1$ and $\vartheta(0) = 1$ such that $\forall u \in \mathcal{D}(\mathcal{E})$, $\forall v \in \mathcal{D}(\mathcal{E}) \cap L^\infty(\Omega; I\!R^N)$, $\forall c > 0$ we have

$$\lim_{h \to \infty} \left[\sup_{\substack{\|z-u\|_p \leq c \\ \mathcal{E}(z) \leq c}} \mathcal{E}\left(\vartheta\left(\frac{z}{h}\right) v \right) \right] = \mathcal{E}(v);$$

$(\mathcal{E}_2.2)$ For all $u \in \mathcal{D}(\mathcal{E})$:

$$\lim_{h \to \infty} \mathcal{E}\left(\vartheta\left(\frac{u}{h}\right) u \right) = \mathcal{E}(u) .$$

Moreover, let $G : \Omega \times I\!R^N \to I\!R$ be a function such that

(G_1) $G(\cdot, s)$ is measurable for every $s \in I\!R^N$;

(G_2) for every $t > 0$ there exists $\alpha_t \in L^1(\Omega)$ such that

$$|G(x, s_1) - G(x, s_2)| \leq \alpha_t(x)|s_1 - s_2|$$

for a.e. $x \in \Omega$ and every $s_1, s_2 \in I\!R^N$ with $|s_j| \leq t$; for a.e. $x \in \Omega$ we set

$$G^\circ(x, s; \hat{s}) = \gamma^\circ(s; \hat{s}) , \quad \partial_s G(x, s) = \partial\gamma(s) ,$$

where $\gamma(s) = G(x, s)$ (see Definition 1.13);

(G_3) there exist $a_0 \in L^1(\Omega)$ and $b_0 \in I\!R$ such that

$$G(x, s) \geq -a_0(x) - b_0|s|^p \quad \text{for a.e. } x \in \Omega \text{ and every } s \in I\!R^N ;$$

(G_4) there exist $a_1 \in L^1(\Omega)$ and $b_1 \in I\!R$ such that

$$G^\circ(x, s; -s) \leq a_1(x) + b_1|s|^p \quad \text{for a.e. } x \in \Omega \text{ and every } s \in I\!R^N .$$

Because of $(\mathcal{E}_1)$ and (G_3), we can define a lower semicontinuous functional $f : L^p(\Omega; \mathbb{R}^N) \to \mathbb{R} \cup \{+\infty\}$ by

$$f(u) = \mathcal{E}(u) + \int_\Omega G(x, u(x))\, dx\,.$$

Remark 11.1 According to $(\mathcal{E}_1)$, the functional $\mathcal{E}$ is lower semicontinuous. Condition $(\mathcal{E}_2)$ ensures that $\mathcal{E}$ is continuous at least for some particular restrictions.

Remark 11.2 If $s \mapsto G(x,s)$ is of class C^1 for a.e $x \in \Omega$, the estimates in (G_2) and in (G_4) are respectively equivalent to

$$|s| \le t \Longrightarrow |D_s G(x,s)| \le \alpha_t(x)\,,$$

$$D_s G(x,s) \cdot s \ge -a_1(x) - b_1|s|^p\,.$$

Because of (G_2), for a.e. $x \in \Omega$ and any $t > 0$ and $s \in \mathbb{R}^N$ with $|s| < t$ we have

$$\forall \hat{s} \in \mathbb{R}^N : |G^\circ(x, s; \hat{s})| \le \alpha_t(x)|\hat{s}|\,; \tag{11.1}$$

$$\forall s^* \in \partial_s G(x,s) : |s^*| \le \alpha_t(x)\,. \tag{11.2}$$

In the following, we set $\vartheta_h(s) = \vartheta(s/h)$, where ϑ is a function as in $(\mathcal{E}_2)$, and we fix $M > 0$ such that $\vartheta = 0$ outside $B_M(0)$. Therefore

$$\forall s \in \mathbb{R}^N : |s| \ge hM \Longrightarrow \vartheta_h(s) = 0\,. \tag{11.3}$$

Our first result concerns the connection between the notions of generalized directional derivative and subdifferential in the functional space $L^p(\Omega; \mathbb{R}^N)$ (in the sense of Definitions 1.13 and 1.14) and the more concrete setting of hemivariational inequalities, which also involves the notion of generalized directional derivative, but in $\mathbb{R}^N$.

If $u, v \in L^p(\Omega; \mathbb{R}^N)$, we can define $\int_\Omega G^\circ(x, u; v)\, dx$ if we agree, as in [27], that

$$\int_\Omega G^\circ(x, u; v)\, dx = +\infty$$

whenever

$$\int_\Omega [G^\circ(x, u; v)]^+\, dx = \int_\Omega [G^\circ(x, u; v)]^-\, dx = +\infty\,.$$

With this convention, $\{v \mapsto \int_\Omega G^\circ(x, u; v)\, dx\}$ is a convex functional from $L^p(\Omega; \mathbb{R}^N)$ into $\overline{\mathbb{R}}$.

Theorem 11.3 (Degiovanni, Marzocchi and Rădulescu [8]) Let $u \in \mathcal{D}(f)$. Then the following facts hold:

(a) for every $v \in \mathcal{D}(\mathcal{E})$ there exists a sequence $\{v_h\}$ in $\mathcal{D}(\mathcal{E}) \cap L^\infty(\Omega; \mathbb{R}^N)$ satisfying $[G^\circ(x, u; v_h - u)]^+ \in L^1(\Omega)$, $\|v_h - v\|_p \to 0$ and $\mathcal{E}(v_h) \to \mathcal{E}(v)$;

(b) for every $v \in \mathcal{D}(\mathcal{E})$ we have

$$f^\circ(u; v - u) \leq \mathcal{E}(v) - \mathcal{E}(u) + \int_\Omega G^\circ(x, u; v - u)\, dx\,; \tag{11.4}$$

(c) if $\partial f(u) \neq \emptyset$, we have $G^\circ(x, u; -u) \in L^1(\Omega)$ and

$$\mathcal{E}(v) - \mathcal{E}(u) + \int_\Omega G^\circ(x, u; v - u)\, dx \geq \int_\Omega u^* \cdot (v - u)\, dx \tag{11.5}$$

for every $u^* \in \partial f(u)$ and $v \in \mathcal{D}(\mathcal{E})$ (the dual space of $L^p(\Omega; \mathbb{R}^N)$ is identified with $L^{p'}(\Omega; \mathbb{R}^N)$ in the usual way);

(d) if $N = 1$, we have $[G^\circ(x, u; v - u)]^+ \in L^1(\Omega)$ for all $v \in L^\infty(\Omega; \mathbb{R}^N)$.

Proof. (a) Let us set $\mathcal{G}(v) = \int_\Omega G(x, v)dx$. Given $\varepsilon > 0$, by $(\mathcal{E}_2.2)$ we have $\|\vartheta_h(v)v - v\|_p < \varepsilon$ and $|\mathcal{E}(\vartheta_h(v)v) - \mathcal{E}(v)| < \varepsilon$ for h large enough. Then, by $(\mathcal{E}_2.1)$ we get $\|\vartheta_k(u)\vartheta_h(v)v - v\|_p < \varepsilon$ and $|\mathcal{E}(\vartheta_k(u)\vartheta_h(v)v) - \mathcal{E}(v)| < \varepsilon$ for k large enough. Of course $\vartheta_k(u)\vartheta_h(v)v \in L^\infty(\Omega; \mathbb{R}^N)$, and by (11.1) we have

$$\begin{aligned} G^\circ(x, u; \vartheta_k(u)\vartheta_h(v)v - u) &\leq \vartheta_k(u)\vartheta_h(v)G^\circ(x, u; v - u) \\ &\quad + (1 - \vartheta_k(u)\vartheta_h(v))G^\circ(x, u; -u) \\ &\leq (h + k)M\alpha_{kM}(x) + [G^\circ(x, u; -u)]^+\,. \end{aligned}$$

From (G_4) we infer that $[G^\circ(x, u; -u)]^+ \in L^1(\Omega)$ and assertion (a) follows.

(b) Without loss of generality, we may assume that $[G^\circ(x, u; v - u)]^+ \in L^1(\Omega)$. Suppose first that $v \in \mathcal{D}(\mathcal{E}) \cap L^\infty(\Omega; \mathbb{R}^N)$ and take $\varepsilon > 0$.

We claim that for every $z \in L^p(\Omega; \mathbb{R}^N)$, $t \in]0, 1/2]$ and $h \geq 1$ with $hM > \|v\|_\infty$, we have

$$\begin{aligned} &\frac{1}{t}(G(x, z + t(\vartheta_h(z)v - z)) - G(x, z)) \\ &\leq 2\,(\|v\|_\infty\, \alpha_{hM} + a_1 + b_1(|z| + |v|)^p)\,. \end{aligned} \tag{11.6}$$

In fact, for a.e. $x \in \Omega$, by Lebourg's mean value theorem, there exist $\bar{t} \in]0, t[$ and $u^* \in \partial_s G(x, z + \bar{t}(\vartheta_h(z)v - z))$ such that

$$\frac{1}{t}(G(x, z + t(\vartheta_h(z)v - z)) - G(x, z)) = u^* \cdot (\vartheta_h(z)v - z)$$

$$= \frac{1}{1-\bar{t}} \left[\vartheta_h(z) u^* \cdot v - u^* \cdot (z + \bar{t}(\vartheta_h(z) v - z)) \right] .$$

By (11.2) and (11.3), it easily follows that

$$\frac{|\vartheta_h(z) u^* \cdot v|}{1-\bar{t}} \le 2 \, \|v\|_\infty \, \alpha_{hM} \, .$$

On the other hand, from (G_4) we deduce that for a.e. $x \in \Omega$

$$\frac{u^* \cdot (z + \bar{t}(\vartheta_h(z) v - z))}{1-\bar{t}}$$

$$\ge -\frac{1}{1-\bar{t}} G^\circ(x, z + \bar{t}(\vartheta_h(z) v - z); -(z + \bar{t}(\vartheta_h(z) v - z))$$

$$\ge -\frac{1}{1-\bar{t}} (a_1 + b_1 |z + \bar{t}(\vartheta_h(z) v - z)|^p) \ge -2 \, (a_1 + b_1 (|z| + |v|)^p) \, .$$

Then (11.6) is readily established.

For a.e. $x \in \Omega$ we have

$$G^\circ(x, u; \vartheta_h(u) v - u) \le \vartheta_h(u) G^\circ(x, u; v - u) + (1 - \vartheta_h(u)) G^\circ(x, u; -u)$$

$$\le [G^\circ(x, u; v - u)]^+ + [G^\circ(x, u; -u)]^+ \, .$$

Furthermore, for a.e. $x \in \Omega$ and every $s \in \mathbb{R}^N$, (G_2) implies $G^\circ(x, s; \cdot)$ to be Lipschitz continuous, so in particular

$$\lim_{h \to \infty} G^\circ(x, u; \vartheta_h(u) v - u) = G^\circ(x, u; v - u) \quad \text{a.e. in } \Omega \, .$$

Then, given

$$\lambda > \int_\Omega G^\circ(x, u; v - u) \, dx \, ,$$

by Fatou's Lemma there exists $\overline{h} \ge 1$ such that

$$\int_\Omega G^\circ(x, u; \vartheta_h(u) v - u) \, dx < \lambda \quad \text{and} \quad \|\vartheta_h(u) v - v\|_p < \varepsilon, \tag{11.7}$$

for all $h \ge \overline{h}$.

By the lower semicontinuity of $\mathcal{G}$, there exists $\overline{\delta} \in]0, 1/2]$ such that for every $z \in B_{\overline{\delta}}(u)$ one has $\mathcal{G}(z) \ge \mathcal{G}(u) - \frac{1}{2}$. Then for every $(z, \mu) \in B_{\overline{\delta}}(u, f(u)) \cap \operatorname{epi}(f)$ it follows

$$\mathcal{E}(z) \le \mu - \mathcal{G}(z) \le \mu + \frac{1}{2} - \mathcal{G}(u) \le f(u) + \overline{\delta} - \mathcal{G}(u) + \frac{1}{2} \le \mathcal{E}(u) + 1 \, .$$

Let now $\sigma > 0$. By assumptions $(\mathcal{E}_1)$ and $(\mathcal{E}_2.1)$ there exist $h \ge \overline{h}$ and $\delta \le \overline{\delta}$ such that

$$\|v\|_\infty < hM \, ,$$

$$\mathcal{E}(z) > \mathcal{E}(u)-\sigma\,, \quad \mathcal{E}(\vartheta_h(z)v) < \mathcal{E}(v)+\sigma\,, \quad \|(\vartheta_h(z)v-z)-(v-u)\|_p < \varepsilon\,,$$

for any $z \in B_\delta(u)$ with $\mathcal{E}(z) \leq \mathcal{E}(u) + 1$.

Taking into account (1.44), (11.6) and (11.7), we deduce by Fatou's Lemma that, possibly reducing δ, for any $t \in]0, \delta]$ and for any $z \in B_\delta(u)$ we have

$$\int_\Omega \frac{1}{t}(G(x, z + t(\vartheta_h(z)v - z)) - G(x, z))\,dx < \lambda\,.$$

Now let $\mathcal{V} : (B_\delta(u, f(u)) \cap \mathrm{epi}(f)) \times]0, \delta] \to B_\varepsilon(v - u)$ be defined setting

$$\mathcal{V}((z, \mu), t) = \vartheta_h(z)v - z\,.$$

Since $\mathcal{V}$ is evidently continuous and

$$f(z + t\mathcal{V}((z, \mu), t)) = f(z + t(\vartheta_h(z)v - z))$$

$$\leq \mathcal{E}(z) + t\,(\mathcal{E}(\vartheta_h(z)v) - \mathcal{E}(z)) + \mathcal{G}(z + t(\vartheta_h(z)v - z)$$

$$\leq \mathcal{E}(z) + (\mathcal{E}(v) - \mathcal{E}(u) + 2\sigma)t + \mathcal{G}(z) + \lambda t = f(z) + (\mathcal{E}(v) - \mathcal{E}(u) + \lambda + 2\sigma)t\,,$$

we have

$$f^\circ_\varepsilon(u; v - u) \leq \mathcal{E}(v) - \mathcal{E}(u) + \lambda + 2\sigma\,.$$

By the arbitrariness of $\sigma > 0$ and $\lambda > \int_\Omega G^\circ(x, u; v - u)\,dx$, it results

$$f^\circ_\varepsilon(u; v - u) \leq \mathcal{E}(v) - \mathcal{E}(u) + \int_\Omega G^\circ(x, u; v - u)\,dx\,.$$

Letting $\varepsilon \to 0^+$, we get (11.4) when $v \in \mathcal{D}(\mathcal{E}) \cap L^\infty(\Omega; \mathbb{R}^N)$.

Let us now treat the general case. If we set $v_h = \vartheta_h(v)v$, we have $v_h \in L^\infty(\Omega; \mathbb{R}^N)$. Arguing as before, it is easy to see that

$$G^\circ(x, u; v_h - u) \leq [G^\circ(x, u; v - u)]^+ + [G^\circ(x, u; -u)]^+\,,$$

so that

$$\limsup_h \int_\Omega G^\circ(x, u; v_h - u)\,dx \leq \int_\Omega G^\circ(x, u; v - u)\,dx\,.$$

On the other hand, by the previous step it holds

$$f^\circ(u; v_h - u) \leq \mathcal{E}(v_h) - \mathcal{E}(u) + \int_\Omega G^\circ(x, u; v_h - u)\,dx\,.$$

Passing to the lower limit as $h \to \infty$ and taking into account the lower semicontinuity of $f^\circ(u, \cdot)$ and $(\mathcal{E}_2.2)$, we get (11.4).

(c) We already know that $[G^\circ(x,u;-u)]^+ \in L^1(\Omega)$. If we choose $v=0$ in (11.4), we obtain

$$f^\circ(u;-u) \leq \mathcal{E}(0) - \mathcal{E}(u) + \int_\Omega G^\circ(x,u;-u)\,dx\,.$$

Since $\partial f(u) \neq \emptyset$, it is $f^\circ(u;-u) > -\infty$, hence

$$\int_\Omega [G^\circ(x,u;-u)]^-\,dx < +\infty\,.$$

Finally, if $u^* \in \partial f(u)$ we have by definition that

$$f^\circ(u;v-u) \geq \int_\Omega u^* \cdot (v-u)\,dx$$

and (11.5) follows from (11.4).
(d) From (11.1) it is clear that $G^\circ(x,u;v-u)$ is summable where $|u(x)| \leq \|v\|_\infty$. On the other hand, where $|u(x)| > \|v\|_\infty$ we have

$$G^\circ(x,u;v-u) = \left(1 - \frac{v}{u}\right) G^\circ(x,u;-u)$$

and the assertion follows from (G_4). ■

Since f is only lower semicontinuous, we are interested in the verification of the condition $(epi)_c$. For this purpose, we consider an assumption (G_3') on G stronger than (G_3).

Theorem 11.4 Assume that

(G_3') there exist $a \in L^1(\Omega)$ and $b \in I\!R$ such that

$$|G(x,s)| \leq a(x) + b|s|^p \quad \text{for a.e. } x \in \Omega \text{ and every } s \in I\!R^N\,.$$

Then for every $(u,\lambda) \in \mathrm{epi}(f)$ with $\lambda > f(u)$ it is $|d\mathcal{G}_f|\,(u,\lambda) = 1$ (see Definition 1.8 and the lines below it). Moreover, if $\mathcal{E}$ and $G(x,\cdot)$ are even, for every $\lambda > f(0)$ we have $\left|d_{Z\!\!Z_2}\mathcal{G}_f\right|(0,\lambda) = 1$.

Proof. Let $\varrho > 0$. Since

$$\forall \tau \in [0,1] : G^\circ(x,u;\tau u - u) = (1-\tau)G^\circ(x,u;-u) \leq [G^\circ(x,u;-u)]^+\,,$$

by $(\mathcal{E}_2.2)$ and (G_4) there exists $\overline{h} \geq 1$ such that

$$\|\vartheta_{\overline{h}}(u)u - u\|_p < \varrho\,, \quad \mathcal{E}(\vartheta_{\overline{h}}(u)u) < \mathcal{E}(u) + \varrho\,,$$

$$\forall h \geq \overline{h} : \int_\Omega G^\circ(x,u;\vartheta_h(u)\vartheta_{\overline{h}}(u)u - u)\,dx < \varrho\,.$$

Set $v = \vartheta_{\overline{h}}(u)u$.

By $(\mathcal{E}_2.1)$ there exist $h \geq \overline{h}$ and $\delta \in]0,1]$ such that

$$\|\vartheta_h(z)v - z\|_p < \varrho, \quad \mathcal{E}(\vartheta_h(z)v) < \mathcal{E}(u) + \varrho,$$

whenever $\|z-u\|_p < \delta$ and $\mathcal{E}(z) \leq \lambda + 1 - \mathcal{G}(u) + \varrho$.

By decreasing δ, from (G_3'), (11.6) and (1.44) we deduce that

$$|\mathcal{G}(z) - \mathcal{G}(u)| < \varrho, \quad \int_\Omega \frac{1}{t}(G(x, z + t(\vartheta_h(z)v - z)) - G(x,z))\,dx < \varrho$$

whenever $\|z-u\|_p < \delta$ and $0 < t \leq \delta$.

Define a continuous map

$$\mathcal{H} : \{z \in B_\delta(u) : f(z) < \lambda + \delta\} \times [0,\delta] \to X$$

by $\mathcal{H}(z,t) = z + t(\vartheta_h(z)v - z)$. It is readily seen that $\|\mathcal{H}(z,t) - z\|_p \leq \varrho t$.

If $z \in B_\delta(u)$, $f(z) < \lambda + \delta$ and $0 \leq t \leq \delta$, we have

$$\mathcal{E}(z) = f(z) - \mathcal{G}(z) < \lambda + \delta - \mathcal{G}(u) + \varrho \leq \lambda + 1 - \mathcal{G}(u) + \varrho,$$

hence, taking into account the convexity of $\mathcal{E}$,

$$\begin{aligned}\mathcal{E}(z + t(\vartheta_h(z)v - z)) &\leq \mathcal{E}(z) + t(\mathcal{E}(\vartheta_h(z)v) - \mathcal{E}(z)) \\ &\leq \mathcal{E}(z) + t(\mathcal{E}(u) - \mathcal{E}(z) + \varrho).\end{aligned}$$

Moreover, we also have

$$\mathcal{G}(z + t(\vartheta_h(z)v - z)) \leq \mathcal{G}(z) + t\varrho \leq \mathcal{G}(z) + t(\mathcal{G}(u) - \mathcal{G}(z) + 2\varrho).$$

Therefore

$$f(z + t(\vartheta_h(z)v - z)) \leq f(z) + t(f(u) - f(z) + 3\varrho).$$

and the first assertion follows by Corollary 1.6.

Now assume that $\mathcal{E}$ and $G(x,\cdot)$ are even and that $u = 0$. Then, in the previous argument, we have $v = 0$, so that $\mathcal{H}(-z,t) = -\mathcal{H}(z,t)$ and the second assertion also follows. ■

We provide in what follows a criterion which is useful in the verification of the Palais-Smale condition. For this purpose, we consider further assumptions on $\mathcal{E}$, which ensure a suitable coerciveness, and a new condition (G_4') on G, stronger than (G_4), which is a kind of one-sided subcritical growth condition.

Theorem 11.5 (Degiovanni, Marzocchi and Rădulescu [8]) Let $c \in \mathbb{R}$. Assume that

($\mathcal{E}_3$) for every $\{u_h\}$ bounded in $L^p(\Omega; \mathbb{R}^N)$ with $\{\mathcal{E}(u_h)\}$ bounded, there exists a subsequence $\{u_{h_k}\}$ and a function $u \in L^p(\Omega; \mathbb{R}^N)$ such that

$$\lim_{k\to\infty} u_{h_k}(x) = u(x) \quad \text{for a.e. } x \in \Omega\,;$$

($\mathcal{E}_4$) if $\{u_h\}$ is a sequence in $L^p(\Omega; \mathbb{R}^N)$ weakly convergent to $u \in \mathcal{D}(\mathcal{E})$ and $\mathcal{E}(u_h)$ converges to $\mathcal{E}(u)$, then $\{u_h\}$ converges to u strongly in $L^p(\Omega; \mathbb{R}^N)$;

(G_4') for every $\varepsilon > 0$ there exists $a_\varepsilon \in L^1(\Omega)$ such that

$$G^\circ(x, s; -s) \le a_\varepsilon(x) + \varepsilon |s|^p \quad \text{for a.e. } x \in \Omega \text{ and every } s \in \mathbb{R}^N\,.$$

Then any $(PS)_c$-sequence $\{u_h\}$ for f bounded in $L^p(\Omega; \mathbb{R}^N)$ admits a subsequence strongly convergent in $L^p(\Omega; \mathbb{R}^N)$.

Proof. From (G_3) we deduce that $(\mathcal{G}(u_h))$ is bounded from below. Taking into account $(\mathcal{E}_1)$, it follows that $(\mathcal{E}(u_h))$ is bounded. By $(\mathcal{E}_3)$ there exists a subsequence, still denoted by $\{u_h\}$, converging weakly in $L^p(\Omega; \mathbb{R}^N)$ and a.e. to some $u \in \mathcal{D}(\mathcal{E})$.

Given $\varepsilon > 0$, by $(\mathcal{E}_2.2)$ and (G_4) we may find $k_0 \ge 1$ such that

$$\mathcal{E}(\vartheta_{k_0}(u)u) < \mathcal{E}(u) + \varepsilon\,,$$

$$\int_\Omega (1 - \vartheta_{k_0}(u)) G^\circ(x, u; -u)\, dx < \varepsilon\,.$$

Since $\vartheta_{k_0}(u)u \in \mathcal{D}(\mathcal{E}) \cap L^\infty(\Omega; \mathbb{R}^N)$, by $(\mathcal{E}_2.1)$ there exists $k_1 \ge k_0$ such that

$$\forall h \in \mathbb{N}: \quad \mathcal{E}(\vartheta_{k_1}(u_h)\vartheta_{k_0}(u)u) < \mathcal{E}(u) + \varepsilon\,, \tag{11.8}$$

$$\int_\Omega (1 - \vartheta_{k_1}(u)\vartheta_{k_0}(u)) G^\circ(x, u; -u)\, dx < \varepsilon\,.$$

It follows that $\vartheta_{k_1}(u_h)\vartheta_{k_0}(u)u \in \mathcal{D}(\mathcal{E})$. Moreover, from (11.1) and (G_4') we get

$$\begin{aligned} &G^\circ(x, u_h; \vartheta_{k_1}(u_h)\vartheta_{k_0}(u)u - u_h) \\ &\le \vartheta_{k_1}(u_h) G^\circ(x, u_h; \vartheta_{k_0}(u)u - u_h) + (1 - \vartheta_{k_1}(u_h)) G^\circ(x, u_h; -u_h) \\ &\le \alpha_{k_1 M}(x)(k_0 M + k_1 M) + a_\varepsilon(x) + \varepsilon |u_h|^p\,. \end{aligned}$$

From (1.45) and Fatou's Lemma we deduce that

$$\limsup_{h\to\infty} \int_\Omega [G^\circ(x, u_h; \vartheta_{k_1}(u_h)\vartheta_{k_0}(u)u - u_h) - \varepsilon |u_h|^p]\, dx$$

$$\le \int_\Omega [G^\circ(x, u; \vartheta_{k_1}(u)\vartheta_{k_0}(u)u - u) - \varepsilon |u|^p]\, dx$$

$$\leq \int_\Omega (1 - \vartheta_{k_1}(u)\vartheta_{k_0}(u))G^\circ(x, u; -u)\, dx < \varepsilon\,,$$

hence

$$\limsup_{h\to\infty} \int_\Omega G^\circ(x, u_h; \vartheta_{k_1}(u_h)\vartheta_{k_0}(u)u - u_h)\, dx < \varepsilon \sup_h \|u_h\|_p^p + \varepsilon\,. \quad (11.9)$$

Since $\{u_h\}$ is a $(PS)_c$-sequence, by Theorem 1.7 there exists $u_h^* \in \partial f(u_h)$ with $\|u_h^*\|_{p'} \leq |df|\,(u_h)$, so that $\lim_{h\to\infty} \|u_h^*\|_{p'} = 0$. Applying (*c*) of Theorem 11.3, we get

$$\mathcal{E}(\vartheta_{k_1}(u_h)\vartheta_{k_0}(u)u) \geq \mathcal{E}(u_h) - \int_\Omega G^\circ(x, u_h; \vartheta_{k_1}(u_h)\vartheta_{k_0}(u)u - u_h)\, dx$$

$$+ \int_\Omega u_h^* \cdot (\vartheta_{k_1}(u_h)\vartheta_{k_0}(u)u - u_h)\, dx\,.$$

Taking into account (11.8), (11.9) and passing to the upper limit, we obtain

$$\limsup_{h\to\infty} \mathcal{E}(u_h) \leq \mathcal{E}(u) + 2\varepsilon + \varepsilon \sup_h \|u_h\|_p^p\,.$$

By the arbitrariness of $\varepsilon > 0$, we finally have

$$\limsup_{h\to\infty} \mathcal{E}(u_h) \leq \mathcal{E}(u)$$

and the strong convergence of $\{u_h\}$ to u follows from $(\mathcal{E}_4)$. ■

2. Area Type Functionals

Let $n \geq 2$, $N \geq 1$, Ω be a bounded open subset of $\mathbb{R}^n$ with Lipschitz boundary and let

$$\Psi : \mathbb{R}^{nN} \to \mathbb{R}$$

be a convex function satisfying

$$(\Psi) \quad \begin{cases} \Psi(0) = 0,\ \Psi(\xi) > 0 \text{ for any } \xi \neq 0 \text{ and} \\ \text{there exists } c > 0 \text{ such that } \Psi(\xi) \leq c|\xi| \text{ for any } \xi \in \mathbb{R}^{nN}. \end{cases}$$

We want to study the functional $\mathcal{E} : L^{\frac{n}{n-1}}(\Omega; \mathbb{R}^N) \to \mathbb{R} \cup \{+\infty\}$ defined by

$$\mathcal{E}(u) = \begin{cases} \displaystyle\int_\Omega \Psi(Du^a)\, dx + \int_\Omega \Psi^\infty\left(\frac{Du^s}{|Du^s|}\right) d|Du^s|(x) \\ \displaystyle + \int_{\partial\Omega} \Psi^\infty(u \otimes \nu)\, d\mathcal{H}^{n-1}(x) \quad \text{if } u \in BV(\Omega; \mathbb{R}^N), \\ +\infty \quad \text{if } u \in L^{\frac{n}{n-1}}(\Omega; \mathbb{R}^N) \setminus BV(\Omega; \mathbb{R}^N), \end{cases}$$

where $Du = Du^a\,dx + Du^s$ is the Lebesgue decomposition of Du, $|Du^s|$ is the total variation of Du^s, $Du^s/|Du^s|$ is the Radon-Nikodym derivative of Du^s with respect to $|Du^s|$, Ψ^∞ is the recession functional associated with Ψ, ν is the outer normal to Ω and the trace of u on $\partial\Omega$ is still denoted by u.

Theorem 11.6 The functional $\mathcal{E}$ satisfies conditions $(\mathcal{E}_1)$, $(\mathcal{E}_2)$, $(\mathcal{E}_3)$ and $(\mathcal{E}_4)$.

The Section will be devoted to the proof of this result. We begin by establishing some technical lemmas.

In $BV(\Omega; I\!R^N)$ we consider the norm

$$\|u\|_{BV} = \int_\Omega |Du^a|\,dx + |Du^s|(\Omega) + \int_{\partial\Omega} |u|\,d\mathcal{H}^{n-1}(x),$$

which is equivalent to the standard norm of $BV(\Omega; I\!R^N)$.

Lemma 11.1 For every $u \in BV(\Omega; I\!R^N)$ and every $\varepsilon > 0$ there exists $v \in C_c^\infty(\Omega; I\!R^N)$ such that

$$\|v - u\|_{\frac{n}{n-1}} < \varepsilon, \qquad \left| \int_\Omega |Dv|\,dx - \|u\|_{BV} \right| < \varepsilon,$$

$$|\mathcal{E}(v) - \mathcal{E}(u)| < \varepsilon, \qquad \|v\|_\infty \le \operatorname{ess\,sup}_\Omega |u|.$$

Proof. Let $\delta > 0$, let $R > 0$ with $\overline{\Omega} \subseteq B_R(0)$ and let

$$\vartheta_h(x) = 1 - \min\left\{\max\left\{\frac{h+1}{h}[1 - h\,d(x, I\!R^n \setminus \Omega)], 0\right\}, 1\right\}.$$

Define $\hat{u} \in BV(B_R(0); I\!R^N)$ by

$$\hat{u}(x) = \begin{cases} u(x) & \text{if } x \in \Omega, \\ 0 & \text{if } x \in B_R(0) \setminus \Omega. \end{cases}$$

Thus, if h is sufficiently large, then $\vartheta_h u \in BV(\Omega; I\!R^N)$, $\|\vartheta_h u - u\|_{\frac{n}{n-1}} < \delta$ and

$$\begin{aligned} &\int_\Omega \sqrt{1 + |D(\vartheta_h u)^a|^2}\,d\mathcal{L}^n + |D(\vartheta_h u)^s|(\Omega) \\ &< \int_\Omega \sqrt{1 + |Du^a|^2}\,d\mathcal{L}^n + |Du^s|(\Omega) + \int_{\partial\Omega} |u|\,d\mathcal{H}^{n-1} + \delta \\ &= \int_{B_R(0)} \sqrt{1 + |D\hat{u}^a|^2}\,d\mathcal{L}^n + |D\hat{u}^s|(B_R(0)) + \delta. \end{aligned}$$

Moreover, $\vartheta_h u$ has compact support in Ω and

$$\operatorname{ess\,sup}_\Omega |\vartheta_h u| \leq \operatorname{ess\,sup}_\Omega |u|.$$

If we regularize $\vartheta_h u$ by convolution, we easily get $v \in C_c^\infty(\Omega; I\!R^N)$ with

$$\|v\|_\infty \leq \operatorname{ess\,sup}_\Omega |u|, \quad \|v - u\|_{\frac{n}{n-1}} < \delta$$

and

$$\int_\Omega \sqrt{1 + |Dv|^2}\, d\mathcal{L}^n < \int_{B_R(0)} \sqrt{1 + |D\hat{u}^a|^2}\, d\mathcal{L}^n + |D\hat{u}^s|(B_R(0)) + \delta.$$

Since

$$\|u\|_{BV} = \int_{B_R(0)} |D\hat{u}^a|\, dx + |D\hat{u}^s|(B_R(0)),$$

$$\mathcal{E}(u) = \int_{B_R(0)} \Psi(D\hat{u}^a)\, dx + \int_{B_R(0)} \Psi^\infty\left(\frac{D\hat{u}^s}{|D\hat{u}^s|}\right) d|D\hat{u}^s|,$$

our conclusion follows from [2], Fact 3.1. ■

Lemma 11.2 The following facts hold:

(*a*) $\Psi : I\!R^{nN} \to I\!R$ is Lipschitz continuous of some constant $\mathrm{Lip}(\Psi) > 0$;

(*b*) for any $\xi \in I\!R^{nN}$ and $s \in [0,1]$ we have $\Psi(s\xi) \leq s\Psi(\xi)$;

(*c*) for every $\sigma > 0$ there exists $d_\sigma > 0$ such that

$$\forall \xi \in I\!R^{nN}: \quad \Psi(\xi) \geq d_\sigma(|\xi| - \sigma);$$

(*d*) $\mathcal{E} : BV(\Omega; I\!R^N) \to I\!R$ is Lipschitz continuous of constant $\mathrm{Lip}(\Psi)$;

(*e*) if σ and d_σ are as in (*c*), we have

$$\forall u \in BV(\Omega; I\!R^N): \quad \mathcal{E}(u) \geq d_\sigma\Big(\|u\|_{BV} - \sigma\mathcal{L}^n(\Omega)\Big).$$

Proof. Properties (*a*) and (*b*) easily follow from the convexity of Ψ and assumption (Ψ).

To prove (*c*), assume by contradiction that $\sigma > 0$ and $\{\xi_h\}$ is a sequence with $\Psi(\xi_h) < \frac{1}{h}(|\xi_h| - \sigma)$. If $|\xi_h| \to +\infty$, we would have

$$\Psi\left(\frac{\xi_h}{|\xi_h|}\right) \leq \frac{\Psi(\xi_h)}{|\xi_h|} < \frac{1}{h}\left(1 - \frac{\sigma}{|\xi_h|}\right).$$

Up to a subsequence, $\{\xi_h/|\xi_h|\}$ is convergent to some $\eta \neq 0$ with $\Psi(\eta) \leq 0$, which is impossible. Since $|\xi_h|$ is bounded, up to a subsequence we have $\xi_h \to \xi$ with $|\xi| \geq \sigma$ and $\Psi(\xi) \leq 0$, which is again impossible.

Finally, (d) is a direct consequence of (a) and the definition of $\|\cdot\|_{BV}$, while (e) follows from (c). ∎

Let now $\vartheta \in C_c^1(\mathbb{R}^N)$ with $0 \le \vartheta \le 1$, $\|\nabla\vartheta\|_\infty \le 2$, $\vartheta(s) = 1$ for $|s| \le 1$ and $\vartheta(s) = 0$ for $|s| \ge 2$. Define $\vartheta_h : \mathbb{R}^N \to \mathbb{R}$ and $T_h, R_h : \mathbb{R}^N \to \mathbb{R}^N$ by

$$\vartheta_h(s) = \vartheta\left(\frac{s}{h}\right), \quad T_h(s) = \vartheta_h(s)s, \quad R_h(s) = (1 - \vartheta_h(s))s.$$

Lemma 11.3 There exists a constant $c_\Psi > 0$ such that

$$\mathcal{E}\left(\vartheta\left(\frac{u}{h}\right)v\right) \le \mathcal{E}(v) + \frac{c_\Psi}{h}\|v\|_\infty\|u\|_{BV},$$

$$\mathcal{E}(T_h \circ u) \le \mathcal{E}(u) + c_\Psi|Du|(\{x \in \Omega \setminus S_u : |\tilde{u}(x)| > h\})$$
$$+c_\Psi \int_{\{x\in S_u:|u^+(x)|>h \text{ or } |u^-(x)|>h\}} |u^+ - u^-|\, d\mathcal{H}^{n-1}(x)$$
$$+c_\Psi \int_{\{x\in\partial\Omega:|u(x)|>h\}} |u|\, d\mathcal{H}^{n-1}(x),$$

$$\mathcal{E}(T_h \circ w) + \mathcal{E}(R_h \circ w) \le \mathcal{E}(w) + c_\Psi \int_{\{x\in\Omega:h<|w(x)|<2h\}} |Dw|\, dx$$

whenever $h \ge 1$, $u \in BV(\Omega; \mathbb{R}^N)$, $v \in BV(\Omega; \mathbb{R}^N) \cap L^\infty(\Omega; \mathbb{R}^N)$ and $w \in C_c^\infty(\Omega; \mathbb{R}^N)$.

Proof. Suppose first that $u, v \in C_c^\infty(\Omega; \mathbb{R}^N)$. Then, since

$$D\left[\vartheta\left(\frac{u}{h}\right)v\right] = \vartheta\left(\frac{u}{h}\right)Dv + \frac{1}{h}v \otimes \left[D\vartheta\left(\frac{u}{h}\right)Du\right],$$

by (Ψ) and Lemma 11.2 it follows that

$$\mathcal{E}\left(\vartheta\left(\frac{u}{h}\right)v\right) \le \mathcal{E}(v) + \mathrm{Lip}(\Psi)\frac{\|D\vartheta\|_\infty}{h}\|v\|_\infty \int_\Omega |Du|\, dx. \tag{11.10}$$

In the general case, we consider two sequences $\{u_k\}$, $\{v_k\}$ in $C_c^\infty(\Omega; \mathbb{R}^N)$ converging to u, v in $L^1(\Omega; \mathbb{R}^N)$ with $\int_\Omega |Du_k|\, dx \to \|u\|_{BV}$, $\mathcal{E}(v_k) \to \mathcal{E}(v)$ and $\|v_k\|_\infty \le \|v\|_\infty$. Passing to the lower limit in (11.10), we obtain the first inequality in the assertion.

To prove the second inequality, we first observe that by Lemma 11.2 we have

$$\mathcal{E}(T_h \circ u) \le \mathcal{E}(u) + \mathrm{Lip}(\Psi)\|R_h \circ u\|_{BV}. \tag{11.11}$$

In order to estimate the last term in (11.11), we apply the chain rule. Since $R_h(s) = 0$ if $|s| \le h$ and $\|DR_h\|_\infty \le k_\vartheta$ for some $k_\vartheta > 0$, we have

$$\int_\Omega |D(R_h(u))^a|\,dx \le \int_{\Omega\setminus S_u} |DR_h(\tilde{u})||Du^a|\,dx$$

$$\le k_\vartheta \int_{\{x\in\Omega\setminus S_u:|\tilde{u}(x)|>h\}} |Du^a|\,dx\,,$$

$$\Big|D(R_h(u))^s\Big|(\Omega) \le \int_{\Omega\setminus S_u} |DR_h(\tilde{u})|\,d|Du^s|(x)$$

$$+\int_{S_u} |R_h(u^+) - R_h(u^-)|\,d\mathcal{H}^{n-1}(x)$$

$$\le k_\vartheta\Big(|Du^s|\left(\{x\in\Omega\setminus S_u : |\tilde{u}(x)| > h\}\right)$$

$$+\int_{\{x\in S_u:|u^+(x)|>h \text{ or } |u^-(x)|>h\}} |u^+ - u^-|\,d\mathcal{H}^{m-1}(x)\Big)$$

and

$$\int_{\partial\Omega} |R_h(u)|\,d\mathcal{H}^{n-1}(x) \le k_\vartheta \int_{\{x\in\partial\Omega:|u(x)|>h\}} |u|\,d\mathcal{H}^{n-1}(x)\,.$$

Combining these three estimates, we get

$$k_\vartheta^{-1}\|R_h\circ u\|_{BV} \le \int_{\{x\in\Omega\setminus S_u:|\tilde{u}(x)|>h\}} |Du^a|\,dx$$

$$+|Du^s|(\{x\in\Omega\setminus S_u : |\tilde{u}(x)| > h\})$$

$$+\int_{\{x\in S_u:|u^+(x)|>h \text{ or } |u^-(x)|>h\}} |u^+ - u^-|\,d\mathcal{H}^{m-1}(x)$$

$$+\int_{\{x\in\partial\Omega:|u(x)|>h\}} |u|\,d\mathcal{H}^{n-1}(x)\,. \tag{11.12}$$

Then the second inequality follows from (11.11) and (11.12).

Again, since Ψ is Lipschitz continuous, we have

$$\left|\int_\Omega \Psi(D(T_h\circ w))\,dx - \int_\Omega \Psi(\vartheta_h(w)Dw)\,dx\right|$$

$$\le \frac{\mathrm{Lip}(\Psi)}{h}\int_\Omega \left|D\vartheta\left(\frac{w}{h}\right)Dw\right||w|\,dx$$

$$\le 2\,\mathrm{Lip}(\Psi)\|\nabla\vartheta\|_\infty \int_{\{h<|w|<2h\}} |Dw|\,dx\,.$$

In a similar way, it is also

$$\left|\int_\Omega \Psi(D(R_h \circ w))\,dx - \int_\Omega \Psi((1-\vartheta_h(w))Dw)\,dx\right|$$

$$\leq 2\,\mathrm{Lip}(\Psi)\|\nabla\vartheta\|_\infty \int_{\{h<|w|<2h\}} |Dw|\,dx\,.$$

Hence, combining the last two estimates and taking into account (b) of Lemma 11.2, we deduce

$$\int_\Omega \Psi(D(T_h \circ w))\,dx + \int_\Omega \Psi(D(R_h \circ w))\,dx$$

$$\leq \int_\Omega \Psi(Dw)\,dx + 4\,\mathrm{Lip}(\Psi)\|\nabla\vartheta\|_\infty \int_{\{h<|w|<2h\}} |Dw|\,dx$$

and the proof is complete. ∎

Lemma 11.4 Let $\{u_h\}$ be a sequence in $C_c^\infty(\Omega; I\!R^N)$ and assume that $\{u_h\}$ is bounded in $BV(\Omega; I\!R^N)$.

Then for every $\varepsilon > 0$ and every $\overline{k} \in I\!N$ there exists $k \geq \overline{k}$ such that

$$\liminf_{h\to\infty} \int\limits_{\{k<|u_h|<2k\}} |Du_h|\,dx < \varepsilon\,.$$

Proof. Let $m \geq 1$ be such that

$$\sup_h \int_\Omega |Du_h|\,dx \leq \frac{m\varepsilon}{2}$$

and let $i_0 \in I\!N$ with $2^{i_0} \geq \overline{k}$. Then, since

$$\sum_{i=i_0}^{i_0+m-1} \int_{\{2^i<|u_h|<2^{i+1}\}} |Du_h|\,dx \leq \int_\Omega |Du_h|\,dx \leq \frac{m\varepsilon}{2}\,,$$

there exists i_h between i_0 and $i_0 + m - 1$ such that

$$\int_{\{2^{i_h}<|u_h|<2^{i_h+1}\}} |Du_h|\,dx \leq \frac{\varepsilon}{2}\,.$$

Passing to a subsequence $\{i_{h_j}\}$, we can suppose $i_{h_j} \equiv i \geq i_0$, and setting $k = 2^i$ we get

$$\forall j \in I\!N: \int\limits_{\{k<|u_{h_j}|<2k\}} |Du_{h_j}|\,dx \leq \frac{\varepsilon}{2}\,.$$

This proves the assertion. ■

Lemma 11.5 Let $\{u_h\}$ be a sequence in $C_c^\infty(\Omega; \mathbb{R}^N)$ and let $u \in BV(\Omega; \mathbb{R}^N)$ with $\|u_h - u\|_1 \to 0$ and $\mathcal{E}(u_h) \to \mathcal{E}(u)$.

Then for every $\varepsilon > 0$ and every $\overline{k} \in \mathbb{N}$ there exists $k \geq \overline{k}$ such that

$$\liminf_{h\to\infty} \|R_k \circ u_h\|_{BV} < \varepsilon \, .$$

Proof. Given $\varepsilon > 0$, let $d > 0$ be such that

$$\forall \xi \in \mathbb{R}^{nN} : \quad \Psi(\xi) \geq d\left(|\xi| - \frac{\varepsilon}{3\mathcal{L}^n(\Omega)}\right),$$

according to Lemma 11.2. Let also $c_\Psi > 0$ be as in Lemma 11.3. By (11.12) and Lemma 11.4, there exists $k \geq \overline{k}$ such that

$$\|R_k \circ u\|_{BV} < \frac{d\varepsilon}{3\mathrm{Lip}(\Psi)},$$

$$\liminf_{h\to\infty} \int_{\{k<|u_h|<2k\}} |Du_h| \, dx < \frac{d\varepsilon}{3c_\Psi} \, .$$

From Lemma 11.3 we derive that

$$\mathcal{E}(T_k \circ u) + \liminf_{h\to\infty} \mathcal{E}(R_k \circ u_h) \leq \liminf_{h\to\infty} \mathcal{E}(T_k \circ u_h) + \liminf_{h\to\infty} \mathcal{E}(R_k \circ u_h)$$

$$\leq \liminf_{h\to\infty}\Big(\mathcal{E}(T_k \circ u_h) + \mathcal{E}(R_k \circ u_h)\Big) \leq \mathcal{E}(u) + c_\Psi \liminf_{h\to\infty} \int_{\{k<|u_h|<2k\}} |Du_h| \, dx$$

$$< \mathcal{E}(u) + \frac{d\varepsilon}{3} \leq \mathcal{E}(T_k \circ u) + \mathrm{Lip}(\Psi)\|R_k \circ u\|_{BV} + \frac{d\varepsilon}{3} < \mathcal{E}(T_k \circ u) + \frac{2}{3} d\varepsilon \, ,$$

whence

$$\liminf_{h\to\infty} \mathcal{E}(R_k \circ u_h) < \frac{2}{3} d\varepsilon \, .$$

On the other hand, by Lemma 11.2 we have

$$\mathcal{E}(R_k \circ u_h) \geq d\left(\|R_k \circ u_h\|_{BV} - \frac{\varepsilon}{3}\right)$$

and the assertion follows. ■

Now we can prove the main auxiliary result that we need for the proof of Theorem 11.6. It is a property of the space BV which could be interesting also in itself.

Theorem 11.7 (Degiovanni, Marzocchi and Rădulescu [8]) Let $\{u_h\}$ be a sequence in $BV(\Omega; I\!R^N)$ and let $u \in BV(\Omega; I\!R^N)$ with $\|u_h - u\|_1 \to 0$ and $\mathcal{E}(u_h) \to \mathcal{E}(u)$. Then $\{u_h\}$ is strongly convergent to u in $L^{\frac{n}{n-1}}(\Omega; I\!R^N)$.

Proof. By Lemma 11.1 we may find $v_h \in C_c^\infty(\Omega; I\!R^N)$ with

$$\|v_h - u_h\|_1 < \frac{1}{h}, \quad \|v_h - u_h\|_{\frac{n}{n-1}} < \frac{1}{h}, \quad |\mathcal{E}(v_h) - \mathcal{E}(u_h)| < \frac{1}{h}.$$

Therefore it is sufficient to treat the case in which $u_h \in C_c^\infty(\Omega; I\!R^N)$.

By contradiction, up to a subsequence we may assume that there exists $\varepsilon > 0$ such that $\|u_h - u\|_{\frac{n}{n-1}} \geq \varepsilon$. Let $\tilde{c}$ be a constant such that $\|w\|_{\frac{n}{n-1}} \leq \tilde{c}\|w\|_{BV}$ for any $w \in BV(\Omega; I\!R^N)$. According to Lemma 11.5, let $k \in I\!N$ be such that

$$\|R_k \circ u\|_{\frac{n}{n-1}} < \frac{\varepsilon}{2}, \quad \liminf_{h\to\infty} \|R_k \circ u_h\|_{\frac{n}{n-1}} \leq \tilde{c} \liminf_{h\to\infty} \|R_k \circ u_h\|_{BV} < \frac{\varepsilon}{2}.$$

Then we have

$$\|u_h - u\|_{\frac{n}{n-1}} \leq \|R_k \circ u_h\|_{\frac{n}{n-1}} + \|T_k \circ u_h - T_k \circ u\|_{\frac{n}{n-1}} + \|R_k \circ u\|_{\frac{n}{n-1}}. \quad (11.13)$$

Since $T_k \circ u_h \to T_k \circ u$ in $L^{\frac{n}{n-1}}(\Omega; I\!R^N)$ as $h \to \infty$, passing to the lower limit in (11.13) we get

$$\liminf_{h\to\infty} \|u_h - u\|_{\frac{n}{n-1}} < \varepsilon,$$

whence a contradiction. ■

Proof of Theorem 11.6. It is well known that $\mathcal{E}$ satisfies condition $(\mathcal{E}_1)$. Conditions $(\mathcal{E}_2)$ are an immediate consequence of Lemma 11.3. From (e) of Lemma 11.2 and Rellich's theorem it follows that $\mathcal{E}$ satisfies condition $(\mathcal{E}_3)$. To prove $(\mathcal{E}_4)$, let $\{u_h\}$ be a sequence in $L^{\frac{n}{n-1}}(\Omega; I\!R^N)$ weakly convergent to $u \in BV(\Omega; I\!R^N)$ such that $\mathcal{E}(u_h)$ converges to $\mathcal{E}(u)$. Again by (e) of Lemma 11.2 and Rellich's theorem we deduce that $\{u_h\}$ is strongly convergent to u in $L^1(\Omega; I\!R^N)$. Then the assertion follows from Theorem 11.7. ■

3. A Result of Clark Type

Let $n \geq 2$ and Ω be a bounded open subset of $I\!R^n$ with Lipschitz boundary, let $\Psi : I\!R^{nN} \to I\!R$ be an even convex function satisfying (Ψ) and let $G : \Omega \times I\!R^N \to I\!R$ be a function satisfying (G_1), (G_2), (G_3'), (G_4')

with $p = \frac{n}{n-1}$ and the following conditions:

$$\begin{cases} \text{there exist } \tilde{a} \in L^1(\Omega) \text{ and } \tilde{b} \in L^n(\Omega) \text{ such that} \\ G(x,s) \geq -\tilde{a}(x) - \tilde{b}(x)|s| \quad \text{for a.e. } x \in \Omega \text{ and every } s \in I\!R^N ; \end{cases} \tag{11.14}$$

$$\lim_{|s|\to\infty} \frac{G(x,s)}{|s|} = +\infty \quad \text{for a.e. } x \in \Omega ; \tag{11.15}$$

$$\{s \mapsto G(x,s)\} \text{ is even for a.e. } x \in \Omega . \tag{11.16}$$

Our purpose in this Section is to prove the following multiplicity result.

Theorem 11.8 (Degiovanni, Marzocchi and Rădulescu [8]) For every $k \in I\!N$ there exists Λ_k such that for any $\lambda \geq \Lambda_k$ the problem

$$\begin{cases} u \in BV(\Omega; I\!R^N) \\ \mathcal{E}(v) - \mathcal{E}(u) + \int_\Omega G^\circ(x,u;v-u)\,dx \\ \geq \lambda \int_\Omega \frac{u}{\sqrt{1+|u|^2}} \cdot (v-u)\,dx, \ \forall v \in BV(\Omega; I\!R^N) \end{cases}$$

admits at least k pairs $(u,-u)$ of distinct solutions.

For the proof we need the following preliminary result.

Lemma 11.6 Let $\{u_h\}$ be a bounded sequence in $L^{\frac{n}{n-1}}(\Omega; I\!R^N)$, which is convergent a.e. to u, and let $\{\varrho_h\}$ be a positively divergent sequence of real numbers.

Then we have

$$\lim_h \int_\Omega \frac{G(x,\varrho_h u_h)}{\varrho_h}\,dx = +\infty \qquad \text{if } u \neq 0 ,$$

$$\liminf_h \int_\Omega \frac{G(x,\varrho_h u_h)}{\varrho_h}\,dx \geq 0 \quad \text{if } u = 0 .$$

Proof. If $u = 0$, the assertion follows directly from (11.14). If $u \neq 0$, we have

$$\int_\Omega \frac{G(x,\varrho_h u_h)}{\varrho_h}\,dx$$

$$\geq \int_{\{u\neq 0\}} \frac{G(x,\varrho_h u_h)}{\varrho_h}\,dx - \frac{1}{\varrho_h}\int_{\{u=0\}} \tilde{a}\,dx - \int_{\{u=0\}} \tilde{b}|u_h|\,dx .$$

From (11.14), (11.15) and Fatou's Lemma, we deduce that

$$\lim_h \int_{\{u\neq 0\}} \frac{G(x,\varrho_h u_h)}{\varrho_h}\,dx = +\infty ,$$

thereby the result. ∎

Proof of Theorem 11.8. First of all, set

$$\widetilde{G}(x,s) = G(x,s) - \lambda(\sqrt{1+|s|^2} - 1)\,.$$

It is easy to see that also $\widetilde{G}$ satisfies (G_1), (G_2), (G_3'), (G_4'), (11.14), (11.15), (11.16) and that

$$\widetilde{G}^\circ(x,s;\hat{s}) = G^\circ(x,s;\hat{s}) - \lambda\frac{s}{\sqrt{1+|s|^2}}\cdot\hat{s}\,.$$

Define a lower semicontinuous functional $f : L^{\frac{n}{n-1}}(\Omega;I\!R^N) \to I\!R\cup\{+\infty\}$ by

$$f(u) = \mathcal{E}(u) + \int_\Omega \widetilde{G}(x,u)\,dx\,.$$

By (11.16), f is even and satisfies condition $(epi)_c$, by Theorem 11.4. We claim that

$$\lim_{\|u\|_{\frac{n}{n-1}}\to\infty} f(u) = +\infty\,. \tag{11.17}$$

To prove it, let $\{u_h\}$ be a sequence in $BV(\Omega;I\!R^N)$ with $\|u_h\|_{\frac{n}{n-1}} = 1$ and let $\varrho_h \to +\infty$. By (e) of Lemma 11.2, there exist $\tilde{c} > 0$ and $\tilde{d} > 0$ such that

$$\forall u \in BV(\Omega;I\!R^N): \quad \mathcal{E}(u) \geq \tilde{d}\Big(\|u\|_{BV} - \tilde{c}\mathcal{L}^n(\Omega)\Big)\,.$$

If $\|u_h\|_{BV} \to +\infty$, it readily follows from (11.14) that $f(\varrho_h u_h) \to +\infty$. Otherwise, up to a subsequence, u_h is convergent a.e. and the assertion results from Lemma 11.6 and the inequality

$$f(\varrho_h u_h) \geq \varrho_h\left[\tilde{d}\left(\|u_h\|_{BV} - \frac{\tilde{c}}{\varrho_h}\mathcal{L}^n(\Omega)\right) + \int_\Omega \frac{\widetilde{G}(x,\varrho_h u_h)}{\varrho_h}\,dx\right]\,.$$

Since f is bounded below on bounded subsets of $L^{\frac{n}{n-1}}(\Omega;I\!R^N)$, it follows from (11.17) that f is bounded below on all $L^{\frac{n}{n-1}}(\Omega;I\!R^N)$. Furthermore, it also turns out from (11.17) that any $(PS)_c$ sequence is bounded, hence f satisfies $(PS)_c$ by Theorem 11.5.

Finally, let $k \geq 1$, let $w_1, \ldots, w_k$ be linearly independent elements of $BV(\Omega;I\!R^N)$ and let $\psi : S^{k-1} \to L^{\frac{n}{n-1}}(\Omega;I\!R^N)$ be the odd continuous map defined by

$$\psi(\xi) = \sum_{j=1}^k \xi_j w_j\,.$$

Because of (G_3'), it is seen that

$$\sup\left\{\mathcal{E}(u)+\int_\Omega G(x,u)\,dx : u\in\psi(S^{k-1})\right\}<+\infty$$

and

$$\inf\left\{\int_\Omega(\sqrt{1+|u|^2}-1)\,dx : u\in\psi(S^{k-1})\right\}>0\,.$$

Therefore there exists $\Lambda_k>0$ such that $\sup_{\xi\in S^{k-1}} f(\psi(\xi))<0$ whenever $\lambda\geq\Lambda_k$.

By Theorem 11.1, we infer that f admits at least k pairs $(u_k,-u_k)$ of critical points. Therefore, by Theorem 1.7, for any u_k it is possible to apply Theorem 11.3 (with $\widetilde{G}$ in place of G), whence the assertion. ■

4. An Inequality Problem with Superlinear Potential

Let $n\geq 2$ and Ω be a bounded open subset of $\mathbb{R}^n$ with Lipschitz boundary, let $\Psi:\mathbb{R}^{nN}\to\mathbb{R}$ be an even convex function satisfying (Ψ) and let $G:\Omega\times\mathbb{R}^N\to\mathbb{R}$ be a function satisfying (G_1), (G_2), (G_3'), (G_4'), (11.16) with $p=\frac{n}{n-1}$ and the following condition:

$$\left\{\begin{array}{l}\text{there exist } q>1 \text{ and } R>0 \text{ such that}\\ G^\circ(x,s;s)\leq qG(x,s)<0\\ \text{for a.e. } x\in\Omega \text{ and every } s\in\mathbb{R}^N \text{ with } |s|\geq R\,.\end{array}\right. \tag{11.18}$$

Define $\mathcal{E}$ as before and an even lower semicontinuous functional $f:L^{\frac{n}{n-1}}(\Omega;\mathbb{R}^N)\to\mathbb{R}\cup\{+\infty\}$ by

$$f(u)=\mathcal{E}(u)+\int_\Omega G(x,u)\,dx\,.$$

Theorem 11.9 (Degiovanni, Marzocchi and Rădulescu [8]) There exists a sequence $\{u_h\}$ of solutions of the problem

$$\left\{\begin{array}{l}u\in BV(\Omega;\mathbb{R}^N)\\ \mathcal{E}(v)-\mathcal{E}(u)+\displaystyle\int_\Omega G^\circ(x,u;v-u)\,dx\geq 0\quad\forall v\in BV(\Omega;\mathbb{R}^N)\end{array}\right.$$

with $f(u_h)\to+\infty$.

Proof. According to (11.1), we have

$$|s|<R\quad\Longrightarrow\quad|G^\circ(x,s;s)|\leq\alpha_R(x)|s|\,.$$

Combining this fact with (11.18) and (G_3'), we deduce that there exists $a_0 \in L^1(\Omega)$ such that

$$G^\circ(x, s; s) \leq qG(x, s) + a_0(x) \text{ for a.e. } x \in \Omega \text{ and all } s \in \mathbb{R}^N. \quad (11.19)$$

Moreover, from (11.18) and Lebourg's mean value theorem it follows that for every $s \in \mathbb{R}^N$ with $|s| = 1$ the function $\{t \to t^{-q}G(x, ts)\}$ is nonincreasing on $[R, +\infty[$. Taking into account (G_3') and possibly substituting a_0 with another function in $L^1(\Omega)$, we deduce that

$$G(x, s) \leq a_0(x) - b_0(x)|s|^q \quad \text{for a.e. } x \in \Omega \text{ and all } s \in \mathbb{R}^N, \quad (11.20)$$

where

$$b_0(x) = \inf_{|s|=1} (-R^{-q}G(x, Rs)) > 0 \quad \text{for a.e. } x \in \Omega.$$

Finally, since $\{\hat{s} \to G^\circ(x, s; \hat{s})\}$ is a convex function vanishing at the origin, we have $G^\circ(x, s; s) \geq -G^\circ(x, s; -s)$. Combining (11.19) with (G_4'), we deduce that for every $\varepsilon > 0$ there exists $\tilde{a}_\varepsilon \in L^1(\Omega)$ such that

$$G(x, s) \geq -\tilde{a}_\varepsilon(x) - \varepsilon|s|^{\frac{n}{n-1}} \quad \text{for a.e. } x \in \Omega \text{ and all } s \in \mathbb{R}^N. \quad (11.21)$$

By Theorem 11.4 we have that f satisfies $(epi)_c$ for any $c \in \mathbb{R}$ and that $\left|d_{\mathbb{Z}_2}\mathcal{G}_f\right|(0, \lambda) = 1$ for any $\lambda > f(0)$.

We also recall that, since Ψ is Lipschitz continuous, there exists $M \in \mathbb{R}$ such that

$$(q+1)\Psi(\xi) - \Psi(2\xi) \geq \frac{q-1}{2}\Psi(\xi) - M, \quad (11.22)$$

$$(q+1)\Psi^\infty(\xi) - \Psi^\infty(2\xi) \geq \frac{q-1}{2}\Psi^\infty(\xi). \quad (11.23)$$

We claim that f satisfies the condition $(PS)_c$ for every $c \in \mathbb{R}$. Let $\{u_h\}$ be a $(PS)_c$-sequence for f. By Theorem 1.7, there exists a sequence $\{u_h^*\}$ in $L^n(\Omega; \mathbb{R}^N)$ with $u_h^* \in \partial f(u_h)$ and $\|u_h^*\|_n \to 0$. According to Theorem 11.3 and (11.19), we have

$$\begin{aligned}
\mathcal{E}(2u_h) &\geq \mathcal{E}(u_h) - \int_\Omega G^\circ(x, u_h; u_h)\, dx + \int_\Omega u_h^* \cdot u_h\, dx \\
&\geq \mathcal{E}(u_h) - q\int_\Omega G(x, u_h)\, dx + \int_\Omega u_h^* \cdot u_h\, dx - \int_\Omega a_0(x)\, dx.
\end{aligned}$$

By the definition of f, it follows

$$qf(u_h) + \|u_h^*\|_n \|u_h\|_{\frac{n}{n-1}} + \int_\Omega a_0(x)\, dx \geq (q+1)\mathcal{E}(u_h) - \mathcal{E}(2u_h).$$

Finally, applying (11.22) and (11.23) we get

$$qf(u_h) + \|u_h^*\|_n \|u_h\|_{\frac{n}{n-1}} + \int_\Omega a_0(x)\, dx \geq \frac{q-1}{2} \mathcal{E}(u_h) - M\mathcal{L}^n(\Omega)\,.$$

By (e) of Lemma 11.2 we deduce that $\{u_h\}$ is bounded in $BV(\Omega; \mathbb{R}^N)$, hence in $L^{\frac{n}{n-1}}(\Omega; \mathbb{R}^N)$. Applying Theorem 11.5 we get that $\{u_h\}$ admits a strongly convergent subsequence and $(PS)_c$ follows.

Lemma 11.7 (Marzocchi [19]) There exist a strictly increasing sequence $\{W_h\}$ of finite-dimensional subspaces of $BV(\Omega) \cap L^\infty(\Omega)$ and a strictly decreasing sequence $\{Z_h\}$ of closed subspaces of $L^p(\Omega)$ such that $L^p(\Omega) = W_h \oplus Z_h$ and $\bigcap_{h=0}^\infty Z_h = \{0\}$.

Proof. Let $m > n/2$ and let $\{e_h\}$ be a Hilbert basis in $W^{-m,2}(\Omega)$ constituted by elements of $W_0^{m,2}(\Omega)$. Let W_h be the linear space spanned by $e_0, \ldots, e_h$ and $\tilde{Z}_h$ the closed linear space spanned by $e_{h+1}, e_{h+2}, \ldots$, so that $W^{-m,2}(\Omega) = W_h \oplus \tilde{Z}_h$. It is obvious that $\bigcap_{h=0}^\infty \tilde{Z}_h = \{0\}$. If we set $Z_h = \tilde{Z}_h \cap L^p(\Omega)$, the spaces W_h and Z_h have the requested properties. ∎

By Lemma 11.7, there exist a strictly increasing sequence $\{W_h\}$ of finite-dimensional subspaces of $BV(\Omega; \mathbb{R}^N) \cap L^\infty(\Omega; \mathbb{R}^N)$ and a strictly decreasing sequence $\{Z_h\}$ of closed subspaces of $L^{\frac{n}{n-1}}(\Omega; \mathbb{R}^N)$ such that $L^{\frac{n}{n-1}}(\Omega; \mathbb{R}^N) = W_h \oplus Z_h$ and $\bigcap\limits_{h=0}^{\infty} Z_h = \{0\}$. By (e) of Lemma 11.2 there exists $\varrho > 0$ such that

$$\forall u \in L^{\frac{n}{n-1}}(\Omega; \mathbb{R}^N): \quad \|u\|_{\frac{n}{n-1}} = \varrho \quad \Longrightarrow \quad \mathcal{E}(u) \geq 1\,.$$

We claim that

$$\lim_h \Big(\inf\{f(u) : u \in Z_h, \|u\|_{\frac{n}{n-1}} = \varrho\}\Big) > f(0)\,.$$

For showing this assume by contradiction that $\{u_h\}$ is a sequence with $u_h \in Z_h$, $\|u_h\|_{\frac{n}{n-1}} = \varrho$ and

$$\limsup_h f(u_h) \leq f(0)\,.$$

Taking into account (G_3') and Lemma 11.2, we derive that $\{\mathcal{E}(u_h)\}$ is bounded, so that $\{u_h\}$ is bounded in $BV(\Omega; \mathbb{R}^N)$. Therefore, up to a subsequence, $\{u_h\}$ is convergent a.e. to 0. From (11.21) it follows that

$$\liminf_h \int_\Omega \Big(G(x, u_h) + \varepsilon |u_h|^{\frac{n}{n-1}}\Big)\, dx \geq \int_\Omega G(x, 0)\, dx\,,$$

hence

$$\liminf_h \int_\Omega G(x, u_h)\, dx \geq \int_\Omega G(x, 0)\, dx$$

by the boundedness of $\{u_h\}$ in $L^{\frac{n}{n-1}}(\Omega; \mathbb{R}^N)$ and the arbitrariness of ε. Therefore

$$\limsup_h \mathcal{E}(u_h) \leq \mathcal{E}(0) = 0$$

which contradicts the choice of ϱ.

Now, fix $\overline{h}$ with

$$\inf\{f(u) : u \in Z_{\overline{h}}, \|u\|_{\frac{n}{n-1}} = \varrho\}\Big) > f(0)$$

and set $Z = Z_{\overline{h}}$ and $V_h = W_{\overline{h}+h}$. Then Z satisfies assumption (a) of Theorem 11.2 for some $\alpha > f(0)$.

Finally, since V_h is finite-dimensional,

$$\|u\|_G := \left(\int_\Omega b_0 |u|^q dx\right)^{\frac{1}{q}}$$

is a norm on V_h equivalent to the norm of $BV(\Omega; \mathbb{R}^N)$. Then, combining (11.20) with (d) of Lemma 11.2, we see that also assumption (b) of Theorem 11.2 is satisfied.

Therefore there exists a sequence $\{u_h\}$ of critical points for f with $f(u_h) \to +\infty$ and, by Theorems 11.3 and 11.4, the result follows. ■

References

[1] A. Ambrosetti and P. H. Rabinowitz, Dual variational methods in critical point theory and applications, *J. Funct. Anal.* **14** (1973), 349-381.

[2] G. Anzellotti, The Euler equation for functionals with linear growth, *Trans. Amer. Math. Soc.* **290** (1985), 483-501.

[3] M. F. Bocea, Multiple solutions for a class of eigenvalue problems involving a nonlinear monotone operator in hemivariational inequalities, *Appl. Anal.* **65** (1997), 395-407.

[4] H. Brézis, *Opérateurs maximaux monotones et semigroupes de contractions dans les espaces de Hilbert*, North-Holland Mathematics Studies, **5**, Notas de Matemàtica (50), North-Holland, Amsterdam-London, 1973.

[5] K. C. Chang, Variational methods for non-differentiable functionals and their applications to partial differential equations, *J. Math. Anal. Appl.* **80** (1981), 102-129.

[6] D. C. Clark, A variant of the Ljusternik-Schnirelmann theory, *Indiana Univ. Math. J.* **22** (1972), 65-74.

[7] E. De Giorgi, A. Marino and M. Tosques, Problemi di evoluzione in spazi metrici e curve di massima pendenza, *Atti Accad. Naz. Lincei Rend. Cl. Sci. Fis. Mat. Natur.* **68** (1980), 180-187.

[8] M. Degiovanni, M. Marzocchi and V. Rădulescu, Multiple solutions of hemivariational inequalities with area-type term, *Calc. Var. P.D.E.* **10** (2000), 355-387.

[9] M. Degiovanni and F. Schuricht, Buckling of nonlinearly elastic rods in the presence of obstacles treated by nonsmooth critical point theory, *Math. Ann.* **311** (1998), 675-728.

[10] I. Ekeland and R. Temam, *Convex Analysis and Variational Problems*, Studies in Mathematics and its Applications, **1**, North-Holland Publishing Co., Amsterdam-Oxford, 1976.

[11] F. Gazzola and V. D. Rădulescu, A nonsmooth critical point theory approach to some nonlinear elliptic equations in $\mathbb{R}^n$, *Differential Integral Equations* **13** (2000), 47-60.

[12] D. Goeleven, D. Motreanu and P. D. Panagiotopoulos, Multiple solutions for a class of eigenvalue problems in hemivariational inequalities, *Nonlinear Anal.* **29** (1997), 9-26.

[13] D. Goeleven, D. Motreanu and P. D. Panagiotopoulos, Multiple solutions for a class of hemivariational inequalities involving periodic energy functionals, *Math. Methods Appl. Sci.* **20** (1997), 547-568.

[14] D. Goeleven, D. Motreanu and P. D. Panagiotopoulos, Semicoercive variational-hemivariational inequalities, *Appl. Anal.* **65** (1997), 119-134.

[15] D. Goeleven, D. Motreanu and P. D. Panagiotopoulos, Eigenvalue problems for variational-hemivariational inequalities at resonance, *Nonlinear Anal.* **33** (1998), 161-180.

[16] D. Kinderlehrer and G. Stampacchia, *An introduction to variational inequalities and their applications*, Pure and Applied Mathematics, **88**, Academic Press, New York-London-Toronto, Ont., 1980.

[17] A. Marino, The calculus of variations and some semilinear variational inequalities of elliptic and parabolic type, *Partial Differential Equations and the Calculus of Variations. Volume II. Essays in Honor of Ennio De Giorgi*, Birkäuser, Boston, 1989.

[18] A. Marino and D. Scolozzi, Geodetiche con ostacolo, *Boll. Un. Mat. Ital. B* (6) **2** (1983), 1-31.

[19] M. Marzocchi, Multiple solutions of quasilinear equations involving an area-type term, *J. Math. Anal. Appl.* **196** (1995), 1093-1104.

[20] D. Motreanu and Z. Naniewicz, Discontinuous semilinear problems in vector-valued function spaces, *Differential Integral Equations* **9** (1996), 581-598.

[21] D. Motreanu and P. D. Panagiotopoulos, A minimax approach to the eigenvalue problem of hemivariational inequalities and applications, *Appl. Anal.* **58** (1995), 53-76.

[22] D. Motreanu and P. D. Panagiotopoulos, Double eigenvalue problems for hemivariational inequalities, *Arch. Rational Mech. Anal.* **140** (1997), 225-251.

[23] Z. Naniewicz and P. D. Panagiotopoulos, *Mathematical Theory of Hemivariational Inequalities and Applications*, Monographs and Textbooks in Pure and Applied Mathematics, **188**, Marcel Dekker, Inc., New York, 1995.

[24] P. D. Panagiotopoulos, *Hemivariational Inequalities. Applications in Mechanics and Engineering*, Springer-Verlag, Berlin, 1993.

[25] P. D. Panagiotopoulos and V. D. Rădulescu, Perturbations of hemivariational inequalities with constraints and applications, *J. Global Optim.* **12** (1998), 285-297.

[26] P. H. Rabinowitz, *Minimax Methods in Critical Point Theory with Applications to Differential Equations*, CBMS Reg. Conf. Series Math. **65**, Amer. Math. Soc., Providence, RI, 1986.

[27] R. T. Rockafellar, Integral functionals, normal integrands and measurable selections, *Nonlinear Operators and the Calculus of Variations* (Bruxelles, 1975), 157-207, *Lecture Notes in Math.*, **543**, Springer, Berlin-New York, 1976.

[28] M. Struwe, *Variational Methods*, Springer-Verlag, Berlin, 1990.

Nonconvex Optimization and Its Applications

1. D.-Z. Du and J. Sun (eds.): *Advances in Optimization and Approximation.* 1994 ISBN 0-7923-2785-3
2. R. Horst and P.M. Pardalos (eds.): *Handbook of Global Optimization.* 1995 ISBN 0-7923-3120-6
3. R. Horst, P.M. Pardalos and N.V. Thoai: *Introduction to Global Optimization* 1995 ISBN 0-7923-3556-2; Pb 0-7923-3557-0
4. D.-Z. Du and P.M. Pardalos (eds.): *Minimax and Applications.* 1995 ISBN 0-7923-3615-1
5. P.M. Pardalos, Y. Siskos and C. Zopounidis (eds.): *Advances in Multicriteria Analysis.* 1995 ISBN 0-7923-3671-2
6. J.D. Pintér: *Global Optimization in Action.* Continuous and Lipschitz Optimization: Algorithms, Implementations and Applications. 1996 ISBN 0-7923-3757-3
7. C.A. Floudas and P.M. Pardalos (eds.): *State of the Art in Global Optimization.* Computational Methods and Applications. 1996 ISBN 0-7923-3838-3
8. J.L. Higle and S. Sen: *Stochastic Decomposition.* A Statistical Method for Large Scale Stochastic Linear Programming. 1996 ISBN 0-7923-3840-5
9. I.E. Grossmann (ed.): *Global Optimization in Engineering Design.* 1996 ISBN 0-7923-3881-2
10. V.F. Dem'yanov, G.E. Stavroulakis, L.N. Polyakova and P.D. Panagiotopoulos: *Quasidifferentiability and Nonsmooth Modelling in Mechanics, Engineering and Economics.* 1996 ISBN 0-7923-4093-0
11. B. Mirkin: *Mathematical Classification and Clustering.* 1996 ISBN 0-7923-4159-7
12. B. Roy: *Multicriteria Methodology for Decision Aiding.* 1996 ISBN 0-7923-4166-X
13. R.B. Kearfott: *Rigorous Global Search: Continuous Problems.* 1996 ISBN 0-7923-4238-0
14. P. Kouvelis and G. Yu: *Robust Discrete Optimization and Its Applications.* 1997 ISBN 0-7923-4291-7
15. H. Konno, P.T. Thach and H. Tuy: *Optimization on Low Rank Nonconvex Structures.* 1997 ISBN 0-7923-4308-5
16. M. Hajdu: *Network Scheduling Techniques for Construction Project Management.* 1997 ISBN 0-7923-4309-3
17. J. Mockus, W. Eddy, A. Mockus, L. Mockus and G. Reklaitis: *Bayesian Heuristic Approach to Discrete and Global Optimization.* Algorithms, Visualization, Software, and Applications. 1997 ISBN 0-7923-4327-1
18. I.M. Bomze, T. Csendes, R. Horst and P.M. Pardalos (eds.): *Developments in Global Optimization.* 1997 ISBN 0-7923-4351-4
19. T. Rapcsák: Smooth Nonlinear Optimization in R^n. 1997 ISBN 0-7923-4680-7
20. A. Migdalas, P.M. Pardalos and P. Värbrand (eds.): *Multilevel Optimization: Algorithms and Applications.* 1998 ISBN 0-7923-4693-9
21. E.S. Mistakidis and G.E. Stavroulakis: *Nonconvex Optimization in Mechanics.* Algorithms, Heuristics and Engineering Applications by the F.E.M. 1998 ISBN 0-7923-4812-5

Nonconvex Optimization and Its Applications

22. H. Tuy: *Convex Analysis and Global Optimization.* 1998 ISBN 0-7923-4818-4
23. D. Cieslik: *Steiner Minimal Trees.* 1998 ISBN 0-7923-4983-0
24. N.Z. Shor: *Nondifferentiable Optimization and Polynomial Problems.* 1998 ISBN 0-7923-4997-0
25. R. Reemtsen and J.-J. Rückmann (eds.): *Semi-Infinite Programming.* 1998 ISBN 0-7923-5054-5
26. B. Ricceri and S. Simons (eds.): *Minimax Theory and Applications.* 1998 ISBN 0-7923-5064-2
27. J.-P. Crouzeix, J.-E. Martinez-Legaz and M. Volle (eds.): *Generalized Convexitiy, Generalized Monotonicity: Recent Results.* 1998 ISBN 0-7923-5088-X
28. J. Outrata, M. Kočvara and J. Zowe: *Nonsmooth Approach to Optimization Problems with Equilibrium Constraints.* 1998 ISBN 0-7923-5170-3
29. D. Motreanu and P.D. Panagiotopoulos: *Minimax Theorems and Qualitative Properties of the Solutions of Hemivariational Inequalities.* 1999 ISBN 0-7923-5456-7
30. J.F. Bard: *Practical Bilevel Optimization.* Algorithms and Applications. 1999 ISBN 0-7923-5458-3
31. H.D. Sherali and W.P. Adams: *A Reformulation-Linearization Technique for Solving Discrete and Continuous Nonconvex Problems.* 1999 ISBN 0-7923-5487-7
32. F. Forgó, J. Szép and F. Szidarovszky: *Introduction to the Theory of Games.* Concepts, Methods, Applications. 1999 ISBN 0-7923-5775-2
33. C.A. Floudas and P.M. Pardalos (eds.): *Handbook of Test Problems in Local and Global Optimization.* 1999 ISBN 0-7923-5801-5
34. T. Stoilov and K. Stoilova: *Noniterative Coordination in Multilevel Systems.* 1999 ISBN 0-7923-5879-1
35. J. Haslinger, M. Miettinen and P.D. Panagiotopoulos: *Finite Element Method for Hemivariational Inequalities.* Theory, Methods and Applications. 1999 ISBN 0-7923-5951-8
36. V. Korotkich: *A Mathematical Structure of Emergent Computation.* 1999 ISBN 0-7923-6010-9
37. C.A. Floudas: *Deterministic Global Optimization: Theory, Methods and Applications.* 2000 ISBN 0-7923-6014-1
38. F. Giannessi (ed.): *Vector Variational Inequalities and Vector Equilibria.* Mathematical Theories. 1999 ISBN 0-7923-6026-5
39. D.Y. Gao: *Duality Principles in Nonconvex Systems.* Theory, Methods and Applications. 2000 ISBN 0-7923-6145-3
40. C.A. Floudas and P.M. Pardalos (eds.): *Optimization in Computational Chemistry and Molecular Biology.* Local and Global Approaches. 2000 ISBN 0-7923-6155-5
41. G. Isac: *Topological Methods in Complementarity Theory.* 2000 ISBN 0-7923-6274-8
42. P.M. Pardalos (ed.): *Approximation and Complexity in Numerical Optimization: Concrete and Discrete Problems.* 2000 ISBN 0-7923-6275-6
43. V. Demyanov and A. Rubinov (eds.): *Quasidifferentiability and Related Topics.* 2000 ISBN 0-7923-6284-5

Nonconvex Optimization and Its Applications

44. A. Rubinov: *Abstract Convexity and Global Optimization*. 2000 ISBN 0-7923-6323-X
45. R.G. Strongin and Y.D. Sergeyev: *Global Optimization with Non-Convex Constraints*. 2000 ISBN 0-7923-6490-2
46. X.-S. Zhang: *Neural Networks in Optimization*. 2000 ISBN 0-7923-6515-1
47. H. Jongen, P. Jonker and F. Twilt: *Nonlinear Optimization in Finite Dimensions*. Morse Theory, Chebyshev Approximation, Transversability, Flows, Parametric Aspects. 2000 ISBN 0-7923-6561-5
48. R. Horst, P.M. Pardalos and N.V. Thoai: *Introduction to Global Optimization*. 2nd Edition. 2000 ISBN 0-7923-6574-7
49. S.P. Uryasev (ed.): *Probabilistic Constrained Optimization*. Methodology and Applications. 2000 ISBN 0-7923-6644-1
50. D.Y. Gao, R.W. Ogden and G.E. Stavroulakis (eds.): *Nonsmooth/Nonconvex Mechanics*. Modeling, Analysis and Numerical Methods. 2001 ISBN 0-7923-6786-3
51. A. Atkinson, B. Bogacka and A. Zhigljavsky (eds.): *Optimum Design 2000*. 2001 ISBN 0-7923-6798-7
52. M. do Rosário Grossinho and S.A. Tersian: *An Introduction to Minimax Theorems and Their Applications to Differential Equations*. 2001 ISBN 0-7923-6832-0
53. A. Migdalas, P.M. Pardalos and P. Värbrand (eds.): *From Local to Global Optimization*. 2001 ISBN 0-7923-6883-5
54. N. Hadjisavvas and P.M. Pardalos (eds.): *Advances in Convex Analysis and Global Optimization*. Honoring the Memory of C. Caratheodory (1873-1950). 2001 ISBN 0-7923-6942-4
55. R.P. Gilbert, P.D. Panagiotopoulos† and P.M. Pardalos (eds.): *From Convexity to Nonconvexity*. 2001 ISBN 0-7923-7144-5
56. D.-Z. Du, P.M. Pardalos and W. Wu: *Mathematical Theory of Optimization*. 2001 ISBN 1-4020-0015-4
57. M.A. Goberna and M.A. López (eds.): *Semi-Infinite Programming. Recent Advances*. 2001 ISBN 1-4020-0032-4
58. F. Giannessi, A. Maugeri and P.M. Pardalos (eds.): *Equilibrium Problems: Nonsmooth Optimization and Variational Inequality Models*. 2001 ISBN 1-4020-0161-4
59. G. Dzemyda, V. Šaltenis and A. Žilinskas (eds.): *Stochastic and Global Optimization*. 2002 ISBN 1-4020-0484-2
60. D. Klatte and B. Kummer: *Nonsmooth Equations in Optimization*. Regularity, Calculus, Methods and Applications. 2002 ISBN 1-4020-0550-4
61. S. Dempe: *Foundations of Bilevel Programming*. 2002 ISBN 1-4020-0631-4
62. P.M. Pardalos and H.E. Romeijn (eds.): *Handbook of Global Optimization, Volume 2*. 2002 ISBN 1-4020-0632-2
63. G. Isac, V.A. Bulavsky and V.V. Kalashnikov: *Complementarity, Equilibrium, Efficiency and Economics*. 2002 ISBN 1-4020-0688-8
64. H.-F. Chen: *Stochastic Approximation and Its Applications*. 2002 ISBN 1-4020-0806-6

Nonconvex Optimization and Its Applications

65. M. Tawarmalani and N.V. Sahinidis: *Convexification and Global Optimization in Continuous and Mixed-Integer Nonlinear Programming*. Theory, Algorithms, Software, and Applications. 2002 ISBN 1-4020-1031-1
66. B. Luderer, L. Minchenko and T. Satsura: *Multivalued Analysis and Nonlinear Programming Problems with Perturbations.* 2002 ISBN 1-4020-1059-1

KLUWER ACADEMIC PUBLISHERS – DORDRECHT / BOSTON / LONDON